Decentralised Sanitation and Reuse

Concepts, Systems and Implementation

Edited by

Piet Lens, Grietje Zeeman and Gatze Lettinga
Department of Environmental Technology,
University of Wageningen, The Netherlands

Published by IWA Publishing, Alliance House, 12 Caxton Street, London SW1H 0QS, UK

Telephone: +44 (0) 20 7654 5500; Fax: +44 (0) 20 7654 5555; Email: publications@iwap.co.uk
www.iwapublishing.com

First published 2001

Printed by TJ International (Ltd), Padstow, Cornwall, UK

British Library Cataloguing in Publication Data
A CIP catalogue record for this book is available from the British Library

ISBN 1 900222 47 7

Front cover photographs

Left: Biogas latrine shortly before completion: biogas plant for domestic wastewater, organic kitchen and household waste (Ethiopia). Photograph courtesy of H. Euler/C. Kellner (TBW GmbH, Frankfurt am Main, Germany).

Right: Use of treated sewage for irrigation in broad bean (*Vicia faba*) crops (Egypt). Photograph courtesy of J.B. van Lier (EP&RC, Wageningen, The Netherlands).

Faded background: Modified UASB pilot plant (0.86 m^3) for the treatment of black or grey water of two houses (9 persons) in Bandung (Indonesia).

Contents

List of contributors

Andrea Angelakis
National Agricultural Research Foundation, Institute of V.V.F. of Iraklio, Water Resources and Environment Division, PO Box 1841, 71110 Iraklio, Crete, Greece

Raf Bellers
Aquafin, Dijkstraat 8, 2630 Aartselaar, Belgium

Lailach Ben-David
Ben-Gurion University, Environmental Water Resources Centre, The Institute for Desert Research, Kiryat Sde-Boker 84990, Israel

Markus Boller
Swiss Federal Institute for Environmental Science and Technology (EAWAG), Ueberlandstrasse 109, CH-8600 Duebendorf, Switzerland

Paul Cooper
The Ladder House, Cheap Street, Chedworth, Cheltenham, Gloucestershire GL54 4AB, UK

Claire Diaper
School of Water Sciences, Cranfield University, Cranfield MK43 0AL, UK

Fatma El-Gohary
Water Pollution Control Department, National Research Centre, Tahreer Street, Dokki, 12622 Cairo, Egypt

Eva Eriksson
Department of Environmental Science and Engineering, Technical University of Denmark, Building 115, 2800 Lyngby, Denmark

Hartlieb Euler
TBW GmbH, Baumweg 10, 60316 Frankfurt am Main, Germany

Leonid Gillerman
Ben-Gurion University, Environmental Water Resources Centre, The Institute for Desert Research, Kiryat Sde-Boker 84990, Israel

Tova Halmuth
Central Virological Laboratory, Sheba Medical Centre, Tel-HaShomer 52621, Israel

Lennert Heip
Aquafin, Dijkstraat 8, 2630 Aartselaar, Belgium

Mogens Henze
Department of Environmental Science and Engineering, Technical University of Denmark, Building 115, 2800 Lyngby, Denmark

Nigel Horan
Department of Civil Engineering, University of Leeds, Leeds LS2 9JT, UK

Look Hulshoff Pol
Department of Environmental Technology, University of Wageningen, PO Box 8129, 6700 EV Wageningen, The Netherlands

Bruce Jefferson
School of Water Sciences, Cranfield University, Cranfield MK43 0AL, UK

Simon Judd
School of Water Sciences, Cranfield University, Cranfield MK43 0AL, UK

Youssouf Kalogo
Laboratory Microbial Ecology, University of Gent, Coupure L. 653, 9000 Gent, Belgium

Ludmilla Kats
Ben-Gurion University, Environmental Water Resources Centre, The Institute for Desert Research, Kiryat Sde-Boker 84990, Israel

Katarzyna Kujawa-Roeleveld
Department of Environmental Technology, University of Wageningen, PO Box 8129, 6700 EV Wageningen, The Netherlands

Tove Larsen
Swiss Federal Institute for Environmental Science and Technology (EAWAG), Ueberlandstrasse 109, CH-8600 Duebendorf, Switzerland

Jon Kristensson
Kristinsson BV Architect & Ir. Bureau, Noordenbergsingel 10, 7411 SE Deventer, The Netherlands

Katsuhiko Kuniyasu
Japan Education Centre of Environmental Sanitation, 3-33-2 Haramachi, Shinjuku-ku, Tokyo, Japan

Anna Ledin
Department of Environmental Science and Engineering, Technical University of Denmark, Building 115, 2800 Lyngby, Denmark

Piet Lens
Department of Environmental Technology, University of Wageningen, PO Box 8129, 6700 EV Wageningen, The Netherlands

Gatze Lettinga
Department of Environmental Technology, University of Wageningen, PO Box 8129, 6700 EV Wageningen, The Netherlands

Jorgen Logstrup
DRT-TransForm, Borgervede 6, 1st floor, 1300 Copenhagen, Denmark

Alexia Luisings
Delft University of Technology, Faculty of Architecture, Environmental Design, Berlageweg 1, Postbus 5043, 2600 GA Delft, The Netherlands

Yossi Manor
Central Virological Laboratory, Sheba Medical Centre, Tel-HaShomer 52621, Israel

Harri Mattila
Tampere University of Technology (TUT), Institute of Water and Environmental Engineering (IWEE), PO Box 541, FIN-33101 Tampere, Finland. Also: Finnish Environment Institute, FEI, PO Box 140, FIN-00251 Helsinki, Finland

Janusz Niemczynowicz
Department of Water Resources Engineering, University of Lund, Box 118, 22100 Lund, Sweden.

Hallvard Odegaard
Department of Hydraulic and Environmental Engineering, Norwegian University of Science and Technology, N-7491 Trondheim, Norway

Hideaki Ohmori
Japan Education Centre of Environmental Sanitation, 3-33-2 Haramachi, Shinjuku-ku, Tokyo, Japan

Gideon Oron
Ben-Gurion University, Environmental Water Resources Centre, The Institute for Desert Research, Kiryat Sde-Boker 84990, Israel

Ralf Otterpohl
Technische Universität Hamburg-Harburg, Arbeitsbereich Abfallwirtschaft, Eißendorferstr. 42, 21071 Hamburg, Germany

Erik Poppe
Aquafin, Dijkstraat 8, 2630 Aartselaar, Belgium

Lucas Reijnders
Anna van den Vondelstraat 10, 1054 GZ Amsterdam, The Netherlands

Japp Schiere
AqN-Consult, Oudeweg 63, 9201 EK Drachten, The Netherlands

Miquel Salgot
Laboratori d'Edafologia, Facultat de Farmàcia, Universitat de Barcelona, Joan XXIII s/n. 08028, Barcelona, Spain

Friedhelm Streiffeler
Humbolt/dt-Universitat zu Berlin, LGF-WISOLA, Unter den Linden 6, 10099 Berlin, Germany

Anke Stubsgaard
Department of Monitoring and Information Technology, DHI, Gustav Wieds Vej 10, 8000 Arhus C, Denmark

Paul Terpstra
De Dreijenborch, Ritzema bosweg 32a, 6703 AZ Wageningen, The Netherlands

Willy Verstraete
Laboratory Microbial Ecology, University Gent, Coupure L. 653, 9000 Gent, Belgium

Michael von Hauff
Volkswirtschaftslehre und Wirtschaftspolitik, Universitaet Kaiserslautern, Gottlieb-Daimler-Strasse, 67663 Kaiserslautern, Germany

Rony Wallach
Faculty of Agriculture, The Hebrew University, Rechovot 76100, Israel

Madeleen Wegelin-Schuringa
IRC International Water and Sanitation Centre, PO Box 2869, 2601 CW Delft, The Netherlands

Peter Wilderer
Institute of Water Quality Control and Waste Management, Technical University of Munich, Am Coulombwall, D-85748 Garching, Germany

Takeshi Yahashi
Japan Education Centre of Environmental Sanitation, 3-33-2 Haramachi, Shinjuku-ku, Tokyo, Japan

Xinmi Yang
Japan Education Centre of Environmental Sanitation, 3-33-2 Haramachi, Shinjuku-ku, Tokyo, Japan

Grietje Zeeman
Department of Environmental Technology, University of Wageningen, PO Box 8129, 6700 EV Wageningen, The Netherlands

Preface

The technological achievements in our society during the last century are enormous in practically all fields. And this process of innovation and invention still continues, sometimes at an ever-increasing speed. Where will this process lead? Can society accommodate all the technical achievements and innovations; do they always comprise improvements for mankind? An increasing number of people are concerned about how society is developing in our world. It appears that the unbelievably rapid progress made in all kind of technologies is not sufficiently balanced by improved social conditions and welfare. The most fundamental human needs of millions of people are not being met; people are faced with serious pollution of their environment; and mineral resources are wasted by a small fraction of mankind living in prosperity. The gap between North and South, between the poor and the rich, is rapidly increasing. Something is seriously wrong in our society: we should learn to use our knowledge and our technologies for the benefit of everyone instead of for a small privileged fraction, so that everyone can enjoy life in a clean environment and in a sustainable way.

Although sanitation issues play a relatively modest role when compared to various other problems we are facing, they are important because sanitation directly affects quality of life. Poor sanitation represents a serious risk for human health, and for that reason sanitation is in most countries a governmental responsibility. Governments do all that they can to limit environmental pollution and to reduce the public health risks to citizens of waste and wastewaters. For that reason, it is common practice in many countries to convey human excreta and other household wastes away from residential sites as quickly as possible, generally even without considering its content, its potential value, or its impact to the neighbours. Although the way we handle our excreta is full of taboos, one may wonder why humanity creates such a big discrepancy between feeding and excreting. It is common practice to flush away excreta and most people do not want to be responsible for its impact. This behaviour is a key to solving the current sanitation dilemma. One may wonder why so many people, even in developing countries, prefer to spend their money on cars, videos or portable telephones instead of investing in sanitation and, thus, a clean environment. Proper sanitation solutions need to develop the same type of appeal as modern gadgets, it seems, in order to allow our generation and future generations to be able to live in a harmonious environment.

As a matter of fact, current sanitation technologies are very similar to those developed 100 years ago: transport the problem out of the residential area. They do not consider resource preservation or the reuse of residues and wastes. This book critically overviews and evaluates current approaches to sanitation, not in order to reject technologies, but to attempt to find ways of modifying current systems to develop and introduce new processes and concepts which meet the needs of sustainable development. Many different aspects need to be considered in addition to technical know-how, including architecture, town planning and socio-economics. This book pays special attention to (de)centralisation in sanitation, an aspect that integrates all these elements.

Our main objective in assembling the various chapters has been to present major, up-to-date reviews. Each separate chapter is presented on a standalone basis, so that the reader will find it most helpful to consider only the theme of each chapter. There are nevertheless many connections between what may at first seem to be quite different subjects. It was our intention to draw out and emphasise interdisciplinary linkages. For this reason, a comprehensive index is included to facilitate cross-reference. We hope that the work described in this book will inspire those already working in the field and encourage those who are starting to explore this field.

We wish to thank all the contributors to this book for their enthusiastic and prompt compilation of their contributions. The book is based on the Euro Summer School 'Decentralised sanitation and reuse', held from 18–23 June

2000, in Wageningen, The Netherlands. This summer school was financially supported by the EU Program 'Improving the Human Potential' (IHP-1999-0060). In addition to most of the oral presentations, a few invited contributions are also included in this volume. We are also grateful to Alan Click and Alan Peterson of IWA Publishing for their help and support in publishing this book.

Piet Lens
Grietje Zeeman
Gatze Lettinga
Wageningen, March 2001

Part I

The DESAR concept for environmental protection

1

Environmental protection technologies for sustainable development

G.Lettinga, P. Lens and G. Zeeman

1.1 SUSTAINABLE DEVELOPMENT AND ENVIRONMENTAL PROTECTION

The quality of our environment is a matter of eminent concern for all of us, especially for future generations. It is increasingly felt that the time has come for some drastic changes in the way our environment is protected from pollution, i.e. how to maintain a high diversity of life, at a local and a global level, and how to prevent the exhaustion of resources. Our society urgently needs a sustainable lifestyle, which also adheres to the environmental protection

 Decentralised Sanitation and Reuse: Concepts, Systems and Implementation.
Edited by P. Lens, G. Zeeman and G. Lettinga. ISBN: 1 900222 47 7

technologies adopted. This implies a holistic, multidisciplinary approach to current global problems, such as overpopulation, malnutrition, desertification, water quality deterioration, and so on. From ecology, one knows that uncontrolled exponential growth should be prevented in an ecosystem if a population wants to colonise that ecosystem in a sustainable way. Harmony and balance are needed around the carrier capacity of an ecosystem, and one can question if sustainability and growth can really co-exist. Sustainability and exponential growth certainly cannot. With respect to human society this not only applies to the prevention of the exhaustion of resources, but also to social justice; that is, the prevention of extreme prosperity of a few at the expense of hopeless poverty of the vast majority.

The concept of sustainability is certainly not new. However, it is difficult to be understood and implemented worldwide. The lack of parameters to quantify sustainability contributes to the vagueness of the concept. This leads to insufficiently clearly defined targets or actions proposed by politicians and/or policy makers, although this can be on purpose on certain occasions. Even when international committees are involved, such as the well-known Brundlandt Committee, the definition given to the concept is so open that it is easy for people, institutions and governments to avoid taking proper measures. For instance, when governments come up with extremely stringent standards for protecting, for example, the aquatic environment from pollution, the question arises, 'What sense does it make to pursue a paradisaical natural environment in a single country or region, when at the same time little if any money or technology is made available to contribute to highly needed environmental improvement in less prosperous countries?' This is a type of environmental tunnel vision which presumably mainly serves to give the citizens the impression that politicians are concerned about having a clean environment. But in fact it mainly serves the short-sighted objectives which have little to do with sustainable and robust environmental protection. As a result, developments frequently move in the opposite direction to that which was originally planned.

1.2 SUSTAINABILITY IN ENVIRONMENTAL PROTECTION TECHNOLOGIES

1.2.1 Centralised sanitation and sustainability

Of all the wastewater in the world, 95 per cent is released to the environment without treatment (Niemczynowics 1997). In 1997, three billion people on earth lacked adequate sanitation. If sanitation provisions continue to be installed based on the current standard, up to 5.5 billion people will be without sanitation by the year 2035, many of whom will be living in crowded urban settlements

(Niemczynowics 1997). As a consequence of this lack of sanitation, 3.3 million people die annually from diarrhoeal diseases, out of 3.5 billion infected. In Africa alone, 80 million people are at risk from cholera, and the 16 million cases of typhoid infections each year are a result of lack of adequate sanitation and clean drinking water (WHO 1996). Although inadequate sanitation is less of a problem in European countries, regular epidemic breakthroughs (for example, of *Cryptosporidium*, *Gardia* and *Legionella* and even cholera) indicate that developed countries also face problems of improper sanitation.

One of the main reasons for this situation is the high cost of current water-borne sanitation techniques and methods. It is obvious that the established sanitary engineering world in the public sanitation sector emphasises the implementation of very expensive (both in investment and operation) high-tech centralised systems. This applies to the processing and distribution of drinking water and to the collection, transport and treatment of solid waste and wastewater. A huge number of such centralised urban sanitation (CUS) systems have been developed and implemented in the last century, especially in the industrialised world (Harremoes 1999). Huge investments have been made to install the sewerage systems required, and the maintenance of these systems is also expensive (see Chapter 30). According to Grau (1994), countries with an average annual per capita gross national product (GNP) of below US$1,000 not only lack the resources to construct treatment plants, but also cannot maintain them, even if these plants were constructed free of charge. Moreover, as the lifetime of a sewer is only in the order of 50-70 years, these investments have to be made again and again.

CUS systems are based on the collection and transport of wastewater via an extended sewer system to centralised treatment systems. These systems use clean water (mainly tap water) as the transport medium of domestic wastes which are frequently relatively concentrated, for example, faeces. Moreover, very little, if any, recovery of useful by-products such as fertilisers (nitrogen, phosphorous or potassium) is achieved. On the contrary, huge amounts of poorly stabilised and polluted sludges are generated which have to be disposed of because they are not acceptable for agricultural reuse. Thus, the CUS approach is far from sustainable. The proper functioning of CUS systems depends on energy supply, computer hardware, and so on, making them vulnerable to theft, sabotage and military attack in poor and politically unstable countries.

1.2.2 Decentralised sanitation and sustainability

One can not ignore the role that modern CUS systems have played in the efficient protection of the environment over the last century and the great increase in public comfort they have offered to the industrial Western world. However, many aspects of the CUS concept conflict with sustainable sanitation (see Table 1.1). Enormous amounts of clean water are wasted by using it as a transport medium. Since the waste is highly diluted in this way, expensive, energy-consuming and technically complex wastewater treatment technologies have to be applied. Thus, CUS systems constitute a heavy financial burden on society, particularly in less prosperous countries. Moreover, many sewerage systems cannot cope with stormwater and, during periods of heavy rainfall, untreated wastewater is released into the environment via sewer overflow. CUS concepts are not sustainable because resources are consumed and – except the treated water – not recovered.

Table 1.1. Criteria for sustainable sanitation

- Little, if any, dilution of high strength domestic (and industrial) residues with clean water
- Maximisation of recovery and reuse of treated water and by-products, e.g. for irrigation, fertilization and soil conditioning
- Application of efficient, robust and reliable wastewater collection and transport systems and treatment technologies, which require few resources and which have a long lifetime

The lack of sustainability of the CUS concept becomes obvious in the case of diffuse pollution. At present, full coverage with sanitation and treatment is achieved only in rich countries, serving only 6 per cent of the world's population. No matter how high the levels of industrialisation and modernisation are, there is still a major part of the population living in the rural countryside. Even in very industrialised regions, such as Western Europe, the US and Japan, the percentage of the population living in rural areas can still be up to 20-40% (Watanabe and Iwasaki 1997). Human activities in these rural areas, i.e. isolated resorts and communities, hotels and camping sites, generate diffusive pollution by wastewater and solid wastes. A variety of land use practices, such as farming, timber harvesting, construction, mining and land disposal also contribute to diffuse pollution (Boller 1997).

The generally accepted concept that water can be obtained from nature in any quantities by the use of suitable technology has strongly contributed to this situation (Niemczynowics 1997). The question of how much water we really need and what quality this should be was not quantified until recently. Water of the highest quality is needed only for drinking and cooking, making up about 5 per cent of current total water consumption. However, at present, all water delivered is of the same quality since there is only one water network. Furthermore, all delivered water will be contaminated if water closets are used, so that all the water expelled from our houses after use is called 'wastewater', of which a large part is actually 'wasted water'.

Many of the drawbacks of the CUS approach can be overcome when applying decentralised sanitation concepts (see Table 1.2). These concepts put an emphasis on prevention (for example, little if any use of clean water for transport, but separation of concentrated and diluted wastewater in the house, and separate treatment of each (see Table 1.3), treatment in or near the community, application of low cost and sustainable treatment systems, recovery and reuse of useful by-products, also at or nearby the site (for example, water and nutrients for agricultural purposes, energy in the form of biogas for domestic purposes (Lettinga 1996; MacKenzie 1996; Van Lier and Lettinga 1999) (Figure 1.1). Decentralised urban sanitation (DUS) systems are in principle much less vulnerable, because their operation is independent of complex infrastructures such as energy and water supply; they are simple and robust (van Riper and Geselbracht 1998).

Table 1.2. Some criteria for robust urban sanitation

- Little dependency on complex infra-structural services, for example, power and/or water supply
- High self-sufficiency in construction, operation and maintenance of systems (independent of highly specialised people/companies)
- Low vulnerability to sabotage, destruction, etc.
- High public participation; acceptable to all social actors
- Applicable at any site and scale

Table 1.3. Prevention of environmental pollution problems

- Complete utilisation of all possible waste resources
- No pollution of water, soil or air
- Finding a proper final destination for any type of residue

1.2.3 Centralised versus decentralised sanitation

Thus far, the established sanitary engineering world remains reluctant to developments and new technologies which could lead to a more sustainable and robust alternative to the CUS concept, for example, Decentralised Sanitation and Reuse (DESAR) concepts. The statements in Table 1.4 illustrate the doubts the established civil engineering-oriented sanitation world has in the development and implementation of low-cost, simple and decentralised solutions for environmental protection. It should also be noted that established groups working on so-called low-cost wastewater treatment systems, such as constructed wetlands and/or lagoons, also have prejudices against both decentralised and centralised treatment. Apparently, each specialised group sticks to their own system(s), advertises them wherever possible, for their own commercial interests, and sometimes perhaps for scientific prestige. Unfortunately, as in many other fields of human development, disagreements among specialists take place at the expense of those who really need the solution and, of course, nature itself.

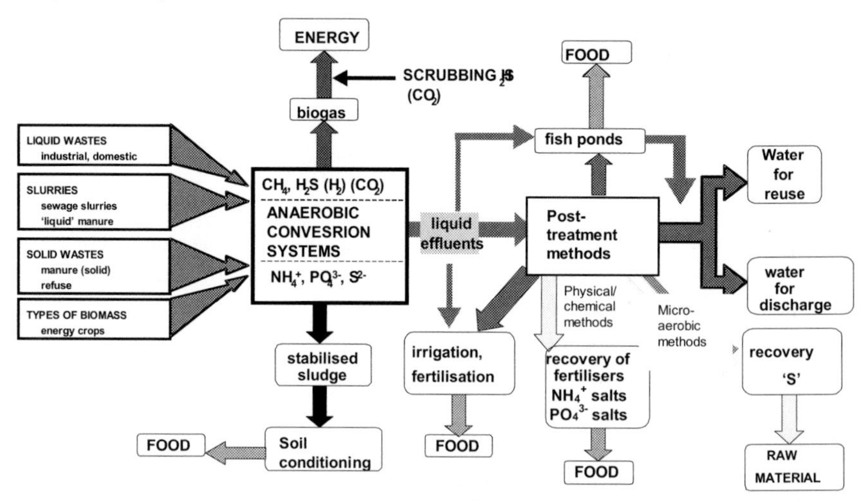

Figure 1.1. The potential for anaerobic conversion in the recovery of resources (water, food, new materials and energy) from waste.

According to Harremoës (1997, 1999), 'There is no miracle "low tech" solution in sight, because it [environmental pollution] is a social rather than a technical problem.' One can wonder if the same holds for high-tech solutions: do miracle high-tech solutions exist? The crucial point to be recognised here is that good low-tech systems can be developed and even – to some extent – are already available (Figure 1.1.). Undoubtedly the origin of environmental pollution is more social than technical, but one of the major reasons for current environmental pollution is the continued application of non-sustainable CUS solutions. These modern Western concepts may appear to be (at least for the time being) economically and technically achievable in the prosperous industrialised world, but they are far too expensive and complex for poor countries (Grau 1996). This, however, does not stop established consultants, contractors and scientists from attempting to implement these systems in countries that lack the financial resources and expertise to operate and maintain them.

Table 1.4. The evaluation of DUS as an alternative to CUS methods, from the viewpoint of the established sanitary engineering world (Harremoës 1997)

'Local wastewater treatment is not a viable solution in cities, because the approach is either "low tech", which does not live up to established hygienic requirements and risk assessment, or it is "high tech", which suffers from energy consumption.'

'Presently available "low tech" centralised urban sanitation (CUS) approaches are not simple and not easy solutions. Consequently they are not better than available "high tech" CUS solutions.'

'The present decentralised urban sanitation (DUS) systems lack adaptability to the urban environment, manageability and control (maintenance of standards).'

In the less prosperous world there exists a growing demand for integrated decentralised sanitation systems providing opportunities to save and reuse resources, thus for DESAR-type of environmental protection solutions. It should be emphasised here that DESAR systems do not mean small scale systems. It means the redirection of the water and nutrient cycles within the community, and this can still be done using relatively large scale systems. They are also not low-tech either, although the core parts of the system are essentially low-tech, sustainable and robust. For the recovery of resources such as fertilisers a high-tech system and a more centralised approach will undoubtedly be needed. Each situation is unique and has its own optimal solution.

The potentials of the decentralised approach have been clearly demonstrated in the industry in recent decades with the implementation of integrated anaerobic and physical/chemical treatment, mainly in European countries. The question we face when we want to apply the decentralised approach to the sanitation sector is 'What are the best DESAR systems for the various situations prevailing in urban regions, and how to developed them?' Even when stimulating the DESAR approach in the public sector, it should not be suggested that the available CUS systems should be abandoned immediately. However, attempts should be made to move step by step to more sustainable and robust sanitation systems, and to limit the extent of centralisation to reach a rational optimum, in stead of implementing CUS as the only available sanitation solution. A lot remains to be done to define such a rational optimum, and it is clear that there may exist several 'optima' for a specific situation, each with its own typical characteristics.

The purpose of this book is to contribute to the selection and implementation of sustainable environmental protection in the public sanitation sector. This is a very broad and multidisciplinary field. It is obviously impossible to go into detail of all related aspects. Therefore, the sanitary solutions available are reviewed and critically evaluated. Moreover, information is given on the many non-sanitary engineering aspects (for example, sociological and economical aspects, environmental and public health, architecture and town planning) which are often decisive in the successful implementation of sanitary systems.

1.3 REFERENCES

Boller, M. (1997) Small wastewater treatment plants – a challenge to wastewater engineers. *Wat. Sci. Tech.* **35**(6), 1–12.

Grau, P. (1994) What next? *Wat. Qual. Int.* **4,** 29–32.

Grau, P. (1996) Low cost wastewater treatment. *Wat. Sci. Tech.* **33**(8), 39–46.

Harremoës, P. (1997) Integrated water and waste management. *Wat. Sci. Tech.* **35**(9), 11–20.

Harremoes, P. (1999) Water as a transport medium for waste out of towns. *Wat. Sci. Tech.* **39**(5), 1–8.

Lettinga, G. (1996) Sustainable integrated biological wastewater treatment. *Wat. Sci. Tech.* **33**(3), 85–98.

MacKenzie, K. (1996) On the road to a biosolids composting plant. *Biocycle* **37**, 58–61.

Niemczynowics, J. (1997) The water profession and agenda 21. *Wat. Qual. Int.* **2**, 9–11.

Van Lier, J.B. and Lettinga, G. (1999) Appropriate technologies for effective management of industrial and domestic wastewaters: the decentralised approach. *Wat. Sci. Tech.* **40**(7), 171–183.

van Riper, C. and Geselbracht, J. (1998) Water reclamation and reuse. *Wat. Environ. Res.* **70**, 586–589.

WHO (1996) Water supply and sanitation sector monitoring. Report 1996: "Sector status as of 31 December 1994". In: WHO/EOS/96.15. Geneva, Switzerland.

Watanabe, Y. and Iwasaki, Y. (1997) Performance of hybrid small wastewater treatment system consisting of jet mixed separator and rotating biological contactor. *Wat. Sci. Tech.* **35**(6), 63–70.

2

Historical aspects of wastewater treatment

P. F. Cooper

2.1 INTRODUCTION

This chapter does not attempt to give a detailed history of wastewater treatment but instead to give an overview, point to the most significant developments and describe why we are where we are in the treatment of sewage. The paper is written from the viewpoint of a practising wastewater process engineer rather than that of an academic historian. It is assumed that the majority of readers are from a similar area of experience and so detailed explanations of the wastewater treatment unit processes are not included. A great deal of reference is made to history of developments in the UK, particularly to those in London. This is largely because the UK was one of the first industrialised countries and hence

Edited by P. Lens, G. Zeeman and G. Lettinga. ISBN: 1 900222 47 7

experienced the problems which result from very densely populated cities before many other countries. Mention is made of potable water supply and sewerage systems since they are intimately associated with the development of wastewater treatment, but they are not discussed in detail.

2.2 EARLY HISTORIC TIMES

The use of sewers is not new. In the Mesopotamian empire (3500 to 2500 BC) some homes were connected to a stormwater drain system to carry away wastes. In Babylon there were latrines which were connected to 18 inch (450 mm) diameter vertical shafts lined with perforated clay pipes leading to cesspools. However most people in Babylon threw debris including garbage and excrement on to the unpaved streets. The streets were periodically covered with clay, eventually raising the street levels to the extent that stairs had to be built down into houses.

In the Indus city of Mohenjo-daro (located in Pakistan) the wealthy as well as some of the peasants used latrines and cesspools. These were connected to drainage systems in the streets from whence the liquid flowed to cesspools or through drains to the nearest river. In some cases terracotta pipes were used to connect second-floor bathrooms to street sewers.

Archaeologists have found four separate drainage systems at King Minos' Royal Palace at Knossos (Crete), which dates from 1700 BC. The wastewater drained through terracotta pipes which were joined with cement into stone sewers. Rainwater-fed cisterns and stone aqueducts tapped available water sources to deliver a continuous flow of water through the bathrooms and latrines which eventually discharged to the Kairatos River. From 2000 BC the island of Crete had a drainage system made up of terracotta pipes with bell and spigot joints sealed with cement. The system conveyed mainly stormwater but also some human waste. Water stored in large jars was used to fill the system periodically. Wolfe (1999) states that many of the drains are still in use today.

There was a recent discovery of a stone lavatory with running water in a royal tomb from the Western Han dynasty (206 BC to AD 24) in the central province of Henan, China (Rennie 2000).

The Ancient Greeks (300 BC to 500 AD) tackled the problem of waste in a different way. They had public latrines which drained into sewers which conveyed the sewage and stormwater to a collection basin outside the city. From there brick-lined conduits took the wastewater to agricultural fields which used the wastewater for irrigation and to fertilise crops and orchards. The sewers were periodically flushed with wastewater.

A good review of the very earliest uses of sewers and waste disposal is given by Wolfe (1999) in the special issue of *World of Water 2000*. The reader is referred to that review for more detailed information.

2.3 ROMAN TIMES: 800 BC TO 450 AD

In about 800 BC the Romans constructed the Cloaca Maxima, the central sewer system, to drain the marsh upon which Rome was later built. The system took surface water to the River Tiber. By 100 AD the system was almost complete and connections had been made to some houses. The streets were still open sewers and, although many Romans used the public latrines, human wastes were still thrown into the streets. Water was supplied by an aqueduct system which carried away sewage and wastewater from the public baths and latrines thence to the sewers beneath the city and finally into the Tiber. The streets were regularly washed with water from the aqueduct system and the waste washed into the sewers (Wolfe 1999).

The Romans knew of the need for clean water and the need to dispose of wastewater away from the source of drinking water. In the UK they built their villas on the sides of hills where springs emerged from the hillside, and disposed of their wastewater to streams away from their villas. It has long been known that the Romans built brick-lined sewers in London (which they called Londinium). However, it has recently been discovered that these were preceded by wood-lined sewers which drained the water from the city to the River Thames. Pieces of the brick-lined sewers still exist.

2.4 THE SANITARY DARK AGES: 450 TO 1750

When the Roman empire collapsed their sanitary approach collapsed with it, since it depended upon far-reaching aqueducts and these needed effective government and the protection of a powerful army (Wolfe 1999).

During this period the main form of waste disposal (solid or liquid) in European cities such as Paris and London was simply to dispose of it in the streets. The terms 'Tout a la rue' (Paris), 'All in the road', 'Gare de l'eau' (Edinburgh) and 'Gardyloo' (Glasgow) come from that period. Often it was just thrown from windows and God help anyone who happened to be passing. This is the basis of the custom for the gentleman to walk on the side of the pavement closest to the road so that he could shield the lady from the splashing of passing carts and coaches and chamber pots of human waste which were flung from the second-storey windows which overhung the pavement.

Paris was founded upon the ruins of the Roman city Lutece in 360 AD (Wolfe 1999). Waste went into the streets where rainfall and heavy traffic helped it to decompose and it was picked over by pigs and wild dogs or collected by scavengers for fertiliser. In the thirteenth century King Phillipe Augustus ordered the city's roads to be paved to reduce the stench of the mixed garbage and sewage. However, once paved, the wastes could not break down to mud and in 1348 King Phillipe VI formed the first corps of sanitation workers to clean the streets. He also issued an ordinance that required all citizens to sweep in front of their houses and dispose of garbage to dumps. The first covered sewer was built in 1370 which dumped sewage into the River Seine near the Louvre. The French monarchy only took action over the sewers if affected by the smell. King Francois I moved his mother to the Tuilleries to escape the stench. In 1539, when plagues swept Europe, King Francois I ordered houseowners to build cesspools (indoor pit toilets) for sewage collection in new houses. These were constructed so that they leaked and did not have to be emptied often. These continued to be used until the late 1700s.

In London cesspools were in existence in 1189. The first Mayor of London, Henry Fitzalwyn, ruled that they be located no less than 2.5 feet (75 cm) from neighbouring buildings if made of stone or 3.5 feet (105 cm) if constructed of other materials (Wolfe 1999). Cistercian monks in the south of Scotland built stone-lined sewers to drain latrines in the monks' cells to the nearby watercourse. The clay pipes and brick-lined sewers put in place by the Romans in London were still in use but they were originally intended to take surface waters. Stephen Halliday, in his recently published book *The Stink of London* details the work done by Sir Joseph Bazalgette in providing sewers to cleanse Victorian London in the second half of the 1800s. In this he quotes a statement from *The Builder* journal written in 1884 which pointed out that as late as 1800 it was a penal offence to discharge sewage or any noxious matter to sewers that were meant for surface drainage only. The sewage of the city was to be collected in cesspools and their contents conveyed into the countryside for application to land. This was done in medieval times by 'rakers' or 'gong-fermors' who removed the foul sewage from the cesspools and sold it to farmers just outside the city walls. By the 1300s the city of Norwich, at that time the second largest city in England after London, was selling 'night soil' to farmers outside the walls of the city as fertiliser (Campbell 2000). Cesspools were built to drain into the street by a crude culvert, but when these became blocked the sewage spread under buildings and contaminated shallow wells and waterways that supplied drinking water. Several hundred thousand Londoners died from cholera, typhoid, plague and pestilence before the city realised that its own waste was causing the problems (Wolfe 1999). Overflowing cesspools could drain into neighbouring dwellings, causing poor families to live in houses saturated by their neighbours' excrement. Entire families were killed by asphyxiation from hydrogen sulphide coming from sewage collecting in or below their cellars.

In 1596 Sir John Harington had designed two water closets (called The Necessary) for Queen Elizabeth I but these did not achieve popularity until adopted by Londoners late in the 1700s. (Thomas Crapper in 1861 achieved long-standing fame for inventing a better flushing mechanism than his predecessors.)

2.5 THE AGE OF SANITARY ENLIGHTENMENT AND THE INDUSTRIAL REVOLUTION: 1750 TO 1950

2.5.1 The age of miasmas, disease, a shortage of safe water and development

This period is characterised by a high population growth in the new industrial cities, leading to high population densities (see Figures 2.1 and 2.2 and Table 2.1). The growth rate in London was extremely high, increasing from just under 1 million in 1801 to 2.8 million in 1861 (see Figure 2.2 and Halliday 1999) and to 6.5 million by 1900 (Lee 1997). The increasing death rates (Table 2.2) are now known to be related to water and waste-borne disease.

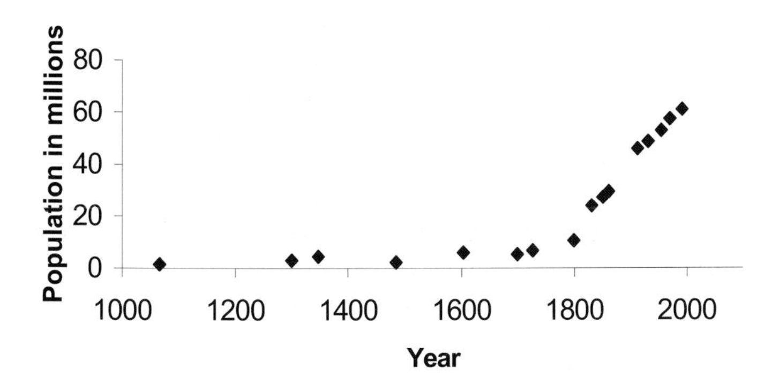

Figure 2.1. Increase in population of the British Isles over the last millennium (Lee 1997, 1999).

Table 2.1. Growth of population in new industrial towns. Yorkshire Wool Industry Towns (Stanbridge 1976)

	1801	**1831**
Huddersfield	15000	34000
Bradford	29000	77000
Halifax	63000	110000
Leeds	53000	123000

Table 2.2. Disease in the Industrial Revolution in Great Britain. Deaths per 1000 people (Stanbridge 1976)

	1811	**1841**
Birmingham	15	27
Leeds	20	27
Bristol	17	31
Manchester	30	34
Liverpool	21	35

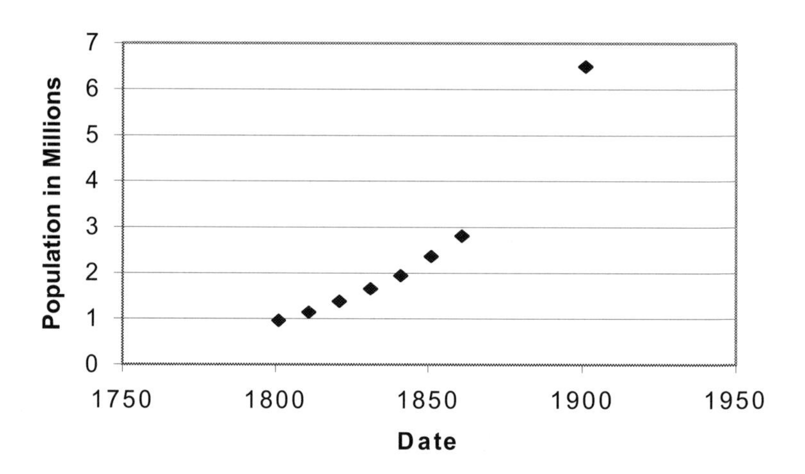

Figure 2.2. Growth of population in London in the 19th century (Halliday 1999; Lee 1997).

The early part of the nineteenth century (1820 to 1850) saw great debate as to how diseases like cholera and typhus was spread and what could be done to prevent it in the rapidly expanding cities and towns as the Industrial Revolution gathered pace. Outbreaks of cholera took place in other large European cities. This was the era of Miasmas. The belief was that that miasmas (noxious gaseous emanations and infections derived from the rotting waste which abounded in streets and public places) led to disease. In other words people were made ill as a result of poisoned air. This idea was put forward by the "Miasmatists", who included Florence Nightingale and Edwin Chadwick. Another rival group, the "Contagionists" believed that disease was passed by physical contact, whether from human to human, or through the consumption of infected food or water. Drs John Snow and William Budd were amongst those who saw that infected drinking water seemed to be the likeliest source of disease and particularly cholera (Chartered Institute of Environmental Health 1998). The link was not established until later in the century but the developments in Europe and most particularly in the US were influenced by the English 'Sanitary Idea' in the 1840s (Melosi 2000) Filth and foul smells were thought to be responsible for epidemics. Whilst the miasmatic theory did not show the cause of disease, it did place a great deal of emphasis on the need for sanitation to combat the disease.

Melosi (2000) pays fulsome tribute to 'the nineteenth century English civil engineers and sanitarians who became leaders in setting standards for water and wastewater systems throughout Europe and north America'. In particular he pays considerable attention to the work of Sir Edwin Chadwick (1800-1890), referring to the period up to 1830 as 'Pre-Chadwickian'. Chadwick was a lawyer and journalist who was associated with Jeremy Bentham and other Philosophical Radicals known as 'Utilitarians' in the 1820s. He developed an interest in the condition of the London slums and whilst doing this contacted typhus, from which he recovered. He was appointed to the commission enquiring into the state of the Poor Law, which resulted in the 1834 Poor Law report. Chadwick took on the role of Secretary to the Commission and in 1842 produced the *Report on the Sanitary Conditions of the Labouring Population of Great Britain*. The report made the following recommendations:

- Provision of water supply to every house
- Use of water-closets over older systems (earth closets and privies)
- Discharge of domestic wastewater direct to sewer rather than to cesspools
- Sewers to also take solid refuse from streets
- Sewers, instead of discharging to watercourse, to convey sewage to an agricultural area away from town where its manurial value could be utilised. (This is now called land treatment.)

The report contained much background material and thinking. Here are a few snippets:

- It was 'not customary to provide sanitary accommodation in poor areas and very few privies existed in crowded courts (yards)'
- 33 privies for 7905 persons in Liverpool
- 2 privies for 80 persons in Manchester
- waste in yards 6 inches (15 cm) deep

One major result that came from this report was the 1848 Public Health Act which set up Local Boards of Health and gave them the power to construct sewers.

Despite the success of the measures that he advocated, Chadwick was not a popular man. He was very determined and this meant that he created friction. One of the main reasons for his unpopularity was that he was the Secretary to the new Poor Law Commission, which had been set up to make the Poor Law Amendment Act 1834 work (Chartered Institute of Environmental Health 1998). The Poor Law was the first official form of social welfare provision in the UK but it was extremely unpopular. As an assistant commissioner, Chadwick had been instrumental in writing the report on which the act was based and hence it was natural that people should regard him as the architect of the monolithic, all-purpose workhouses designed to deter people from entering them. He had been against these huge establishments which could house up to 2,000 miserable people. He had wanted specialised workhouses for different needs and was concerned that children should be properly looked after. Charles Dickens was one of his chief critics and attacked him in newspaper articles and in *Oliver Twist*. Later in 1851 Dickens became a supporter of the Chadwickian reforms since his brother-in law, Henry Austin, a public health engineer and Secretary of the General Board of Health, was able to show the benefits of the Chadwick recommendations.

Chadwick proposed a hydraulic (or arterial-venous) system that would bring potable water into homes equipped with water closets and then carry effluent out to public sewer lines to be deposited as 'liquid manures' on to neighbouring agricultural fields (Melosi 2000). He also proposed the 'backyard tubular drainage' system in which sewage was drained from the backs (where the privies, latrines and water closets were placed) of back-to-back houses (being built in the poorer areas in the rapidly expanding cities) rather than putting the sewer connection through the fronts of the houses as was usual (General Board of Health 1852). He claimed that it would reduce the cost of sewer runs by two-thirds to four-fifths and allow the use of smaller sewers. This idea was taken up successfully 130 years later in Brazil (Mara 1999).

The water closet which began to be adopted by Londoners in the late 1700s gained tremendous popularity in the 1800s because of its ability, once connected to the sewer, to immediately remove human waste from the house, thus making cesspools no longer necessary. This improved the living conditions in homes but transformed the River Thames, from which most of the city's water supply was drawn, into a virtual cesspool. The water volume in the London drainage system almost doubled in the six years from 1850 to 1856 as a result of increased use of water closets (Halliday 1999).

Chadwick did not get everything his own way. In 1855, after another cholera epidemic, Parliament passed an act that established the Metropolitan Board Of Works to develop an adequate sewerage system for London. Joseph Bazalgette became the chief engineer. He was opposed to the Chadwickian idea of collecting sewage and using it on farm land. Instead he proposed a series of main intercepting sewers running east–west which collected discharge before it got to the Thames. He proposed that the discharge should all go to outfalls downriver from the city. His original proposal in 1856 was rejected because of the outfall location. Two years later the government reversed their decision as a result of the Great Stink of 1858 (Halliday 1999; Melosi 2000). Hot weather and the use of thousands of water closets created an ungodly stench lasting two years, caused by the putrefaction of sewage caught in the tidal reach of the river. Sessions of Parliament (located at the riverside) were only made bearable by hanging sheets soaked with lime of chlorine from each open window (Melosi 2000). The construction of the Bazalgette sewer system started in 1858 and was essentially complete by 1865. A total of 83 miles (133 km) of sewers were laid to drain an area of about 100 square miles (256 km^2). This was one of the examples of the principle 'the solution to pollution is dilution' which had been applied by the Greeks and Romans, but it was not until later in the 1800s that it was realised that it was not good enough to dilute and disperse, and that something would have to be done to remove the pollutants.

The Chadwickian ideas greatly influenced thinking in the US, especially in the east coast cities, including New York, which were growing at similar rates to some of the European cities. In 1845 Dr John Griscom, the New York City Inspector, produced a study, *The sanitary conditions of the laboring population of New York*. Over the next century there was a free exchange of ideas between the east coast cities and the large European cities.

Another major contribution at this time was made by Dr John Snow who was able to provide the link between disease and sanitary conditions, the solution to the link with miasmas. In 1849 he wrote an article 'On the mode of Transmission of Cholera'. He believed that it was transmitted by water contaminated by the vomit and faecal matter of cholera patients. He was able to prove the theory in 1854 when a severe bout of cholera occurred in London (Binnie 1999). He carefully documented the number of cholera deaths occurring in houses served by two of the city's water companies (which served a total of about 300,000 people (BBC 2001)). The two companies supplied water to people in the same areas of the city but derived their water from different sources. He showed that there were 315 deaths per 10,000 houses in the area served by the Southwark Water Company which drew its water from the heavily contaminated lower reaches of the River Thames. In the same period there were only 37 deaths per 10,000 houses served by the Lambeth Water Company which took its water from the upper reaches of the Thames. In particular he showed that in one area near the intersection of Cambridge and Broad Streets more than 500 people died of cholera in 10 days in 1854. After investigation he concluded that these were linked with water taken from the Broad Street pump. He had the handle removed from the pump and the epidemic was contained. In this study he also made use of microscopy and the work of Dr Arthur Hill Hassall (Bingham 1999). Figure 2.3 shows the course of the Broad Street outbreak. In fact it is clear from this figure that the epidemic was almost over. Dr Snow was well aware of this and the real significance of the removal of the pump handle was to prevent a second epidemic because there was a new cholera case in the house (which was later found to have been the source of the contamination of the pump) on the day that the handle was removed (BBC 2001). It is now known that cholera is caused by the bacillus *Vibrio cholerae* which thrives in warm and humid conditions. The whole progress of the disease can take as little as 5 to 12 hours but is usually 3 to 4 days. The incubation period is thought to be a minimum of 24 hours and a maximum of 5 days (Evans 1987).

At this period industrialisation was gathering pace in mainland Europe, in particular in the German states. Epidemics of cholera had periodically caused heavy loss of life in the large European cities for the same reasons as in London (Evans 1987). The first comprehensive sewer network in Europe was begun in Hamburg in 1848 by the English engineer William Lindley (Melosi 2000; Evans 1987). Lindley had gone to Germany in 1838 to construct a railway and then stayed to construct public bath- and wash-houses and then later the sewer network. He was a disciple of both Isambard Kingdom Brunel (railways and civil engineering, such as bridges) and Edwin Chadwick. Lindley was involved in the reconstruction of the city following the Great Fire of Hamburg and this allowed him to get agreement on the construction of a centralised sewer network. In November 1842 he travelled to London to discuss the latest ideas with Chadwick and to examine the progress made. He proposed a sewerage network for the central city of Hamburg in 1843 and this was finally accepted, with construction starting in 1848. The system came into operation in 1853 and the district of St Pauli was connected in 1859. By 1860 there were 48 km of sewers in the city but this network did not cover the new suburbs (Evans 1987). It was not until the 1890s that the municipal authorities could claim that all of the city was completely sewered. The last cholera epidemic was in 1893.

Lindley had included that the idea of wastes being sold to farmers as fertiliser was impractical. This system allowed tidal flow to flush out the main sewers once a week. It made a big impact and was the model for other European cities where English engineers were employed and in the US where it was the model for the New York and Chicago sewerage systems. Lindley was also involved in potable water supply and sanitation projects in Budapest, Warsaw, St Petersburg, Basle and Frankfurt, amongst others.

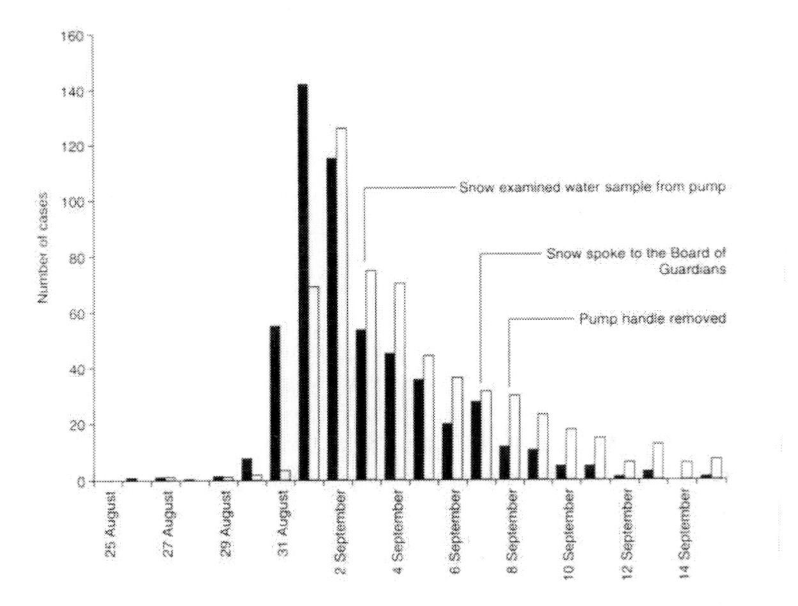

Figure 2.3. Onset of fatal cases of and numbers of deaths in the Broad Street outbreak.

2.5.3 Land treatment

Following the rapid expansion of the cities and towns the first treatment process applied was land treatment, a process which went back to Roman times and even into pre-historic times (Wolfe 1999). One of the first organised users was James Smith, a Stirlingshire cotton mill owner. He found that taking the excrement from his factory privies to his farm improved crop yields (Stanbridge 1976). In 1842 he moved to London and adopted the ideas of James Vetch for distributing sewage on the land by hoses and jets. These ideas were enthusiastically followed by Edwin Chadwick. He was greatly encouraged by Justus von Liebig, the eminent German chemist, who argued that the fertiliser value (particularly the phosphate content) should be used on agricultural land (Stanbridge 1976). Smith was appointed to the Commission on Health of Towns. A whole range of ideas modifications and process designs were used over the next fifty years. The large towns and cities bought more and more land for their sewage 'farms'. To this day many sewage treatment works are referred to in common parlance as sewage farms. The use of land treatment continued into the twentieth century and the last system in the UK continued to be used until the 1980s. The systems were gradually abandoned because:

(1) They used large areas of land which became more difficult and hence expensive to purchase around the expanding towns and cities.
(2) The land suffered from clogging and waterlogging.
(3) They were unable to achieve the higher hygiene standards required.

2.5.4 Chemical treatment

Chemical treatment of sewage discharges had been used in Paris in 1740 using lime as the precipitant (Wardle 1893).

Between 1850 and 1910 there were several hundred patent applications for recipes for treating sewage. The purpose was twofold: (a) to treat the sewage to remove enough of the pollutant concentration to get it to the point at which the effluent could be safely disposed, and (b) to produce artificial guano. In the early 1800s the UK imported large quantities of the bird droppings from South America and the Galapagos Islands to fertilise farmland and improve cropping yields. It had been shown that sewage could improve fertility but the sewage was diluted and it needed large areas to apply. By using chemicals to enhance the settlement rate and recover more of the solids in a much smaller volume the sludge produced could provide the concentrated fertiliser. Hence it was thought that two problems could potentially be solved at once.

Chemical treatment was helpful in removing some of the polluting load but it had two main disadvantages: it could only remove suspended pollutants and this left about one third of the total pollutant load present in the treated sewage, and it produced a much larger quantity of sludge, which was difficult to dispose of.

When the biological processes (which removed dissolved pollutants as well as the suspended matter) came along at the end of the 1800s then chemical treatment gradually went out of use. It underwent a revival in the 1970s for the removal of phosphates, and continues in this role today (Culp and Culp 1971).

2.6 DEVELOPING THE BASIC TREATMENT PROCESSES: 1870 TO 1914

2.6.1 Primary settlement

When it first became practice to use farm land to treat sewage, trenches or pits were sometimes dug to remove the heavier solids prior to application, thereby reducing the load on the land. When they were filled they were covered over (Stanbridge 1976) and others dug. Possibly the first use of this was at Craigentinny Meadows in Edinburgh in 1829.

The next development consisted of flat-bottomed tanks which were sometimes clay-lined. It seems that these were operated on a fill-and-draw basis, with the removal of water by siphoning. In his patent of 1846 for lime as a precipitant William Higgs mentioned 'tanks or reservoirs in which the contents of sewers and drains from cities, towns and villages are to be collected and the solid animal or vegetable matters therein contained, solidified and dried' (Stanbridge 1976). Horizontal-flow tanks seem to have been invented in the 1850s and radial-flow tanks in 1905. Many of these systems had to be manually desludged with scrapers and squeegees. There were some bucket-and-winch operated systems for desludging in the 1850s and 1860s but true power-operated mechanical systems did not make an appearance until the 1900s.

In 1860, L.H. Mouras of Vesoul in France designed a cesspool in which the inlet and outlet pipes dipped below the water surface thus forming a water seal. This so-called 'fosses Mouras' was described by the Abbe Moigno in 'Cosmos les Mondes' in 1881 as it had been found that liquefaction of the solids took place, which was attributed to anaerobic action (Stanbridge 1976). This is a pre-cursor of modern septic tanks. In 1895 Donald Cameron, the city Surveyor of Exeter, and F.J. Cummins patented a similar, but improved, system and Cameron called it a 'septic tank'. The process gained great popularity and one observer commented, 'Since the septic tank idea gained favour every designer of sewage tank has used the name septic for his tank, and apparently with good reason, for originally the word septic meant simply bacterial, just as the word anti-septic means anti-bacterial' (Melosi 2000).

The Imhoff tank, designed in 1906 by Karl Imhoff of the Emscher Drainage Board in Germany, was a further advance. This improved upon the design of septic tanks by using two chambers which allowed the separation of the settlement and sludge digestion processes. The system was so successful that Imhoff tanks comprised nearly half the total treatment works in the US by the end of the 1930s (Wolfe 1999), and it is still in worldwide use.

2.6.2 Biological filters

Up to 1900 virtually all the sewage treatment, where it existed, was carried out by land treatment. The farms were not always successful, as waterlogging was a major problem (Nicoll 1988). As the population continued to expand it became more and more difficult to find sufficient areas of land on the fringes of the towns and cities. The idea that there might be better ways, using 'organisms', gradually began to emerge. In 1870 Sir Edward Frankland established the fundamental principles of filtration through soil on which much of future developments depended (Second Royal Commission on Rivers Pollution 1870). In one of his experimental filters containing coarse porous gravel at Beddington Sewage Farm in Croydon, south of London, it was found that a rate of application of 0.045 m^3/m^3 of bed per day produced a well-nitrified effluent and that the 'filter' showed no signs of clogging after four months of operation (Stanbridge 1976). In 1882 Warington wrote that, 'sewage contains the organisms for its own destruction, and these may be so cultivated as to effect the purpose.' He went on to suggest the first idea of a filter bed which would have 'a greater oxidising power than would be possessed by an ordinary soil' (Nicoll 1988). He also suggested the use of a filter containing a more porous medium than natural soil (Stanbridge 1976). In 1887 William Dibdin (the chief chemist of the London Metropolitan Board of Works and later the London County Council from 1882 to 1897) stated that,

> ...in all probability the true way of purifying sewage...will be first to separate the sludge, and then turn into neutral effluent a charge of the proper organism, whatever that may be, specially cultivated for the purpose; retain it for a sufficient period, during which time it should be fully aerated, and finally discharge it into the stream in a purified condition. This is indeed what is aimed at and imperfectly accomplished on a sewage farm.

This is probably the first statement of what is achieved by modern primary and secondary treatment. The idea that there might be a way of biologically treating sewage was revolutionary at the time, but the sewage farm did demonstrate that if sewage was passed through a sandy, gravelly soil it became less polluting and from this came the idea of 'artificial ground' which led on to the 'contact bed', and eventually to the modern biological filter (Nicoll 1988). After Warington's suggestion (Warington 1882), Baldwin Latham installed 'artificial filters' at Merton, south of London, that contained alternating layers of burnt clay and soil (Stanbridge 1976). Between 1885 and 1891, various artificial filters were constructed across the UK.

The dramatic breakthrough in biological filter design for more reliable performance was made in the US at the Lawrence Experimental Station of the

Massachusetts State Board of Health (MSBH) which had been established in 1886. Local Boards of Health were set up in the US in the 1880s along similar lines as in the UK twenty years earlier. They were set up to combat disease in the rapidly growing cities.

Table 2.3 shows the very rapid rate of growth in population in the USA and the population served by sewage treatment in the period 1880 to 1920.

Table 2.3. Urban population in the USA 1880 to 1920 (from Melosi 2000)

Year	US population (million)	Urban population (million)	Population with sewage treatment (million)
1880	50.15	14.13	0.005
1890	62.95	22.11	0.100
1900	75.99	30.16	1.000
1910	91.97	41.99	4.450
1920	105.71	54.16	9.500

When it was set up, the Lawrence experimental station had been intended to conduct chemical analysis, but the association of drinking water with typhoid led the station to concentrate upon bacteriology (Melosi 2000) and then carry out tests on sewage treatment. They were evaluating the suitability of Massachusetts soil for oxidising organic matter in sewage. They confirmed Frankland's finding that gravel was the best filtering medium and in November 1890 the first 'trickling filter' was commissioned (Stanbridge 1976). Following on from this breakthrough, there was rapid progress in the US and the UK. At first the systems installed were intermittent filtration and contact beds but soon they developed as continuous flow filters as we know them today. The contact beds developed in the 1890s were:

...essentially tanks containing broken stones, slate or other coarse inert substances which provided a relatively large specific surface area for microbial growth. They were operated on a "fill-and-draw" basis, and bacteria on the filter bed decomposed the organic matter in the sewage. When the filter was empty, bacterial growth would be stimulated by the flow of air through the voids in between filter material' (Wolfe 1999; American Public Works Association 1976).

One of the earliest biological filters was used at Salford near Manchester in the UK, in 1893 whilst the first in the US was used at Madison, Wisconsin in 1901. Between 1895 and 1920 many were installed to treat sewage from towns and cities in the UK. This rapid application had a negative effect upon the later implementation of the activated sludge process in the UK after it was invented in 1913. City and town councillors were reluctant to spend money on another new-fangled process when they had already committed taxpayers' money to the biological filter process !

From that time on it was a case of gradual development of the biological filter process which does not look too different today than it did in 1900. Many of the early 20th century systems are in operation throughout the world.

2.6.3 The Royal Commission on Sewage Disposal

In 1898 an important event occurred in the formation of the Royal (Iddesleigh) Commission on Sewage Disposal by the UK government. This commission was to write a series of ten reports between 1901 and 1915. The Royal Commission's eighth report in 1912 (Royal Commission on Sewage Disposal 1912) had significant effects since it was concerned with the standards (and testing methods) to be applied to the sewage and effluent being discharged to rivers. It recommended the so-called '20:30 standard', 'Royal Commission Standard' or 'general standard', which was copied by many other countries. This is a general standard of 20mg BOD_5/litre, 30 mg suspended solids/litre for effluent discharges from sewage treatment works. What is often forgotten is that this standard is specific to a dilution of at least eight-fold being achieved in the receiving water !

2.7 THE AGE OF PROCESS DEVELOPMENT: 1914 TO 1965

2.7.1 Activated sludge

Since about 1882 experiments had been carried out on the aeration of settled sewage but in the last two decades of the nineteenth century research efforts had concentrated on treatment by the promising biological filtration theories. In November 1912 Dr Gilbert Fowler of the University of Manchester visited the US in connection with the pollution of New York harbour (Cooper and Downing 1997, 1998). He was also employed as the consultant chemist to Manchester Corporation. On his return he described to his colleagues, Edward Ardern and William Lockett, some experiments that he had seen at the Lawrence Experimental Station of the Massachusetts State Board of Health, in which sewage was aerated in a bottle which had been internally coated with green algae. Tests had also been carried out in an aerated tank containing slabs of slate spaced 25mm apart. Fowler suggested to his colleagues that similar tests should be carried out in Manchester. He was keen on finding a clotting mechanism and had in 1913 worked with Mumford on the M7 mechanism (Ardern and Lockett 1914; Coombs 1992). This was a bacterium found in colliery workings which could help to precipitate organic matter in the presence of low concentrations of iron salts. During 1913 and 1914 they aerated sewage continuously for several weeks and achieved complete nitrification. Lockett allowed the treated liquid to settle and decanted off the supernatant liquid leaving behind the first activated sludge. The bottles were covered with brown paper to cut out light and prevent the growth of algae. Whereas other workers undertaking similar work had discarded the sewage in its entirety after the aeration, the Manchester workers then added further portions of sewage and aerated this in contact with the original settled solids. They found that after each of these aeration periods the amount of solids, now called sludge, had increased and that the period needed for oxidation of the matter in the sewage reduced until it was eventually possible to achieve complete oxidation in 24 hours (Institute of Water Pollution Control 1987). These tests were all done at the Davyhulme Sewage Works in Manchester using sewage from four different districts of Manchester plus a sample from Macclesfield. The results were discussed in the classic paper by Ardern and Lockett which was presented to the society of the Chemical Industry at the Grand Hotel, Manchester on 3 April 1914. During 1914 the process was scaled up to pilot plant scale at Davyhulme Sewage Works. Some of the tests were continuous-flow experiments and some used the fill-and-draw technique (which was a precursor of the modern sequencing batch reactors). The initial Davyhulme work was done with coarse-bubble aeration and later with fine-bubble aeration. Two years later the first full-scale continuous-flow was installed at Worcester (Coombs 1992; Institute of Water Pollution Control 1987).

In 1914 a large-scale test had been carried out at Salford using the fill-and-draw technique. It is interesting to note that these fill-and-draw plants achieved full nitrification and there was no problem with sludge settlement or bulking. By the time that the first book was written on the activated sludge process (Martin 1927) the process was being used in the US, Denmark, Germany, Canada, the Netherlands and India (Professor Fowler had gone to work at the Indian Institute of Technology).

The first British city to fully apply the activated sludge process was Sheffield in 1920. By contrast, its application in the US was far more rapid. The reason for this was because following the First World War capital for investment was very limited in the UK and because all the major cities had already invested in sewage works based on the biological filter process in the period between 1890 and 1910. Hence the major activated sludge works at Mogden in London (which served 1.25 million people), Davyhulme in Manchester and Coleshill in Birmingham were not built until 1934 or 1935. In the US, by contrast, many of the activated sludge plants were the first form of sewage treatment ever used.

Large-scale tests (500m^3/day) took place at San Marcos in Texas in 1916. This was followed by full-scale tests at Houston, Texas (40,000m^3/day) in 1917, Des Plaines, Illinois (20,000m^3/day) in 1922, Milwaukee (170,000m^3/day) and Indianapolis (190,000m^3/day) in 1925 and then in Chicago North (660,000m^3/day) in 1927.

The process was first applied in Europe in Denmark in the Soelleroed Municipality in 1922 (Henze *et al.* 1997). Work commenced in Germany in 1924 when the first experimental plant was built at Essen by Imhoff (von der Emde 1964, 1997). This was followed by the first full-scale system in Germany at Essen-Rellinghausen in 1926. In 1927 Kessener treated an abattoir effluent at Apeldoorn, in the Netherlands (Institute of Water Pollution Control 1987) using an activated sludge process equipped with a brush aerator.

In 1938 Mohlman, reviewing the first twenty-five years of the activated sludge process for the Federation of Sewage Works Association in the US, wrote:

> In 1913, activated sludge was discovered and recognised by W.T. Lockett in the course of some bottle experiments in the laboratory of the Manchester sewage treatment works. In 1938, the activated sludge process is in operation in hundreds of full-scale sewage treatment works and more than a billion gallons of sewage are treated every day. Activated sludge plants are now operated all over the world, extending from Helsinki, Finland to Bangalore, India; from Flin Flon, Manitoba, Canada to Glenelg, Australia; and from Golden Gate Park, San Francisco to Johannesburg, South Africa. Huge plants are in operation at London, New York, Chicago, Cleveland and Milwaukee. This astounding growth in the past twenty-five years is unparalleled in the history of sewage treatment, and must be ascribed to the fact that the activated sludge process is in harmony with the speed of and science of modern life. Sewage treatment works in our modern cities can no longer be obnoxious or inefficient. They must be free from odour, occupy limited area, and be amenable to scientific control.

The Second World War held up development of the process until about 1948 when the search for a way to better control plant performance began. This search was to occupy many workers over the next forty years in many different countries.

The activated sludge process and its many variants is now the main engine of secondary sewage treatment and has probably had the biggest impact of all processes upon environmental improvement in the past century.

Progress in the rest of Europe had not been as rapid as in the UK and the US. In Finland, progress was delayed by the Russian War and occupation, and in 1910 Finand only had three sewage treatment works. This rose to seven by 1950 (Katko 1997) and grew quickly in the 1960s after the Water Act had been passed. The effect of this upon the health of the population in contrast with Sweden, Switzerland and England and Wales is seen in Table 2.4. There are now 110 modern sewage treatment works in Finland (Katko 1997).

Table 2.4. Mortality rates from typhus and paratyphoid fever in selected European countries from 1930 to 1959 (from Katko 1997)

Country	Period	Mortality rate per million people per year
England and Wales	1941–1950	1.5
Sweden	1941–1947	4.0
Switzerland	1941–1949	5.3
Finland	1931–1940	25.0
Finland	1941–1950	43.0

2.8 PROCESS REFINEMENT TOWARDS STANDARDS DICTATED BY ENVIRONMENTAL PROTECTION: 1965 TO 2000

In this period the emphasis has been on:

- more widespread application of known techniques for BOD and TSS removal;
- environmental protection and improvement by the removal of nitrate, phosphate and ammoniacal nitrogen; and
- disinfection.

More fixed film process variants of the original biological filters have gradually been developed. Examples of these such as submerged aerated biological filters and plastic media biological filter systems are now common.

2.8.1 Nutrient removal

Nutrient removal processes to help prevent eutrophication and to protect water sources from high nitrate concentrations have developed rapidly in this period.

By the 1960s the main engine of secondary treatment was the activated sludge process. One of the major problems with activated sludge systems in the period up to the early 1960s was that the oxidation of ammoniacal nitrogen (nitrification) was not reliable or predictable. The solution to this was discovered by an investigation by Downing *et al.* (1964) at WPRL (later part of WRc) at Stevenage. The results of that work are now incorporated into design methods and computer models. Biological denitrification had been known about since the late 1800s but denitrification first took place in sewage treatment in the late 1930s (Edmondson and Goodrich 1947). They used the nitrate as a source of oxygen for an overloaded biological filter. In 1962, in the US, Ludzack and Ettinger put forward the use of anoxic zones to achieve biological denitrification in an activated sludge process. This concept is now standard practice in all AS processes and some fixed film processes.

The problem of how to remove phosphorus in activated sludge processes was solved by James Barnard (1974) and his colleagues in South Africa. This technique is now applied worldwide. In the second half of the twentieth century the South African water industry has protected its water resources very carefully and developed recycling processes because of a water shortage and a rapidly growing population. As a result, some of the most advanced sewage treatment processes have been developed here.

2.8.2 Standards

In the 1970s, a move started to raise standards and improve environmental protection, to some extent driven by public opinion and greater public awareness. The first step in this direction was the Clean Water Act in the US in 1972. As the European Union expanded from the original five states to the present fifteen there have been a series of directives aimed at the prevention of water pollution and protection of cross-border water resources. This began with the Surface Water Directive in 1975 followed by the Bathing Water Directive in 1976, the Fishing Waters Directive in 1978, the Shellfish Water Directive in 1979 and the Drinking Water Directive in 1980. The Urban Waste Water Treatment Directive (CEC 1991) has had a very significant impact upon operators in the last five years since it provides European-wide standards and introduces more stringent standards for nitrogen and phosphorus levels.

2.8.3 Sludge treatment and disposal

Little has been said about sludge treatment and disposal in the earlier periods. It has become a more significant problem in the last twenty years as easy disposal routes have been gradually closed. It is no longer permissible in Europe to discharge sewage sludge to sea; a common practice until the 1990s. Standards for disposal on agricultural land have also become tighter. Many new processes have been proposed and developed. The most common use in the UK is still on agricultural land but incineration and drying/pelletisation are becoming more popular. The sludge treatment and disposal route should be considered at the earliest stage in any process design.

2.8.4 Computer modelling and control

The advent of industrial electronic computers (and electrically controlled valves) in the late 1970s made automatic control of process units a possibility for the first time, and this has progressed apace since that time. In the late 1980s when the first affordable personal computers (PCs) became available, there was another change with respect to the development of computer models of the treatment processes, in particular the activated sludge process, which had previously required powerful mainframe computers. The IAWPRC model (based on COD) (Olsson and Newell 1999) and the WRc STOAT model (based on BOD) (Smith and Dudley 1998) have led the way. They are particularly helpful in allowing a 'dry run' of weather conditions and checking how outside factors, such as storm conditions, will affect the treatment process (Smith *et al.* 1998).

2.8.5 Reed beds/constructed wetlands

Over the past twenty years there has been a rise in interest in less sophisticated drainage and treatment systems such as pond and wetland treatment systems. This has been driven in Europe by the desire to provide safe treatment at a lower cost.

The use of reed beds (also known as constructed wetlands) came in the 1980s (Cooper and Findlater 1990; Cooper *et al.* 1996). These systems are particularly useful for small rural decentralised wastewater treatment systems. There are tales which indicate their use long ago in Italy, even that the Romans may have known of their use.

2.8.6 Anaerobic treatment of wastewaters

Over the past fifty years a number of attempts have been made to apply anaerobic processes to the treatment of wastewaters. There has been considerable success in treatment of agricultural and industrial wastewaters largely based upon the UASB (upflow anaerobic sludge blanket) reactors pioneered in the Netherlands in the 1970s (Zeeman *et al.* 2001). These process systems have been successful because these agricultural and industrial wastewaters are usually warm or concentrated (or both) organic wastes. Municipal domestic sewage is usually cold and weak and so efforts to apply anaerobic treatment have not yet been successful. Recently, considerable research effort has been devoted to the anaerobic treatment of the concentrated wastewaters that result from the separation of 'grey' and 'black' waters in domestic homes (Zeeman *et al.* 2001). This looks to be a promising possibility for localised decentralised treatment, but will not be the solution to treatment of the present weak domestic sewage.

2.8.7 Membrane systems

One of the most important developments relates to the use of membranes. This is possibly the most novel process of the past forty years. Tertiary or quaternary treatment using membranes for removing bacteria is already carried out in Europe, Australia and the US. The potential for use of membranes in reverse osmosis (RO), micro-filtration (MF) and ultra-filtration (UF) has been known since the 1960s (they were used in the American missions to the moon) but research and development has only recently resulted in membranes that are cheap enough to allow for their use with concentrated wastes such as sewage. Total operating costs have dropped four-fold, that is, by 75 per cent since 1992 (and are probably around a hundred times cheaper than they were in the early 1970s). Probably the most exciting application is in membrane biological reactors (MBRs) such as the Kubota system from Japan. In this the membrane panels are inserted directly into the activated sludge aeration tank. This has several advantages:

(1) It allows for operation without a settlement stage (always the most unreliable part of any AS process).
(2) It eliminates a substantial amount of piping.
(3) A tertiary treated effluent is produced in two stages or even one stage.
(4) A disinfected effluent with no TSS is produced, which could be reused for a secondary purpose.
(5) An automated AS process unit could be operated as a package unit for much smaller populations. In the past, poor sludge settlement has hindered this application.
(6) It is possible in this system to allow the biomass concentration to increase to more than 15,000 mg/litre which means that the size of the aeration reactor can be dramatically reduced.

Kubota systems are already being used for small populations in Japan and the UK (Yates 2000). Two village/town systems for 4,000 and 23,000 people have been installed in the UK but it should be noted that the system was developed in the Kubota business group which had responsibility for populations up to 50 people. It may thus be a system that has great potential for a small decentralised sewerage system and for some reuse of treated effluent. A membrane system is also in use at the Millennium Dome at Greenwich, London for grey-water recycling to provide water for toilet flushing.

2.9 CONCLUDING REMARKS

History shows that change comes in cycles and that ideas and processes come back into use when developments in other fields make the improvements needed to allow them to succeed. A good example of this is the sequencing batch reactor (SBR). This was the original form of activated sludge process, the fill-and-draw process. It made a comeback in the 1990s because there is a need for a process which avoids bulking sludge. Interest in it began to revive in the 1970s but has developed strongly in recent years because the invention of computers and electronically controlled valves allows SBRs to operate automatically whereas, in the 1920s, everything had to be done by manual labour.

Another example of this cyclic situation is the current interest in Brazil in backyard drainage/condominial sewerage, a process originally proposed in the 1850s. Yet another example is the widespread use of chemicals which are now used for phosphorus removal rather than enhanced suspended solids removal, as in the previous century.

International cooperation and the free exchange of ideas has been very influential in accelerating development, particularly between 1850 and 1950. At this time there was a considerable exchange of ideas between London and east coast cities in the US such as New York and Boston which were experiencing rapid growth and problems in controlling sewage-linked diseases. A similar exchange of ideas has been seen within Europe and continues to this day.

The water-carriage system is very old. It began in around 2000 BC in Greece and then took hold in the UK as a result of the work of Edwin Chadwick in the 1840s. It is of course now the main form of sewage treatment in developed countries. Care will need to be taken in designing systems to treat sewage from low water use or vacuum systems since the high concentrations of ammoniacal nitrogen in these concentrated wastes may be toxic to the nitrifiers. I find it difficult to see this being usurped as the main form of sewerage system but I can also see the benefit of decentralised systems for small populations and rural areas far from large treatment works. The treatment of sewage from these systems is already being carried out in pond systems and reed beds/constructed wetland systems worldwide. These systems have huge potential for developing countries since they are cheap and can be constructed by local people using simple techniques and equipment (Cooper and Pearce 2000; Mara 2000).

The trend in Europe over the last thirty years has been to organise water and wastewater treatment on a river basin basis by using river basin authorities rather than by municipal councils, as happened in earlier times. This has benefited the areas and population by improving environmental protection and possibly also by lowering costs.

2.10 TIMELINE FOR WASTEWATER TREATMENT

3500-2500 BC	Mesopotamian empire stormwater drainage system. In Babylon clay pipes led to cesspools.
1700 BC	Four separate drainage systems in King Minos' palace. In Knossos, Crete, terracotta pipes drained to stone sewers.
c. 800 BC	Cloaca Maxima central sewer system built in Rome.
c. 100 AD	Sewer network in Rome connected to houses.
c. 400 AD	Brick sewers in London.
c 1100	Cistercian monasteries in Scotland locate next to watercourses and flush latrines via sewers to watercourse.
1189	Regulations in London on placement of cesspools.
1370	First covered sewer in Paris dumps sewage into the River Seine near the Louvre.
1531	Commission on Sewers in London.
1596	Sir John Harington builds two water closets for Queen Elizabeth I. Called the 'Necessary', this is the first water closet flushed by a valve system.
1740	First recorded mention of chemical treatment of sewage. Lime used in Paris.
1776	Magistrate John Shortbridge requires Glasgow tenants to drain water from kitchens via lead pipes, and excreta to be taken to middens.
1790	First sewer built in Glasgow.
1793	First water closet in Glasgow, 200 years after its invention. Edwin Chadwick publishes the landmark report to the Poor Law Commissioners, Report on the Sanitary Condition of the Labouring Population of the Great Britain. Health of Towns Association formed.
1844	Commission on Health of Towns adopted Chadwick's Proposals.
1846	First British patent on chemical treatment is granted to W. Higgs for the use of lime.
1848	Public Health Act in the UK masterminded by Edwin Chadwick. Set up local Boards of Health and gave them rights to construct sewers.
1849	Metropolitan Commission of sewers for London.
1848-54	Dr John Snow proves link between cholera outbreak and water supply polluted by sewage.

1853	First comprehensive sewerage system completed in Hamburg, Germany. System designed by William Lindley serves as model for US and European cities.
1850-1910	Many patents applied for in the UK and US for chemical treatment of sewage. Four hundred and seventeen patents granted in the UK between 1856 and 1876.
1860	Overflowing cesspool (precursor of septic tank) designed in France by L.H. Mouras.
1862-65	More soldiers die from typhoid and cholera than combat in US Civil war.
1866	Medical Officer of the Privy Council (advisers to Queen Victoria) reported that death rates had dropped considerably where the Chadwick report recommendations were followed.
1868-70	Frankland's tests on filtration of sewage through soil and gravel (an extension of land treatment). Nitrification achieved.
1870-90	Many tests in the UK and US on filtration of sewage through various media.
1887	Dibdin suggests basis for biological treatment by organisms and describes modern primary and secondary treatment
1890	First true biological filter at Lawrence Experimental Station, Massachusetts State Board of Health, US.
1890-1900	Many tests and designs in the UK follow up American work on biological filters.
1895	Cameron and Cummins (Exeter) patent septic tank.
1898	1st Royal Commission on Sewage Disposal in the UK.
1906	Imhoff tank designed in Germany.
1912	8th Royal Commission on Sewage Disposal defines the 20 mg BOD/litre; 30 mg SS/litre 'Royal Commission Standard'.
1913	First laboratory experiments on activated sludge by Fowler, Ardern and Lockett at University of Manchester, UK.
1916	First full-scale activated sludge plant at Worcester. Large-scale tests in the US. First full-scale AS plant in US at Houston, Texas.
1922	Activated sludge plant built at Soelleroed, Denmark.
1924	Pilot AS plant in Germany at Essen.
1926	Full-scale AS plant at Rellinghausen, Germany.
1927	Kessener brush aeration, Apeldoorn, the Netherlands.
1936	Denitrification used in Sheffield.
1964	Development of basis for consistent nitrification by Downing, Painter and Knowles, WPRL, Stevenage, UK.

1972	Biological phosphorus removal described by Barnard in South Africa.
1970s	Development of dynamic process computer models by WRc and IAWPRC.
1990s	Membrane biological reactors developed in Japan.

2.11 REFERENCES

American Public Works Association (1976) *History of Public Works in the United States, 1776-1976,* p. 403.

Ardern, E. and Lockett, W.T. (1914) Experiments in the oxidation of sewage without the aid of filters. *Journal of the Society of the Chemical Industry* **33**(10), 524.

Barnard, J.L. (1974) Cut P and N without chemicals. Part 1. *Water and Wastes Engineering* **72**(6), 705.

BBC (2001) BBC Radio 4 Radio Series. *Disease Detectives; Cholera and Dr John Snow,* 24 January, 2001, British Broadcasting Corporation, UK.

Bingham, P. (1999) Dr Arthur Hill Hassall 1817-1894: microscopist to the Broad Street outbreak. *Health and Hygiene* **20**, 106–108.

Binnie, C. (1999) The Present London. In World of Water 2000 (full reference given below), pp. 40–51.

Campbell, B. (2000) Britain 1300. *History Today* **50**(6), June, 10–17.

Chartered Institute of Environmental Health (1998) For the common good: 150 years of Public Health. CIEH, London, UK.

Coombs, E.P. (1992) *Activated Sludge Ltd. – The Early Years*, published privately by C.R. Coombs, Bournemouth, Dorset, UK.

Commission of the European Communities (1991) Urban Waste Water Treatment Directive.

Cooper, P.F. and Findlater, B.C. (eds) (1990) *Constructed Wetlands in Water Pollution Control*, Pergamon Press, Oxford, UK, p. 605.

Cooper, P.F., Job, G.D., Green, M.B. and Shutes, R.B.E. (1996) *Reed Beds and Constructed Wetlands for Wastewater Treatment.* Water Research Centre, Medmenham, UK, p. 208.

Cooper, P.F. and Downing, A.L. (1997) Milestones in the development of the activated sludge process over the past eighty years. Paper presented to the CIWEM Conference, Activated Sludge into the 21st Century, Manchester, UK, 17–19 September.

Cooper, P.F. and Downing, A.L. (1998) Milestones in the development of the activated sludge process over the past eighty years. *Journal of the Chartered Institution of Water and Environmental Management* **12**(5), 303–313. [This is an abridged version of the paper above and the pre-1950 material has been drastically shortened.]

Cooper, P.F. and Pearce, G. (2000) The potential for the application of constructed wetlands in the village situation in arid developing countries. Paper presented to the CIWEM Aqua Enviro conference, Wastewater Treatment: Standards and Technologies to meet the Challenge of the 21st Century, Leeds, 4–6 April.

Culp, R.L. and Culp, G.L. (1971) *Advanced Waste Water Treatment,* Van Nostrand Reinhold, New York.

Dibdin, W.J. (1887) Sewage sludge and its disposal. *Proceedings of the Institution of Civil Engineers*, p. 155.

Downing, A.L., Painter, H.A. and Knowles, G. (1964) Nitrification in the activated sludge process. *Journal of the Institute of Sewage Purification* **2**, 130.

Edmondson, J.H. and Goodrich, S.R. (1947) Experimental work leading to increased efficiency in the bio-aeration process of sewage purification and further experiments on nitrification and recirculation in percolating filters. *Journal of the Institute of Sewage Purification* **2**, 17–43.

Evans, R.J. (1987) *Death in Hamburg: Society and politics in the cholera years 1830–1910,* Oxford University Press, Oxford, UK, p. 673.

General Board of Health (1852) Minutes of information collected with reference to works for the removal of soil, water or drainage of dwelling houses and public edifices and for sewerage and cleansing of the sites of towns, HMSO, London, p. 144.

Halliday, S. (1999) *The Great Stink of London: Sir Joseph Bazalgette and the Cleansing of the Victorian Metropolis*, Sutton Publishing, Stroud, Gloucestershire, p. 210.

Henze, M., Harremoes, P., la Cour Jansen, J. and Arvin, E. (1997) *Wastewater Treatment – Biological and Chemical Processes*, 2nd edn, Springer Verlag, Berlin, p. 54.

Institute of Water Pollution Control (1987) *Unit Processes: Activated Sludge*, IWPC, London, UK. [Reprinted and updated in 1997 by Chartered Institution of Water and Environmental Management.]

Katko, T.S. (1997) *Water: Evolution of water supply and sanitation in Finland from the mid-1800s to 2000.* Finnish Water and Waste Water Works Association, Helsinki, Finland.

Lee, C. (1997) *This Sceptred Isle: 55 BC–1901*, BBC Books and Penguin Books, London, p. 616.

Lee, C. (1999) *This Sceptred Isle: Twentieth Century,* BBC Worldwide and Penguin Books, London, p. 497.

Mara, D. (1999) Condominial sewerage in Victorian England. *Water and Environmental Manager*, September, 4–6.

Mara, D. (2000) A long way to go: Wastewater treatment in developing countries in the twenty-first century. Paper presented to the CIWEM and Aqua Enviro conference, Wastewater Treatment: Standards and Technologies to meet the Challenge of the 21st Century, Leeds, 4–6 April.

Melosi, M V. (2000) *The Sanitary City: Urban infrastructure in America from colonial times to the present,* Johns Hopkins University Press, Baltimore, Maryland, p. 578.

Mohlman, F.W. (1938) Twenty-five years of activated sludge, Chapter VI in *Modern Sewage Disposal*, anniversary book of the Federation of Sewage Works Association, USA, p 38.

Nicoll, E.H. (1988) *Small Water Pollution Control Works: Design and Practice*, Ellis Horwood, Chichester, UK.

Olsson, G. and Newell, R. (1999) *Wastewater Treatment Systems: Modelling, Diagnosis and Control*, IWA Publishing, London.

Rennie, D (2000) China flushed with pride over loo find. The *Daily Telegraph*, 27 July, p 18.

Royal Commission on Sewage Disposal Eighth Report (1912) Standards and tests for sewage and sewage effluent discharging to rivers and streams. Cd 6464, London.

Second Royal Commission on Rivers Pollution (1870) First report. London.

Stanbridge, H.H. (1976) *History of Sewage Treatment in Britain* (in 12 volumes), Institute of Water Pollution Control, Maidstone, UK.

Smith, M., Cooper, P.F., McMurchie, J., Stevenson, D., Mann, B., Stocker, D., Bayes, C. and Clark, D. (1998) Nitrification trials at Dunnswood Sewage Treatment Works and process modelling using WRc STOAT. *Water and Environmental Management* **12**(3), 157–162.

Smith, M and Dudley, J (1998) Dynamic process modelling of activated sludge plants. *Water and Environmental Management* **12**(5), 346–356.

von der Emde, W. (1964) 50 Jahre Schlammbelebungsverfahren, *Gas-u-Wasser Fach* **105**(28), 755–760. [In German.]

von der Emde, W. (1997) The history of the activated sludge process. Review for ATV, Germany.

Wardle, T. (1893) *On Sewage Treatment and Disposal*, John Heywood, London, p 43.

Warington, R. (1882) Some practical aspects of recent investigations on nitrification. *Journal of the Royal Society of Arts*, 532–544.

Wolfe, P. (1999) History of wastewater. In *World of Water 2000* (full reference given below) pp. 24–36.

Wolfe, P. (ed) *World of Water 2000 – The Past, Present and Future* (1999). Water World/Water and Wastewater International Supplement to PennWell Magazines, Tulsa, OH, USA, pp. 167.

Yates, J. (2000) Membrane process systems are establishing a significant position for the treatment of both municipal and industrial wastewaters. Paper presented to the CIWEM and Aqua Enviro conference, Wastewater Treatment: Standards and Technologies to meet the Challenge of the 21st Century, Leeds, 4–6 April.

Zeeman, G., Kujawa-Roeleveld, K. and Lettinga, G. (2001) Anaerobic treatment systems for high-strength domestic waste (water) streams. Chapter 12 in this book.

3

Decentralized versus centralized wastewater management

P. A. Wilderer

3.1 INTRODUCTION

The ministers of the environment of the EU member states decided at their annual meeting in Weimar in May 1999 to adopt and enforce a concept called 'Integrated Product Policy (IPP)'. They declared this concept a common basis of the European environmental legislation. IPP was defined as a 'governmental means to gradually improve products and services with respect to their environmental impacts taking into account the entire life span of the products and services', from cradle to grave, so to speak. This concept is aimed at overcoming the so-called 'end-of-the-pipe' technology and replacing it by an overall material flux management approach (see Figure 3.1) that includes minimization of the material flux, recovery of valuable materials and returning

Edited by P. Lens, G. Zeeman and G. Lettinga. ISBN: 1 900222 47 7

them into the material cycle. Environmentally and economically sound production of goods (that is, clean technology) and the application of advanced recycling technologies are of concern here, but equally important are the environmentally and economically sound distribution and use of goods and services (for instance, minimization of fuel consumption and CO_2 emissions from motor vehicles). Thus, IPP takes an holistic approach. IPP aims for low consumption of resources, long-lasting technology and advanced treatment requirements. It may be understood as the consequence of life cycle analysis that favours the sustainable development of economy, society and environment.

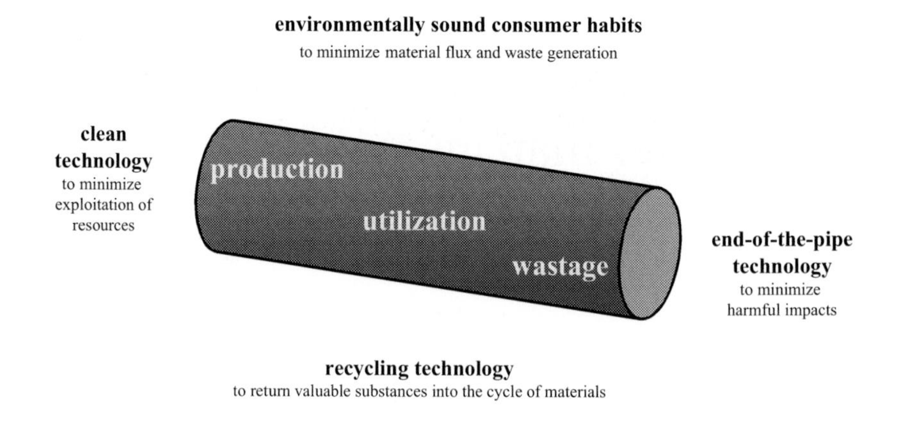

Figure 3.1. 'The pipe' as a model of the traditional linear sequence of production, utilization and waste discharge, overlaid by the holistic approach inherent in the IPP concept.

What has all this to do with water and wastewater, with centralized and decentralized water and wastewater management? One may assume that water is not a product but a natural good. It is taken from rivers, reservoirs or from the ground and delivered to customers, for instance to private households, to enterprises or to industrial plants. However, to be able to meet the quality requirements of the customers water has to be purified. As a result, water becomes a product that has to be purchased for further use by customers, and it eventually becomes wastewater. The linear sequence of production, usage and wastage typical for any industrial products applies also for water.

Treatment and distribution of water and wastewater requires the supply of building and piping materials, pumps and energy as well as a variety of chemicals, for example, chlorine. Usage of water is directly or indirectly associated with environmental impacts, in particular when the water is used as process water in industry. The treated wastewater is discharged to surface water bodies and subsequently reused by municipalities and industries located downstream. IPP with respect to water management means that water uptake and distribution, the use of water as well as the collection, treatment and disposal of wastewater have to be seen holistically as a system to be optimized in order to minimize environmental impacts (Figure 3.2).

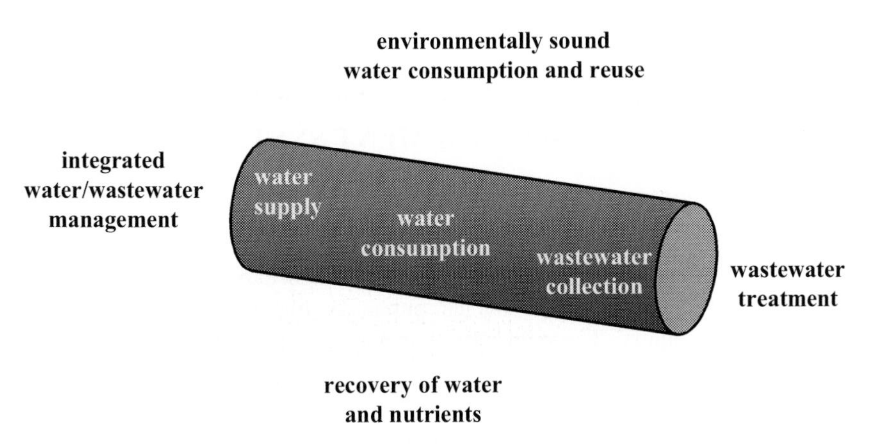

Figure 3.2. According to the IPP concept, the linear sequence of water supply, consumption and wastewater treatment has to be embedded into a closed-loop of water reclamation and reuse. Efforts have to be made to reduce water uptake, the introduction of harmful substances by users, leakage of sewer pipes, and the consumption of materials and energy for wastewater treatment.

When assessing the concept of IPP, it becomes clear that the aims of the concept can only be approached when the technical systems as they have been developed over the years are critically reviewed. When taking the IPP concept as a measure, further improvement and optimization of the traditional urban water/wastewater management systems becomes questionable. But are there any alternatives? To properly answer this question, economic and legal conditions as well as climatic and cultural differences in both industrialized and developing countries have to be taken into account.

It is estimated that within the next fifty years the percentage of people without access to safe drinking water will dramatically increase unless water supply and wastewater treatment systems are built and improved on a large scale (World Bank 2000). Already today, billions of people have no access to proper sanitation. Collection and treatment of solid waste is provided to far under half the earth's population. Millions of people are killed every year by water-borne diseases (Kalbermatten *et al.* 1999). Disease is spreading in many developing regions in Africa, Asia and Latin America. The quality of freshwater resources is deteriorating in many countries, caused by the discharge of untreated or poorly treated domestic and industrial wastewater, runoff from extensively fertilized agricultural land, leachates from mining pits and landfills. Not only developing countries are affected, but also industrial countries where water treatment regulations are improperly enforced.

3.2 EVALUATION OF CENTRALIZED WATER/ WASTEWATER MANAGEMENT SYSTEMS

3.2.1 Achievements and benefits

Supply of high quality drinking water, collection of sewage and stormwater and advanced treatment of the collected water prior to discharge into natural water bodies are major prerequisites of the public health, industrial growth and prosperity of a country. The relevance of this statement becomes evident when looking back in history. The rise of the industrialized countries was closely related to advances in and implementation of water supply technology, urban sanitation and wastewater treatment technology.

The municipal water and wastewater management systems that have developed over the years in industrialized countries are characterized by:

- acquisition of fresh water from protected areas (groundwater protection zones, protected water reservoirs);
- controlled purification of raw water and safe distribution of high quality water in sufficient quantity all year round;
- collection of sewage and stormwater by means of sewers;
- transport of the collected wastewater out of the urban area;
- advanced treatment of wastewater and stormwater;
- control of the treated water before discharge into natural surface water bodies; and
- treatment and utilization or controlled disposal of the waste sludge.

This method is commonly called 'central' water/wastewater management (Figure 3.3) because all the water to be distributed in the urban area is purified at discrete locations, and the wastewater collected in the area is sent to a discrete plant for treatment and discharge.

The advantages of this method are many. In particular, the treatment plants (water and sewage works) can be reliably and efficiently managed and controlled to the benefit of both the consumers and the environment. Besides, it is assumed that one large treatment plant is less expensive both in terms of capital and operating costs than many small plants serving the same urban area.

The key factor in this centralized treatment concept is 'control'. Violation of quality and quantity standards may have serious consequences. People may become ill when pathogenic organisms are not efficiently removed from water prior to distribution. Long-term health problems may develop when harmful substances such as heavy metals, chlorinated hydrocarbons, pharmaceuticals or hormone-like compounds are allowed to enter the distribution system in excess amounts. With the deterioration of the health of a population the economy of the region is threatened, and public welfare may decrease. Similarly, non-compliance with quantitative requirements may cause shortcomings in industry. Of equal importance is control of efficient operation of the wastewater treatment plant. Violation of the discharge limits would cause deterioration of the surface water quality. Downstream, where the water is used as a source of drinking water, the purification of the water would require more effort, greater care and increased costs.

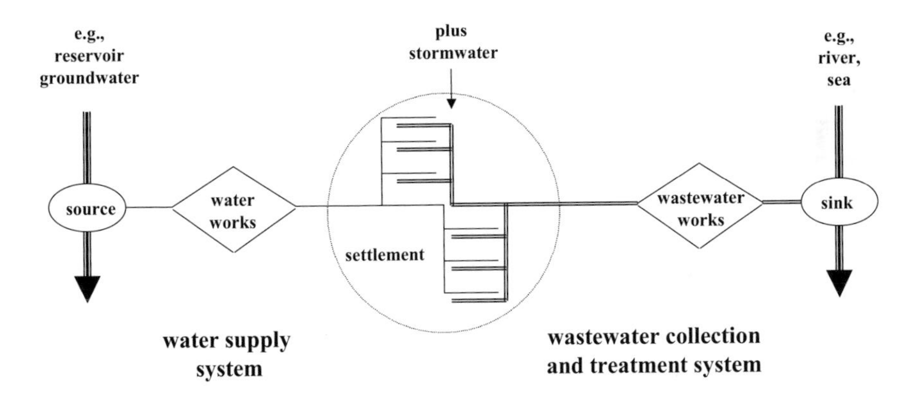

Figure 3.3. Traditional centralized system of water supply/wastewater collection and treatment.

3.2.2 Inconsistencies

Despite of all these positive aspects the central water/wastewater management system has disadvantages which should not be neglected, especially in view of the IPP concept mentioned above.

The cost benefits of central systems diminish when the costs of building and maintaining the distribution and collection system are taken into account. The cost of the installation of the water supply net and the sewer system are almost one order of magnitude higher than the cost of building the treatment facilities. Surveys conducted in many cities have revealed extensive leakage causing infiltration of groundwater or exfiltration of wastewater and subsequent groundwater pollution. The rehabilitation cost for the piping system in Germany, for instance, is estimated to be in the range of 100 billion euros.

Estimating the cost of worldwide implementation of centralized systems, it becomes evident that the capacity of the global money market would not be sufficient to cover the need for investment capital. It is hard to believe that centralized systems are seen as the best solution for problems in developing countries, especially in mega-cities.

There are further problems to be considered when assessing centralized systems:

- Only a small fraction of the high quality water distributed in urban areas is used for drinking, washing and cooking. The majority is used for cleaning, flushing and for watering plants and lawns. A significant amount of the drinking water is required just as a means to transport the pollutants to the wastewater treatment plant. Why should we use high quality drinking water for nothing but for transportation of waste?
- Taking water from a discrete location, sending it to a widely sealed urban area, and discharging it eventually to a distant surface water body may negatively effect the water balance in that area. The groundwater table may drop, increasing the need for artificial irrigation of gardens, parks and agricultural land. As the result, more water has to be taken from the natural water sources, thus making the situation even worse.
- Sewer pipes may leak, causing pollution of soil and groundwater by ex-filtration. Vice versa, infiltration will result in an increase of the hydraulic loading of the treatment plant. Higher capacity of pipes and tanks are needed to cope with an extra hydraulic load. Moreover, the wastewater becomes diluted which makes treatment more difficult.

- Implementation of a dual water supply system to allow distribution of fresh and reclaimed water is not possible on a large scale because of cost and hygiene reasons.
- Combining all kinds of wastewater and stormwater leads to a highly complex variety of pollutants that fluctuates heavily in composition and concentration. Thus, effective removal of the pollutants becomes very difficult.
- Wastewater contains components such as phosphorus which could be used as fertilizer provided the product is not spoiled by problematic substances such as heavy metals (Larsen and Gujer 1996a). Since the valuable components of wastewater are polluted and diluted, it is impossible to reclaim them for further use.
- Likewise, the sludge gained from wastewater treatment facilities is highly polluted but contains, potentially at least, a variety of valuable substances. Sludge can be converted into compost usable as a soil conditioner and as fertilizer. Unfortunately, sludges from central municipal treatment plants are often loaded with harmful materials including pathogenic organisms, household chemicals, pharmaceuticals and heavy metals making it difficult to convert the sludge into a useful product.

3.2.3 Conclusions and questions

Taking all these arguments into account it becomes obvious that the centralized system – as beneficial as it is in many respects – obviously does not comply with the demands of the IPP concept. Moreover, it cannot be applied, for financial reasons, to solve the problems of the developing countries that still have to install proper sanitation systems.

But what are the alternatives? The following questions should all be asked in the search for a workable system:

- Are decentralized systems the better approach, or a combination of both, centralized and decentralized systems?
- How small may the treatment plant of a decentralized system be?
- Is a decentralized system financially affordable?
- Can decentralized systems provide the same degree of safety and reliability as centralized systems?
- Who would own the individual plants and who would be responsible for their operation and maintenance?

Question after question springs up but satisfactory answers have not yet been found. In recent years, an increasing number of research teams and companies have become active in the field of decentralized wastewater treatment. In the rest of this chapter, an attempt is made to provide an overview of the various approaches, and the current state of technology.

3.3 DECENTRALIZED WATER/WASTEWATER MANAGEMENT SYSTEMS

3.3.1 The traditional approach

Decentralized sanitation systems already exist in many parts of the world, mostly in rural areas. They consist of many small wastewater treatment facilities designed and built locally. The current state of technology has been summarized and assessed by Crites and Tchobanoglous (1998), Loetscher *et al.* (1997) and Loetscher (1999). In general, the treatment results achieved in practice are not satisfactory when IPP-relevant criteria (low consumption of resources, long-lasting technology, advanced treatment requirements) are used as a measure. There are three major concerns:

- the effluent quality is mostly low and rarely allows safe reuse of water;
- treatment plants are not properly operated; and
- plants are difficult to supervise and control by water authorities.

In contrast to modern approaches, the methods applied are either primitive (for example, aqua privies or pit latrines) or low-tech (one, two or three chamber septic tanks, for instance). Close-to-nature systems such as ponds, constructed wetlands or vertical infiltration plants are also used. A further category of decentralized systems use advanced technical solutions, mostly small versions of technologies normally applied at large treatment plants (for instance, trickling filters or sequencing batch reactors (SBRs)). The problem is that even if the technology applied has a high treatment potential it is insufficiently exploited since the plants receive little to no attention by the owners.

Typically, the different wastewater streams generated in households (Figure 3.4), shops or industries are combined and sent to the treatment plants disregarding their specific composition, the hazards contained in the specific wastewater stream, and any reuse potentials. Only rarely is the purified wastewater considered as a valuable source for flushing or cleaning purposes (Asano *et al.* 1996).

3.3.2 Novel ideas and concepts

To overcome the obvious shortcomings of the centralized approach and to lead the way towards an integrated, ecologically and economically sound water/wastewater management system, various researchers have recently presented ideas and innovative concepts (for example, Larsen and Gujer 1996b; Otterpohl *et al.* 1997; Venhuizen 1997; Zeeman *et al.* 2000). These proposals have in common:

- integration of water, wastewater and household waste management systems;
- separate collection and treatment of the various categories of waste streams generated in the catchment area (house, dwelling, settlement; factory, industrial park);
- recovery of valuable substances for further and mostly direct use (for example, water, compost, biogas and fertilizer) based on the concept of 'industrial ecology' (Jelinski *et al.* 1992; Raha *et al.* 1999).

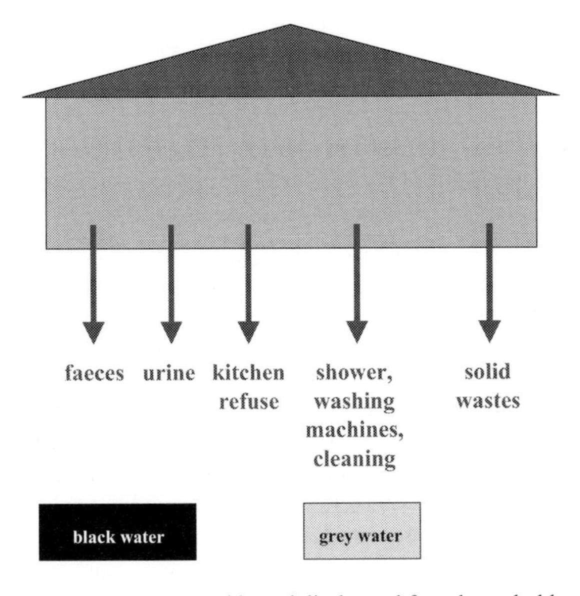

Figure 3.4. Waste streams generated in and discharged from households.

By integration of water and wastewater management systems on a local scale (Figure 3.5) the need for a fresh water supply should be minimized.

Subsequently, exploitation of freshwater sources (groundwater or surface water reservoirs) can be kept at a minimum. This allows the maintenance of reasonably high groundwater levels and undisturbed growth of plants, especially trees. In water-scarce areas reduction in fresh water supply rates is of real economical importance. Costs of desalination, for instance, can be minimized.

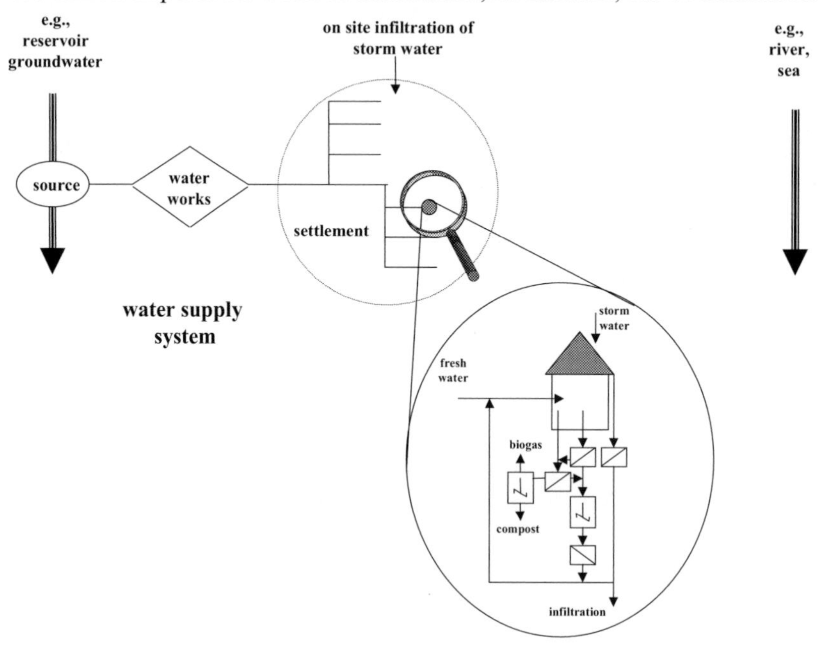

Figure 3.5. Concept of a decentralized water/wastewater system where stormwater and treated wastewater are used as sources of water for flushing, cleaning and watering gardens.

By collecting wastewater with a particular composition, purification and material conversion processes can be applied which are adjustable to the composition of the wastewater as well to specific requirements for further utilization of the treatment products. With respect to the municipal applications (family and apartment houses, cafeterias, restaurants and hotels) five categories of wastewater streams can be discriminated (see Figure 3.4);

- wastewater containing faeces;
- wastewater containing mainly urine;
- wastewater from washing machines, washing bowls, showers, baths or cleaning (grey water);

- wastewater from kitchen sinks (including kitchen refuse cut into pieces by means of sink waste disposal units, i.e. comminutors); and
- solid waste not suitable for on site treatment and reuse (for example, paper, plastics, metals).

These categories of wastes differ greatly in composition and concentration of the various components. The amount of hazardous substances they contain is very different.

Black water (faeces plus urine) is of major concern with respect to health risks. It may contain pathogenic organisms as well as pharmaceutical residuals. Since the concentration of organic material is high, conversion into biogas appears to be attractive. Urine contains high amounts of nitrogen and phosphorous and could be used as a source of fertilizer production. Kitchen refuse is high in organic load and can be converted into biogas and compost, potentially in combination with the black water constituents. Grey water is low in concentration of organic compounds. Most, but not all, of the grey water components are easily biodegradable. The concentration of inorganic substances in grey water is very low. This type of wastewater can be purified relatively easily and used, thereafter, for various purposes, for instance instead of drinking water for flushing toilets, for cleaning and irrigation. No matter what purpose the treated water is dedicated to, it should be converted into water of close-to-drinking-water-quality to avoid any negative health impacts.

The proposals currently discussed in relation to decentralized systems differ in the following ways:

- separate collection of the four major types of household wastewater;
- inclusion of stormwater collected on roofs and driveways in the water collection and reuse system;
- tailored technologies applied to treat the various wastewater fractions; and
- reintroduction of the treatment products into the material cycle.

The proposed systems differ also in scale. Solutions are discussed focusing on the treatment of wastewater from single houses, apartment complexes, industrial parks or entire residential areas.

The separate collection of urine is recommended by Larsen and Gujer (1996a,b). For this specific purpose, specially designed toilet bowls are required. Otterpohl *et al.* (1997) advocate the collection of black water plus urine by means of vacuum toilets since they are widely used in ships, trains and aircraft. Wilderer and Schreff (2000) proposed traditional flushing toilets operated with treated grey water. The black water is combined with urine and organic waste from kitchen sinks and sent to a high rate digester for biogas production (Figure 3.6) while the grey water is aerobically treated in a biofilm reactor.

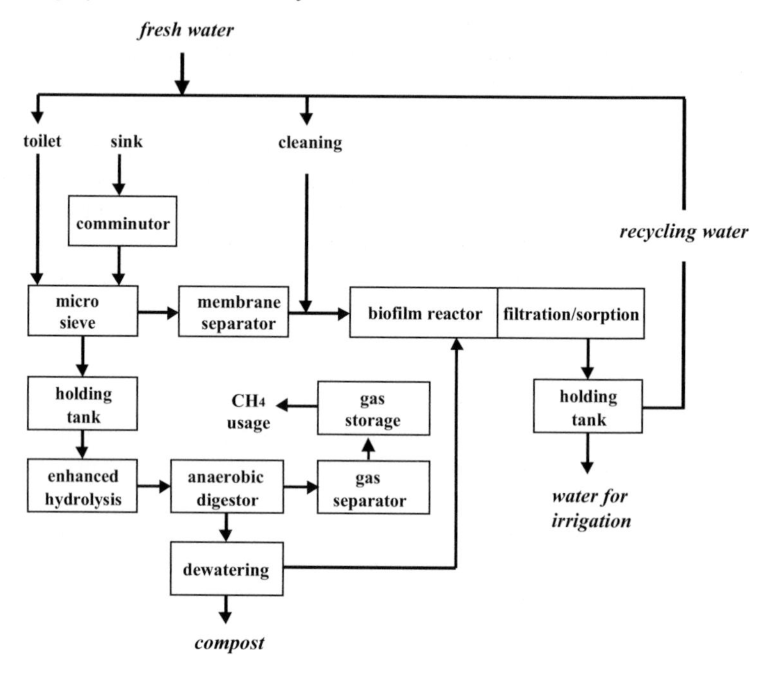

Figure 3.6. Flow diagram of a system treating black water, including organic waste, and grey water from washing machines etc., proposed by Wilderer and Schreff (2000).

As a technologically promising method, most of the proponents of new sanitation systems use membrane separation for solid–liquid separation, and anaerobic treatment of black water, including kitchen refuse, for converting organic materials into biogas and compost.

3.3.3 Cost considerations

Wastewater treatment systems, both small and large, are designed to protect the environment from excess discharges of pollutants and nutrients. If a city adopts the concept of decentralization, it should be accepted that the same degree of treatment is to be achieved by the small decentralized plants as by the single large plant.

$\sum$ discharge of small decentralized plants $\leq$ discharge of the respective large centralized plant

This concept translates into treatment goals and effluent quality parameters which are significantly tougher than the ones applied to traditional small wastewater treatment plants. The following minimum requirements must be met by the new types of small plants:

(1) The treated effluents, both water and solids, must be safe and hygienic since it is unavoidable that people, pets and other animals will come into contact with treated water and compost. Health risks are to be minimized as far as possible.
(2) The treated water must meet drinking water standards, even if the water is not to be used for drinking or cooking.
(3) The solids produced must be fully stabilized and free of noxious odours.
(4) The biogas must be treated so that it can be used as an energy source without damaging burners, machines or catalytic converters (for example, by corrosion).

It is obvious that a treatment plant that meets these high quality criteria is technically very complex, and requires a high degree of control, maintenance work and operator skills. At first glance the cost of building, implementing and operating such a small plant are enormously high. Who would be able and willing to pay for such a plant? Would they be affordable by people living in the mega-cities of developing countries who are desperately in need of a safe water supply and organized waste management?

Characteristically, modern technical systems of all kinds are complex. A computer, for instance, is a technically advanced, complicated and highly sophisticated device. So is a cellular phone or a car. Surprisingly, computers, cell phones and cars are used worldwide, even in developing countries, by ‘ordinary’ people with limited technical skills. They are surprisingly low in cost despite of their high degree of complexity, and they are reliable. Why is this? The simple answer is that all these high-tech systems are designed by teams of highly specialized engineers and they are mass produced. Why not apply this concept to small wastewater treatment systems as well?

A modern car is designed by a team of experts and built to a high degree by robots. It is user-friendly so that people can drive it without needing to know about the technical details that make the car run. The owner of a car, a computer or a cell phone does not need to maintain the devices he uses; any maintenance or repair work required is taken care of by specialized service companies. Again, why not apply this concept to small wastewater treatment systems?

It is not the local consulting bureau, builder or plumber who will be able to design and build a high-tech wastewater treatment system that produces potable water for a reasonable price. To make a modern decentralized sanitation system work, to make it affordable and to make it reliable, the design, construction and servicing of treatment plants must be industrialized. After several decades of intensive research, we have the knowledge needed to design efficient and compact 'machines' that convert wastewater and solid wastes into economically valuable products. Thus, the basis for proper design and fabrication is available. The market for high-tech, compact treatment systems already exists. This market needs to be cultivated. Simultaneously, a network of supervision and service stations should be established which ensures that the plants run smoothly. They may be run by private companies or by the government. In either case, independent quality control must be provided.

3.3.4 Problems yet to be solved

The evaluation presented above might be too optimistic when critically assessed in more detail. The problem is that wastewater and kitchen refuse contain not only readily biodegradable substances but recalcitrant ones as well. Wastewater and kitchen wastes – the input into the treatment system – are highly variable in composition, volumetric flow and mass flux. Techniques to make use of treated water and biogas are not well developed. In densely populated areas treated water cannot be used for irrigation or infiltration because of the low hydraulic conductivity of the soil, and there may be no need for locally produced compost. New legislation is necessary, and regulations must be strictly enforced to avoid the outbreak of disease and health problems caused by inadequately operated systems.

Of major concern are the chemicals used in households, including pharmaceutical products, especially chemicals such as pesticides which may accumulate in the water that is recycled in the house and which may enter into the food chain by irrigation of crops in the garden. Hormone-like substances, waste medicines and metabolites of medicines will form part of the constituents of the wastewater and the solid waste stream. The effects these substances may have on human health in the long term is currently unknown. To make integrated decentralized treatment systems feasible, advanced research is necessary, as well as the combined actions of the chemical and pharmaceutical industry.

The variability of composition and flux of wastewater and solid waste could be handled by providing volumetric buffer capacity in the system. Since the system under consideration must be compact to be acceptable, a volume-based buffer concept has only a limited applicability. Research and *in situ* investigations are necessary to identify better alternatives.

The applicability of washing machines and dishwashers to accept treated instead of fresh water has not yet been properly investigated. Of particular concern are humic substances which may develop and accumulate in the recycled water. Their effect on the washing process is unclear and needs investigation. The use of biogas is also problematic. Elimination of hydrogen sulphide, humidity and odour is necessary to be able to use the gas in households. It would be ideal if the gas could be fed into a fuel cell and used for the production of electricity. This development is a long way off, however.

For densely populated areas the concept of the reuse of water and solid materials needs to be considered. The basic question is to what extent the treated water can be reused on site, and to what degree the treated water must be substituted by fresh water. Again, investigation over a long period of time must be carried out in order to reach accurate answers. With respect to solid materials to be taken away from the site, solutions may be found relatively easily (e.g., collection by truck). Alternatively, the solid materials could be liquefied and converted into methane gas. The technology needed for the realization of such a concept is not yet available.

With respect to legislation and enforcement, new regulations have to be defined. Ideally, the decentralized treatment plants should be implemented and owned by individuals. However, the proper functioning of the plants is in the interest of the society as a whole. That means that the city or state should have the power to control the operational status of the plants, and to force property owners, if necessary, to make the system work properly. Robust and reliable sensor systems are required to allow remote control of the individual plants. There is still much research and development work to do on this topic.

3.4 SUMMARY AND CONCLUSIONS

The centralized water supply/sanitation system developed and implemented in industrial states has its merits. It is not an universally applicable solution, however.

As an alternative, integrated water and solid waste management concepts are being discussed by various groups of researchers. The basic idea is to treat the wastewater (together with kitchen refuse) on site in small, compact, highly efficient treatment systems, and make direct use of the treatment products (water, compost, biogas).

The technology needed to build such a system is largely available. By mass production of such treatment systems and by the establishment of a service system, decentralized wastewater/waste management concepts can be assumed to be low-cost as well as efficient in operation.

However, various problems have to be solved before such systems can be implemented on a global scale. Of concern are chemicals and pharmaceuticals used in households, humic substances produced in biological reactors, and hydrogen sulphide and odours contained in the biogas.

New regulations have to be defined to clarify ownership and responsibilities, and to force plant owners to comply with regulations and ensure that the treatment systems run properly.

3.5 REFERENCES

Asano, T., Maeda, M. and Takaki, M. (1996) Wastewater reclamation and reuse in Japan: Overview and implementation examples. *Wat. Sci. Tech.* **34**(11), 219–226.

Crites, R. and Tchobanoglous, G. (1998) *Small and Decentralized Wastewater Management Systems,* McGraw Hill, Boston, USA.

Jelinski, L.W., Graedel T.E., Laudise, R.A., McCall, D.W. and Patel, C.K.N. (1992) Industrial Ecology: Concepts and approaches. *Proceedings of the National Academy Science USA* **89**, 793–797.

Kalbermatten, J.M., Middelton, R. and Schertenleib, R. (1999) *Household-Centered Environmental Sanitation,* EAWAG News, Duebendorf, Switzerland.

Larsen, T. and Gujer, W. (1996a) Separate management of anthropogenic nutrient solutions. *Wat. Sci. Tech.* **34**(3–4), 87–94.

Larsen, T. and Gujer, W. (1996b) The concept of sustainable urban water management. *Wat. Sci. Tech.* **35**(9), 3–10.

Loetscher, T. (1999) Appropriate sanitation in developing countries – the development of a computerised decision aid. Ph.D. thesis, the University of Queensland, Brisbane, Australia.

Loetscher, T., Keller, J. and Greenfield, P. (1997) Appropriate sanitation in developing countries. *World Water* **20**(9), 16–20.

Otterpohl, R., Grottker, M. and Lange, J. (1997) Sustainable water and waste management in urban areas. *Wat. Sci. Tech.* **35**(9), 121–133.

Raha, S., Ghose, A.K. and Allen, N.W. (1999) Industrial ecology – analogy, or the technical core of sustainability. *Green Business Opportunities* **5**(2), 5–10.

Venhuizen, D. (1997) Paradigm shift: Decentralized wastewater systems may provide better management at less cost. *Water Environment & Technology*, Water Environment Federation, 49–52.

Wilderer, P.A. and Schreff, D. (2000) Decentralized and centralized wastewater management: a challenge for technology developers. *Wat. Sci. Tech.* **41**(1), 1–8.

World Bank (2000) *Korrespondenz Abwasser* **47**(6), 807.

Zeeman, G., Sanders, W. and Lettinga, G. (2000). Feasibility of the on-site treatment of sewage and swill in large buildings. *Wat. Sci. Tech.* **41**(1), 9–16.

Part II

Waste and wastewater characteristics and its on-site collection

4

Types, characteristics and quantities of classic, combined domestic wastewaters

Mogens Henze and Anna Ledin

4.1 INTRODUCTION

The production of waste from human activities is unavoidable. However, not all humans produce the same amount of waste. The amount and type of waste produced in households is influenced by the behaviour, lifestyle and standard of living of the population as well as the technical and juridical framework by which people are surrounded. In households most waste will end up as solid and liquid waste, and there are significant possibilities for changing the amounts and composition of the two waste streams generated. In relation to the sustainablilty

 Decentralised Sanitation and Reuse: Concepts, Systems and Implementation. Edited by P. Lens, G. Zeeman and G. Lettinga. ISBN: 1 900222 47 7

aspects of household waste generation, flexibility and adaptivity of the chosen solutions are essential to obtain long-lasting sustainable solutions. What we believe is the most sustainable solution today may not be so tomorrow due to new findings and the development of new technologies.

The objective of this chapter is to present the characteristics of the classic combined domestic wastewater, and to discuss the elements of which it consists. This will let us then consider changes to the present technologies used in households.

2.2 WASTEWATER COMPONENTS

Wastewater components can be divided into different main groups as shown in Table 4.1. In the following the composition of different types of wastewater is shown, based on domestic wastewater and municipal wastewater with no industrial influence.

Table 4.1. Components present in domestic wastewater. (Based on Henze *et al.* 2001)

Component	Of special interest	Environmental effect
Microorganisms	Pathogenic bacteria, virus and worms eggs	Risk when bathing and eating shellfish
Biodegradable organic materials	Oxygen depletion in rivers, lakes and fjords	Fish death, odours
Other organic materials	Detergents, pesticides, fat, oil and grease, colouring, solvents, phenols, cyanide	Toxic effect, aesthetic inconveniences, bio accumulation in the food chain
Nutrients	Nitrogen, phosphorus, Ammonium	Eutrophication, oxygen depletion Toxic effect
Metals	Hg, Pb, Cd, Cr, Cu, Ni	Toxic effect, bioaccumulation
Other inorganic materials	Acids, for example hydrogen sulphide, bases	Corrosion, toxic effect
Thermal effects	Hot water	Changing living conditions for flora and fauna
Odour (and taste)	Hydrogen sulphide	Aesthetic inconveniences, toxic effect
Radioactivity		Toxic effect, accumulation

4.2.1 Domestic wastewater/municipal wastewater

The concentrations found in wastewater are a combination of pollutant load and the amount of water with which the pollutant is mixed. The daily or yearly polluting load may thus form a good basis for an evaluation of the composition of wastewater. Table 4.2 shows the figures for different countries. Many of the figures are estimates. The composition of domestic wastewater and municipal wastewater varies significantly both in terms of place and time. This is partly due to variations in the discharged amounts of substances. However, the main reasons are variations in water consumption in households and infiltration and exfiltration during transport in the sewage system. The composition of typical domestic wastewater/municipal wastewater is shown in Tables 4.3–4.7. Concentrated wastewater represents cases with low water consumption and/or infiltration. Dilute wastewater represents high water consumption and/or infiltration. Stormwater will further dilute the wastewater as most stormwater components have lower concentrations than very diluted wastewater.

Table 4.2. Pollution load related to persons. (Based on Treibel 1982; Henze 1977, 1982, 2001; Andersson 1978; EPA 1977; Lønholdt 1973)

	Pollutant	Denmark	Brazil	Egypt	Italy	Sweden	Turkey	USA
kg/(capita·y)	BOD	20–25	20–25	10–15	18–22	25–30	10–15	30–35
	SS	30–35	20–25	15–25	20–30	30–35	15–25	30–35
	N-total	5–7	3–5	3–5	3–5	4–6	3–5	5–7
	P-total	1.5–2	0.6–1	0.4–0.6	0.6–1	0.8–1.2	0.4–0.6	1.5–2
	Detergents	0.8–1.2	0.5–1	0.3–0.5	0.5–1	0.7–1.0	0.3–0.5	0.8–1.2
g/(capita ·y)	Hg	0.1–0.2		0.01–0.2	0.02–0.04	0.1–0.2	0.01–0.02	
	Pb	5–10		5–10	5–10	5–10	5–10	
	Zn	15–30		15–30	15–30	10–20	15–30	
	Cd	0.2–0.4				0.5–0.7		

COD = (2–2.5) × BOD; VSS = (0.7–0.8) × SS; NH_3–N = (0.–0.7) × N-total

Tables 4.3 and 4.4 give an overview of the content of organic matter in domestic/municipal wastewater.

Table 4.5 gives the concentration of nutrients in domestic/municipal wastewater. These concentrations are sensitive to changes in the handling of urine and use of detergents with or without phosphorus.

Table 4.3. Typical average contents of organic matter (g O_2/m^3 = mg/l = ppm) in domestic wastewater (Henze 1982, 1992; Ødegaard 1992). From Henze et al. 2001

Analysis parameters	Wastewater type			
	Concentrated	Moderate	Diluted	Very diluted
Biochemical oxygen Demand, BOD				
Infinite	530	380	230	150
7 days	400	290	170	115
5 days	350	250	150	100
Dissolved	140	100	60	40
Dissolved, very easily degradable	70	50	30	20
After 2h settling	250	175	110	70
Chemical oxygen demand with dichromate, COD				
Total	740	530	320	210
Dissolved	300	210	130	80
Suspended	440	320	190	130
After 2h settling	530	370	230	150
Inert, total	180	130	80	50
Dissolved	30	20	15	10
Suspended	150	110	65	40
Degradable, total	560	400	240	160
Very easily degradable	90	60	40	25
Easily degradable	180	130	75	50
Slowly degradable	290	210	125	85
Heterotrophic biomass	120	90	55	35
Denitrifying biomass	80	60	40	25
Autotrophic biomass	1	1	0,5	0,5
Chemical oxygen demand with permanganate, COD_P				
Total	210	150	90	60

Table 4.4. Typical average contents of organic matter in domestic wastewater (Henze 1982, 1992; Ødegaard 1992). (From Henze et al. 2001)

Analysis parameters	Unit	Wastewater type			
		Concentrated	Moderate	Diluted	Very diluted
Total organic carbon	g C/m^3	250	180	110	70
Carbohydrate	g C/m^3	40	25	15	10
Proteins	g C/m^3	25	18	11	7
Fatty acids	g C/m^3	65	45	25	18
Fats	g C/m^3	25	18	11	7
Fats, oil and grease	g/m^3	100	70	40	30
Phenol	g/m^3	0.1	0.07	0.05	0.02
Phtalate, DEHP[1]	g/m^3	0.3	0.2	0.15	0.07
Phtalates, DOP[2]	g/m^3	0.6	0.4	0.3	0.15
Nonylphenoles, NPE[3]	g/m^3	0.08	0.05	0.03	0.01
Detergents, anion[4]	g LAS/m^3	15	10	6	4

[1] Di(2-ethylhexyl)phthalate
[2] Di-n-octylephthalate
[3] Nonylphenoles and nonylphenoleethoxylates
[4] LAS = Lauryl Alkyl Sulphonate

Table 4.5. Typical content of nutrients in domestic wastewater. (Based on Henze 1982, 1992; Ødegaard 1992. From Henze et al. 2001).

Analysis parameters	Unit	Wastewater type							
		Concentrated		Moderate		Diluted		Very diluted	
Total nitrogen	g N/m^3	80		50		30		20	
Ammonia nitrogen[1]	GN/m^3	50		30		18		12	
Nitrite nitrogen	g N/m^3	0.1		0.1		0.1		0.1	
Nitrate nitrogen	g N/m^3	0.5		0.5		0.5		0.5	
Organic nitrogen	g N/m^3	30		20		12		8	
Kjeldahl nitrogen[2]	g N/m^3	80		50		30		20	
Total phosphorous	g P/m^3	23	(14)[3]	16	(10)	10	(6)	6	(4)
Orthophosphate	g P/m^3	14	(10)	10	(7)	6	(4)	4	(3)
Polyphosphate	g P/m^3	5	(0)	3	(0)	2	(0)	1	(0)
Organic phosphate	g P/m^3	4	(4)	3	(3)	2	(2)	1	(1)

[1] $NH_3 + NH_4^+$
[2] organic nitrogen + $NH_3 + NH_4^+$
[3] for detergents without phosphorus

Table 4.6. Typical content of metals in domestic wastewater, mg metal/m^3. (Data from Henze 1982; Henze et al. 2001)

Analysis parameter	Wastewater type			
	Concentrated	Moderate	Diluted	Very diluted
Aluminium	1000	650	400	250
Arsenic	5	3	2	1
Cadmium	4	2	2	1
Chromium	40	25	15	10
Cobalt	2	1	1	0.5
Copper	100	70	40	30
Iron	1500	1000	600	400
Lead	80	65	30	25
Manganese	150	100	60	40
Mercury	3	2	1	1
Nickel	40	25	15	10
Silver	10	7	4	3
Zinc	300	200	130	80

Table 4.7. Different parameters in domestic wastewater. (Data from Henze 1982).

Analysis/matter	Unit	Wastewater type			
		Concentrated	Moderate	Diluted	Very diluted
Suspended solids	g SS/m^3	450	300	190	120
Suspended solids, volatile	g VSS/m^3	320	210	140	80
Precipitate after 2h	ml/l	10	7	4	3
Precipitate after 2h, suspended solids	g/m^3	320	210	140	80
Precipitate suspended solids, volatile	g/m^3	220	150	90	60
Suspended solids after 2h	g SS/m^3	130	90	50	40
Absolute viscosity	kg/(m· s)	0.001	0.001	0.001	0.001
Surface tension	dyn/cm^2	50	55	60	65
Conductivity	mS/m[1]	120	100	80	70
PH		78	78	78	78
Alkalinity (TAL)	eqv/m^3 [2]	37	37	37	37
Sulphide[3]	g S/m^3	0.100	0.100	0.100	0.100
Cyanide	g/m^3	0.050	0.035	0.020	0.015
Chloride[4]	g Cl/m^3	500	360	280	200
Boron	g B/m^3	1.0	0.7	0.4	0.3

[1] mS/m = 10 μS/cm = 1 m mho/m
[2] 1 eqv/m^3 = 1 m eqv/l = 50 mg $CaCO_3$/l
[3] $H_2S + HS^- + S^{--}$
[4] with 100 gCl/m^3 in the water supply

The metals in wastewater can influence the possibilities for recycling of the wastewater treatment sludge to farmland. Typical values are given in Table 4.6.

Table 4.7 gives a range of hydrochemical parameters for domestic/municipal wastewater.

Table 4.8 gives an idea of the concentration of micro-organisms in raw domestic wastewater.

Table 4.8. Concentrations of micro-organisms in wastewater (number of micro-organisms per 100 ml). (Based on Henze et al. 2001)

	High	Low
E.Coli	$5 \cdot 10^8$	10^6
Coliforms	10^{13}	10^{11}
Cl.perfringens	$5 \cdot 10^4$	10^3
Fecal streptococcae	10^8	10^6
Salmonella	300	50
Campylobacter	10^5	$5 \cdot 10^3$
Listeria	10^4	$5 \cdot 10^2$
Staphyllococus aureus	10^5	$5 \cdot 10^3$
Coliphages	$5 \cdot 10^5$	10^4
Giardia	10^3	10^2
Roundworms	20	5
Enterovirus	10^4	10^3
Rotavirus	100	20

It is not only the wastewater from the catchment that a treatment plant has to handle. The bigger the plant, the more internal wastewater recycles and external deliveries there are to be dealt with. In relation to decentralised wastewater handling, septic tank sludge is of interest.

Table 4.9 shows the typical composition of septic sludge. Septic tank wastes are often driven to the treatment plant and mixed with the incoming wastewater. This extra waste stream can contribute significantly to the total waste load of the wastewater.

Table 4.9. Composition of septic sludge. (From Henze et al. 2001)

Compound	Septic sludge		Unit
	High	Low	
BOD total	30000	2000	g/m^3
BOD soluble	1000	100	g/m^3
COD total	90000	6000	g/m^3
COD soluble	2000	200	g/m^3
Total nitrogen	1500	200	gN/m^3
Ammonia nitrogen	150	50	gN/m^3
Total phosphorus	300	40	gP/m^3
Orthophosphate	20	5	gP/m^3
Suspended solids	100000	7000	g/m^3
Volatile suspended solids	60000	4000	g/m^3
Precipitate after 2h	900	100	Ml/l
Chloride	300	50	g/m^3
Sulphide	20	1	g/m^3
PH	8.5	7.0	
Alkalinity	40	10	ekv/m^3
Lead	30	10	mg/m^3
Total iron	200	20	g/m^3
Faecal coliforms	10^8	10^6	No./100 ml

4.3 POPULATION EQUIVALENT AND PERSON LOAD

Sometimes the volume of wastewater is expressed in units of population equivalent (PE). Population equivalent can be expressed in water volume or BOD. The following two definitions are used worldwide:

1 PE = 0.2 m^3/d

1 PE = 60 g BOD/d

These two definitions are based on fixed figures. They are used for administrative purposes and in some cases for simple design. The actual contribution from a person living in a sewer catchment, the person load (PL) can vary considerably, as indicated in Table 4.10. The reasons for the variation can vary due to lifestyle reasons.

Table 4.10. Variations in person load. (From Henze *et al.* 2001)

Component	Level
BOD g/(capita·d)	15–80
COD g/(capita·d)	25–200
Nitrogen g/(capita·d)	2–15
Phosphorus g/(capita·d)	1–3
Wastewater m^3/(capita·d)	0.05–0.40

PE and PL are often confused, so care should be taken when using them. They are both based on average contributions, and used to give an impression of the loading of wastewater treatment processes. They should not be calculated from data based on short time intervals.

The person load varies from country to country, as seen in the annual values given in Table 4.2.

4.4 COLOURED WASTEWATER

The wastewater described in Tables 4.3–4.8 is mixed wastewater from households. Different colours of wastewater can also be defined, as in Table 4.11.

Table 4.11. The wastewater palette (definition of wastewater fractions from households).

Type	Content
Classic	toilet, bath, kitchen, wash
Black	toilet
Grey	bath, kitchen, wash
Light grey	bath, wash
Yellow	urine
Brown	faeces

4.5 WASTES FROM HOUSEHOLDS

A tool for the detailed analysis of waste composition is mass flow analysis. In the case of household waste, the composition of wastewater and solid wastes from households is a result of contributions from various sources within the household. It is possible to change the amount and the composition of the waste streams. The amount of a given waste stream can be decreased or increased, depending on the optimal solution. For example, a reduction in the amount of waste present in the wastewater can be achieved by two means:

(1) Reduction of waste generated in the household.
(2) Diversion of certain waste loads to the solid waste of the household.

The amount of organic waste and nutrients produced in households in developed countries is shown in Table 4.12. From this table, it is easy to get an idea of the potential for changing the wastewater composition.

4.5.1 Physiological wastes

Options for reducing the physiologically generated amount of waste are not obvious, although human diet influences the amount of waste produced. Thus we have to accept this waste generation as a natural result of human activity. Separating toilet waste (physiological waste or anthropogenic waste) from the waterborne route means a significant reduction in the nitrogen, phosphorus and organic load in the wastewater. Waste generated after the separation has taken place must still be transported out of the household, and in many cases out of the city.

There are many different technologies for handling this sort of waste, including:

(1) The night soil system, used worldwide.
(2) Compost toilets, mainly used in individual homes in agricultural areas (preferably with urine separation in order to optimise the composting process).
(3) Septic tanks followed by infiltration or transport by a sewer system.

Recently a significant interest in separating urine has developed, due to the high nutrient content in urine (Sundberg 1995). Table 4.12 shows that urine is the main contributor to nutrients in household wastes.

Table 4.12. Sources for household wastewater components and their values for non-ecological living. (From Henze 1997; Sundberg 1995; Eilersen *et al.* 1999)

	Unit	Toilet		Kitchen	Bath/ laundry	Total
		Total incl. urine	Urine			
Wastewater	m^3/y	19	11	18	18	55
BOD	kg/y	9.1	1.8	11	1.8	21.9
COD	kg/y	27.5	5.5	16	3.7	47.2
Nitrogen	kg/y	4.4	4.0	0.3	0.4	5.1
Phosphorus	kg/y	0.7	0.5	0.07	0.1	0.87
Potassium	kg/y	1.3	0.9	0.15	0.15	1.6

4.5.2 Liquid kitchen waste

Kitchen waste includes a significant amount of organic matter which traditionally ends up in wastewater. It is possible to divert some liquid kitchen wastes to solid waste by cleantech cooking, thus obtaining a significant reduction in the overall organic load of the wastewater (Danish EPA 1993). Cleantech cooking means that food waste is discarded into the waste bin and not flushed into the sewer by water from the tap. The diverted part of the solid organic waste from the kitchen can be disposed together with the other solid wastes from the household (see below). The grey wastewater from the kitchen could be used for irrigation or, after treatment, for toilet flushing. Liquid kitchen wastes also contain household chemicals, the use of which can affect the composition and load of this waste.

4.5.3 Laundry and baths

This wastewater carries a minor pollution load, part of which comes from household chemicals, the use of which can affect the composition and the load of this waste fraction. Waste from laundry and baths could be used together with the traditional kitchen wastewater for irrigation. Alternatively, it can be reused for toilet flushing. In both cases considerable treatment is needed.

4.5.4 Solid kitchen waste

Solid waste generated by the urban population will not be reduced significantly in our lifetime. Thus we need to face this fact and select the optimal waste handling technology. Historical waste handling procedures may not be the best to use today.

The compostable fraction of the solid waste from the kitchen can either be kept separate or combined with traditionally waterborne kitchen wastes, for later composting or anaerobic treatment.

The use of kitchen disposal units grinders for handling the compostable fraction of the solid waste from households is currently being discussed in many countries, especially in the US, Canada, Australia and Northern Europe (Jones 1990; Tabasaran 1984; Nilsson *et al.* 1990; Dircks *et al.* 1997). Sometimes this option is discarded due to the increased waste load to the sewer. However, waste is generated in households, and it must be transported out of households and out of cities by some means. As with most other options discussed in this section, the discharge of solid waste to the sewer does not change the total waste load produced by the household, but it will change the final destination of the waste. Transporting the compostable fraction of household waste by lorry often results in significant occupational health problems (Christensen 1998). Using sewers as a transport system for solid waste can reduce this type of problem.

4.5.5 Pollution loads in coloured domestic wastewater

Table 4.13 illustrates the annual loads of various compounds in different types of domestic wastewater. There is a large variation in annual loads thus wastewater will vary considerably in quality, with respect to handling and reuse.

Table 4.13. Typical pollution loads in domestic wastewater (kg per person per year)

	BOD	COD	Nitrogen	Phosphorus	Potassium
Classic	22	47	5.1	0.9	1.6
Black	9	27	4.4	0.7	1.3
Grey	13	20	0.7	0.2	0.3
Light grey	1.8	3.7	0.4	0.1	0.15
Yellow	1.8	5.5	4.0	0.5	0.9
Brown	7.3	22	0.4	0.2	0.4

4.6 WATER CONSUMPTION

Water consumption by households is an important part of waste generation. Table 4.14 shows water consumption split into fractions according to where the water is used. Note that infiltration into sewers is also considered to be water consumption since it depletes groundwater resources, irrespective of whether this resource is exploited or not. This infiltration may not be groundwater, but may also be drinking water from leaking water supply pipes. The contribution of these areas to water consumption varies with geographical location and local culture/lifestyle.

Table 4.14. Typical water consumption values in Northern Europe, l/(capita·d). (Based on Henze 1997)

Water consumption	Today	With water-saving mechanisms[1]	Primary water consumption with water-saving mechanisms[2]
Toilet	50	25/0	0
Bath	50	25	25
Kitchen	50	25	25
Laundry	10	5	1
Infiltration	80	25	n/a
Total	240	105	51

[1] New installations, sewer rehabilitation; [2] Plus the use of secondary water (for toilets and laundry)

4.6.1 Water-saving mechanisms

Water-saving mechanisms and sewer rehabilitation can help to obtain the figures shown in the middle column of Table 4.14. Only a part of the future water supply has to be drinking water (primary water), as shown in the right-hand column. The remaining part of the water consumption could be covered by secondary water (rainwater or treated grey water).

It is possible to considerably reduce the water consumption in the various areas. Without much effort, water consumption in a household can be reduced by 50 per cent, and by even more with the rehabilitation of sewers. However, it takes time and money to achieve this goal. For households, a realistic strategy is to install water-saving devices when old devices need to be replaced. This means that it will take some 20 to 40 years in developed countries before most installations are replaced.

Water savings have two important implications for society. The first is the reduced amount of energy used in water supply and wastewater treatment. The second is the reduced exploitation of freshwater resources. Water for toilets and washing of clothes can realistically be replaced by secondary water, for example rainwater. It is thus obvious that a small water consumption is more sustainable than a high one, from an energy consumption point of view.

Water savings will increase the concentration of pollutants in the wastewater and decrease treatment costs slightly, due to less water being treated and the quantity of pollutants that should be removed remaining static. Table 4.15 shows the concentrations of various pollutants in domestic wastewater for varying volumes of household wastewater, including that in a sewer system.

Table 4.15. Concentrations of pollutants in traditionally generated raw wastewater by water savings/sewer renovation. (Some data from Henze *et al.* 2001)

Wastewater including infiltration (g/m^3)	250 l/(capita·d)	160 l/(capita·d)	80 l/(capita·d)
COD	520	815	1625
BOD	240	375	750
Nitrogen	50	80	165
Phosphorus	10	16	31

4.7 WASTEWATER DESIGN FOR HOUSEHOLDS

The use of one or more of the waste handling technologies mentioned earlier in households in combination with water-saving mechanisms makes it possible to design wastewater with a specified composition, which will be optimal for its further handling. If the goal is to reduce the pollutant load to the wastewater, Table 4.16 illustrates possible reductions in waste loads by various actions.

Table 4.16. Reduced waste load to wastewater by toilet separation and cleantech cooking g/(capita·d). (From Henze 1997)

Technology	Present	Toilet separation[1]	Cleantech cooking[2]
COD	130	55	32
BOD	60	35	20
Nitrogen	13	2	1.5
Phosphorus	2.5	0.5	0.4

[1] Water closet → dry/compost toilet.
[2] Part of cooking waste not to the sink, but to solid waste.

The coupling of water savings and load reductions gives one more degree of freedom for the design of a given wastewater composition. The results are shown in Table 4.17.

Table 4.17. The concentration of pollutants in raw wastewater with toilet separation and cleantech cooking (assuming use of detergents without phosphorus). (From Henze 1997)

Wastewater production	250 l/(capita·d)	160 l/(capita·d)	80 l/(capita·d)
COD g/m^3	130	200	400
BOD g/m^3	80	125	250
Nitrogen g/m^3	6	9	19
Phosphorus g/m^3	1.6	2.5	5

The changes obtained in the wastewater composition also influence the detailed composition of the COD. This can result in changes between the soluble and the suspended fractions, or changes in degradability of the organic matter, for example, leading to more or less easily degradable organic matter in the given wastewater fraction. Today we know that the composition of wastewater has a significant influence on the treatment processes used (Henze *et al.* 1995).

By changing technology used in households and by diverting as much of the organic waste to the sewer system as possible, it is possible to obtain wastewater characteristics like those shown in Table 4.18.

Table 4.18. Concentration of pollutants in raw wastewater by maximum load of organic waste (assuming detergents without phosphorus). (From Henze 1997)

Wastewater production	250 l/(capita·d)	160 l/(capita·d)	80 l/(capita·d)
COD	880	1375	2750
BOD	360	565	1125
Nitrogen	59	92	184
Phosphorus	11	17	35

The most common separation is the separation of toilet waste from the rest of the wastewater. This will result in grey and black wastewater generation, the characteristics of which can be seen in Table 4.19. For more details on greywater, see Ledin *et al.* 2000.

Table 4.19. Characteristics of grey and black wastewater. Low values can be due to high water consumption; high values due to low water consumption or high pollution load from kitchens. [* Excluding the content in the water supply]. (Based on Henze 1997; Sundberg 1995; Almeida *et. al* 2000; Eilersen *et al.* 1999)

Compound	Grey wastewater		Black wastewater		Unit
	High	Low	High	Low	
BOD total	400	100	600	300	g O_2/m^3
COD total	700	200	1500	900	g O_2/m^3
Total nitrogen	30	8	300	100	g N/m^3
Total phosphorus	7	2	40	20	g P/m^3
Potassium*	6	2	90	40	g K/m^3

4.8 WASTEWATER CHANGES BY WASTE TRANSPORT

All household waste must be transported to a treatment and disposal site. There are three means of transport: (1) By lorry, (2) By sewer, (3) By soil infiltration.

Neither sewers nor soil are mobile as seen from a general viewpoint. But they act as vehicles for the displacement of household waste.

Local infiltration can be used for stormwater disposal and can also take care of some household waste, but part of the waste still has to be transported away from households by long distance transport by either lorry or sewer.

The best means of transport must be selected for the treatment and disposal of the waste. Assuming that the waste cannot be transported by soil, Table 4.20 shows volumes of waste that can be transported by sewer and by lorry. If the compostable part of the solid waste is transported by sewer, the maximum amount of waste that can be transported by this means is shown. The minimum waste transport by sewer is the case when toilet separation is used and part of the liquid kitchen waste ends up as solid waste. The minimum and the maximum amounts of waste that can be transported from households by lorry are also shown in Table 4.20.

Table 4.20. Minimum and maximum transport of waste in g/(capita·d) by lorry and sewer for household waste. (From Henze 1997)

	Minimum				Maximum			
Transport	COD	BOD	N	P	COD	BOD	N	P
by sewer	33	20	1.5	0.4	220	90	14.7	2.8
by lorry	0	0	0	0	188	70	13.2	2.4

4.9 REFERENCES

Almeida, M.C., Butler, D. and Friedler, E. (2000) At source domestic wastewater quality. *Urban Water* **1**, 49–55.

Andersson, L. (1978) Föroreningar i avloppsvatten från hushåll. (Pollutions in Dometic Wastewater). Statens Naturvårdsverk, Stockholm, Sweden.

Christensen, T.H (1998) Affaldsteknologi (Solid waste technology). Teknisk Forlag, Copenhagen, Denmark, p. 46.

Danish Environmental Protection Agency (Danish EPA) (1993) *Husspildevand og renere teknologi (Domestic wastewater and clean technology).* Miljøprojekt Nr. 219, Miljøstyrelsen, Strandgade 29, DK1401 Copenhagen, Denmark. (In Danish.)

Dircks, K., Pind, P. and Henze, M. (1997) Garbage grinders. Environmental improvement or pollution? *Vand & Jord* **4**, 205–209. (In Danish.)

Eilersen, A.M., Nielsen, S.B., Gabriel, S., Hoffmann, B., Moshøj, C.R., Henze, M., Elle, M. and Mikkelsen, P.S. (1999) Assessing the sustainability of wastewater handling in non-sewered settlements. Department of Environmental Science and Engineering, Technical University of Denmark. Accepted for *Ecological Engineering.*

Henze, M. (1977) Approaches and Methods in the Estimation of the Polluting Load from Municipal Sources in the Mediterranean Area. Paper presented at the Meeting of Experts on Pollutants from Landbased Sources, Geneva, 19–24 September. United Nations Environment Programme (UNEP Project Med X).

Henze, M. (1982) Husspildevands sammensætning (The Composition of Domestic Wastewater). *Stads og Havneingeniøren Journal* **73**, 386–387.

Henze, M. (1992) Characterization of wastewater for modelling of activated sludge processes. *Wat. Sci. Tech.* **25**(6), 1–15.

Henze, M. (1997) Waste design for households with respect to water, organics and nutrients. *Wat. Sci. Tech.* **35**(9), 113–120.

Henze, M., Harremoës, P., la Cour Jansen, J. and Arvin, E. (2000) *Wastewater Treatment: Biological and Chemical Processes*, 3rd edn, Springer-Verlag, Berlin.

Jones, P.H. (1990) Kitchen garbage grinders. The effect on sewerage systems and refuse handling. University report, Institute of Environmental Studies, University of Toronto, Canada.

Ledin, A., Eriksson, E. and Henze, M. (2001) Aspects of groundwater recharge using grey wastewater. Chapter 18 of this book.

Lønholdt, J. (1973) Råspildevands indhold af BI5, N og P (The Content in Raw Wastewater of BOD, N and P). *Stads og Havneingeniøren Journal* **64**, 138–144.

Nilsson, P., Hallin, P.O., Johanson, J., Karlén, L., Lijla, G., Petersson, B.Å. and Pettersson, J. (1990) Source separation with garbage grinders in households. Bulletin VA 56. Institute for Water Technology, Lund University, Sweden. (In Swedish.)

Ødegaard, H. (1992) Norwegian experiences with chemical treatment of raw wastewater. Presented at the Management of Waste Waters in Coastal Areas conference, Montpellier, France, 31 March–2 April.

Sundberg, K. (1995) *What Is The Content In Wastewater From Households?* Swedish Environmental Protection Agency, Report 4425, Stockholm, Sweden. (In Swedish.)

Tabasaran, O. (1984) *Study On The Effects Of Home Garbage Grinders On Drainage Systems, Sewage Treatment Plants, Receiving Water Courses And Garbage Disposal Procedures.* Institut für Siedlungswasserwirtschaft, University of Stuttgart, Germany.

Triebel, W. (1982) *Lehr und Handbuch der Abwassertechnik. (Wastewater Techniques: Textbook and Manual)*, 3rd edn, Verlag von Wilhelm Ernst, Berlin. (In German.)

United States Environmental Protection Agency (1977) *Process Design Manual. Wastewater Treatment Facilities for Sewered Small Communities.* US EPA, Cincinnati, Ohio (EPA 625/177009).

5

Types, characteristics and quantities of domestic solid waste

K. Kujawa-Roeleveld

5.1 INTRODUCTION

Getting rid of household rubbish has always been a problem but only recently has it hit epidemic proportions. For thousands of years people managed solid waste by gathering it up, carting it out, and dumping or burying it in an isolated place. Crude as it was, this system worked because most of this waste consisted of biodegradable organic compounds that decomposed easily. In addition, the volume of rubbish was much lower than now because there were fewer people and less packaging material. Over the last fifty years, new synthetic and hazardous materials have been introduced into the waste stream. This has seriously complicated the problem, since many of these materials are not biodegradable and some produce toxic residues that has led to tighter

Edited by P. Lens, G. Zeeman and G. Lettinga. ISBN: 1 900222 47 7

environmental controls on landfills and during incineration. With open space in shortage, many communities are literally drowning in municipal solid waste.

The term ‘solid waste’ means any rubbish, refuse and other discarded solid material, including those resulting from industrial operations and from community activities, but does not include solids or dissolved materials in domestic sewerage or other significant pollutants in water sources such as silt, dissolved or suspended solids in industrial waste water effluents, dissolved material in irrigation return flows or other water pollutants.

The main strategies for managing solid waste streams are:

- landfilling;
- incineration;
- recycling; and
- reduction in waste generations.

Currently most solid wastes are disposed of in landfill sites, which are becoming scarcer, leading to increased costs of waste disposal. So, waste recycling to reduce costs and help the environment becomes imperative. In general, disposal should be the last option selected when dealing with solid waste. Only after any attempts have been made to reduce, reuse and recycle the waste, should it be disposed of.

The separated organic fraction of municipal solid waste, known as biowaste (fruit, vegetable and garden rubbish), is suitable for reuse on-site. Some forms of solid waste are known as hazardous wastes, and require special disposal procedures. Due to these costly procedures, it is important that their generation is reduced, and correct bins are used when disposing of such wastes.

5.2 TYPES OF DOMESTIC SOLID WASTE

5.2.1 Biowaste

Biowaste is composed of organic products collected outdoors and indoors. Indoor biowaste originates mainly from the kitchen but often also contains flowers and houseplants. Outdoor material originates mainly from gardens and consists of leaves, grass clippings, branches, garden topsoil, woody debris etc. The organic fraction of municipal solid waste may make up more than 50% of the total amount of generated waste (Roosmalen and van de Langrijt 1989).

5.2.1.1 Indoor waste – kitchen waste

Kitchen waste (fruits, vegetables, coffee grounds, tea bags, egg shells etc.) is a nutrient-rich source of organic material and can be easily biologically degraded.

The characteristic feature of kitchen waste is that the ratio of volatile solids to total solids is higher than with other biodegradable domestic solid waste. The term 'swill' is often used to define this waste, especially when it originates from units larger than household, for example from restaurants or hospitals (van Duynhoven 1994). Although this waste can be collected and composted at a central municipal facility, there are many advantages to treating kitchen waste on site. Food waste composting on site allows this waste to be converted into a valuable source of organic material – compost.

5.2.1.2 Outdoor waste – garden waste

Garden waste materials constitute in many parts of the world a significant part of the municipal biowaste stream. They may account for nearly 20% of the domestic (household) waste stream (Cooley *et al.* 1999). Almost a third of all summertime waste may be grass cuttings. Garden waste is more resistant to biodegradation processes than indoor biowaste because of the presence of components such as lignocellulose. If co-digestion or co-composting with, for example, kitchen waste is implemented, this will be not stable because of variations in the composition of garden waste during different seasons. If, however, they are treated together, the resulting digested product is an appropriate material to be used as a soil conditioner.

5.2.2 Household hazardous waste

The term 'hazardous waste' means a solid waste or combination of solid wastes that, because of its quantity, concentration or physical, chemical or infectious characteristics may:

- cause or significantly contribute to an increase in mortality or an increase in serious irreversible illness; or
- pose a potential hazard to human health and the environment when improperly treated, stored, transported or disposed of.

Hazardous waste comprises used or leftover products from homes, garages, and gardens that contain potentially hazardous substances. Because of their chemical nature, they can poison, corrode, explode or ignite when handled improperly. Categories and examples are given in Table 5.1. Law often prohibits these wastes from being put in the domestic rubbish or buried at a landfill site. Usually only the empty containers can be disposed of. The following warnings on the labels of products indicate whether products are potentially hazardous:

Danger, Poison, Corrosive, Toxic, Flammable, Combustible, Pesticide, Caustic, Caution or Warning.

Table 5.1. Categories and examples of potentially hazardous products present in solid domestic waste

Paint products	Cleaning products	Lawn and garden care
Oil based and latex paint Paint thinners and removers Stains and varnishes Aerosol cans Epoxies and Adhesives Wood preservatives	Bleach and liquid cleaners Drain openers Oven cleaners Upholstery cleaners Metal and furniture polish Tub and title Spot removers	Weedkillers Insecticides Fertilisers with weedkiller Rodent bait Ant and roach spray
Automotive	**Miscellaneous**	
Motor oil Oil filters Gasoline Solvents Brake fluid Transmission fluid Lead-acid batteries Polishes and waxes	Driveway sealer Roofing tar Small engine/boat fuel Pool chemicals Photo chemicals	

5.2.3 Faeces

When compost toilets or dry toilets are installed in households, faeces can be categorised as solid domestic waste. They are also known as human, physiological solid waste.

5.2.4 Construction and demolition debris

Construction and demolition (C&D) debris results from the construction, renovation and demolition of buildings. The primary components of C&D debris include concrete, wood, asphalt, gypsum drywall, metal, dirt, rocks, cardboard, paper and plastic. Smaller amounts of other materials, some of which may be hazardous, may also be present.

5.2.5 Other solid waste streams

In addition to the categories of solid domestic waste mentioned earlier, the following waste streams are also produced in households:

- paper (newspaper, cardboard, packing materials);
- plastic (mainly packing material);
- metal (tins);
- glass; and
- textiles.

The solid waste not collected separately for recycling is known as grey solid waste. The fraction of recyclable material in grey waste depends on the goodwill and environmental awareness of the population.

5.3 CHARACTERISTICS OF DOMESTIC SOLID WASTE

Domestic solid waste can vary considerably in terms of volume and content. Its composition can depend on local development and economic conditions.

5.3.1 Quantitative characterisation of solid domestic waste

Generally organic material constitutes the largest fraction of total solid domestic waste, followed by paper and plastic. For instance the composition of total solid waste produced in a household in Beirut (Lebanon) is given in Table 5.2 and Figure 5.1. In one of the three residential districts of Beirut examined (the Rauoche sector) 0.77 and 0.63 kg/person/day (kg/p/d) of solid waste is produced in the summer and winter periods respectively. Including the data from other cities, the annual level of waste generated amounts to 274 kg/p/year (Ayoub *et al.* 1996).

Table 5.2. Overall average values for the composition (by weight) of the total amount of solid wastes as determined in three locations of Beirut (Ayoub *et al.* 1996)

Components	Quantity (kg/p/d)
Organic material	0.463
Paper and cardboard	0.103
Plastics	0.083
Metals	0.020
Textiles	0.025
Glass	0.039
Others	0.017
Total	0.75

Solid household waste in the Netherlands in 1991 amounted to 350 kg/p/year (including coarse solid waste) (Haskoning 1992, RIVM 1989a). The RIVM study (1989a) discovered that the production of solid waste in a household remained relatively constant between 1970 and 1987 with a small exception regarding glass (a reduction in waste) and plastic (an increase in waste).

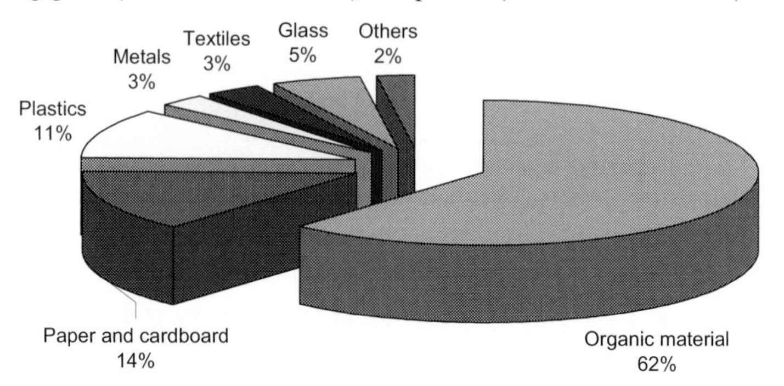

Figure 5.1. A graphical representation of the distribution of different waste categories in the total amount of solid waste in Lebanon, Beirut (Ayoub *et al.* 1996).

The other examples of solid domestic waste distribution as determined in the US and the Netherlands are presented in Figure 5.2 and Table 5.3.

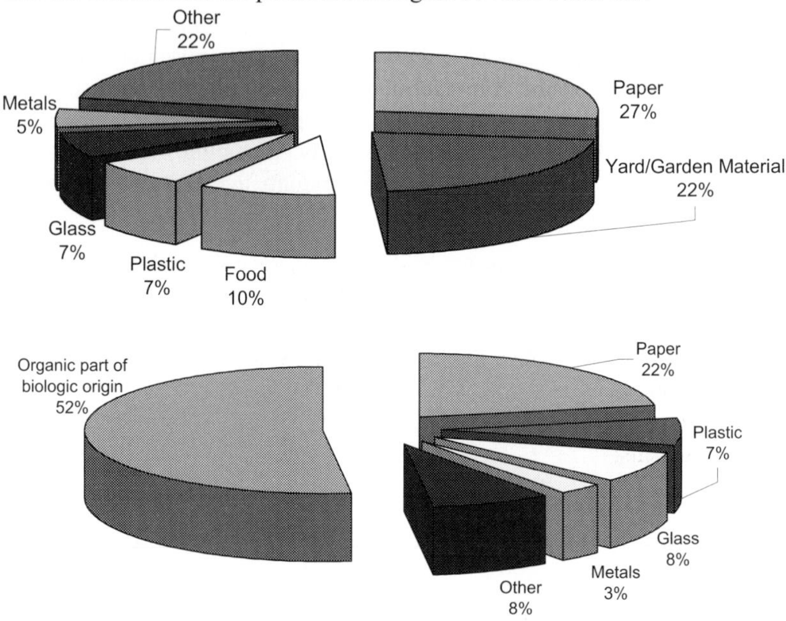

Figure 5.2. Composition of residential refuse in the USA (upper figure) and the Netherlands (lower figure). (Roosmalen and van de Langrijt 1989).

Table 5.3. Overview of waste category distribution (in%) in total solid domestic waste quantity reported in different countries

	Country				
Material type	USA	The Netherlands 1989 (left) and 1991 (right)		Lebanon (1994–1996)	Australia (1995)[2]
Paper	27	22	24.2	13.7	22.3
Organic compostable		52	48[3]	61.7	56.1
Yard, garden	22				
Food	10				
Other organic			2.3[4] 2.1[5]		2.9
Glass	7	8	7.2	5.2	3.8
Plastic	7	7	7.1	11.1	7.3
Ferrous	5		2.6		3.1
Nonferrous			0.6		0.9
Metals		3[1]		2.7	
Textiles			2.1	3.3	
Household hazardous			0.4		0.2
Others	22	8	3.5	2.3	3.2

[1] ferrous; [2] 300 household mobile garbage bins were randomly selected and the contents of each bin were sorted and weighed, [3] VFY and undefined waste (for example, sandy material), [4] bread, [5] animal (pet) waste.

5.3.2 Qualitative characterisation of domestic solid waste

5.3.2.1 Biowaste

Biowaste or selected organic household waste, also termed VFY (vegetable, fruit and yard (garden) waste), generally consists of the following components:

- leaves, peel and the remains of vegetables, fruit and potatoes;
- all food remains: meat, fish, sauce, bones;
- cheese and cheese rind;
- eggshells;
- coffee filters and coffee grounds, tea bags and tea leaves;
- nuts and peanut shells;
- cut flowers, house plants, small quantities of pot ground;
- pet dung (manure);
- grass-clippings, straw and leaves;
- small branches and garden plants; and
- newspaper used when peeling the vegetables.

For the separation and further management of VFY it is very important that it is pure, that is, it has a low fraction of non-compostable (non-biodegradable) components. The pureness of Dutch VFY has been estimated to be between 90–97% (Haskoning 1992) depending on the season. The majority of problematic compostable waste consisted of paper (which is also compostable) while pollutants such as plastic and iron constituted 0.5%. The moisture content of VFY is approximately 55–60% (weighted) while the organic fraction is between 400–600 kg/m^3. The average mass ratio of carbon and nitrogen is approximately 1:30.

A more specific fractionation of the total organic content of biowaste (Table 5.4) was performed in Switzerland by Perringer (1999).

Table 5.4. Biowaste composition expressed as a percentage of the total volatile solids in Switzerland (Perringer 1999)

Cellulose	Hemi-Cellulose	Sugars[1)]	Proteins	Lipids	Lignin
21.4	12.6	32.0	18.7	10.0	5.3

[1] 24.3% soluble

The organic fraction of household waste in Finland has been shown to contain 31% total solids (TS) and 22.5% volatile suspended solids (VSS); 20.6% of TS is carbohydrates while lipids and proteins made up 2.9 and 5.7% respectively (Jokela and Rintala 1999).

A fractionation of VFY was carried out in 1987 and 1988 by RIVM (RIVM 1988 and 1989b). The results are plotted in Figure 5.3.

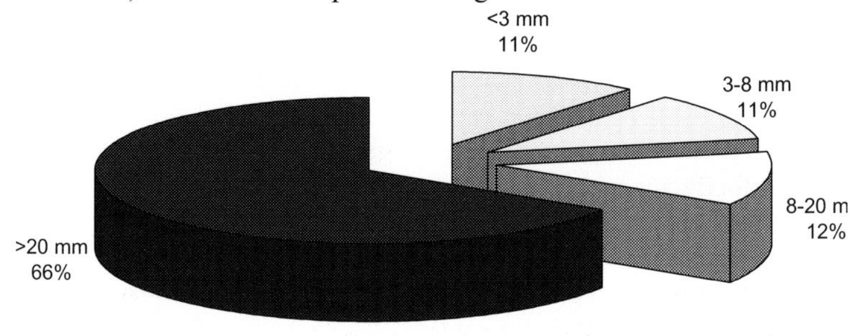

Figure 5.3. Physical fractionation of VFY (RIVM 1988 and 1989b).

The concentration of Ca, Mg and NH_4 in biowaste is relatively high in relation to the phosphorus content. This can contribute to the danger of struvite ($MgNH_4PO_4$) or hydroxyapatite ($Ca(PO_4)_3OH$) formation during anaerobic digestion if this treatment option for biowaste is chosen.

Recycling of biowaste can sometimes be impossible because of the heavy metals that exceed allowable standards (Veeken 1998). A large fraction of biowaste is not organic but is made of soil minerals. This is because approximately 80% of biowaste is collected outdoors and a substantial amount of soil is introduced into the biowaste. When biowaste originates from a village, soil minerals make up more than 50% of the total VFY waste. When the biowaste comes from indoor waste (in towns) there are few soil minerals.

The heavy metal content in fractionated waste corresponds with the natural concentration of these components (Veeken 1998). The heavy metal content of fresh plant material and food residues is very low, but the heavy metal content of partly decayed and humified organic matter from humus layer of soils is higher. These are, however, the normal concentrations in humus. The metal levels found in sandy garden soils are also the natural concentrations. Heavy metal contamination can be caused by the introduction of collected outdoor waste originating from industrial and urban regions or by progressing environmental pollution. An analysis of physical fractions performed by Veeken (1998) (see Table 5.5) showed that 45–60% of heavy metals accumulate in the organo-mineral fraction with particle size smaller than 0.05 mm that constitutes only 17% of biowaste. This fraction consists of humic substances, metal oxides and clay minerals, which all bind heavy metals very strongly.

Table 5.5. The heavy metal content (mg/kg of dry matter) in biowaste in relation to size fraction compared with Dutch legal demands for required compost quality. (SDU 1991)

Metal	Total biowaste	Organic part		BOOM standards	
		>1 mm	<1 mm	Compost	Clean compost
Cd	0.6	0.25	1.0	1	0.7
Cu	20	23	60	60	25
Pb	65	40	150	100	65
Zn	140	110	300	200	75

Grass clippings contain 2 to 4% nitrogen, which is more than for other organic matter sources such as leaves (0.5 to 1%), coffee grounds (0.5 to 2%) and horse manure (1 to 2%). If blended properly with soil, the nitrogen level of the clippings increases the microbial decomposition of the grass.

5.3.2.2 Kitchen waste

A detailed characterisation of kitchen waste is available in the literature. This depends strongly on the geographical location of analysis carried out. Some examples are listed in Tables 5.6 (Italy), 5.7 (Korea), 5.8 (Germany) and 5.9 (the Netherlands).

Table 5.6. Main characteristics of source collected daily from supermarkets and canteens of the city of Treviso (Cecchi *et al.* 1994)

Parameter (unit)	Average	Range
TS (g/kg)	81.8	54.4–132.7
VSS (g/kg)	67	46.8–105.6
COD_T (kgO_2/kgTS))	1.0	0.7–1.5
Total N (%TS)	2.1	1.4–3.3
Total P (%TS)	2.8	1.3–3.3

Table 5.7. Composition of crude kitchen refuse (Paik *et al.* 1999)

Parameter (unit)	Average	Range
TS (g/L)	168	119–235
VSS (g/L)	154	108–216
COD_T (g/L)	180	102–239
BOD (g/L)	108	69–170
Total N (g/L)	5	3.2–7.8
Total P (g/L)	0.63	0.350–0.9
Carbohydrates (g/L)	31	9.5–71.4
Proteins (g/L)	29.4	17–46.7
Lipids (g/L)	27.8	17.6–48.2
C/N ratio (-)	10.9	7.2–21.1

Table 5.8 compares the typical composition in Germany of the organic fraction of municipal solid waste (OFMSW) and kitchen waste. The N- and P-content of both types of biowaste are comparable. The high salt content in kitchen waste can be problematic in composting processes.

Detailed data is also available regarding swill composition for hospitals. This does not, in principle, differ significantly in composition from household kitchen waste but it does differ in quantity. The composition of swill as found in a number of Dutch hospitals is given in Table 5.9. The average volume of swill as measured in 1993 per patient was found to be 0.4 litre/person/day (L/p/d) (Braber 1993; Duynhoven 1994) while for a household this value was found to be 0.2 L/p/d.

Table 5.8. Comparison between the composition of OFMSW and kitchen waste from canteens and restaurants in Germany (Kübler *et al.* 1999)

Parameter (unit)	OFMSW		Kitchen waste	
	Average	Range	Average	Range
TS (% weight)	39	23–35	27	19–37
VSS (%TS)	63	57–70	93	88–96
Salt (%TS)		2.0–2.7		8–11
N (%TS)		2.2–3.4		3.2–4.0
P (%TS)		0.4–0.6		0.5–0.7

Table 5.9. Composition of swill in Dutch hospitals (Braber 1993; Duynhoven 1994)

Parameter	Fraction, (%TS)	Concentration, (g/L)	Concentration, (g/p)
TS		162	
VSS	92.3	149	
COD		278	111.2
Carbohydrates	27.9	98.5[1]	39.4[1]
Protein	20	48.6[1]	19.4[1]
Lipids	55.3	131.1[1]	52.4[1]
P_{tot}	0.6[2]	0.970	
pH		4.4	

[1] in gCZV/L
[2] percentage VSS

Swill has a very high biodegradability. A study by Veeken and Hamelers (1999) on the hydrolysis of compost sub-components showed that biowaste collected indoors (swill) has a significantly higher biodegradability than outdoor waste. For bread, orange peel and coffee filters this factor was as high as 90, 92 and 99% respectively (based on total COD).

According to Vermeulen *et al.* (1993), kitchen derived biowaste contains nitrogen as high as 4.7 gN/kg_{waste} while the concentration of ammonium-nitrogen (NH_4-N) is approximately 3.5 gN/kg_{waste}. A high level of ammonia can inhibit methanogenic bacteria if anaerobic digestion is applied. When kitchen waste is landfilled together with other biowaste without being first sorted out, high levels of ammonia can be solubilised into the landfill leachate (Jokela and Rintala 1999).

5.3.2.3 Faeces

A detailed composition of faeces can be calculated based on the amount of food that is consumed daily and the fraction that is normally found in physiological solid waste (excreted, not digested).

Faeces contains bacteria, digestion liquids and enterocytes. Approximately one-third of faeces consists of food remains, one-third is intestinal bacteria and one-third is from the intestine itself (enterocytes and liquids). As a result, the composition of faeces does not vary much in relation to the eating pattern. The weight of faeces is basically determined by non-biodegradable fibre (for example, from bread and other grain products, vegetables, potatoes and fruit).

According to several researchers the dry weight of faeces can vary between 70 and 170 g/person/day (g/p/d) while the average frequency of excretion oscillates around 1 time/person/day (1/p/d). (Table 5.10).

Table 5.10. Total and dry weight of faeces and the average frequency of passing stools

Reference	Total weight of faeces (g/p/d)	Dry weight of faeces (g/p/d)	Frequency (1/p/d)
Bingham 1979	70-140	19-38	-
Cummings *et al.* 1992	106	-	-
Glatz and Katan 1993	170	44.2	1.2
Cummings *et al.* 1996	138	34	0.9
Belderok *et al.* 1987	100-200		

A detailed composition of faeces is given by van der Wijst and Groot-Marcus (1998). Based on the digestibility and consumption of nutritious matter (Table 5.11), the COD-content of the faeces of an average Dutchman is 41.4 g/p/d.For comparison, the table shows the composition of faeces calculated based on dry matter (Wijn and Hekkens 1985).

The consumed lipids are 90–98% absorbed by human bodies. In faeces, lipids are found as a mixture of fatty acids, Ca- and Mg- fatty acids, cholesterol and vegetable sterols (Stasse-Wolthuis and Fernandes 1991). The presence of lipids slows down the digestive process. In the Netherlands 120 g/p/d lipids and 1–3 g/p/d cholesterol are taken upeach day, and approximately 1 g/p/d lipids and 0.5 g/p/d cholesterol are excreted each day.

Carbohydrates constitute an important energy source for a human being and are present in food as starch (60%), saccharose (30%) and lactose (10%). The digestibility of carbohydrates is approximately 70% (Bingham 1979). The absorption efficiency of carbohydrates in the intestine depends on the type of starch (Stasse-Wolthuis and Fernandes 1991).

Table 5.11. Digestibility, consumption of nutritious matter and COD-value of physiological solid waste of an average healthy Dutchman (van der Wijst and Groot-Marcus 1998)

Component	% not digested	Consumption of components (g/p/d)	Average COD (g/p/d)	Wijn and Hekkens 1985 (g/p/d)
Carbohydrates				7.3
mono-sacharides	2–4	121	3.8	
poli-sacharides	2–22	126	15.81	
Lipids	2–10	92	14.86	13.3
Protein	<5	81	6.93	15.4
Total			41.4	36.0

The proteins originating from plants and meat-products constitute the only source of nitrogen for a human. The majority of nitrogen is excreted via urine and the rest via faeces, sweat, dead hair and skin cells. These proteins contain approximately 16% nitrogen. In healthy people, the same amount of nitrogen is excreted as is taken up. In faeces, approximately 5% of the nitrogen consumed can be measured.

The nitrogen content in faeces was assessed in several types of research and the results are listed in Table 5.12. They vary significantly because of different approaches used in the experiments. For the Dutch situation it is assumed that faecal nitrogen is from 1 to 2 gN/p/d.

Table 5.12. The faecal N concentration as measured in different countries

Faecal N (g/p/d)	Type of experiment	Reference
1.62	In UK 12 people with different eating patterns were examined	Wijn and Hekkens (1985)
5.0–7.0	The average from a review of literature from different countries	Flameling (1994)
2.4	In Brazil 5 men were subjected to a diet consuming 56 g protein per day	Sergio Marchini *et al.* (1996)
0.27	In USA 5 women were subjected to a diet of 70 g protein per day	Fricker *et al.* (1991)
1.13	In Nigeria 12 women were given a diet of 30 g protein per day	Egun and Atinmo (1993)

5.4 MANAGEMENT OF DOMESTIC SOLID WASTE

This section briefly describes current waste management, its advantages and disadvantages and developing trends in relation to DESAR concepts. The overall trend is to reuse and recycle as much as possible. In this context waste management is prioritised as follows:

- a reduction in waste generation;
- recycling, composting and reuse,
- incineration with energy recovery; and
- disposal.

5.4.1 Reduction in waste generation

Solid waste can be significantly reduced (by up to 85%) simply by increasing control over processes and changing purchasing practices (especially applicable to restaurants and cafeterias). Logical and simple examples applicable to households and hospitals, offices, restaurants, bars and so on are as follows:

- purchasing varying sizes of canned/bottled products so only what is needed is used, resulting in fewer leftovers;
- buying supplies in bulk if this means less packaging;
- avoiding single-use, disposable items such as paper plates;
- avoiding using disposable plastic, paper or polystyrene items;
- using water and energy efficiently by planning cooking and dishwashing;
- using cloth towels, reusable food containers and china (not plastic) crockery and cutlery;
- making table napkins from damaged or worn tablecloths;
- using fabric table napkins rather than paper;
- using badly damaged towels and tablecloths as cleaning rags and cloths;
- giving unused food to charity;
- recycling as much as possible – for example, glass, containers, cooking oils, aluminium cans;
- designating a space where people can deposit their empty drink containers for recycling; and
- setting up a (worm) compost bin to compost food waste.

5.4.2 Recycling

Recycling promotes closed cycles of carbon and nutrients and it reduces the emission of greenhouse gases. Sorting waste on a household level is a good starting point for sustainable waste management. The scale and efficiency of waste sorting is encouraged by favourable conditions – local authority equipment, type of housing, age group, income and level of education. Sorting practice varies, since the types of waste material involved will reflect environmental awareness in a given area (Jeunesse 2000).

Local authority policy has a positive impact on waste sorting and thus on recycling. Next to this, the type of housing people live in has a direct effect on recycling trends. A survey carried out by Ademe (Jeunesse 2000) in France showed that 75% of people living in detached houses recycled glass compared with only 46% of people living in flats. Living in a flat discourages individual recycle initiatives. Proportional relations are reported between income, age, level of education and level of environmental awareness. Sorting one form of specific waste material implies sorting others.

The waste material that may be sorted for recycling commonly includes:

- glass;
- paper;
- biowaste;
- batteries and other hazardous material;
- plastics; and
- metal.

Biowaste recycling can be achieved by biological conversion into compost, which can be further reused as soil conditioner or fertiliser. The decomposition and recycling of organic waste is an essential part of soil building and healthy plant growth in forests, meadows and gardens. The pilot programme in Bellport, New York showed that the mixed waste stream could be reduced by 30% simply by home composting (Cooley *et al.* 1999).

In the industrialised world the amount of waste disposed is reduced by recycling materials such as newspaper, glass bottles and aluminum cans. In for instance Bellport (mentioned above) the tipping fee for mixed waste amounts to $66 per ton while the tipping fee for plastic, metal, and glass recyclables is $33 per ton. This indicates that an increasing level of reuse and recycling can create additional savings.

Even with relatively advanced organised separation procedures additional waste is produced. This is sometimes called grey solid waste, and it consists of all other types of waste and smaller or higher quantities of recycling material depending on whether and to what degree local people cooperate with the recycling scheme. In the Netherlands, for example, grey waste is generally brought to a separation plant (Grontmij 1997). The objective of a separation plant is to separate the waste into three streams: a high-energy (or caloric) fraction, a low-energy fraction, and a paper/plastic/iron/non-iron (other metals) fraction. The high-energy fraction is suitable for incineration while the low-energy fraction is subjected to washing and digestion. Metals (ferro- and non-ferro fractions) are reused in the metals industry. The paper/plastic mixture is used for further processes in the paper/plastic industry.

5.4.2.1 Composting

Compost is the product of an aerobic process by which plant and other organic materials decompose under controlled conditions. When biowaste is placed in a pile which contains sufficient nutrients, moisture and oxygen, bacteria and fungi break down the waste as they feed upon it. The finished product, a dark brown, dirt-like, crumbly and earthy-smelling substance formed of decomposing organic matter, is called stabilised compost or humus.

Every organic structure has a carbon to nitrogen ratio (C:N) ranging from 500:1 for sawdust to 15:1 for table scraps. To attain an ideal C:N ratio for the activity of compost microbes in different types of wastes (with different C:N ratios), these should be mixed together. A C:N ratio of 30:1 is ideal for a fast, hot composting, and a higher ratio (e.g. 50:1) will be adequate for a slower composting.

Composting is the most practical and convenient way to handle garden waste. It can be easier and cheaper than bagging this waste or taking it to a local tip. Anything that was once alive can be composted. Garden waste, such as fallen leaves, grass clippings, weeds and dead plants, make excellent compost. Woody garden wastes can be clipped and sawed down to a size useful for a wood stove or fireplace, or they can be run through a shredder for mulching and path-making. Used as mulch or for paths, they will eventually decompose and become compost. Compost can be used to enrich flower and vegetable gardens, to improve the soil around trees and shrubs, as a soil substitute for house plants and planter boxes and, when screened, to grow seeds in or to use as lawn top-dressing.

Compost is a major fertility input. High quality compost supplies nutrients, trace elements, organic matter, humus and biological activity. The abundant beneficial microbes in compost contain most of the nutrients needed by plants. As these microbes die, otherwise unavailable nutrients are provided, resulting in a slow release of nutrients over the course of a growing season. Compost can reduce fertiliser input costs and increase fertiliser efficiency.

Compost encourages the formation of stable soil aggregates. This is vital to soil structure, and to the creation of the soil's ecosystem, which allows beneficial microbes, fungi and plants to flourish. Compost improves the water-holding capacity of sandy soils, and adds structure and permeability to clay soils. A good granular soil structure allows water, air and nutrients to pass freely to the plant's roots, thus promoting soil and plant health. Compost speeds up the rate of seed germination and stimulates plant growth. Horticultural experiments – on everything from sod to container plants – demonstrate that plants and seedlings grow on much more quickly by adding compost to the planting mix. In addition, plants grown in soil mixed in with compost are more able to withstand heat stress, and are less susceptible to attacks by pests.

The addition of nutrients from humus reduces the need for commercial fertilisers. Composting also preserves valuable resources by reducing the amount of waste.

5.4.2.2 Anaerobic digestion of biowaste

During the anaerobic digestion of biowaste (VFY), organic material is converted in anaerobic conditions to methane (CH_4), carbon dioxide (CO_2) and biomass. The fundamentals of the anaerobic digestion process and its application in practice are discussed elsewhere in this book (see Chapters 10–12). The biogas composition after digestion is 55% CH_4 and 45% CO_2. The addition of paper contributes to a higher biogas yield. The addition of paper has two objectives:

- to increase gas production; and
- to reduce the risk of acidification.

The product of digestion, stabilised regarding organic content, but containing significant concentrations of soluble nutrients, can be reused as a soil conditioner or fertiliser.

5.4.2.3 The management of grass-clippings

One way to cut down on the grass clippings generated after mowing the lawn is to leave them on the lawn after mowing, since they provide a natural fertiliser. Another way, if a professional service is used, is to ask for a grass growth retardant. The best of these products breaks down to CO_2 and water and slows down grass growth by up to 50%.

Disposing of grass clippings is very costly and wasteful. Grass clippings make up thousands of tons of solid waste in some American states (for instance New Jersey; see Cooley *et al.* 1999). In fact, nearly one-third of all summer waste handled by rubbish men consists of grass clippings. They represent a waste management cost that is paid either directly to the rubbish collector, or indirectly in taxes. As an example, each ton of grass clippings brought to a landfill site costs between $65 and $100 in disposal fees. Furthermore, landfilled grass clippings decompose slowly, due to a lack of oxygen.

When recycled grass clippings are returned to the soil, nearly 10 grams of valuable nitrogen is recycled to every square metre of lawn each year. That may not be enough to keep the lawn really healthy, so another 7.5 g/m^2 of nitrogen must be added each year.

5.4.2.4 Burying kitchen waste

Food waste can be buried in empty spots in vegetable and flower gardens. Buried food scraps may take from two to six months to decompose, depending on soil moisture, temperature, worm population and food source. In good garden soil, leafy greens will break down in weeks while whole citrus peels may take several months in a loose and fertile garden soil.

5.4.2.5 Hazardous waste management

Hazardous waste management procedures depend on the type of waste. All hazardous materials are sent away to be recycled or incinerated. The following items can be recycled: car batteries, used motor oil, oil filters, antifreeze and latex paint.

5.4.3 Incineration

Incineration is considered to be the most expensive method of solid waste management. Incineration may cause air pollution (for example, of heavy metals, chlorinated dioxins and dibenzofurans) and thus an expensive post-treatment is required (White *et al.* 1995). Incineration also contributes to the greenhouse effect and leads to the production of hazardous by-procucts that have to be landfilled (McBean *et al.* 1995).

Hazardous waste management often includes incineration. Solvents and other flammables are incinerated for energy. Other materials, such as pesticides, are destroyed.

In current policy concerning incineration and co-incineration of hazardous and non-hazardous waste, more and more attention is being paid to preventing and reducing pollution in all aspects of the process, including installation, operation, air and water emissions and residues (Peyret 2000).

5.4.4 Landfill or illegal disposal

The problems related to landfilling are a lack of space, the uncontrolled emission of greenhouse gases and the possibility of groundwater or soil contamination due to leakage (Farquhar and Rovers 1973; Hjelmar 1995). Consistent with current trends, the design and operation of landfills should also reflect the three Rs of waste management, namely reduce, reuse and recycle.

This expresses efforts to prevent or at least minimise the production of leachate and landfill gas. Recent developments forbin solid waste with an organic matter content of higher than 5% to be landfilled.

It is illegal to dispose of hazardous waste in rubbish. Dumped hazardous materials can leach into groundwater, and contaminate creeks and rivers. Unfortunately, illegal dumping in unauthorised areas is a common practice. Materials typically dumped include:

- construction and demolition waste, such as drywall, roofing shingles, wood, bricks and concrete;
- abandoned cars, car parts and scrap tyres;
- appliances, furniture;
- household rubbish; and
- medical waste;

When litter is left on sidewalks or along kerbs, it may get washed into storm drains during a heavy rain, eventually to reach the nearest river or ocean. Litter can be very dirty and, as well as not looking very nice, it may carry germs or toxicants. Since litter is exposed to elements, it may start to decompose and this can result in a foul smell.

5.5 REFERENCES

Australian Waste Database (1997) Composition of solid domestic waste, Mitcham, SA. The University of New South Wales. Sydney, Australia. (http://www.water.civeng.unsw.edu.au/water/awdb/compostn/sa/Mitcham.htm)

Ayoub, G.M., Acra, A., Abdallah, R. and Merhebi, F. (1996) Fundamental aspects of municipal refuse generated in Beirut and Tripoli. Field studies 1994–1996. Department of Civil and Environmental Engineering, Reports of American University of Beirut (http://www.sdnp.org.lb/ump/solid10.html)

Belderok, B., Breedveld, B.C., Douwes, A.C., Fernandes, J., Korver, O., Nagengast, F.M., Smit, G.P.A., Swinkels, J.J.M. and Vandewoude, M.F.J. (1987) Langzame en snelle koolhydraten, serie: Voeding en Gezondheid. Samson Stafleu, Alphen aan den Rijn, the Netherlands (In Dutch)

Bingham, S., (1979) Low-residue diet: a reappraisal of their meaning and content. *Journal of Human Nutrition* **33**, 5–16.

Braber, K. (1993) Anaerobe vergisting van swill, Publicatiecentrum NOVEM, SITTARD (In Dutch)

Cecchi, F., Battistoni, P, Pavan, P., Fava, G. and Mata-Alvarez, J. (1994) Anaerobic digestion of OFMSW (organic fraction of municipal solid waste) and BNR (biological nutrient removal) processes: a possible integration – preliminary results. *Wat. Sci. Tech* **30**(8), 65–72.

Cooley, A., Stravinski, D. and Tripp, J.T.B. (1999) *The Village Of Bellport's Program For The Home Composting Of Kitchen Waste*. Reports of Environmental Defence, New York, 1999. (http://www.edf.org/pubs/Reports/compost.html)

Cummings, J.H., Bingham, S.A., Heaton, K.W. and Eastwood, M.A. (1992) Fecal weight, colon cancer risk, and dietary intake of nonstarch polyssacharides (dietary fiber). *Gastroenterology* **103**(6), 1783–1789.

Cummings, J.H., Beatty, E.R., Kingman, S.M., Bingham, S.A. and Englyst, H.N. (1996) Digestion and physiological properties of resistant starch in the human large bowel. *British Journal of Nutrition* **75**, 733–747.

Duynhoven, van A.H.M. (1994) Verwijdering van organisch keukenafval. De beordeling van drie verwijderingsmethoden voor het Academisch Ziekenhuis Nijmegen. MSc report Milieukunde nr. 79, Katholieke Universiteit Nijmegen, The Netherlands (In Dutch).

Egun, G.N. and Atinmo, T. (1993) Protein requirement of young adult Nigerian females on habitual Nigerian diet at the usual level of energy intake. *British Journal of Nutrition* **70**, 439–448.

Farquhar, G.J. and Rovers, F.A. (1973) Gas production during refuse decomposition. *Water, Soil and Air Pollution* **2**, 483.

Flameling, A.G. (1994) Studies into possibilities of anaerobic treatment of domestic wastewater in order to reduce the greenhouse effect. Doctoraal scriptie, Landbouwuniversiteit Wageningen, The Netherlands (In Dutch).

Fricker, J., Rozen, R., Melchior, J.C. and Apfelbaum, M. (1991) Energy metabolism adaptation in obese adults on a very low calorie diet. *American Journal of Clinical Nutrition* **53**, 826–830.

Glatz, J.F.C. and Katan, M.B. (1993) Dietary saturated fatty acids increase cholesterol synthesis of fecal steroid excretion in healthy men and women. *European Journal of Clinical Investigation* **23**, 648–655.

Grontmij (1997) MER Vagron. Samenvatting, Grontmij Advies en Techniek, De Bilt, The Netherlands (in Dutch)

Haskoning (1992) Report Milieu-effectrapport Vergistingsinstallatie GFT-afval Midden-Brabant, Haskoning, Nijmegen, The Netherlands.

Hjelmar, O. (1995) Composition and management of leachate from landfills within the EU. In: Proceedings Sardinia 1995, Fifth International Landfill Symposium, Calgliari, 243.

Jeunesse, V. (2000) Waste sorting in France. Journal Hors-Serie Environnement & Technique. Salon Paris 2000, 32-33.

Jokela, J.P.Y. and Rintala, J.A. (1999) Long-term anaerobic incubation of source-sorted putrescible household waste: ammonification, methane production and effect of waste characteristics. Proceedings of the Second International Symposium on Anaerobic Digestion of Solid Waste (II ISAD-SW), June, Barcelona, Spain.

Kübler, H., Hoppenheidt, K., Hirsch, P., Kottmair, A., Nimmrichter, R., Nordsieck, H., Mücke, H. and Swerev, M. (1999) Full-scale co-digestion of organic waste. Proceedings of the Second International Symposium on Anaerobic Digestion of Solid Waste (II ISAD-SW), June, Barcelona, Spain.

McBean, E.A., Rovers, F.A. and Farquhar, G.J. (1995) *Solid Waste Engineering and Design*, Prentice Hall, New Jersey, USA.

Paik, B-C., Shin, H-S., Han, S-K., Song, Y-C., Lee, C-Y. and Bae, J-H. (1999) Enhanced acid fermentation of food waste in the leaching bed. Proceedings of the Second

International Symposium on Anaerobic Digestion of Solid Waste (II ISAD-SW), June, Barcelona, Spain.

Péringer, P. (1999) Biomethanation of sorted household waste: experimental validation of a relevant mathematical model. Proceedings of the Second International Symposium on Anaerobic Digestion of Solid Waste (II ISAD-SW), June, Barcelona, Spain.

Peyret, L. (2000) Waste. The future European framework for incineration. Journal Hors-Serie Environnement & Technique. Salon Paris 2000, 33-34.

RIVM (1988) Fysisch en chemisch onderzoek aan huishoudelijk afval van 1987 inclusief batterijen. RIVM report. Bilthoven, The Netherlands (In Dutch).

RIVM (1989a) Afval 2000 – een verkenning van de toekomstige afvalverwijderingsstructuur. RIVM report, Bilthoven, The Netherlands (In Dutch).

RIVM (1989b) Fysisch onderzoek naar de samenstelling van het Nederlandse huishoudelijke afval. Resultaten 1988. RIVM report, Bilthoven, The Netherlands (In Dutch).

Roosmalen, G.R. E.M. van and Langrijt, J.C. van de (1989) Green waste composting in the Netherlands. *Biocycle* **30**, 32–35.

SDU (1991) Besluit overige organische meststoffen (BOOM). *Staatblad* **613**, 1–45. (In Dutch.)

Sergio Marchini, J., Moreira, E.A.M., Moreira, M.Z., Hiramatsu, T., Dutra de Oliveira, J.E. and Vannucchi, H. (1996) Whole body protein metabolism turnover in men on a high or low calorie rice and bean Brazilian diet. *Nutrition Research* **16**(3), 435–441.

Stasse-Wolthuis M. and Fernandes J. (1991) Voeding en spijsvertering. Bohn Stafleu Van Loghum, Houten/Antwerpen (In Dutch).

Veeken, A. (1998) Removal of heavy metals from biowaste. Modelling of heavy metal behaviour and development of removal technologies. Ph.D. thesis, Wageningen University, the Netherlands.

Veeken, A. and Hamelers, B. (1999) Effect of temperature on hydrolysis rates of selected biowaste components. *Bioresource Technology* **69**, 249–254.

Vermeulen, J., Huysmans, A., Crespo, M., van Lierde, A., De Rycke, A. and Verstraete, W. (1993) Processing of biowaste by anaerobic composting to plant growth substrates. *Wat. Sci. Tech.* **27**, 109–119.

White, P.R., Franke, W. and Hindle, P. (1995) Integrated solid waste management – a life cycle inventory, Blackie, London.

Wijn, J.F. de and Hekkens, W.Th.J.M. (1995) Fysiologie van de voeding, 2nd edition, Bohn, Stafleu Van Loghum, Houten, The Netherlands, ISBN 9031310093.

Wijst, van der, M. and Groot-Marcus, A.P. (1998) Huishoudelijk afvalwater. Berekening van de zuurstofvraag, Huishoud en Consumentenstudie, Landbouwuniversiteit Wageningen, The Netherlands, report STOWA 98-40 (In Dutch).

6

The collection and transport of wastewater

L. Heip, R. Bellers and E. Poppe

6.1 INTRODUCTION

Water is probably the most basic of all human needs. Therefore it is not surprising that most human development happened at or close to a river. The river was not only used as a source for drinking water, but also as an easy means to dispose of wastewater. Traditionally the wastewater was disposed of in the simplest way possible, either by discharging it into ditches and brooks, or simply letting it flow over land to the river.

As cities developed, the disposal of wastewater started to become more and more of a problem. From very early on the use of water from the river within the city limits as drinking water was abandoned in most cities. Allowing the

Edited by P. Lens, G. Zeeman and G. Lettinga. ISBN: 1 900222 47 7

wastewater to flow through the streets and in open ditches also caused numerous problems. As early as 4500 BC the Assyrians built sewer networks. This problem was also recognised in ancient Rome. The administrators of the city of Rome built the most famous ancient sewer, namely the Cloaca Maxima, to alleviate this problem (Berlamont 1997).

In most medieval cities, however, the traditional way of disposing of wastewater was still widely practiced. In the city of Antwerp in Belgium, for example, the many brooks were open watercourses until the eighteenth century, causing many problems, some innocent, such as odour or the occasional drunken lout falling into them, and some less innocent, such as waterborne diseases. Therefore, the city of Antwerp persuaded its inhabitants to culvert these watercourses by allowing them to make use of the extra land for building and expanding their homes. These culverted watercourses still form the backbone of the sewerage system in the historical centre of Antwerp, forming a network of underground canals (Figure 6.1).

Figure 6.1. Culverted watercourse as sewer: the Ruien in Antwerp (photo: Aquafin).

With the Industrial Revolution in the nineteenth century the situation worsened dramatically, with outbreaks of cholera as a result. In England the first large modern sewerage systems were built in the middle of the nineteenth century (Stedman 1995). These sewer systems served a dual purpose. They were meant to dispose of wastewater but also to serve as flood-protection. On paved surfaces that form the major part of any city, rainwater does not infiltrate. Thus, with every storm, streets tended to transform into rivers. So the sewers were designed not only to transport wastewater to the river, but to transport rainwater as well. In fact, the sewers were designed to dispose of all water to the river as quickly as possible.

This solution did at first solve the problems, but soon the dramatic decrease in river water quality became a major threat to public health (Martin 1927). The rivers were no longer able to cope with the pollution load since their self-cleaning ability had been surpassed (Hosten 1991).

For a long time the only way domestic wastewater was treated was by spreading it over land. At the end of the nineteenth century a number of discoveries led to the development of the first processes for wastewater treatment. The principle is mostly based on boosting natural self-cleaning ability by adding air to the wastewater before discharging it into the river. The biomass responsible for the breakdown of the pollution can be either suspended (activated sludge) or attached (e.g. trickling filter). These processes, collectively known as 'aerobic treatment', are focused on the removal of organic pollution (Verstraete *et al.* 1999).

An alternative to aerobic treatment is anaerobic treatment, where organic pollution is converted into biogas, allowing energy-recuperation (Lettinga 2001). This process, however, is primarily used for industrial wastewater. Up to now, it has rarely been used in a large public sewer system, but is now increasingly applied in tropical countries.

More recently (after the Second World War) new problems emerged with other components. Nutrients such as nitrogen and phosphorous, combined with organic pollution, led to profuse algae growth, depleting the oxygen content of rivers and lakes at night-time. This in turn prevented higher life forms such as fish from developing. The eutrophication process can be prevented by removing the nutrients down to their natural background concentration. Since the mid-1980s, new processes, mainly variations on the known activated sludge processes, have been developed to remove the nutrients from the wastewater (Wanner *et al.* 1992) (see Figure 6.2).

6.2 TYPES OF SEWER SYSTEMS

Nowadays, the main purpose of sewerage systems is to transport wastewater to a suitable treatment plant. Several different designs are used to serve this purpose.

Historically, the first large sewer networks were combined systems, transporting rainwater and wastewater through the same pipes. More recently dual-pipe systems are more commonly used. In these so-called separate systems one pipe carries wastewater while the other pipe carries rainwater.

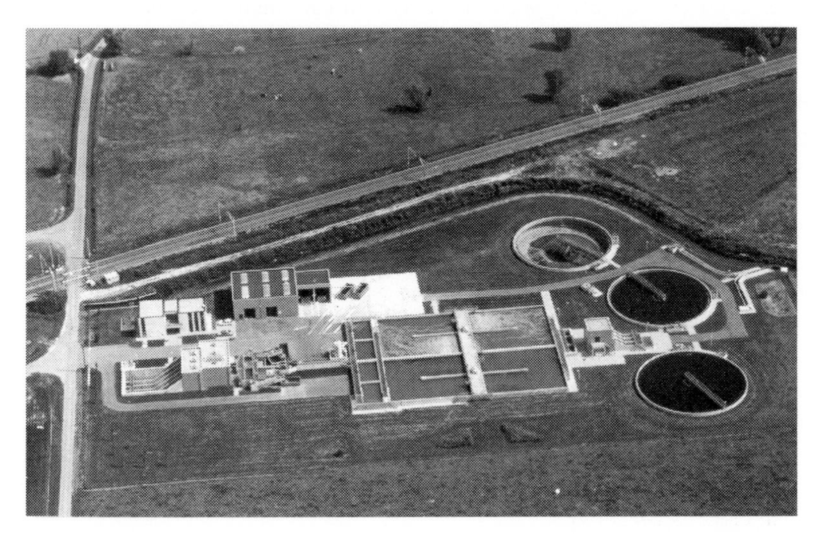

Figure 6.2. Nutrient-removing Krüger-type wastewater treatment plant (WWTP), Tielt (photo: Aquafin).

Both combined and separate sewer systems are based on gravitational flow. This implies that the pipes always need to slope towards the downstream end of the system. If this means that the pipes are too deep – a technical maximum depth at which pipes can still economically be placed is 8–10 metres – a pumping-station is required. In flat regions, such as Flanders and the Netherlands, large sewer networks often have numerous pumping stations. This makes those systems relatively costly and difficult to maintain.

In relatively small sewage catchment areas alternative solutions can be used. One alternative is the vacuum system, in which all pipes are kept under-pressurised by a central pumping-station (Gray 1991; Schinke 1999) (see Figure 6.3). The wastewater is sucked into the pipes and transported to a wastewater treatment plant (WWTP). The pipes do not need to have a specific slope, since the flow is not caused by gravity. All the pipes can be placed near to the surface. Laying the pipes becomes cheaper and repair and maintenance costs if one of the pipes should break are also reduced. Leaks cannot cause pollution to the environment, so this system also has an environmental advantage over gravitational systems. The gravitational systems can become leaky upon ageing, causing groundwater to become polluted or draining the clean groundwater to a WWTP.

A major disadvantage of the vacuum system is its low durability. Vacuum pipes can break easily, especially where they are connected together. If the system fails, the houses connected to that branch of the system can, until a repair is effected, no longer discharge their wastewater. Also the numerous valves (one for every couple of houses) can cause serious operational problems. When one of the valves breaks, the system ceases to function. Thus, the operational cost of such a system tends to be high. New materials have, however, considerably increased the durability of this system, so it may become competitive.

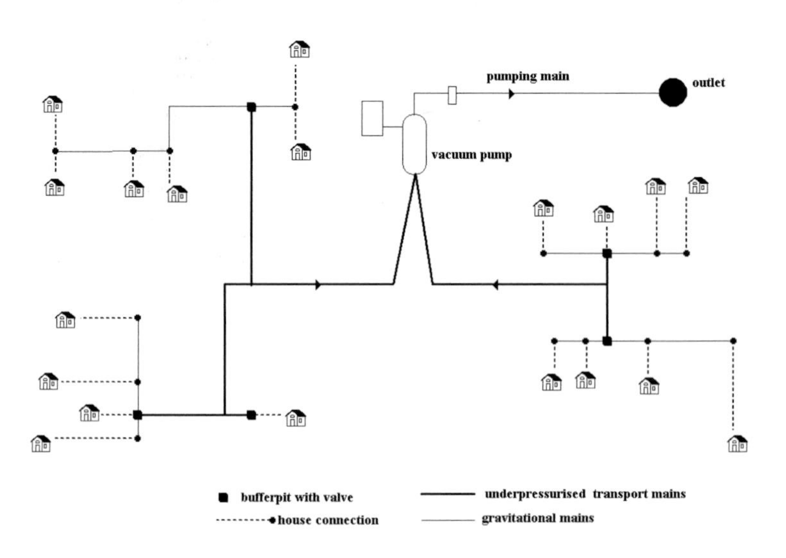

Figure 6.3. Schematic representation of a vacuum system.

Another alternative to the gravitational system is the system of pressurised sewers (Pfeiffer 1996; Dugre 1995). In such a system (Figure 6.4) wastewater from a couple of houses is collected and pumped into a common pressure main that is fed by a number of pumping stations. Again, the pipes connecting the pumping stations can be kept small and do not need a specific slope, so they can be constructed near to ground level, thus reducing the construction cost. The pipes are relatively durable. Even when the pipes break, the discharge of wastewater is still possible. However, a broken pipe will lead to massive pollution, since most of the sewage will be pumped straight into the soil.

The cascading sewerage system is positioned somewhere between a full gravitational system and pressurised sewers (Anonymous 1994). Instead of pumping into a common pressure main, a gravitational sewer connects each pumping-station. The location of each pumping station is determined by the requirement that the gravitational sewers must not become too deep. This system is especially suited for ribbon development.

Vacuum systems, pressurised sewers and cascading sewers are only suited for wastewater. A separate rainwater system is always necessary.

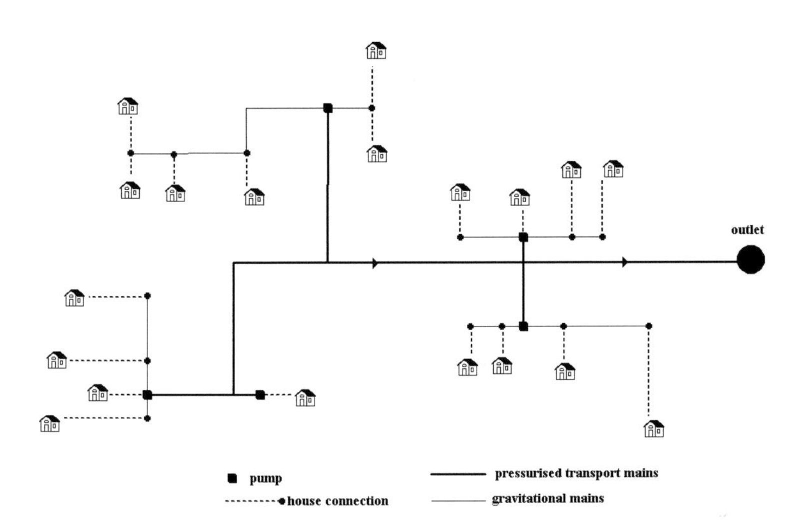

Figure 6.4. Schematic representation of a pressurised sewer system.

As the pressures encountered in any type of sewer system are usually low (usually not higher than 5–10 mH_2O) a large range of different materials can be used. They range from concrete over cast iron to polymers such as high density polyethlene (HDPE), polyvinychloride (PVC) or fibre-reinforced polyester. For gravitational sewers concrete is most common, reinforced for any diameter larger than 900 mm. For smaller pipes (for example, house connections) vitrified clay pipes are also regularly used. All these materials can withstand limited pressures. For pressure mains, cast iron is the most common material, although reinforced concrete can also be used. The choice of material is mostly made on economic grounds.

6.3 COMBINED SEWER SYSTEMS

6.3.1 Why combined sewer systems?

Combined sewer systems were not originally developed for wastewater treatment. Their main purpose was to discharge both wastewater and rainwater to a receiving watercourse as fast as possible.

Often a large portion of the sewer system is already in operation before wastewater treatment is started. In those cases the existing sewer system is almost always a combined system with one or more direct discharge points to one or more receiving watercourses. This existing sewer system represents a big capital investment. Thus, for economic reasons planning wastewater treatment also incorporates this existing collection system.

In urban Western Europe and North America, the building of sewer systems preceded wastewater treatment, so most existing sewers are combined. Table 6.1 describes the sewerage arrangements for the twelve countries in the European Union (EU) (excluding Austria, Finland and Sweden).

Table 6.1. Wastewater collecting systems and treatment provisions in EU member states in 1994 (Henderson 1998) [? means that no information was received from the country in question]

Country	B Belgium	DK Denmark	F France	D Germany
Population (millions)	9.9	5.1	57.8	80.3
% urban	70	85	72	?
%rural	30	15	28	?
Population density (persons/sq. km)	323	119	100	220
Percentage of population connected to collection systems	58	94	74	90
%urban	?	99	90	?
%rural	?	1	50	?
Percentage of wastewater receiving treatment (secondary or better)	25	92	50	78
Percentage of urban area served by combined collecting systems	70	45–50	75–80	67
Age profile of collecting systems (where known)	?	50% built after 1960 20% built after 1980	?	74% built after 1945 60% built after 1963

Country	GR Greece	IRL Ireland	I Italy	LUX Luxembourg
Population (millions)	10.3	3.6	56.7	0.4
% urban	58	56	74	57
%rural	42	44	26	43
Population density (persons/sq. km)	74	50	188	142
Percentage of population connected to collection systems	45	67	82	96
%urban	80	99	93	100
%rural	?	23	50	93
Percentage of wastewater receiving treatment (secondary or better)	18	25	40	84
Percentage of urban area served by combined collecting systems	20	60–80	60–70	80–90
Age profile of collecting systems (where known)	60% built after 1960	?	40% built after 1965	50% built after 1965

Country	NL The Netherlands	P Portugal	E Spain	UK United Kingdom
Population (millions)	14.9	10.5	39.1	57.5
% urban	88	64	75	88
%rural	12	36	25	12
Population density (persons/sq. km)	348	114	78	232
Percentage of population connected to collection systems	97%	62%	82%	96%
%urban	100%	88%	?	99%
%rural	30%	15%	?	85%
Percentage of wastewater receiving treatment (secondary or better)	78%	40–50%	45%	83%
Percentage of urban area served by combined collecting systems	74%	40-50%	96%	70%
Age profile of collecting systems (where known)	50% after 1955	70% after 1960	?	50% after 1945

6.3.2 Design of combined sewer systems

Incorporating the existing combined sewer system in wastewater treatment infrastructure implies connecting the discharge points to a WWTP.

The flow during dry-weather-flow conditions (DWF conditions) is only a fraction of the maximum flow during storm-events. In the temperate climate of central and northern Europe, the peak flows can easily be up to a hundred times higher than the foul flow. In more southern regions this can be even higher. It is not economically feasible to transport all this water to a WWTP. Furthermore, no wastewater treatment process is flexible enough to treat flows ranging between 1 and 100 times the foul flow. Moreover, the contaminant loading also fluctuates because of the diluting effect of stormwater. Therefore a limit must be set for the amount of wastewater to be treated at the WWTP. Typically, this limit is set at between 2 and 10 times dry weather flow (see Table 6.2). When flow exceeds this limit, the surplus flow is spilled to a receiving watercourse.

Table 6.2. Combined sewer overflow (CSO) settings (flow transported above DWF levels before a spill occurs) in EU-member states (Henderson 1998)

Country	CSO setting
Belgium	2-5 × mean DWF
Denmark	5 × peak DWF (8–10 × mean DWF)
France	3 × peak DWF (4–6 × mean DWF)
Germany	7 × DWF (2 × mean DWF + infiltration to treatment) and ATV Guideline A128 (90% load to treatment)
Greece	3–6 × mean DWF
Ireland	6 × DWF (recently formula A, see UK)
Italy	3–5 × mean DWF
Luxembourg	3 × peak DWF (4–6 × mean DWF); now ATV Guideline A128
Netherlands	Site-specific; minimal storage 7 mm of runoff from impervious area
Portugal	6 – mean DWF
Spain	3–5 × mean DWF
UK	Historically 6 × mean DWF Now Formula A: DWF + 1360P + 2^E P: population, E: industrial effluent

In some countries (UK, Denmark, the Netherlands, Ireland) the design of combined sewer overflows (CSOs) is based more and more on Environmental Quality Objectives/Environmental Quality Standards (EQO/EQS). In this approach, the effect of the drainage system on the receiving watercourse, including dilute pollution and wastewater treatment, is compared to the quality objectives of this watercourse. This approach requires the modelling of all system components, that is, sewerage and river (Van Assel *et al.* 1997).

Figure 6.5. CSO (Photo: Objectief).

Once a choice has been made about the carry-on flow for overflow structures, the design of a combined sewer system is based solely on its function as a rainwater carrier. The guidelines for the design of combined sewers set a flooding frequency that is considered to be acceptable (Table 6.3).

The size of the combined sewers is then determined in accordance with these guidelines. Since the flow in combined sewers is caused by rainfall, and since rainfall is a stochastic phenomenon, an assumption needs to be made as to the intensity of rainfall used for design. The most accurate approach makes use of a historical series of rainfall. A full record of historical rainfall, known as a time series of rainfall, can however not be used with every design method. When this cannot be used, a design storm (that is, a synthetic storm with certain statistical properties, used to calculate flows in sewers) which has predetermined statistical properties is used.

Historical rainfall series are country-specific and can usually be provided by the meteorological office of the country in question. The second step in the process is to determine the amount of rainfall that is transported to the sewers. This is a very complex process that is not yet fully understood. Phenomena such as evapo-transpiration, puddle formation, and so on, can only be approximated. This complex process is often highly simplified in order to be manageable in a design procedure.

Table 6.3. EU norm EN 752-2; 1996 – Drain and sewer systems outside buildings (EU-norm EN 752-2; 1996, Part 2: Performance requirements)

Location	Surcharge-frequency	Flooding-frequency
Rural areas	1 in 1 year	1 in 10 years
Residential areas	1 in 2 years	1 in 20 years
City centres/industrial areas/commercial areas		
- with flooding check	1 in 2 years	1 in 30 years
- without flooding check	1 in 5 years	
Underground railway/underpasses	1 in 10 years	1 in 50 years

Notes
Surcharge-frequency indicates the time period in which no storms should cause any pipes in the sewer system to become pressurised.
Flooding frequency indicates the time period in which no flooding should occur.
Any national norm always takes precedence even if the EU norm is stricter (e.g. the Flemish norm is that storms with a return period of 1 in 2 years, even when pipes become pressurised the pressure should never rise higher than 50 cm below ground level, whereas with storms with a return period of 1 in 5 years; no flooding should occur.

Several different methods exist for calculating the necessary pipe size given a certain inflow into the sewer system (Berlamont 1997). They range from very simple manual methods, such as the rational method (Brown 1993), to very complex numerical models solving the full set of St-Venant equations. The latter requires a powerful computer and appropriate software (Crabtree *et al.* 1994; Long 1995).

6.3.3 Ancillary structures

At certain strategic locations in a combined sewer system, overflows limit the carry-on flow to the WWTP to a preset flow. The amount and location of those overflows is determined by ecological and economical criteria. The further downstream in a sewer system an overflow is located, the bigger the upstream pipes must be to prevent flooding. On the other hand, every overflow will pollute its receiving watercourse. Limiting the number of overflows to the locations where they cause the least harm ecologically is also a criterion to be taken into account.

A correct design both from an economical and an ecological viewpoint requires a very complex process incorporating EQO/EQS criteria. The tools available for predicting the loads spilled by overflows and the quality of effluent discharges of WWTPs and industrial treatment plants are currently not very accurate. Nevertheless, different design methods that take this into account already exist and are increasingly used, both for drafting sewer masterplans in developing areas and for finding solutions where existing sewer systems cause environmental and/or flooding problems.

In a combined sewer system, eliminating combined sewer overflows (CSOs) altogether is not feasible. As more and more direct discharges are eliminated, the pollution caused by these overflows becomes increasingly important. However, a number of measures are available to limit the pollution of an overflow to a minimum.

The biggest pollution associated with a spill event is related to sediments. Thus, by allowing for settlement in the design of an overflow structure the amount of pollutants spilled can be minimised. These improved overflow structures can be divided into two categories, high-sided weir overflows and vortex overflows. Recent research has provided new design criteria for both kinds of overflows (Luyckx 1997; Van Poucke 1998). These design criteria can result in a maximum settlement efficiency of as much as 75%. Overflow-structures designed according to these guidelines will however be two to three times the size of traditional overflow structures and will cost up to three times more.

Another possible way of minimising pollution is to reduce the number of spill events that take place. This is best done by providing off-line storage (Berlamont 1997). If this storage is provided with a settlement tank, this kills two birds with one stone. The spill frequency will reduce and when a spill occurs, the water will have settled, reducing the spilled load.

Providing some kind of treatment after the overflow structure is another possibility to reduce CSO pollution. Reed-beds are one option. As overflows tend to operate on a very irregular basis, it is not always easy to keep the plants alive. Floating mats with plants can provide a solution here (Figure 6.6). This system has been tested successfully in Germany (Janssen 1998; Van Authaerden 1999) and will be tested shortly in Belgium.

The end-point of any combined sewer system should be a WWTP. The maximum flow for which a WWTP is designed is usually equal to the CSO setting (see Table 6.2). Traditionally the WWTP will not biologically treat all the water that is pumped during wet weather. Usually half is treated biologically and mechanically. The other half is only treated mechanically. The mechanical treatment usually consists of a screen, a sand trap and sometimes a fat trap. The water that is only treated mechanically is temporarily stored in a settlement basin. Although this basin is emptied to the biological treatment after the storm, overflow of this basin always discharges directly into the receiving watercourse. The design of the basin, based on retention time, causes the overflow to operate

Figure 6.6. Floating mat with plants (photo: Bitumar).

regularly. Since the water spilled over this overflow is only treated mechanically, it can cause significant pollution in the receiving watercourse. A way around this, which is becoming common practice in Flanders, is treating the full flow biologically. This requires the process to be flexible enough to cope with flows ranging from one to six times the foul flow. The results in Flanders showed that with a good design and smooth operation, overall pollution is greatly reduced when the full flow is treated biologically. The efficiency of the WWTP decreases when it receives full flow, but the effluent results are still better than the effluent of the normal treatment combined with the spill of the settlement basin following rain events (Carrette *et al.* 1999).

6.3.4 The sewer as chemical and biological reactor

In combined systems the size of the pipes is determined by the rainwater that it needs to transport. This implies that under DWF conditions the residence times in those pipes can be quite long and the flow conditions very slow. Furthermore, due to the slow flow, the pipes will act as large settlement basins. This is the ideal environment for biomass to develop and biodegrade the wastewater (Cao and Alaerts 1995). Both attached and suspended biomass will be able to grow in sufficient numbers to have a significant effect. This effect is usually unwanted. First, the biomass in sewers will degrade the easily degradable substrate. This substrate is necessary to remove nutrients in a WWTP but, in the sewer, the degradation of this substrate does not result in any nutrient removal. Second, under anaerobic conditions, which occur regularly in sewers, hydrogen sulphide (H_2S) will be formed. This H_2S will be oxidised again in the sewer with the formation of sulphuric acid as a consequence (Hvitved-Jacobsen and Nielsen 2000). The acid in its turn will damage concrete, the most commonly used material for sewers (Boon 1995) (Figure 6.7).

Figure 6.7.Damaged sewer as a result of sulphuric acid (Photo: Aquafin).

Settlement in the sewers is also undesirable. A large sediment layer in a sewer diminishes its hydraulic capacity. Flooding will become more likely. Furthermore, if the sediment is resuspended during storm events, there is a distinct possibility that it will be discharged directly to surface water through CSOs (Heip and Ockier 1997).

6.4 SEPARATE SEWER SYSTEMS

Pollution from combined sewer systems originates from the CSOs. The function of a CSO is to allow the excess rainwater that cannot technically and economically be transported to and treated at a WWTP to be discharged to a receiving watercourse. An obvious solution to this problem is separating both flows into a foul flow that can be transported to the WWTP without CSOs and a rainwater flow that can be discharged into surface water.

A completely separate system consists of a dual set of pipes that are not interconnected (see Figure 6.8).

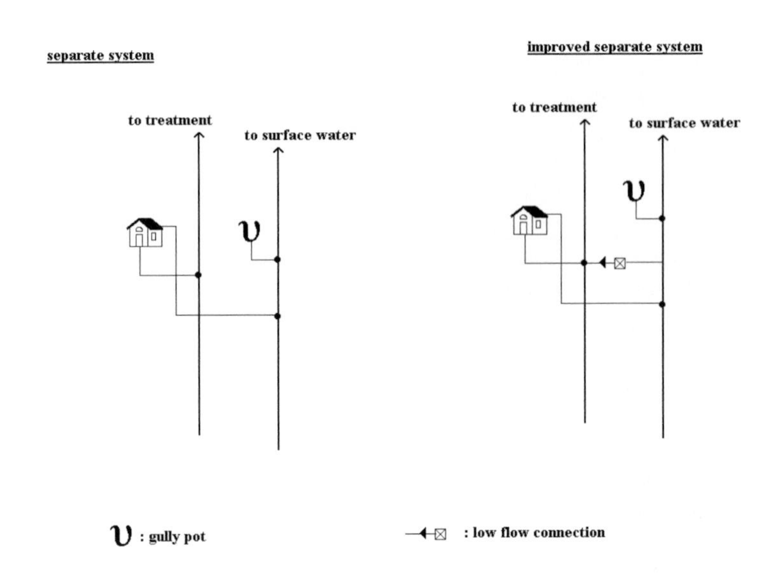

Figure 6.8. Completely separate system and improved separate sewer system.

One set of pipes transports the wastewater to a WWTP, and the other transports the rainwater to surface water. The wastewater system needs to form a network that is connected to the WWTP, but the rainwater system can be made up out of small stretches of sewer connected to surface water. Thus, the rainwater system will consist of a large number of discharges to surface water.

The wastewater system is usually designed by calculating the pipe size needed to transport twice the mean DWF. The rainwater system is designed in exactly the same way as a combined system with direct discharges to the receiving watercourse instead of CSOs.

A good measure for flood protection in the rainwater system is to create the rainwater system as a series of open ditches. Thus the flow from peak storm events will be attenuated and infiltration will be allowed to take place. Another advantage of an open ditch system is the reduced cost.

One of the major disadvantages of a completely separate system is the fact that surface runoff resulting from the rainfall is not always clean, especially after a long dry-spell (Wiggers 1996). Several pollutants may be encountered in the first run-off, e.g. Pb, dust, PAH, salts in winter, rubber, and so on. Pollution prevention might require separate treatment at the discharge points of the rainwater as well. In order to reach a reasonable water quality the treatment that is needed for polluted rainwater is a lot less demanding than the treatment needed for rainwater mixed with wastewater. Simpler and less expensive treatment facilities will suffice, but since a rainwater system tends to have many discharge points, several of those facilities may be required. Overall, the same reduction of pollution may be achieved by properly treating water from spill events in combined systems at a reduced cost.

A possible solution to this problem is an improved separate system (see Figure 6.8). This also consists of two sets of pipes, but the rainwater system is connected to the wastewater system so that the first flow entering the rainwater system will be transported to the WWTP and only the excess flow will be discharged directly to surface water. There are numerous ecological advantages to this system. However, such a system is very expensive and difficult to operate, so it is rarely used. One ecological disadvantage, however, is the fact that the sludge produced at the WWTP will be polluted by components such as Pb, hydrocarbons, and so on that are not normally found in domestic sewage. Where sludge resulting from the treatment of purely domestic wastewater can be easily treated and even reused in agriculture, sludge resulting from the treatment of wastewater from an improved separate sewer system may be too heavily polluted to be reused.

Another major disadvantage of completely separate systems is the danger of wrong connections. If wastewater from no more than 5% of households is wrongly connected to the rainwater system, the overall pollution will be equal to that from the CSOs of a combined system serving the same population. This implies that connections to completely separate systems need to be monitored carefully, which means higher maintenance costs and greater vigilance from the population and the workmen installing the pipes.

Where flooding is concerned, separate systems, if they consist of pipes, may only offer a small improvement where the rainwater system is concerned, since they will behave in a similar way as the combined systems. However, the rainwater system will usually be connected to surface water by numerous outlets. The rainwater flow will thus be spread over a longer stretch of the receiving watercourse, making much better use of the buffering capacity of the receiving watercourse. The risk of flooding will be reduced by such measures.

Furthermore, flooding by relatively clean rainwater is less of a nuisance than flooding by dirty mixed water, although when a house is flooded the quality of the water is usually the least of the householder's worries.

6.5 SEDIMENTATION

Since the flow in the wastewater pipes of a separate sewer system is usually quite small, sedimentation is likely. Particularly in countries with limited water supply or where kitchen disposal units are in common use, the concentration of suspended solids may become high. Regular cleaning will be necessary, but separate sewer systems, when well designed, can still be used.

In well-designed pressurised or vacuum sewer systems the pumps increase the velocity of the water in order to prevent sedimentation, so these systems tend to be less sensitive to sedimentation.

The velocity in combined sewer systems during storm events is usually sufficiently high to prevent sedimentation. During dry weather conditions the flow is small, thus sedimentation is unavoidable. When the system has low gradients, the velocities even in extreme storm conditions may not be high enough to prevent sedimentation. Although these sewer systems may be considered to be poorly designed, in flat regions this cannot always be avoided. Regular monitoring and cleaning will be necessary to maintain these types of sewer system.

6.6 SOURCE CONTROL

For combined sewer systems, it is necessary to use ancillary structures as a solution to diminish the pollution of overflows, since such sewer systems will remain in operation for a very long time. These solutions however are all end-of-pipe solutions. Source control may provide a better, proactive answer to problems caused by CSOs (De Jong *et al.* 1998). In essence, source control implies disconnecting paved areas from the sewer system.

Limiting the amount of rainwater that enters the sewers will reduce the spill frequency and thus pollution. Several systems can be designed for this.

On an individual basis (that is, for one house or a series of buildings), a good solution is a rainwater storage tank with an overflow not connected to the sewer system and where the rainwater is used for domestic purposes (for example, flushing the toilet, washing the car, watering the garden, etc.). Since the overflow is not connected to the sewer system, no rainwater collected in the rainwater tank will ever enter the sewer system. This water will not contribute to any spill event. This measure will also directly influence flooding, provided that the receiving watercourse has enough capacity to cope with the additional inflow of rainwater from the overflow of rainwater storage tanks. Furthermore, drinking water is saved by using rainwater for domestic purposes (Mikkelsen *et al.* 1998). Rainwater storage tanks which have the overflow connected to the sewer system will only limit pollution if the rainwater is reused. If not, the rainwater tank will always be full and will not provide any storage capacity. Even when rainwater is reused, these facilities will only have a limited influence on flooding, since the storage capacity of the tanks is small compared to the volume of water that causes flooding during storm events (Vaes and Berlamont 1998).

For larger areas (for example, industrial estates), a collective storage facility can be used. If both its overflow and outlet are connected to a receiving watercourse, the influence on flooding can be significant. Bearing in mind that reuse of the rainwater stored in such a facility is less feasible because of possible contamination, this can still provide some pollution control when the facility is connected to the sewer system and if the facility is emptied through an outlet in a controlled fashion. Since the volumes involved are larger than those of individual rainwater storage tanks, the influence on flooding will also be higher. The volume of rainwater experienced in flood situations is, however, usually much greater than the storage capacity. Furthermore, buffering the rainwater in rainwater tanks on an individual household basis or even on a larger scale for whole areas will only influence flooding if the storage tank is empty prior to the storm.

Allowing rainwater to infiltrate into the soil will reduce the number of spill events and will also increase flood-protection (see Figure 6.9). This will however be very difficult to budget for, since the influence of infiltration on the reduction of peak flows is not known. As with rainwater storage tanks, any infiltration facility needs to be protected by an overflow, as the peak flows from storm events are likely to exceed the infiltration capacity. As for rainwater storage, the level of protection will be determined according to where the overflow is connected. Protection against both pollution and flooding will be greatest when the overflow is not connected to the sewer system. It should however be emphasised that the infiltration of rainwater pollute the groundwater, due to the pollutants typically found in run-off, such as Pb, Cu, PAH and rubber.

Infiltration will of course only work when the groundwater table is sufficiently low. Some examples of infiltration facilities are:

- infiltration-ditches;
- infiltration-basins, e.g. with honey-comb plastic blocs;
- percolation trench, i.e. a trench filled with a gravel-like material in which the rainwater is allowed to infiltrate;
- overland infiltration, i.e. allowing the rainwater to flow over a meadow thus allowing it to infiltrate.

Figure 6.9. Infiltration ditch under construction (Photo: Haskoning).

Since the flooding behaviour of sewer systems is determined by peak flows during extreme storm events, limiting the mean inflow of rainwater will not necessarily reduce the peak flows. Some beneficial effect on flooding will result from source control, but the magnitude of this effect is hard to discern. With the present state of knowledge no guaranteed added protection can be offered by source control. Although every source control measure will offer some extra protection for flooding, it is difficult to accurately calculate the effect, so it would be unwise to greatly reduce pipe sizes because of source control.

On the other hand, the effect on pollution prevention will be more obvious. Every spill event, no matter how small, that can be prevented by source control reduces pollution. Source control can lead to a smaller design for, for example, settlement retention basins.

Source control is a good measure both in pollution control and flood protection. On its own it will however neither solve flooding nor pollution.

When source control measures are proposed it is essential to take all boundary conditions into account. Infiltration will not work when the groundwater table is too high. Overflows of source control facilities directly connected to a receiving watercourse will not alleviate flooding if the watercourse cannot cope with the added flow (Figure 6.10). Finally, even with source control the design of a combined sewer system may not be significantly smaller, since the effect of source control on peak flows is difficult to determine. Facilities to prevent pollution can on the other hand be designed significantly smaller when source control is used.

Figure 6.10. Flooding in Antwerp in 1998 (Photo: G. Coolens).

6.7 REFERENCES

Anonymous (1994) Mechanische Riolering. Aanbevelingen beheer. Final report of research project 92-03, RIONED, stichting RIONED – Ede, the Netherlands. (In Dutch).

Berlamont, J. (1997) *Rioleringen*. Acco Leuven/Amersfoort.

Boon, A. (1995) Septicity in sewers: causes, consequences and containment. *Wat. Sci. Tech.* **31**(7), 237–253.

Brown, A. (1993) Rational design. Lecture notes, Aquafin.

Cao, Y.S. and Alaerts, G.J. (1995) Aerobic biodegradation and microbial population of a synthetic wastewater in a channel with suspended and attached biomass. *Wat. Sci. Tech.* **31**(7), 181–189.

Carrette, R., Bixio, D., Thoeye, C. and Ockier, P. (1999) Storm operational control: High flow activated sludge process operation. *Wat. Sci. Tech.* **41**(9), forthcoming.

Crabtree, R., Grasdal, H., Gent, R., Mark, O. and Dorge, J. (1994) Mousetrap – deterministic sewer flow quality model. *Wat. Sci. Tech.* **30**(1), 107–115.

De Jong, S.P., Geldof, G.D. and Dirkzwager, A.H. (1998) Sustainable solutions for urban water management. *European Water Management* **1**(5), 47–55.

Dugre, P. (1995) Alternative wastewater collection systems. *Vecteur Environment* **28**(1), 33–42.

Gray, D.D. (1991) Prospects for vacuum sewers. *Wat. Env. Tech.* **3**(7), 47–49.

Heip, L. and Ockier, P. (1997) Vuilvrachtreductie in rioolstelsels: een literatuuroverzicht. *Water* **93**, 51–54. (In Dutch.)

Henderson, R. (1998) Wastewater collecting systems and treatment provisions in EU member states in 1994. Report for the European Wastewater Group.

Hosten, L. (1991) Technologie van de zuivering van water. Lecture notes, Technical Chemistry Laboratory, Faculty of Applied Sciences, University of Gent, Belgium. (In Dutch.)

Hvitved-Jacobsen, T. and Nielsen, P.H. (2000) Sulfur transformations during sewage transport. Chapter 6 in *Environmental Technologies to Treat Sulfur Pollution* (eds P.N.L. Lens and L. Hulshoff), IWA Publishing, London.

Janssen, V. (1998) Optimalisatie van een alternatieve kleinschalige modelwaterzuivering. Graduation thesis, BME-CTL, Gent. (In Dutch.)

Lettinga, G. (2001) *Potentials Of Anaerobic Treatment Of Domestic Sewage Under Temperate Climate Conditions* (Chapter 11 in this book).

Long, R. (1995) Water model is a Derby winner. *Surveyor* **182**(5326), 16–18.

Luyckx, G. (1997) Fysische modelstudie van een hoge zijdelingse overstort. Riooloverstorten: randvoorzieningen (fase 2). Chapter 6 of a study performed by the Universities of Leuven, Brussels, Gent and Antwerp, commissioned by AMINAL and VMM, 6.1–6.38. (In Dutch.)

Martin, A.J. (1927) *The Activated Sludge Process*, MacDonald and Evans, London.

Mikkelsen, P.S., Adeler, O.F., Albrechtsen, H.-J. and Henze, M. (1998) Collected rainfall as a water source in Danish households: What is the potential and what are the costs? Proceedings Options for closed water systems – sustainable water management, International WIMEK congress, 11–13 March, the Netherlands.

Pfeiffer, W. (1996) Requirements for sewerage systems using pressurised and reduced pressure drainage facilities. *3R Internationals* **35**(3/4), 157–165.

Schinke, R. (1999) Vacuum-operated sewer systems – a process offering many often unrealised opportunities. *Korrespondenz Abwasser* **46**(4), 506–513.

Stedman, L. (1995) A journey through time. *Water Resources* **677**, 8–9.

Vaes, G. and Berlamont, J. (1998) Het effect van berging in regenwaterputten, fase 2: het effect op de dimensionering van riolen. Study performed by the University of Leuven, commissioned by Aquafin. (In Dutch.)

Van Assel, J., Dierickx, M. and Heip, L. (1997) Case study Tielt UPM. WaPUG Autumn Meeting, Blackpool, Paper no 7.

Van Authaerden, M. (1999) Optimalisatie van een kleinschalige plantenzuivering bestaande uit een hydrobotanische geul, twee percolatierietvelden en een naklaringsvijver. Graduation thesis, KUL, Leuven. (In Dutch.)

Van Poucke, L. (1998) Terreinmetingen en fysische modelstudie van een omtrekoverstort. Riooloverstorten: randvoorzieningen (fase 3). Chapter 1 of a study performed by the Universities of Leuven, Brussels, Gent and Antwerp, commissioned by AMINAL and VMM, 1.1–1.11.

Verstraete, W., Van Vaerenbergh, E., Bruyneel, B., Poels, J., Gellens, V., Grusenmeyer, S. and Top, E. (1999) Biotechnological processes in environmental technology. Lecture notes, Laboratory Microbial Ecology, Faculty of Agricultural and Applied Biological Sciences, University of Gent, Belgium.

Wanner, J., Cech, J.S. and Kos, M. (1992) New process design for biological nutrient removal. *Wat. Sci. Tech.* **25**, 4–5.

Wiggers, J. (1996) Riolering in de toekomst, duurzame stedelijke waterkringloop in het jaar 2040. Personal note. (In Dutch.)

7

The urban sanitation dilemma

J. Niemczynowicz

7.1 INTRODUCTION

The level of sustainability of any society may depend, among other things, on how it handles water, sanitation and household residuals. The problem of how to handle organic residuals from sanitation and organic wastes has gradually grown worldwide to a dilemma that is discussed in academic, political and economical circles. It has become clear that sanitation and sanitary systems as well as solid waste disposal systems should not only safely evacuate and dispose of human residuals but also deliver the option of reusing nutrients in agriculture. Simultaneously, over the long term, processing of human waste should not bring any risks for the human population and the natural environment.

Mixed sewage coming from a city via sanitary sewers to the treatment plant is obviously not directly applicable for use on agricultural land due to bacterial and chemical pollution. Conclusions drawn by scientists as well as politicians

Edited by P. Lens, G. Zeeman and G. Lettinga. ISBN: 1 900222 47 7

including the governments in several European countries were that, in order to make household nutrients available for recycling in agriculture, sanitary systems must be changed to allow decentralisation, possibly to the level of a single family house or a group of single family houses. Following this idea decentralised sanitation solutions, including composting or urine-separating toilets, were developed and installed in many experimental houses called 'ecological villages' in the late 1970s and 1980s in Sweden. Later, these were installed in thousands of normal houses and public buildings in Sweden. Several new residential areas that were constructed during the 1990s are based on the above. However, to change decentralised systems in the older residential areas would be very difficult, mostly due to the high cost of the necessary re-constructions, but also due to the lack of adequate infrastructure for local processing and transporting the resulting large volumes of organic material from urban areas to agricultural land. Adequate and safe agricultural routines for including the organic material in agricultural practices must be also developed.

Similar developments and the same dilemma as described above have taken place in many towns in developed countries, as well as in many developing countries, especially in mega-cities. The question that all city planners must find the answer to is: Are the new principles and methods of handling sanitation and nutrient flows in a society so compelling that they should entail major changes to current city-, and even country-wide, planning and management?

New sanitation technologies as well as new methods of handling organic residuals that are gradually being introduced into an increasing number of locations also bring several new problems. If these new technologies were carried out without adequate safety measures, they could, in the long term, increase health risks for the human population and the environment. The question of how to handle increasing risk levels should be treated with care and should influence current legislation.

The twentieth and twenty-first centuries can be viewed as a time of increasingly rapid and powerful changes in the relations between mankind and nature. It is necessary to protect the environment and sustain natural resources for future generations. These goals require organisation of all sectors in society. The results of such changes have already contributed to important changes in water management worldwide, including management methods and technologies used in drinking water supply and use, water treatment and sanitation. These changes influence not only technical facilities but also organisational structures, social interactions and lifestyle.

7.2 THE RATIONALE BEHIND THE NEED FOR CHANGE IN PRESENT SANITARY SYSTEMS

The basic motivation behind the need to reshape the management of sanitation nutrients and other streams of organic residuals in society may be found in the so- called 'basic system conditions for sustainable development', formulated in Agenda 21 (UNCED 1992). The following conditions should be adhered to for all future developments including water and sanitation management:

(1) The withdrawal of finite natural resources should be minimised.
(2) The release of non-biodegradable substances to the environment must be stopped.
(3) Physical conditions for circular flows of matter should be maintained.
(4) The withdrawal of renewable resources should not exceed the pace of their regeneration.

The management of wastewater flows is well organised in Swedish society. Practically all Swedish cities are equipped with tree-stage wastewater treatment plants, and implementation for nitrogen reduction will be soon installed across the country. Even the majority of stormwater in Swedish cities undergoes some treatment in a variety of dry or wet ponds, infiltration facilities and wetlands (Niemczynowicz 1999). However, not all problems with pollution in wastewater sludge are resolved in this process, thus hampering the agricultural reuse of nutrients present in wastewater sludge.

Thus, despite the present applications of top technology in sanitation, stormwater and wastewater management in Sweden, as well as several other European countries, current sanitation practice does not entirely agree with conditions for sustainable development.

7.3 THE WASTEWATER SLUDGE PROBLEM

Recently, an intensive debate about the possible use of sewage sludge has been taking place within the Swedish and international research community. For example, Priesnitz (2000) states that sewage sludge has mutagenic effects, i.e. it causes inheritable genetic changes of organisms. The sludge contains heavy metals such as Cadmium (Cd) that accumulate in the human body during the lifetime and can potentially, in the long term, cause kidney disease. Because there is evidence that heavy metals and persistent organic pollution can build up in sludge-treated soils, governments have issued numeric standards for permissible concentrations of metals in soils. These standards limit permissible concentrations of Cd in sludge but plant Cd uptake also depends on the type of soil, rainfall volume and distribution in time, type of crops, time of growth and several other factors. The cadmium content in commercial fertilisers is about 2-3 mg Cd per kg Phospherous while the Cd content is about 50 mg Cd per kg Phospherous in sludge (Lindgren 2000). There are large differences between permissible levels of Cd in sludge between countries. For example in Sweden the standard is 2 mg Cd per dry ton of sludge, while in USA 50 mg/dry ton is allowed.

Sludge also contains many other compounds such as flame-protective chemicals, drugs, antibiotics, substances similar to hormones, bromines, dioxins, furans, PCB and others, many of them unknown. However for the majority of these substances, there are no standards for their regulation in sludge, or information about their potential toxicity and pace of accumulation.

All the compounds mentioned above and other persistent hazardous substances found in sewerage sludge will, sooner or later, enter surface water bodies and groundwater. In the long term these substances may also enter agricultural products. A discussion in 1999 listed hormonal substances found in many rivers and lakes in European countries. No one can say what the long-term ecological and human health effects of this will be.

The movement of metals and other toxins from soil into groundwater, surface water bodies, plants and wildlife is poorly understood. Soil acidity is considered to be a key factor in promoting or retarding the movement of toxic metals into groundwater and accumulation in soils and crops. The National Research Council (NERC) of the US National Academy of Sciences allows using sludge on agricultural land in the short term, 'as long as soils are agronomically used'. However, according to Priesnitz,

> ...research clearly shows that under some conditions (which are not fully understood), toxic organic industrial poisons can be transferred from sludge-treated soils into crops: lettuce, spinach, cabbage, Swiss chard and carrots have been shown to accumulate toxic metals and/or toxic chlorinated hydrocarbons when grown on soils treated with sewage sludge (Priesnitz 2000).

Thus, there is good reason to believe that livestock grazing on plants treated with sewage sludge will ingest the pollutants either through the grazed plants, or by eating sewage sludge along with the plants. Priesnitz goes on to say that: 'Sheep eating cabbage grown on sludge developed lesions of the liver and thyroid gland'. It seems that there is yet no risk-proof method of handling and using wastewater sludge. Incineration, deposition, pelleting and other new methods will sooner or later result in potentially harmful substances entering ecological systems and, later, human bodies. However, perhaps the main problem is that these systems have no ability to safely recycle organic biogenic residuals from human settlements to agriculture.

Handling of wastewater sludge is no trivial problem in large cities and especially in mega-cities i.e. cities with over 10 million inhabitants. In order to highlight the magnitude of the wastewater sludge problem outside Sweden, it is enough to say that 14 European countries together produce 6,631,000 tons of sludge per year. The cost of sludge disposal is an average £200 per dry ton, which gives a total yearly cost of about £1,326,200,000 (Davis 1992). These costs indirectly imply that any investment in sanitation technology that will, instead of producing wastewater sludge, allow nutrients to be directly recycled in agriculture, would quickly pay off.

7.4 HUMAN URINE AS FERTILISER

The urine leaving a healthy human body is sterile. It contains the following weight proportions of nutrients: nitrogen (N): 11, phosphorous (P): 1, potassium, (K): 3. Nitrogen is predominantly (> 80%) in the form of ammonia and is easily accessible for nitrification and use by plants (Hoglund *et al.* 1999, Kärrman *et al.* 1999). Concentration in ca 100 ml water that is usually used for flushing urine from a separating toilet down to the storage container varies between 2.4 to 3.6 g/l N, and ca 0.18 to 0.38 g/l of P.

In order to avoid the leakage of nutrients, urine should be stored in air-free containers that are usually placed under the soil or street surface. In order to avoid nitrogen leakage, storage containers are equipped with airtight covers. Human urine, if stored in containers without contact with air, does not lose its fertilising value for about a year after storage.

The urine produced by one adult per year contains about 5.6 kg nitrogen, 0.5 kg phosphorous and 1.0 kg potassium (Wolgast 1996). The proportions of these nutrients are very similar to the proportions in many commercial fertilisers. The nutrients excreted by one person during one year are enough to produce an amount of grain that has a sufficient nutritional value to cover the nutritional needs of one person for a year. Thus, logically, it is possible to say that, theoretically, there is no reason for anybody in the world to go hungry, independent of location, stage of development or climate, because agricultural nutrients are, at least partially, recyclable through the animal and human food chains.

All the facts stated above are closely related to the issue of equity among people and nations. Developing countries with increasing populations must constantly increase their food production. The questions of which sanitary systems will be chosen in these countries and from which sources will the nutrients necessary to agriculture come, are closely related to the possibility of increasing food production.

7.5 PRESENT SANITATION PRACTICES AND SUSTAINABLE DEVELOPMENT

Facts given in Section 7.2 suggest that management of urban water flows, including stormwater, sanitation, wastewater and sludge in Sweden and in other countries using separation toilets is well organised according to up-to-date standards. Therefore, few ecological problems connected with such management should be expected in these countries. However, *if we come back to the principles of Agenda 21 and the four basic system conditions for sustainable development, it can be shown that this is not the case.* First, wealthy countries have, in the process of developing water-related infrastructure, used a larger share of natural resources than many other countries; the withdrawal of these resources was not minimised, and resources were certainly used faster than they could renew themselves. Second, the cost in terms of money and resources spent on construction, upgrading and maintenance of these facilities is high and cannot be found in developing countries. Third, referring to the current Swedish and international discussion described in Section 7.3, if wastewater sludge is to be used as a fertiliser in agriculture, non-biodegradable substances will accumulate in the soil and will sooner or later enter the human food chain.

Problems of present infrastructure for water and wastewater management are, at least partially, a result of the large scale of these solutions. Centralised sewerage systems collect sewage from large cities, which contains a mixture of useful and harmful substances. Users do not see the connection between their behaviour and the results for the environment. For example, a user pouring chemicals down the toilet does not reflect or know what the result of this might be for the treatment plant and, ultimately, for the environment.

It seems that one rather radical solution to this problem in the future would be to change the scale of the solution from a central to a local one. Such a solution would, in turn, lead to other problems such as higher total costs, higher vulnerability and dependency on the personal behaviour of users. Change of scale cannot be done simultaneously for the whole system. But it can be carried out gradually by changing the system in old buildings during renovations and by constructing sanitation and sewerage systems differently in new housing. A general application of local scale solutions puts obvious obligations on the users. It also requires responsibility and changes in individual behaviour. Poisonous fluids, household chemicals, medicines and so on should not be poured into urine-separating toilets. In order to achieve responsible behaviour by users it would probably be necessary to create new economical incentives for acting in the required manner. Full-scale experiments with new sanitation carried out in Sweden and in other countries bring the hope that such changes in behaviour are possible on a wide scale (Frittschen and Niemczynowicz 1997).

7.6 PRESENT TRENDS IN SUSTAINABLE MANAGEMENT OF ORGANIC MATERIALS

7.6.1 Present trends in Sweden

The Swedish government has recently expressed its serious intention to put Sweden on a course towards sustainable development. The good management of organic material flows is seen as an important element of such development. The Swedish population has been urged by the government to formulate and begin to realise specific plans, known as 'Agenda 21 plans'. A statement by the Swedish Prime Minister Göran Person in 1998 announced that Sweden will lead the way towards sustainable development. According to his statement, Sweden should reach a recycling level of 75 per cent of all organic materials, including organic solid wastes and toilet effluents, from households to agriculture within the next 10 years without increasing the current levels of conservative pollutants in the soil. Technically, this means that present wastewater, sewerage and treatment plants must be adapted to these requirements, or disappear. In practice this will not be as radical as it sounds, but it means that water sanitation must change to adapt to new requirements. It does mean that traditional sanitation with water closets must eventually be exchanged for new so-called 'dry sanitation' solutions such as composting or urine-separating toilets. These ideas have already been tested over several years in 'ecological villages', as mentioned earlier, where alternative sanitation was used. To begin with, composting toilets were used. However, experience showed that composting works poorly in the cold Swedish climate. The next generation of 'ecological sanitation' used in Sweden was based on the separation of urine from faeces directly at source, i.e. in a specially constructed toilet bowl (Figure 7.1(a)). These toilets use a very low amount of water. For example, a toilet system known as 'Dubbletten' reduces a household's water consumption by up to 80 per cent. Some of the water used is evacuated together with the urine to the underground container where it is stored in an airtight container (Fig 7.1(b)). When the container is filled it is emptied by the farmer and, after being diluted with water to a ration of one part of urine to ten parts of water, it is used as agricultural fertiliser. Solid parts are transported, using small amounts of water, to the composting chamber. After about six months the compost is also ready for use in agriculture.

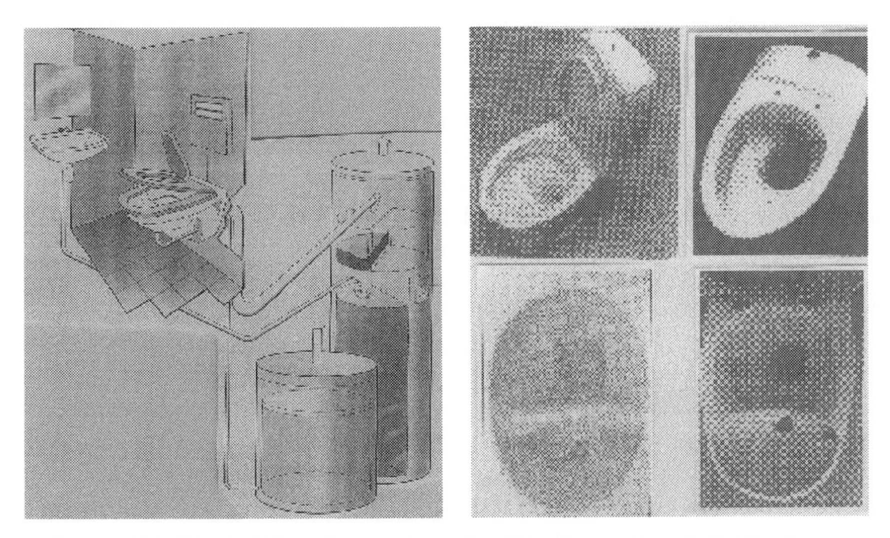

Figure 7.1(a). The 'Dubbletten' separating toilet, (b) and examples of toilet bowls.

7.6.2 Examples of houses using alternative water and sanitation systems in Sweden

Urine-separating toilets are installed in several hundred buildings in Sweden, from family houses to a large exhibition hall in Marienfred in Stockholm, which has up to 25,000 visitors per day. There are many other examples: 44 flats in the residential area of Understenshöjden; 40 flats for 100 people in Falberg; an international student house in Lund which has 175 rooms; an 8-storey building in Stockholm, and many others. There are no official statistics on this because the situation is changing from day to day. Suffice it to say that many new residential areas are equipped with separating sanitation. A similar development in the installation of separating or other 'dry' sanitation solutions is observed in other European cities, notably in Germany, Denmark, Holland and Norway.

7.6.3 A short history of the early 'ecological villages' in Sweden

Composting toilets were first used in Sweden in the early 1980s in so-called 'ecological villages', i.e. residential areas constructed so that self sufficiency with regard to water supply, sanitation, solid waste management could be achieved. At the beginning of this movement, 'ecological living' attracted a small group of ecologically conscious people who wanted to live in harmony with the environment and were willing to sacrifice some of their comfort to achieve this goal. Composting toilets used in the first ecological villages required a lot of maintenance from residents and worked poorly in the cold Swedish climate. Gradually, in newer ecological villages, composting toilets were replaced by urine-separating toilets, that require much less effort from users and do not impact upon standards of living. In several eco-villages urine and compost are used in private gardens and common green areas. In newer residential areas equipped with separation toilets, urine is usually pumped up twice a year by a farmer from a nearby farm, and after dilution with water, it is used as fertiliser. Composted solids are also used in agriculture or, in some city centres, disposed of in sewers.

The trend of ecological living has spread to large parts of Swedish society, and, at present, many houses, multi-storey buildings, residential areas, schools, public buildings, and so on are equipped with urine-separating sanitation.

One of the first eco-villages in Sweden, Toarp eco-village, was constructed in 1992 near Malmö, in the south-west of Sweden. It was a poor start. The housing area of Toarp accommodates about 150 residents in 37 houses. Residents had no influence on the construction of the houses or on the installation of the sanitation equipment. Composting toilets were installed in the houses and composting chambers were placed in cellars. In order to enter the composting chamber for maintenance it was necessary to open an iron lid of 80 × 80cm in size, placed on the street in front of the house. Thus, inspecting the composting chamber was so difficult that residents did not do it, with the consequence that, instead of compost, a smelly wet sludge was produced and had to be removed by hand. Some residents decided to change from composting toilets to urine-separating toilets and have used these up to now. Some more ambitious residents improved their composting toilets and still use them. No one has left the village, however, despite all the mistakes in planning and construction. The lesson learned from this is that people building alternative housing must learn to better plan and construct such houses. The second conclusion is that ecologically minded people can put up with a lot in order to prove their determination to live in an ecologically friendly manner (Fittschen and Niemczynowicz 1997).

A second example, this time of a more successful development, is of the 'Ostra Tom' secondary school in Lund, constructed in 1997. The school is made of recycled tiles, uses a passive energy conservation system with sun panels, and is equipped with a 'Dubbletten' sanitation system, a closed water system with wastewater and stormwater recycling. Wastewater treatment and the composting of organic garbage is carried out locally. The school, with the exception of drinking water, electricity and telephone, does not use and is not connected to any of the other municipal services. Urine and composted solids are collected by a farmer and used locally for agriculture. Everything functions as planned and the students do not understand why their 'normal' school is constantly visited by interested groups and individuals.

7.6.4 Increased risk level caused by decentralisation

The introduction of water closets and increased sanitation took place at the beginning of the nineteenth century in England and later in the century for the remainder of Europe and other developed countries. This brought a drastic improvement in public health in cities manifested, among other ways, by a decrease in epidemics. It was to be expected that the decentralisation of sanitary systems, as well as open handling of urine and solid sanitary wastes in new residential areas equipped with composting and/or separating sanitation could increase health risks for the local population. However, to my knowledge, no investigations have been carried out that would prove the case.

A survey carried out among inhabitants of Toarp eco-village, residents of the International Student House in Lund, and teachers and pupils in Lund's Östra Torn school did not reveal any tendencies to a higher frequency of contagious or other diseases than the rest of the population. Similar results were obtained from an survey carried out among guests at the International Student House in Lund. A tentative conclusion may be that people's all-round hygiene level is more important for health than the type of sanitation used.

7.7 A GLOBAL PERSPECTIVE

According to WMO (Simpson-Hébert 1996), 37% of the world's population did not have adequate sanitation in 1996. Pollution of water sources, often due to contamination from poor sanitation, is responsible for the death of around 25 million people each year. Half of the world's diseases are transmitted by or through water, often in connection with inadequate sanitation. It is estimated that about 50% of the world's population lacked safe drinking water in 1996 and by the 2050 an estimated 65% of people will live in areas that suffer from water shortage (Milburn 1996). More recent sources (Knight 1998) say that the pace of population growth is slowing and, if this trend continues, 25–40% of people will face a shortage of fresh water. Between 1900 and 1995, water use in the world has increased six-fold; more than double the population growth rate (WMO 1998). The continuing provision of water sanitation aggravates water shortage problems worldwide.

Faced with these facts, decision makers in the developing world have a real dilemma on their hands as to how to proceed with the development of water supply and sanitation. Traditional methods used in the development of water resources and in the supply of sanitation are unable to satisfy the fast-growing needs of developing countries, leading to a huge and pressing environmental problem (Milburn 1996). Water and sanitation problems, especially in cities in the developing world, are recognised as one of the greatest obstacles in the process toward sustainable development. Solving these problems depends on research and on introducing innovative technologies in the water sector and, especially, on the types of sanitation that will be included in long-term national development strategies.

7.7.1 Water-saving technologies

An understanding that future sanitation should not be based on water is growing among scientists and practitioners dealing with water and sanitation provision. It is now understood that water-borne sanitation creates long term obligations. It is costly and resource-consuming to construct, maintain and upgrade sewerage systems and treatment plants to satisfy ever more stringent water quality and sludge quality requirements. Simultaneously, the water closet is the largest consumer of water in a household. Simply by the use of dry or separation sanitation, water use in a household can be decreased by as much as 70–80%. However, the most important reason for choosing other solutions than water-born sanitation is that sustainable sanitation systems must deliver the possibility of the safe reuse of nutrients present in human organic residuals in food production. Wastewater contains both highly useful organic material and polluted non-organic material heavily contaminated with pathogens. Once mixed, it is difficult to utilise wastewater in food production. The choice of type of sanitation and sewerage for cities in future has become a central issue in a complex area of future human needs, and a fundamental part of the water management challenge, connecting several issues that have seldom been connected before. It is clear that the issue of sanitation is much wider than previously anticipated: wastewater nutrients or, in general, all organic material from human settlements may constitute valuable material that may end our present dependence on artificial fertilisers in agriculture and enable food production to be boosted without increasing the use of fossil fertilisers. At the same time, organic waste from households, farms, some industries and agriculture may, via biodigesters, deliver a source of clean energy.

7.7.2 Reusing nutrients for agriculture

The new goal of water- and sanitation is not only the safe disposal of human residuals but also the reuse of nutrients from sanitary systems and organic parts of solid wastes in agriculture, including agricultural food production, without a harmful accumulation of toxic substances in soil, surface water bodies and groundwater taking place. Such an approach creates new opportunities in water and sanitation management, especially in newly built residential areas. The wide application of dry sanitation solutions in new residential areas makes expensive investments in water infrastructure and wastewater treatment plants unnecessary. Agriculture may gain a new source of fertilisers produced from organic residuals that contain fewer pollution elements than commercial fertilisers.

The motivation for a new type of water management, including exchanging water closets for dry toilets and, especially, urine-separating toilets is that urine and composted faeces constitute good agricultural fertilisers with a nutrient content similar to that of commercial fertilisers, and with a very low cadmium content. Thus, farmers in Sweden as well as in other European countries are campaigning for the wider use of separation sanitation solutions.

On the global scale, especially in developing countries and in former Eastern and Central European countries where water resources are limited, applying sanitation solutions that do not use water as a means of transportation will bring large cost savings. It will not be necessary to construct sewers and treatment plants. In the long term, the creation of closed nutrient cycles between cities and agricultural land will contribute to an increase of agricultural production without detrimental environmental effects. New thinking and the application of new technology in the whole sanitation–agriculture sector may bring possibilities for more rapid technical and economical development in all countries, including developing countries (Karl 2000).

Surveys carried out did not reveal any observable health problems within populations of residential areas and public buildings equipped with separation sanitation. However, it has not yet been proved that such risks are not connected to handling such sanitary residuals over the long term.

7.7.3 Institutional support

All countries, especially in rapidly growing urban areas in developing countries, are currently facing the major challenge of upgrading their sanitation status and to facilitate the recycling of sanitary nutrients for agriculture. The major dilemma in this context is how to choose the best and most appropriate technology to achieve this goal.

Further development and application of alternative types of sanitation and the improvement of new sanitation systems, as well as the development of methods for utilising sanitary nutrients in agriculture, have been advocated at various prestigious conferences and organisations worldwide. The HABITAT II, UNCHR Conference, Istanbul 1996 (UNCHR 1996) recommended that: 'Governments at the appropriate levels in partnership with other actors should: promote the development and use of efficient and safe sanitary systems such as dry toilets for the recycling of sewage and organic components of domestic waste into useful products such as fertilizers and bio-gas.'

And the World Health Organisation (WHO 1998) said that: 'Aid agencies are encouraged to support research into sanitation systems without water. Educational and training institutions need to adjust their curricula away from sewerage and other water-related sanitation systems and focus on the realities of the world with scarce water resources, growing populations and increasing water shortage.'

7.8 REFERENCES

Agenda 21, UNCED (1992) The Rio Declaration on Environment and Development. The United Nations Conference on Environment and Development, Rio de Janeiro, 3-14 June 1992.

Davis, R.D. (1992) Europe's mountainous problem. *Water Quality International*, 3, 22.

Fittschen, I. and Niemczynowicz, J. (1997) Experiences with dry sanitation and grey water treatment in the ecological village Toarp. *Wat. Sci. Tech.* **35**(9), 161–170.

Habitat Agenda (1996) Recommendations from HABITAT II. United Nations Commission for Habitat Research (UNCHR), Istanbul, Chapter IV, item 141j .

Höglund, C., Stenström, T.A., Vinerås, B. and Jönsson, H. (1999) Chemical and microbiological composition of human urine. In *Proc. Int. Civil and Environmental Eng. Conference*, Bangkok, Thailand, 8–12 November, II 105–112

Karl, D.M. (2000) Phosphorous, the staff of life. *Nature*, **406**, 31–33.

Kärmann, E., Jönsson, H., Sonnesson, U., Gruvberger, C. and Dalemo, M. (1999) System analysis of wastewater and solid organic waste – conventional treatment compared to licit composting, urine separation and irrigation to energy forests. In *Proc. Int. Civil and Environmental Eng. Conference*, Bangkok, Thailand, 8–12 November, V27–V36.

Knight, P. (1998) Environment–population: outlook bleak on water resources. World News Interpress Service, Washington, 17 December.

Lindgren, G. (2000) Kretslopp och Slam hör inte Ihop. (Recycling and sludge do not match). Personal e-mail, 10 January. (In Swedish.)

Milburn, A. (1996) A global freshwater convention – the best means towards sustainable freshwater management. *Proc. Stockholm Water Symposium*, 4–9 August.

Niemczynowicz, J. (1999) Urban hydrology and water management – present and future challenges. *Urban Water Journal* 1, 1–14.

Presnitz, W. (2000) The real dirt on sewage sludge. *Natural Life* magazine, November. http://www.life.ca

Simpson-Hébert, M. (1996) Sanitation myths: obstacles to progress? *Proc. Int. Stockholm Water Symposium*, 4–9 August, 47–53.

UNCHR (1996) Habitat Agenda. Report of UNCHR's Conference on Human Settlements, Habitat II, Istanbul, Chapter IV, Item 141J.

WHO (1998) Collaborative Council Working Group on Sanitation. In *Sanitation Protection* (eds M. Simpson-Hébert and S. Wood), Report of WSCC Working Group.

WMO (1998) World Water Day 1997. http://www.wmo.ch

Wolgast, M. (1995) Recycling system WM ekologen. Stencil WM-Ekologen ABCo, Stockholm, Box 11162, S-10062 Stockholm, Sweden. (In Swedish.)

Part III

Technological aspects of DESAR

A CONCEPTS OF AND TECHNOLOGIES FOR DESAR

B ANAEROBIC PRE-TREATMENT

C LOW STRENGTH WASTEWATER (POST) TREATMENT

D WATER AND MINERAL RESOURCE RECOVERY

E AGRICULTURAL REUSE

8

DESAR treatment concepts for combined domestic wastewater

F.A. El-Gohary

8.1 INTRODUCTION

8.1.1 Background

Water has always been considered to be a critical natural resource on which human survival depends. It is well known that all early civilisations developed and flourished close to rivers, lakes, wetlands, or groundwater resources. Today, although the strategic importance of fresh water is universally recognised more than ever before, and although issues concerning sustainable water management can be found almost in every scientific, social, or political agenda all over the world, water resources face severe quantitative and qualitative threats.

Edited by P. Lens, G. Zeeman and G. Lettinga. ISBN: 1 900222 47 7

Population increase, industrialisation and rapid economic development, followed by political and administrative shortcomings, impose severe risks to the availability and quality of water resources in many areas worldwide.

The problems of water shortage in the Mediterranean region are well documented. Most countries in the Mediterranean area are arid or semi-arid. They have low rainfall, mostly with a seasonal and erratic distribution. Moreover, due to the area's rapid development, which accorded the agricultural sector top priority, conventional water resources have been seriously depleted. This is particularly acute in Southern Mediterranean countries where irrigation accounts for 50% (Algeria) to 90% (Libya) of water use. Furthermore, agriculture has a seasonal demand pattern which often conflicts with other uses such as tourism. UN projections (UN Population Division 1994) show that four Mediterranean countries already have less than the minimum required water availability to sustain their own food production (750 m^3/per capita/yr), by 2025 eight countries will be in virtually the same situation. These countries are essentially all on the southern rim of the Mediterranean (Angelakis *et al.* 1999).

To the scarcity factor must be added that of fragility. Water quality is increasingly endangered by pollution. Consequently, over-exploitation of water resources; lowering of groundwater tables; depletion of surface water bodies; pollution of rivers, lakes and aquifers and natural ecosystem degradation are the main results of the way that water resources have been managed so far.

One of the major sources of water pollution is the uncontrolled discharge of human waste. Although recent years have seen big improvements in access to safe water and adequate sanitation, these gains in many cases have passed by the poor.

The unhealthy conditions of those lacking sanitation cannot be ignored because sanitation-related diseases and polluted water sources often have devastating social, economic; and environmental effects on all urban and rural residents. Waterborne diseases, especially diarrhoeal diseases, are a leading cause of mortality and morbidity among children in the age group 0–14 years. Providing poor urban and rural dwellers with adequate sanitation facilities is therefore, a challenge facing many countries in the Middle East and North Africa.

8.1.2 Small community wastewater system

A major handicap hindering progress in meeting sanitation needs has been the scale of projects addressing sanitation problems. An inevitable result of the implementation of such large-scale, expensive projects is the exclusion of peri-urban and rural areas from such schemes since they cannot afford the high charges. Unbundling of sanitation programmes into smaller-scale projects can bring benefits at an affordable cost to those in greatest need. This does not mean that the macro picture should not be considered. On the contrary, the unbundling should take place after an adaptable strategic macro framework has been defined to sketch out the overall direction for sanitation service provision in the project area. It is within such an overall flexible sketch of the future that the unbundling should take place, with sequencing and details of investments in different service zones driven by demand (Wright 1997). Unbundling is a way of dividing investments into more realistic and more manageable components. There are two forms of unbundling: horizontal and vertical.

In horizontal unbundling, services are subdivided geographically. A large community may be divided into two or more zones, each with its own self-contained sanitation services. Decentralised sewerage is an example of horizontal unbundling that is particularly appropriate in areas with flat terrain and high groundwater tables. Dividing such areas into self-contained zones eliminates the need for expensive pumping stations and interceptor sewers required to serve the whole area with a conventional sewerage system. Another advantage of unbundling is that it reduces the average diameter and depths of sewers as compared to a centralised system. Since these are the two major cost elements (along with the length of sewers), it follows that horizontal unbundling is likely to be economically sound whenever it is technically feasible.

Vertical unbundling is particularly useful in reaching the poor with affordable sanitation services in an incremental way. By separating decisions on in-house improvement from those on neighborhood feeder systems and on city-wide trunk systems, unbundling allows a clear link to be made between immediate benefits and costs. Investment can be made one step at a time, starting with the home. Vertical unbundling consists of the following three technology levels:

(1) In-house infrastructure involves household level systems such as latrines, toilets, septic tanks and house drains. The facilities are located at the point where the waste is generated and the benefit is to the individual householder. In comparison with other levels of investment, in-house systems have the lowest sunk costs. Householders' value judgements are straightforward because benefits are direct. Market forces apply and there is great scope for privatisation of service provision, with competition bringing cost savings. In some peri-urban settlements, lack of secure property rights may be an important issue inhibiting individuals from making investments.

(2) Feeder infrastructure relates to the neighbourhood sewers or collection systems shared among occupants of a street or block of houses. The

users have common interests in ensuring that the systems function properly. Decision-making and payment for feeder systems needs to be shared among the beneficiaries. Sometimes this may come about through a local agency responding to collective demands from groups of users. Incentives may be relevant as a means of stimulating demand, particularly if there is a need to spread the costs of trunk sewerage at a later date. Economies of scale begin to emerge, but sewer systems have higher costs than household sanitation. Market forces and private sector involvement help to keep costs down.

(3) Trunk infrastructure includes mains, sewerage and treatment works serving an entire city or village. The large scale of the operation means high costs and appropriate economies of scale, but savings can be offset by restricted competition. Trunk systems are remote from users, who may not readily appreciate the benefits. Accordingly, user charges may not be the best way to recover investments. Decisions generally need to be made at the local government level, and operations and maintenance may be best funded through general taxation. Privatisation or other forms of private sector involvement are possible, with a need for regulatory safeguards.

8.2 THE DECENTRALISED APPROACH

The decentralised approach is a new means of addressing the wastewater management needs of sewered and unsewered areas in a comprehensive fashion. It allows for the use of individual and shared on-site soil-based systems as long term solutions. By definition, decentralised wastewater management (DWM) employs collection, treatment and disposal/reuse of wastewater from individual homes, clusters of homes, isolated communities, industries or institutional facilities, as well as from portions of existing communities at or near the point of waste generation. Decentralised systems maintain both the solid and liquid fractions of the wastewater near their point of origin, although the liquid portion and any residual solids can be transported to a centralised point for further treatment and reuse (Tchobanoglous 1996).

The elements that DWM systems comprise include: (1) wastewater pre-treatment; (2) wastewater collection; (3) wastewater treatment; (4) effluent reuse or disposal; and (5) biosolids and septage management. Although the components are the same as for large centralised systems, the difference is in the type of technology applied. It should also be noted that not every DWM system will incorporate all of the above elements.

8.2.1 Benefits of decentralised systems

The use of decentralised wastewater treatment systems offers the following advantages (Douglas 1998):

- *Save money* – prevents unnecessary costs by focusing on preventive measures (assessment of community conditions/needs and maintenance of existing systems) instead of reacting to a crisis.
- *Protect the homeowner's investment* – maximises potential for homeowners with existing septic systems to continue to benefit from their original investment.
- *Promote better watershed management* – avoids the potentially large transfers of water from one watershed to another that can occur with centralised treatment.
- *Offer an appropriate solution for low density communities* – in small communities with low population densities (and a smaller tax base), decentralised systems will be the most cost-effective option.
- *Provide a suitable alternative for varying site conditions* – decentralised systems can be designed for sites with shallow water tables, shallow bedrock, low-permeability soils and small property lot sizes.
- *Furnish effective solutions for ecologically sensitive areas* – decentralised systems can provide cost-effective solutions for areas that require advanced treatment, such as nutrient removal or disinfection, while recharging local aquifers and providing reuse opportunities close to points of wastewater generation.

Furthermore, large-scale centralised treatment plants create large amounts of treated wastewater that need to be disposed of in one area. This limits the potential for reuse of the treated effluent and the wastewater is often disposed of directly into the ocean, seas or rivers which in turn can lead to algal blooms and eutrophication.

8.3 TECHNOLOGY OPTIONS

Decentralised wastewater treatment alternatives for small communities can be broadly defined into three categories that represent the basic approaches to wastewater conveyance, treatment, and/or disposal (US EPA 1992).

- Natural systems that utilise soil as a treatment and disposal medium, including land application, constructed wetlands and subsurface infiltration. Some sludge and septage handling systems, such as sand drying beds, land spreading and lagoons, are included.

- Alternative collection system that uses lightweight plastic pipe buried at shallow depths, with fewer pipe joints and less complex access structures than conventional gravity sewers. These include pressure, vacuum, and small-diameter gravity sewer systems.
- Conventional treatment systems that utilise a combination of biological and physical processes, employ tanks, pumps, blowers, rotating mechanisms and/or other mechanical components as part of the overall system. These include suspended growth, fixed growth and combinations of the two. This category also includes some sludge and septage management alternatives, such as digestion, dewatering and composting systems and appropriate disposal information.

It is worth mentioning that the most significant change that has occurred in the past twenty years in the implementation of small and decentralised wastewater management systems is the development of new technology and hardware and the reapplication of old technology using new equipment. Important examples include the use of: (a) developed septic tank systems; (b) high rate anaerobic treatment; (c) alternative wastewater collection technologies; (d) aquatic treatment systems; (e) constructed wetlands; and (f) land treatment systems.

8.3.1 Anaerobic treatment processes

8.3.1.1 Historical background

The first application of anaerobic digestion for sewage treatment can be traced back to about 1860 with the early development of the septic tank by Mouras in France (Dunbar 1908). This was followed by the development of several anaerobic treatment systems, the best known is the Imhoff tank developed by Imhoff in Germany. These systems are primary treatment systems.

Primary treatment of sewage by anaerobic digestion was widely used in Europe between the two world wars. More than 12 million people in Germany were served by anaerobic treatment systems, mostly versions of the Imhoff tank. In the following decades, anaerobic treatment of sewage became less popular than aerobic treatment systems. This was due to the lower removal efficiency obtained in anaerobic systems as compared to the aerobic systems, a situation which could be attributed to a design failure. With the development of a variety of new high rate anaerobic treatment processes, such as the anaerobic filter (AF) (Young and McCarty 1969), the upflow anaerobic sludge blanket reactor

(UASB) (Lettinga *et al.* 1979) and the improvement of the septic systems, a breakthrough has been made in the field of anaerobic treatment of sewage.

8.3.1.2 Sanitation in rural Egypt

The coverage of rural areas in Egypt with appropriate sanitation systems is very low. Unfortunately, the provision with drinking water was not accompanied by programmes for the collection and disposal of the wastewater generated. In view of this design error, the domestic wastewater generated exceeded the capacity of the onsite disposal facilities.

In areas where indoor plumbing has not been installed, or the house is without public water supply, the sanitary pit privy and the concrete vault privy are used.

- The sanitary pit privy or earth pit privy is the most commonly used system in rural Egypt (Figure 8.1). A pit capacity of $1.5m^3$ is considered to be adequate for four or five persons. The pit is lined with masonry to prevent cave-in.

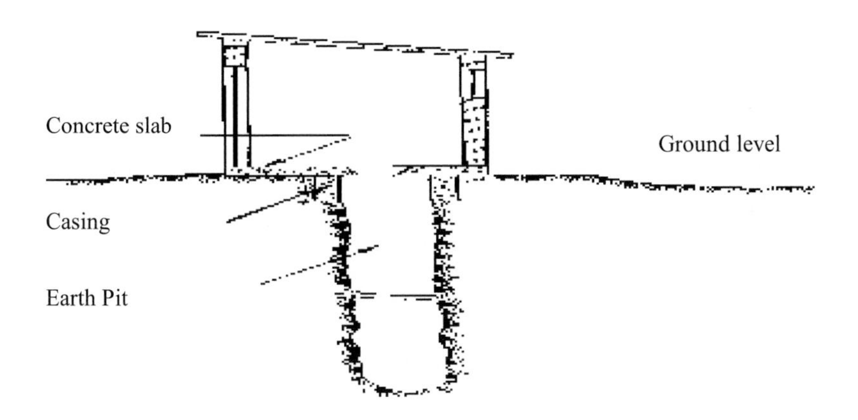

Figure 8.1. Sanitary pit privy.

- In locations where the soil is heavy and impervious or where there is no room to establish the sanitary pit privy at a safe distance from drinking water supply wells, the concrete vault privy is used (Figure 8.2). The vault is constructed of reinforced concrete and has a capacity of 3 cubic feet per person served. The contents of the vault are frequently sprinkled with lime to reduce odours. The vault is emptied when it is about two-thirds full.

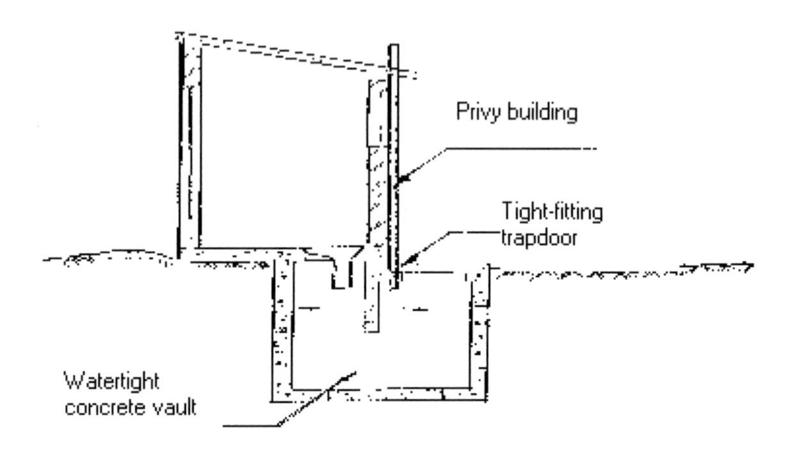

Figure 8.2. Concrete vault privy.

Where indoor plumbing has been installed, and the house is provided with a drinking water supply system, the indoor toilet or privy is connected to an outdoor septic tank and disposal tile field system (septic system; see Figure 8.3). This system has been implemented by the Save the Children Federation's Egypt field office (SCF/EFO) in collaboration with US-AID in rural areas of upper Egypt and Palestine (Kuttab 1993).

Modern conventionally designed septic systems are composed of four basic components:

- building sewer;
- septic tank;
- distribution box; and
- drainfield (or leach field).

There are two basic types of wastewater generated within a home. Blackwater is composed of toilet waste whereas wastewater from sinks, showers and laundry is called greywater. In a conventionally pumped home, these combined wastewaters flow out of the house into the septic tank through a single three or four inch diameter pipe called a building sewer.

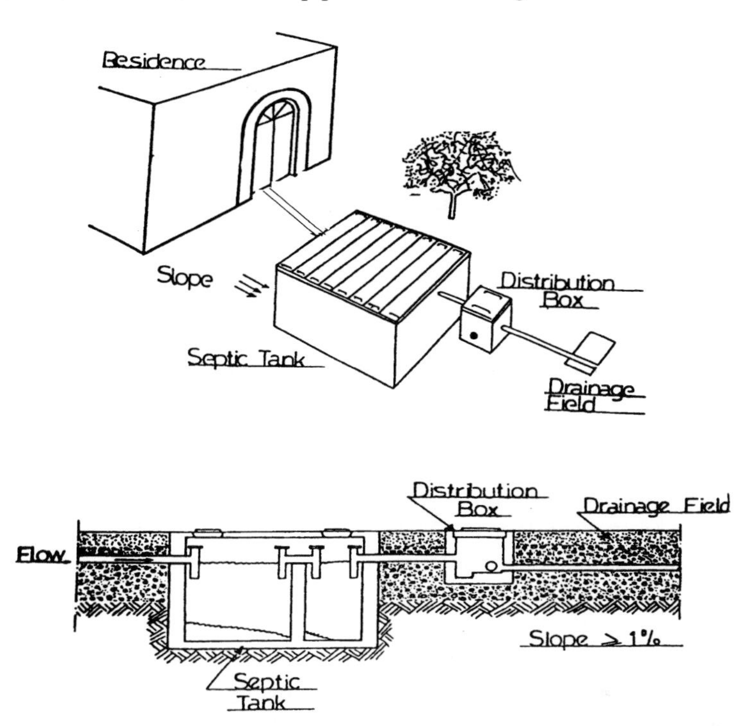

Figure 8.3. Diagram showing a typical subsurface drainage technique.

Typically, septic tanks are made of concrete or fibreglass, although other materials such as steel, redwood and polyethylene have also been used. The use of steel and redwood tanks is no longer accepted by most regulatory agencies. Polyethylene tanks have been used; but their structural integrity is inferior to concrete and fibreglass tanks. Long-term creep, resulting in deformation, has been a problem with polyethylene tanks. Fibreglass tanks, being more expensive, are used in areas inaccessible to concrete tank delivery lorries. Regardless of the construction material, a septic tank must be watertight and structurally sound if it is to function properly, especially where subsequent treatment units such as intermittent and recirculating packed bed filters or pressure sewers are to be used.

Septic tank capacity ranges from 1000 to 2000 gallons, depending on the family size and rate of water consumption. Other applications such as schools, summer camps, parks and hotels may require larger tanks, or more than one tank, often arranged in a series.

Most solids entering the septic tank settle to the bottom and form a sludge layer at the bottom of the tank. Oils and greases and other light materials float to the surface, where a scum layer is formed as floating materials accumulate. The organic material retained in the bottom of the tank undergoes facultative and anaerobic decomposition and is converted to more stable compounds and gases such as carbon dioxide (CO_2), methane (CH_4) and hydrogen sulphide (H_2S). The accumulation of sludge at the bottom and scum at the top of a septic tank gradually reduces the wastewater volume storage capacity of the septic tank. When this happens, incoming solids are not settled efficiently and may be flushed from the tank into the drainfield and cause clogging and premature system failure. Periodic discharge of solids accumulated in the septic tank (every two to three years) helps prevent solids from entering the drainfield. The liquid effluent from the septic tank is discharged via a four inch diameter pipe to a distribution box which separates effluent flow into approximately equal portions to two or more pipe lines leading to the drainfield (Figure 8.3). The typical drainfield consists of a network of four inch perforated plastic pipes surrounded by crushed stone and installed within native soil material. This type of drainfield configuration is a trench design. Once in the drainfield, effluent leaving the pipe network percolates through the crushed stone and moves downward into the underlying soil material.

Wastewater treatment in the soil environment is dependent upon the uptake of nutrients by micro-organisms; mechanical filtering by soil; biological decay; and chemical reactions with soil. All these processes are dependent on retention time. A long retention time means that wastewater is in contact with soil particles and micro-organisms for a longer period of time to promote chemical, microbial and mechanical interactions. Retention time is strongly influenced by soil texture. Coarse-textured sandy soils have a short retention time and pollutant removal efficiencies are correspondingly lower than in medium-textured soils. In contrast, fine-textured silty and clay soils have long retention times. Long retention times maximise chemical, microbial and mechanical wastewater treatment processes. However, because wastewater moves through fine-textured soil so slowly, these soils may not be able to handle the volume of wastewater discharged into them. The most serious operational problem with septic tanks has been the carryover of solids and oils and grease. To overcome this problem, two compartments have been used. A more effective way to eliminate the discharge of untreated solids involves the use of an effluent filter in conjunction with a single compartment tank.

8.3.1.3 Upgrading septic tank effluent

In areas where soil and site conditions are not well-suited for conventional septic systems, alternative systems are often used to guarantee safe sewage treatment and protect public health. The use of small-bore sewer systems and sand filter are such alternatives.

Small-bore sewer system

In a village in Egypt, the World Health Organisation (WHO) and UNICEF have implemented a pilot project. The project consists of the following components:

- improvement of existing pit latrines by the addition of a ventilation pipe;
- construction of additional numbers of ventilated improved pit latrines (VIPL). One pit latrine is allocated for each residential or public building; and
- construction of septic tanks, one tank for each building or for two or more adjacent buildings, depending on the availability of land or width of the streets.

Septic tank effluent is discharged into a small-bore sewerage network where it flows by gravity to a pump station. The accumulated wastewater in the sump of the pump station is pumped to a stabilisation pond for treatment.

Sand filter systems

A typical sand filter system has four working parts:

- the septic tank
- the pump chamber with the pump
- the sand filter
- the disposal component including a drainfield with its replacement area

The pump chamber is a concrete, fibreglass or polyethylene container that collects septic tank effluent. The chamber contains a pump, pump control floats and a high-water alarm float. The control floats are adjustable and are set to pump a specific volume of effluent. When the effluent rises to the level of the ON float, the pump starts to deliver the effluent to the sand filter. The pump lowers the effluent level to the OFF float and stops.

The high-water alarm float in the pump chamber starts an alarm to warn of any pump or system malfunction. The float is set to start when the effluent in the pump chamber rises above the ON float. The alarm should consist of a buzzer and an easily visible light. It should be on an electrical circuit separate from the pump.

The pump discharge pipe is normally provided with a union or other quick-disconnect coupler for easy removal of the pump. A piece of nylon rope or other non-corrodable material should be attached to the pump for taking the pump in and out of the chamber.

The typical sand filter is a concrete or PVC-lined box filled with a specific sand material. A network of small diameter pipes is placed in a gravel-filled bed on top of the sand. The septic tank effluent is pumped under low pressure through the pipes in controlled doses to ensure uniform distribution.

The effluent leaves the pipes, trickles downward through the gravel, and is treated as it filters through the sand. A gravel under-drain collects and moves the treated wastewater to either a second pump chamber for discharge to a pressure distribution drainfield or to a gravity flow drainfield. The second pump chamber may be located in the sand filter (Figure 8.4).

The drainfield receives the treated sand filter effluent for disposal. It has a network of pipes placed in gravel-filled trenches two to three feet wide or beds (over three feet wide) in the soil. The effluent leaves the pipes and trickles downward through the gravel and into the soil.

Every new drainfield is required to have a designated replacement area. This area is similar to the size of the existing drainfield. It must be maintained should the existing system need an addition or repair.

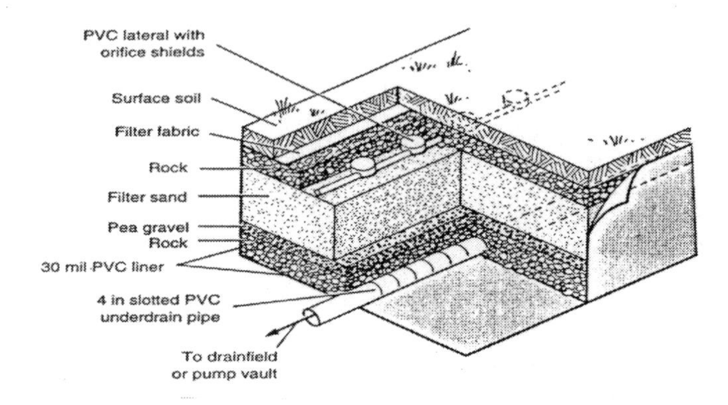

Figure 8.4. Cutaway view of the basic elements of a packed filter (Crites and Tchobanoglous 1998).

8.3.1.4 Upflow anaerobic sludge blanket reactor

The upflow anaerobic sludge blanket (UASB) reactor was developed in the Netherlands (Lettinga *et al.* 1980). The most characteristic device of the UASB reactor is the phase separator. This device is placed at the top of the reactor and divides it into a lower part, the digestion zone, and an upper part, the settling zone (Figure 8.5). The wastewater is introduced as uniformly as possible over the reactor bottom. Due to the inclined walls of the phase separator, the area for the liquid flow in the settling zone increases. Consequently, the upflow velocity of the waste water decreases. As a result flocculation and/or sedimentation of the flocs into the settling zone takes place. At a certain stage the weight of the sludge accumulated in the phase separator will exceed the frictional force that keeps it on the inclined surface and it will slide back into the digestion zone to become, once again, part of the sludge mass that digests the influent organic matter. Thus, the presence of a settler on the top of the digestion zone enables the system to maintain a large sludge mass in the UASB reactor, while an effluent essentially free of suspended solids is discharged (van Haandel and Lettinga 1994). The gases produced under anaerobic conditions (principally methane and carbon dioxide) serve to mix the contents of the reactor as they rise to the surface. The rising gas also helps to form and maintain the flocs. Sludge flocs buoyed up by the gas strike the degassing baffles and settle back down onto the bed from the quiescent settling zone above the sludge blanket. The gas is trapped in a gas collection dome located at the top of the reactor.

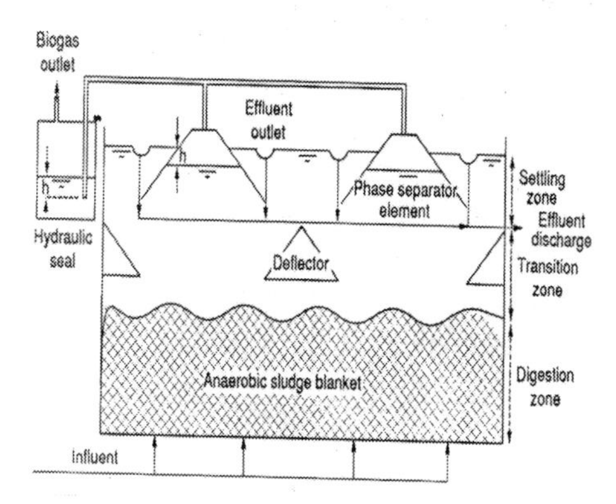

Figure 8.5. Schematic representation of an upflow anaerobic sludge blanket (UASB) reactor (Haandel and Lettinga 1994).

Performance data from Egypt showed the effectiveness of the UASB reactor, which removed up to 85% of the COD and about 85% of the incoming suspended solids at an HRT of 8 hours. Removal of faecal coliforms did not exceed one log. However, the anaerobic sludge bed apparently works as a sieve for all kinds of helminth ovae. None of these ovae were detected in the effluent. Examination of the sludge indicated the presence of *Ascaris spp.* in almost all samples examined; so excess sludge from the UASB reactor should be handled with care. It is worth mentioning that the two-stage system performed better than the one-stage system when operated at the same HRTs of 8 hours (El-Gohary and Nasr 1999).

Upgrading UASB effluent

Anaerobically pre-treated wastewater does not usually comply with the standards required. Therefore, an adequate polishing step is required. Alternative options that can be used to upgrade UASB are as follows:

- aerobic biological treatment (suspended growth or attached growth)
- constructed wetlands
- sand filters
- algal ponds

The use of algal ponds (AP), lemna (duckweed) ponds (LP), fishponds and a compact rotating biological contactor (RBC) to upgrade the UASB effluent has been investigated by El-Gohary *et al.* (1998b). Schematic diagrams of the systems are presented in Figure 8.6. Results showed that a minimum HRT of 10 days was required for the pond system to achieve 99.99% faecal coliform removal.

Corresponding residual count was 1.3×10^3 MPN/100 ml. However, the pond was less effective at removing COD and suspended solids. A sequenced fishpond was able to breed *Tilapia nilotica* with no mortality provided the undissociated NH_3 could be kept below 0.14 mg/l. Fish mortality of 28% was recorded at 0.46 mg/l undissociated ammonia.

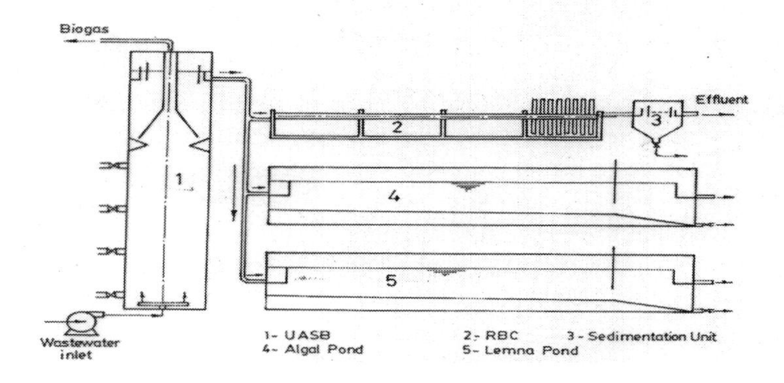

Figure 8.6. An integrated wastewater system (El-Gohary *et al.*1998b).

Much higher COD removal rates were experienced, however, in the LP, which were accompanied by lower pathogen removal rates. A very satisfactory operation was also experienced when combining the RBC with the UASB reactor at a hydraulic loading of 0.06^3 m^3/m^2.day and an average BOD loading of 4.2 g BOD/m^3.d. COD and BOD removal values were 47 and 66% respectively. The faecal coliform count declined by 5 logs. The geometric mean of the residual faecal coliform density averaged 2.8×10^3 MPN/100 ml.

The results obtained indicated that the use of the UASB reactor in combination with the appropriate post-treatment represents an ideal solution for wastewater management in arid Mediterranean countries.

8.3.2 Constructed wetlands

Wetlands are defined as land where the water is above the ground surface or where it can maintain saturated soil conditions for the growth of related vegetation (Reed *et al.* 1988). To avoid interference with natural ecosystems, constructed wetlands in which the hydraulic regime is controlled have been widely used to treat a variety of wastewaters (Hammer 1989).

Constructed wetlands can be considered part of a wastewater treatment system, while most natural wetlands are considered receiving waters and are therefore, subject to applicable laws and regulations regarding discharge. The influent to currently operating constructed wetland systems ranges from septic tanks to secondary effluents.

The two different types of constructed wetlands are characterised by the flow path of the water through the system. The first, called a free-water surface

(FWS) wetland, includes appropriate emergent aquatic vegetation in a relatively shallow bed or channel. The surface of the water in such a system is exposed to the atmosphere as it flows through the area. The second type, called a sub-surface flow (SF) wetland, includes a foot or more of permeable media-rock, gravel, or coarse sand to support the root system of the emergent vegetation. The water in the bed or channel in such a system flows below the surface of the media.

Both types of constructed wetlands are designed to prevent groundwater contamination beneath the bed or channel. Linings range from compacted clay to membrane liners. A variety of water application methods have been used with constructed wetlands, and a number of different outlet structures and methods have been included to control the depth of water in the system.

The principal removal and/or transformation mechanisms in wetland systems are summarised in Table 8.1. Constituents considered are organic matter (for example, BOD), suspended solids, nitrogen, phosphorus, metals, trace organics and pathogens.

Table 8.1. Summary of principal removal and transformation mechanisms in constructed wetlands for the constituents of concern in wastewater. (Source: Crites and Tchobanoglous 1998).

Constituent	Free water system	Subsurface flow
Biodegradable Organics	Bioconversion by aerobic, facultative and anaerobic bacteria on plant and debris surfaces of soluble BOD, adsorption, filtration and sedimentation of particulate BOD	Bioconversion by facultative and anaerobic bacteria on plant and debris surfaces
Suspended solids	Sedimentation, filtration	Filtration, sedimentation
Nitrogen	Nitrification/denitrification, plant uptake, volatilisation	Nitrification/denitrification, plant uptake, volatilisation
Phosphorus	Sedimentation, plant uptake	Filtration, sedimentation, plant uptake
Heavy metals	Adsorption on plant and debris surfaces, sedimentation	Adsorption on plant roots and debris surfaces, sedimentation
Trace organics	Volatilisation, adsorption, biodegradation	Adsorption, biodegradation
Pathogens	Natural decay, predation, UV irradiation, sedimentation, excretion of antibiotics from roots of plants	Natural decay, predation, sedimentation, excretion of antibiotics from roots of plants

8.3.2.1 Free-water-surface constructed wetlands

In a free-water-surface constructed wetland (marsh or swamp), the emergent vegetation is flooded to a depth that ranges from 0.1 to 0.45 m. Typical vegetation for FWS systems includes cattails, reeds, sedges and rushes. A FWS system typically consists of channels or basins with a natural or constructed impermeable barrier to prevent seepage.

Plants in free-water-surface constructed wetlands serve a number of purposes. Stems, submerged leaves and litter serve as support media for the growth of attached bacteria. Leaves above the water surface shade the water and reduce the potential for algal growth. Oxygen is transported from the leaves down into the root zone, which supports the plant growth. A limited amount of oxygen may leak out of the submerged stems to support attached bacterial growth. Pre-treatment for FWS wetlands usually consists of settling (septic tanks or Imhoff tanks), screening with a rotary disk filter, stabilisation lagoons, or UASB reactor. Because major sources of oxygen are surface reaeration in open water from the atmosphere and attached-growth algae, the BOD loading generally needs to be kept below 100 Ib/ac.d. Wastewater is treated as it flows through the vegetation by attached bacteria and by physical and chemical processes.

8.3.2.2 Subsurface-flow constructed wetlands

In a subsurface-flow constructed wetland (Figure 8.7) the wastewater is treated as it flows laterally through the porous medium. Subsurface-flow systems are also known as rock-reed filters, microbial rock plant filters, vegetated submerged beds, marsh beds, and hydrobotanical systems. They have the advantages of smaller land area requirements and avoidance of odour and mosquito problems, compared to FWS systems. Disadvantages of SF systems are their increased cost due to the gravel media and the potential for clogging of the media. Emergent vegetation is planted in the medium, which ranges from coarse gravel to sand. The depth of the bed ranges from 0.45 to 1 m and the slope of the bed is typically 0–0.5%.

The Global Environment Facility (GEF) is currently funding a demonstration wetland project in Egypt. The engineered wetland is designed to treat 25,000 m^3/day of agricultural drainage water. This water is a mixture of water seeping from irrigated fields and partially treated or untreated domestic and industrial wastewater. The main objective of this project is to protect the water quality of Lake Manzala from deterioration.

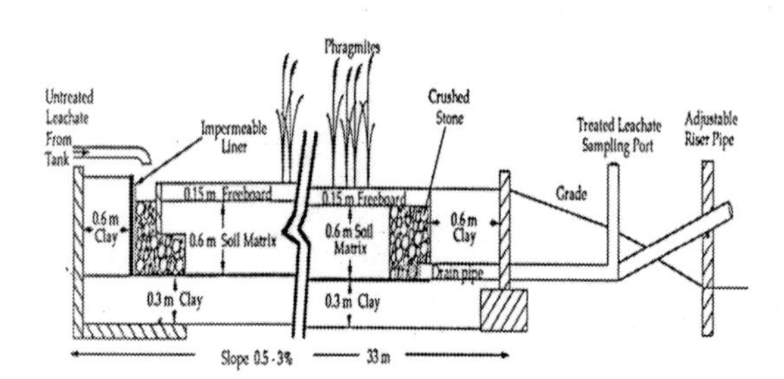

Figure 8.7. A subsurface-flow constructed wetland.

8.3.3 Lagoon treatment

Lagoon treatment systems are earthen basins or reservoirs that are engineered and constructed to treat wastewater. Work on lagoons in the 1940s led to the development of wastewater treatment lagoons as low-cost alternatives (McGauhey 1968; Marais 1970). In practice, the terms 'lagoon' and 'pond' are used interchangeably.

Wastewater treatment lagoons range in depth from shallow to deep, and are often categorised by their aerobic (dissolved oxygen concentration) status and the source of that oxygen for bacterial assimilation of wastewater organics. The four major types of lagoon systems are classified with respect to the presence and source of oxygen in Table 8.2.

Table 8.2. Classification of lagoon systems based on the presence and source of oxygen

Type of lagoon	Presence of oxygen
Aerobic	Photosynthesis provides oxygen for aerobic conditions throughout the water column
Facultative	Surface zone is aerobic. Subsurface zone may be anoxic or anaerobic
Partial-mix aerated	Surface aeration produces aerobic zone that ranges from half depth to total depth depending on oxygen input and lagoon depth
Anaerobic	Entire depth is anaerobic

Lagoon technology is used primarily for small rural communities; however, aerated lagoons and facultative lagoons are often used in medium-sized communities. Lagoon systems are used alone, or in combination with other wastewater treatment systems. The advantages of lagoons include:

- low capital costs
- minimum operational skills needed
- sludge withdrawal and disposal needed only at 10–20 year intervals
- compatibility with land and aquatic treatment processes

The disadvantages of lagoons include:

- large land areas may be required
- high concentrations of algae may be generated, which can be problematic for surface discharge
- non-aerated lagoons often cannot meet stringent effluent limits
- lagoons can impact groundwater negatively if liners are not used, or if they are damaged
- improperly designed and operated lagoons can become odorous
- evaporation of water causes an increase in water salinity

8.3.3.1 Aerobic lagoons

Aerobic lagoons are relatively shallow to allow light to penetrate the full depth of the lagoon. As a result, aerobic lagoons have active photosynthetic activity through the entire water column during daylight hours. Typical lagoon depths range from 0.3–0.6 m. Lagoons designed to maximise the photosynthetic activity of algae are also known as high-rate lagoons. The use of the term 'high-rate' refers to the photosynthetic oxygen production rate of the algae present, not to the metabolic rate, which does not change. The photosynthetic oxygen allows bacteria to aerobically degrade organic material. During daylight hours, dissolved oxygen and pH values rise and peak, followed by a drop in the hours of darkness. Detention times are relatively short, with five days being typical. Aerobic lagoons are used in combination with other lagoons and are limited to warm, sunny climates (Reed *et al.* 1995).

8.3.3.2 Facultative lagoons

Facultative lagoons are the most common and versatile of the different lagoon types. They are 1.5–2.5m deep and are also referred to as oxidation ponds or stabilisation lagoons. Treatment is accomplished by bacterial action in an upper aerobic layer and a lower layer that can be anoxic or anaerobic, depending on wind-induced mixing. Settleable solids are deposited on the lagoon bottom. Oxygen is provided by natural surface aeration and photosynthesis. Facultative lagoons can be used as controlled discharge lagoons, total containment lagoons and as storage lagoons for land treatment systems. To reduce the pond area required, a number of options have been investigated. Increasing pond depth is one example. Research carried out by Oragui has shown that ponds 2–3 m deep can achieve degrees of bacterial and viral removal comparable to those in ponds with a conventional depth of 1–1.5 m (Oragui 1987). Another alternative is to aerate the facultative pond (El-Gohary *et al.* 1994). However, more than one maturation cell is required to allow sedimentation of bio-solids produced.

8.3.3.3 Partial-mix aerated lagoons

Partial-mix aerated lagoons are deeper and more heavily loaded organically than facultative lagoons. Oxygen is supplied typically through floating mechanical aerators or submerged diffused aeration. Aerated lagoons are 2–6 m deep with detention times that range from 3–20 d. The principal advantage of aerated lagoons is that they require less land area than other lagoon systems. Partial-mix aerated lagoons have the same advantages as facultative lagoons (minimal sludge generation and handling) and a lower land requirement.

8.3.3.4 Anaerobic lagoons

Anaerobic lagoons are used for high-strength typically industrial wastewater in remote and rural areas. They have no aerobic zones, are 5–10 m in depth, and have detention times of 20–50 d. Because of the potential for odour production anaerobic lagoons must be covered or sited away from populated areas.

8.3.3.5 Fate of wastewater constituents

BOD removal

Lagoons are low-mass biological reactors. For all but anaerobic lagoons, soluble BOD is reduced by bacterial oxidation. Particulate BOD is removed by sedimentation. In facultative and anaerobic lagoons, anaerobic biological conversion takes place. BOD removal in lagoons depends on the detention time and lagoon water temperature.

Total suspended solids removal

The influent suspended solids are removed by sedimentation in lagoon systems. Algal solids that develop during treatment become the majority of the effluent suspended solids. Effluent suspended solids can range as high as 140 mg/l for aerobic lagoons and 60 mg/l for aerated lagoons. If slow-rate land treatment or reuse follow the lagoon treatment, then algal suspended solids are of little concern. However, because most algal solids are difficult to remove from water and effluent standards often cannot be met, additional processes may be needed to remove the solids.

Nitrogen removal

Nitrogen removal in lagoons appears to be the result of a combination of mechanisms including volatilisation of ammonia (which is pH-dependent), algal uptake, nitrification/denitrification, sludge deposition and adsorption onto bottom soils.

Phosphorus removal

Without the addition of chemicals for precipitation, the removal of phosphorus in lagoons is minimal. Chemical addition using alum or ferric chloride has been used effectively to reduce phosphorus to below 1 mg/l (Reed *et al.* 1995).

Pathogen removal

Significant removal of bacteria, parasites and viruses occurs in multiple cell lagoons with long detention times. Removal of pathogens in lagoons is due to natural die-off, predation, sedimentation and adsorption. Helminths and parasitic cysts and eggs settle to the bottom in the quiescent zone of lagoons. Facultative lagoons with three cells and about twenty days detention time, and aerated lagoons with a separate settling cell prior to discharge, will provide more than adequate helminth and protozoon removal. However, at least 28 days are usually required in hot climates to reduce bacteria numbers to the guide-line level (Mara and Silva 1986).

8.3.3.6 Egyptian experience

In recognition of the acute nature of the sanitation problem in rural areas, the Organisation for Reconstruction and Development of Egyptian Villages and US-AID began the Basic Village Service (BVS) project which covers five different technologiesL extended aeration, oxidation ditch, submerged fixed film reactors, standard stabilisation ponds and modified aerated stabilisation ponds.

A survey carried out by El-Gohary *et al.* (1998) indicated that the two types of oxidation ponds examined receive daily flows above their design values. Consequently, the HRT was reduced by 37% for the aerated pond (Figure 8.8) and by 53% for the standard stabilisation pond (Figure 8.9).

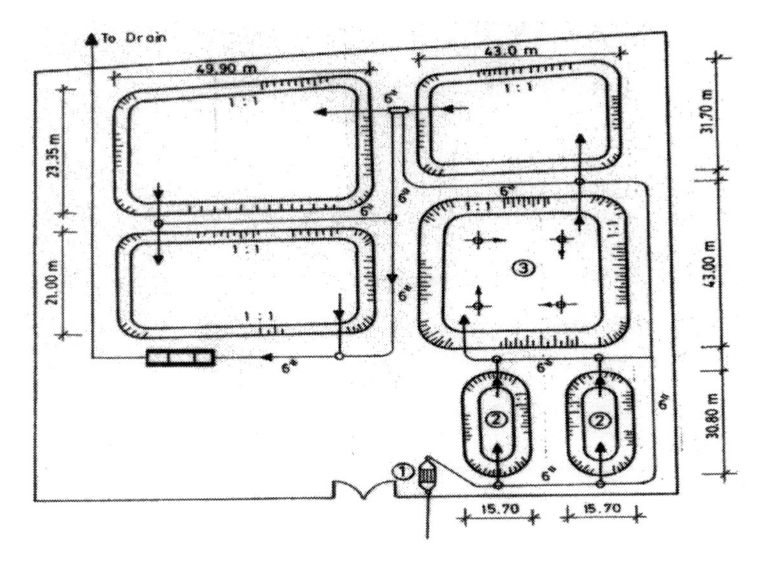

Figure 8.8. Aerated stabilisation pond (El-Gohary *et al.* 1998). 1 = screens; 2 = anaerobic ponds effective depth 3.4 m; 3 = aerated lagoon effective depth 3.5 m; 4 = maturation ponds in series effective depth 1.5 m.

To improve the performance of these treatment plants, expansion of the facilities is required. Since these plants are located in a delta, where land is not available, the replacement of the existing anaerobic ponds with more efficient systems, such as UASB, is recommended.

8.3.3.7 Upgrading lagoon effluent

Lagoon treatment systems are not very efficient at producing effluents with low concentrations of suspended solids. A number of technologies have been used to upgrade lagoon effluents for the removal of suspended solids, including:

- intermittent sand filters
- microstrainers
- rock filters
- dissolved air flotation (DAF)
- floating aquatic plants
- constructed wetlands.

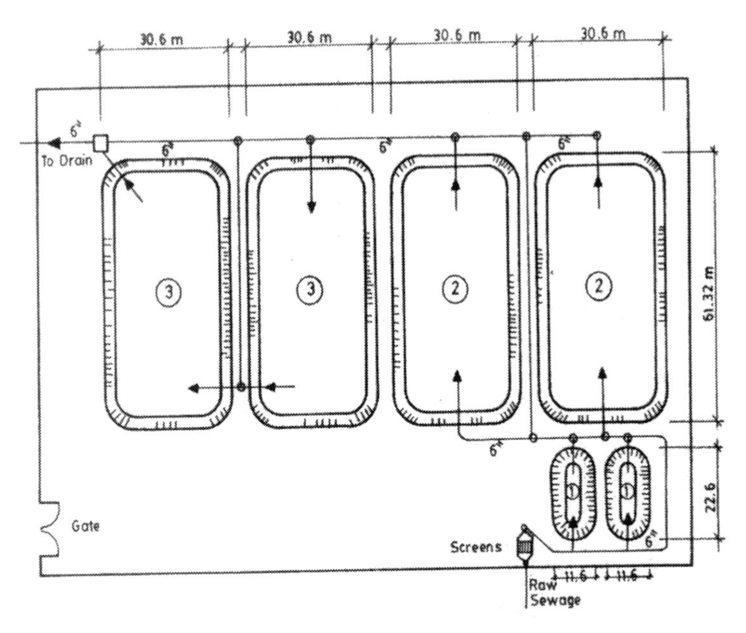

Figure 8.9. Layout of standard stabilisation pond (El-Gohary *et al.* 1998). 1 = anaerobic ponds effective depth 3.2 m; 2 = facultative ponds effective depth 2.05 m; 3 = maturation ponds. effective depth 1.55 m.

Intermittent sand filters (ISFs) are biological and physical treatment units that upgrade lagoon effluent by filtering out suspended solids. Algae collect on the surface of the sand filter as the wastewater is applied and treated. The accumulation of solids occurs in a 50–80 mm layer that must be periodically removed. The depth of sand in the filter must be at least 0.45 m plus a sufficient depth for cleaning cyles once every year (Reed *et al.* 1995). A single cleaning event may remove 25–50 mm of sand. A typical bed depth is 0.9 m. The sand for single-stage ISFs should be between 0.20–0.30 mm with a uniformity coefficient of less than 5.0. Less than 1% of sand particles should be smaller than 0.1 mm.

For lagoon effluent treatment, typical hydraulic loading rates for an ISF range from 0.37–0.56 m/d. For high concentrations of algae, greater than 50 mg/l, the hydraulic loading should be reduced to between 0.19–0.37 m/d to increase the run time between cleaning events. The lower end of the range should be used for filters in cold-weather locations to avoid the possible need for bed cleaning during the winter months.

The total filter area required for an ISF is determined by dividing the average flow rate by the design hydraulic loading rate. One spare filter should be added to ensure continuous operation because it may take several days for a cleaning event to be completed. A minimum of three units is preferred. In small systems that use manual cleaning, the individual bed should be no larger than 90 m^2. In larger systems with mechanical cleaning equipment, the individual beds can be as large as 5000 m^2. A typical ISF is shown in Figure 8.10.

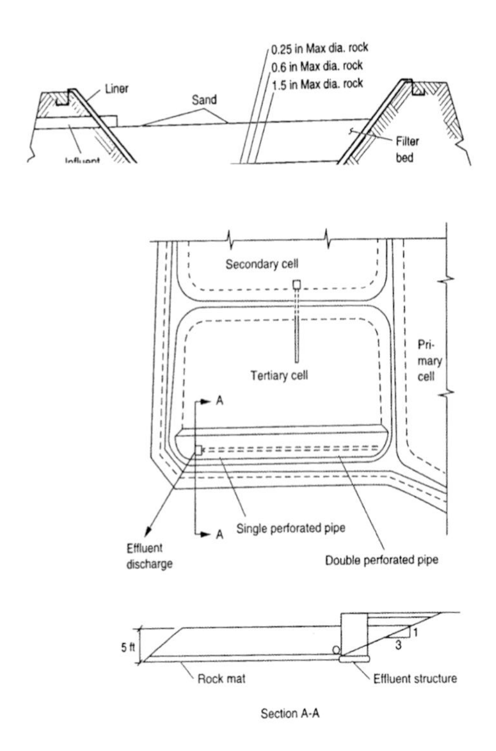

Figure 8.10. Intermittent sand filter (US EPA 1983).

Rock filters remove suspended solids by sedimentation as lagoon effluent flows horizontally through the void spaces in the rocks (Figure 8.11). The accumulated algae are then degraded biologically.

The advantages of the rock filter are its operational simplicity and its relatively low construction cost. Odor problems can occur, especially in wastewater containing significant concentrations of sulfates greater than 50 mg/l.

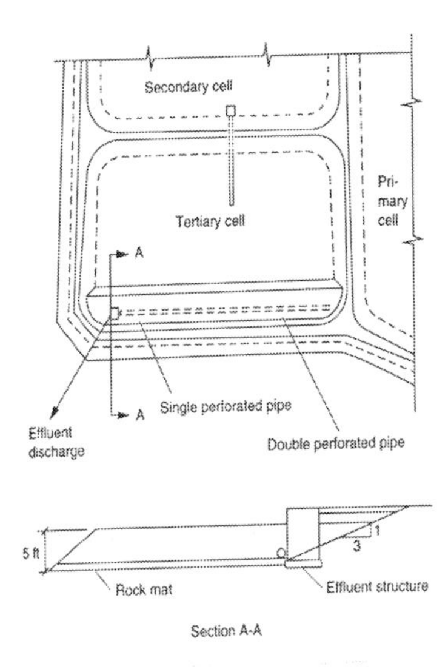

Figure 8.11. A rock filter (US EPA 1983).

8.3.4 Alternative wastewater collection systems

The types of systems used for the collection of wastewater range from conventional gravity sewers to pressure sewers and vacuum sewers. The cost of conventional gravity sewers is high and cannot be afforded by small communities. To avoid these high costs, alternative sewer systems have been developed.

8.3.4.1 Septic tank effluent gravity sewers:

In septic tank effluent gravity (STEG) sewers, a small-diameter (25–50 mm) plastic pipe is used to convey the effluent from a septic tank, equipped with an effluent filter, to a small-diameter collection system (Figure 8.12(a)). Because there are no solids to settle in the collection system, the collection system can be laid at a variable grade, just below the ground surface (e.g. 0.9 m below). As a result, STEG systems are also known as small-diameter variable-grade gravity sewers. Because the collection main is watertight, there is no infiltration in the system. To take advantage of topography, many systems are constructed with a combination of STEG and septic tank effluent pump sewers.

8.3.4.2 Septic tank effluent pump sewers

In modern septic tank effluent pump (STEP) systems, a high-head turbine pump is used to pump screened septic tank effluent into a pressurised collection system (Figure 8.12(b)). The size of discharge line leading from the septic tank is typically 1–1.5 in. The minimum pipe size used for the pressurised collection main is 2 inch diameter plastic pipe. As with the STEG system, infiltration is not an issue because the collection main is watertight. Because the lines are under pressure they can follow the terrain, as a water transmission line does. Because of the shallow burial depth, construction problems resulting from high groundwater and rocky soil can be avoided.

8.3.4.3 Pressure sewers with grinder pumps

In pressure sewer systems with grinder pumps, a septic tank is not used. In its place, the discharge pump, located in a small pump basin, is equipped with chopper blades that cut up the solids in the wastewater so that they can be transported under pressure in a small diameter pipeline (Figure 8.12(b)). As a consequence, higher solids and oil and grease concentrations are encountered. As with the STEP system, infiltration is not an issue because the collection main is watertight. The depth of burial for pressure sewers with grinder pumps is similar to the depth of STEP sewers.

8.3.4.4 Vacuum sewers

In vacuum sewers, a central vacuum source is used to maintain 380–500 m vacuum of mercury on small-diameter collection mains to transport the wastewater from individual homes to a central location (Figure 8.12(c)). As with other alternative collection systems, infiltration is not an issue because the collection vacuum main is watertight.

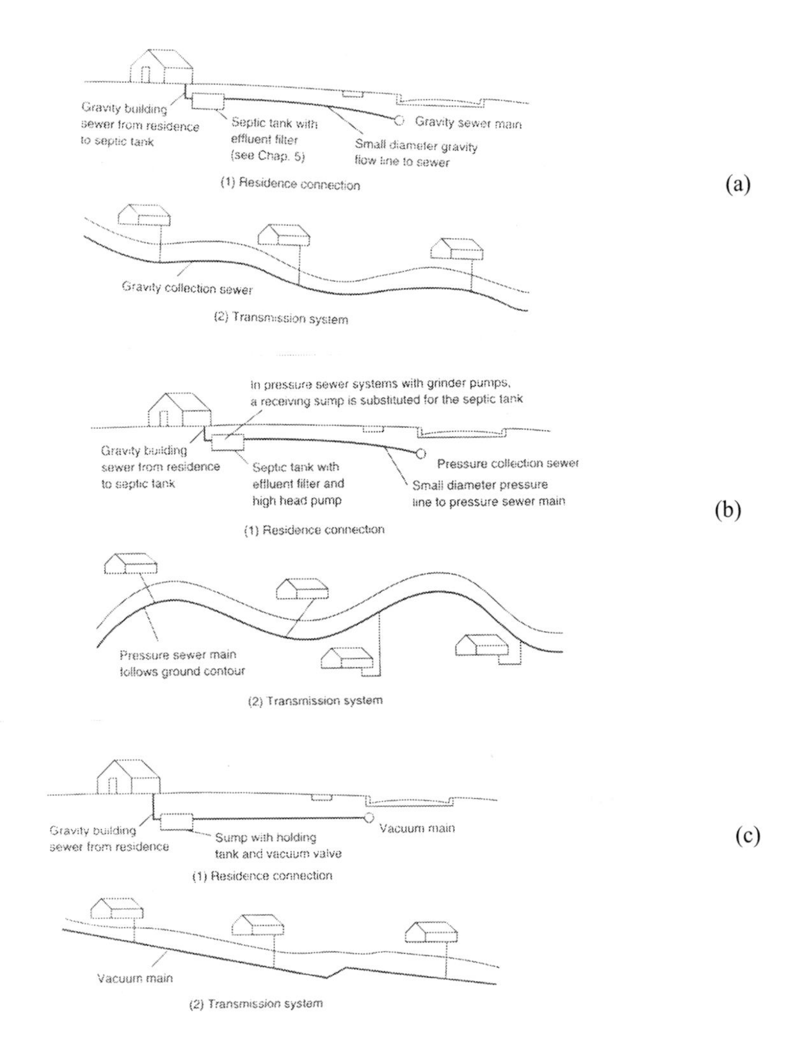

Figure 8.12. Definition sketch for sewer systems: (a) Septic tank effluent gravity (STEG), (b) Septic tank effluent pump (STEP) and pressure sewer with grinder pumps, and (c) vacuum sewer (Crites and Tchobanoglous 1998).

8.3.5 Biosolids and septage management

8.3.5.1 Septage

Septage is the semiliquid material that is pumped out of septic (or interceptor) tanks. It consists of the sludge that has settled to the bottom of the septic tank over a period of years. Septage characteristics are highly variable depending on such factors as tank size and design, user habits, pumping frequency, climate and seasonal weather conditions, and presence of appliances such as garbage grinders, washing machines, and water softeners.

Quantities of septage vary depending on the frequency with which septic tanks are pumped. A typical value for planning purposes is 0.227 m^3.capita/year. If the frequency of septic tank pumping is known, the annual volume of septage can be calculated by:

Annual volume = (number of septic-tanks) (typical volume)/(pumpout interval)

where annual volume = septage quantity, m^3/year; typical volume = typical septic tank volume, m^3; and pumpout interval = time between septic tank pumpings, year.

To determine the maximum and minimum periods of the year, the local septage haulers should be contacted. The major factor that affects septic tank cleaning frequency in cold regions is the weather. The frequency of pumping can be based on the number of people per house.

The most common approach for septage treatment is land application. Co-treatment at a sewage treatment plant is also widely employed. For small communities, independent facilities are usually limited to simple, low-cost approaches such as stabilisation lagoons or lime stabilisation systems.

Land application of septage employs the same techniques used for land application of liquid sewage sludge. Where stabilisation of liquid septage is required prior to land application, lime stabilisation should be considered, since it is one of the simplest and most economical techniques to meet applicable stabilisation criteria. Lime stabilisation involves addition and mixing of sufficient lime to achieve a pH of 12 for at least 30 minutes. To achieve a pH of 12, lime dosages can be expected to be 1500 to 3500 mg/l as Ca $(OH)_2$, which is hydrated lime.

This destroys pathogenic organisms, improves dewaterability, and reduces objectionable odors. Lime can be added in the form of quicklime, Ca0, or $Ca(OH)_2$. In addition, other alkaline materials such as lime kiln dust or cement kiln dust can be used to elevate the pH. Many septage haulers dispose of septage at wastewater treatment plants. This is an acceptable alternative if the plant has sufficient capacity and is equipped to handle the material. The most common approach is to add the septage to the head-works of the plant. In many plants, this allows the septage to undergo screening and grit removal, which is important to remove debris and grit that can wear pump components.

8.3.5.2 Sludge

Characteristics and quantities of sludge vary with the type of wastewater operation or process that produces the sludge as well as the wastewater strength. Important sludge characteristics include the expected quantity, chemical and nutrient content, and heavy metal content.

The first component of many small community sludge management schemes is a holding tank. Depending on the application, holding tanks should be designed to allow periodic sludge settling and decant of supernatant, increasing solids content and reducing the volume for dewatering and/or disposal. This step is followed either by aerobic digestion, anaerobic digestion, sludge lagoons or lime stabilisation.

Aerobic digestion of sewage sludge is commonly used by small communities as a means of stabilisation prior to land application. If application sites are reasonably close to the source of sludge, direct application of liquid sludge is clearly the most economical approach. If not, dewatering via sand drying beds may be used to reduce the volume of sludge to be hauled away.

Aerobic digestion as practised by small communities is typically performed in an open tank. Continuous introduction of air allows biological oxidation of organic matter under aerobic conditions. Aerobic digestion results in a reduction in biodegradable volatile solids, improved dewaterability, and odor reduction. Oxygen may be supplied by mechanical aerators or diffusers.

For small operations, single-tank digesters are commonly used. Periodically, aeration is stopped and solids are allowed to settle for 6 to 12 hours. Clarified supernatant is then decanted and returned to the plant. This procedure maximises the solids capacity of the digester, and also allows a thickened sludge to be removed, thereby reducing the volume to be hauled away. Generally, supernatant quality from an aerobic digester is favorable and impacts on the plant from supenatant decant are minimal.

Anaerobic digestion is nearly as common as aerobic digestion. When dewatering is necessary, sand beds are the most frequent choice.

Modern anaerobic digesters are generally designed as two-stage systems, to maximise the stabilisation of the sludge and the production of biogas (primarily methane) which is considered the most attractive feature of anaerobic digesters.

8.3.5.3 Dewatering options

Dewatering is a physical unit operation in which the moisture content is reduced (solids content increases). A wide variety of operations and equipment can be used to dewater sludge; however, for small plants the principal methods are:

- drying beds
- mechanical dewatering
- reed beds; and
- lagoons.

8.4 CONCLUSIONS

In many countries in the Mediterranean/Middle East region, it has become apparent that it may not be possible, due to both geographic and economic reasons, to provide sewerage facilities for populations in rural and peri-urban areas, either now or in the near future. As a result, the focus of the field of wastewater management should change from the construction and management of regional sewerage systems to the construction and management of decentralised wastewater treatment facilities.

Given the fact that in the arid climate of the Mediterranean, increasing demands are being made on freshwater supplies, it is clear that decentralised systems will increase the opportunities for localised reclamation/reuse. Also, the use of anaerobic treatment as a first step offers good potentials for both on-site and off-site sanitation. Furthermore, anaerobic pre-treatment, complemented by low-cost post-treatment techniques, offers a cost-effective method for reclaiming domestic wastewater for agricultural production.

8.5 REFERENCES

Angelakis, A.N., Marecos, D.M., Bontoux, L. and Asano, T. (1999) The status of wastewater reuse practice in the Mediterranean Basin: the need for guidelines. *Wat. Res.* **33**(10), 2201–2217.

Crites, R. and Tchobanoglous, G. (1998) *Small and Decentralized Wastewater Management Systems*, McGraw Hill, New York.

Douglas, B. (1998) The decentralised approach: an innovative solution to community wastewater management. *WQI* January/February, 29–31.

Dunbar, (1908) *Principles of Sewage Treatment*, Charles Griffen, London.

El-Gohary, F., Abdel Wahaab, R., El-Hawary, S., Shehata, S., Badr, S. and Shalaby, S. (1993) Assessment of the performance of oxidation pond system for wastewater reuse. *Wat. Sci. Tech.* **27**(9), 155–163.

El-Gohary, F.A., Abou-El-Ela, A. El-Hawary, S.A., Shehata, H.M., El-Kamah, H.M. and Ibrahim, H. (1998a) Evaluation of wastewater treatment technologies for rural Egypt. *Intern J. Environmental Studies* **54,** 35–55.

El-Gohary, F.A., Nasr, F.A. and Wahaab, R.A. (1998b) Integrated low-cost wastewater treatment for reuse in irrigation. *Advanced Wastewater Treatment Recycling and Reuse, 2nd International Conference. Resources and Environment: Priorities and challenges,* Milan, 14–16 September.

El-Gohary, F.A. and Nasr, F.A. (1999) Cost-effective pre-treatment of wastewater. *Wat. Sci. Tech.* **30**(5) 97–103.

Hammer, D.A. (1989) Constructed wetlands for wastewater treatment: Municipal, industrial and agricultural, Lewis Publishers, Boca Raton. Florida.

Kuttab, A.S. (1993) Wastewater treatment reuse in rural areas. *Wat. Sci. Tech.* **27**(9), 125–130.

Lettinga, G., Velsen, van L., Zeeuw, de W. and Hobma, S.W. (1979) The application of anaerobic digestion to industrial pollution treatment. *Proc. 1st Int. Symp. on Anaerobic Digestion*, Cardiff, 167–186.

Lettinga, G., van Velsen, A.F.M., Hobma, S.W., de Zeeuw, W.J. and Klapwijk, A. (1980) Use of the upflow sludge blanket (USB) reactor concept for biological wastewater treatment. *Biotechnology and Bioengineering* **22**, 699–734.

Mara, D.D. and Silva, S.A. (1986) Removal of intestinal nematode eggs in tropical waste stabilization ponds. *Journal of Tropical Medicine and Hygiene* **89**(2), 71–74.

Marais, G.V.R. (1970) Dynamic behavior of oxidation ponds. *Proc. 2nd International Symposium on Waste Treatment Lagoons*, FWQA, Kansas City.

McGauhey, P.H. (1968) *Engineering Management of Water Quality*, McGraw Hill, New York.

Oragui, J.I. (1987) The removal of excreted bacteria and viruses in deep waste stabilization ponds in northeast Brazil. *Wat. Sci. Tech.* **19**(12), 569–573.

Reed, S.C, Crites, R.W. and Middlebrooks, E.J. (1995) *Natural Systems for Waste Management and Treatment*, 2nd edn, McGraw Hill, New York.

Tchobanoglous, G. (1996) Appropriate technologies for wastewater treatment and reuse (Australian Water & Wastewater Association), *Water Journal* **23**(4).

US Environmental Protection Agency (1988) Design manual-constructed wetlands and aquatic plant systems for municipal wastewater treatment. EPA/625/1-88/022, Center for Environmental Research Information, Cincinnati, Ohio.

US Environmental Protection Agency (1992) Wastewater treatment/disposal for small communities. EPA/625/R-92/005.

Van Haandel, A.C. and Lettinga, G. (1994) Anaerobic Sewage Treatment. A practical guide for regions with a hot climate. John Wiley & Sons, Chichester.

WPCF (1990) Natural systems for wastewater treatment. Manual of practice No. FD 16, Water Pollut. Cont. Fed., Alexandria, Virginia.

Wright, A.M. (1997) Toward a strategic sanitation approach: Improving the sustainability of urban sanitation in developing countries. UNDP-World Bank.

Young, J.C. and McCarty, P.C. (1969) The anaerobic filter for waste treatment. J.WPCF, **41**(3), 166–171.

9

Design of highly efficient source control sanitation and practical experiences

Ralf Otterpohl

9.1 WELCOME TO THE FUTURE – ZERO EMISSIONS IN MUNICIPAL WASTEWATER MANAGEMENT

If natural processes generated unusable waste, higher forms of life would no longer be possible. We can contribute to the ongoing change from current technology with its excessive waste generation to future no-waste technology. Renewable resources are renewed by the sun and provided by fertile soil and the surface waters (besides direct energy use). Ecological wastewater management will play a key role in the quest for the efficient use and reuse of water, long-

Edited by P. Lens, G. Zeeman and G. Lettinga. ISBN: 1 900222 47 7

term soil fertility and protection of the natural waters. 'Zero emissions' technology aims at 100% reuse of all material. This concept was developed at the UN University in Tokyo in Japan for industrial production (Pauli 2000). The same principles can be applied to municipal wastewater management, ending the concept of 'wastewater'. Sanitation systems can be designed to be more efficient; old and new technology can be applied in source control systems. We can consider sanitation as a production unit that can provide high quality reuse water, safe fertilisers and soil-improving material (including processed biowaste where appropriate). This could be called 'resources management' because there will no longer be any wastewater. Today such approaches exist and can be applied. We are in a fast development phase and many pilot systems are currently being planned, built and operated, which are more economic and more ecologically friendly than 'end-of-the-pipe' systems. Welcome to the future!

9.2 WHAT IS WRONG WITH CONVENTIONAL SANITATION?

The traditional sanitation concept is 'end-of-the-pipe' technology. Acute problems (not long term ones) are solved with appropriate systems instead of being avoided. This is now the standard approach in industrial wastewater treatment and has resulted in technologies of source control with appropriate reuse technology. In the field of municipal wastewater treatment this discussion has just begun (Henze 1997). The first installations of the water- and nutrient-wasting water closet (WC) and sewerage systems were criticised, but alternative systems were not reliable enough at the time (Harremoës 1997; Lange and Otterpohl 1997). Reuse-oriented sanitation ended with the cheap availability of energy and nutrients from fossil sources.

Sanitation concepts should take responsibility for the environment as well as human health. Basic facts for sustainable systems are obvious, nevertheless pilot projects for new approaches are necessary. More extensive and thoughtful planning might end the automatic installation of the water closet-sewerage- wastewater treatment plant (WC-S-WWTP) systems without considering any alternatives.

Agenda 21 of the United Nations includes no accounts of sustainable sanitation concepts (Agenda 21 1992) although water and fertile land are core subjects for the survival of future generations. When planning sanitation, consideration should be given to the worldwide consequences of the implementation of the conventional system. Many experts on sanitation agree on the possibility of resulting disasters even over a short time-span in developing countries.

An assessment of the amazing variety of technical options and their respective economic and social implications will be necessary in order to further develop sanitation. A collection of some source control solutions is given by Henze *et al.* (1997) and Otterpohl *et al.* (1997, 1999a).

Efficient sanitation concepts will have to cooperate with agriculture in order to avoid emissions and allow for reuse of water and nutrients. Sustainable agriculture has to be water-friendly and improve or at least maintain soil quality. Industrial agriculture often results in a rapid degradation of fertile topsoil (Pimentel 1997). Organic fertilisers produced by sanitation and waste management can help to care for and improve the fertile topsoil.

If faeces are mixed with wastewater by the usage of conventional flush toilets this results in a high water demand, the spread of potentially dangerous pathogens and micropollutants (residues of pharmaceuticals) in a large volume of water and also the loss of the option for economic reuse of greywater and to produce fertiliser. The initially small amount of faeces could be hygienised easily and cheaply. For this mixture, known as municipal wastewater, hygienisation is an expensive further treatment step.

Conventional sewerage systems have serious disadvantages: they are a very costly part of the infrastructure (if rehabilitation is carried out). Combined systems emit raw wastewater into receiving water with overflows; storage tanks are very expensive if there are few overflows. Sewerage systems often use a lot of water; even in industrialised countries drainage often amounts to the same volume as the total amount of wastewater. This water dilutes the wastewater and the resulting lower concentrations in the effluent of a treatment plant makes it appear that emissions are low, although loads may be high. In many cases sewerage systems are filtering raw wastewater into the ground, with the resulting potential for pollution.

There has also been discussion recently on hormones; residues of hormones from widely used contraceptive pills have been found in water, showing another weakness of sanitation systems and giving rise to doubt about their effects on men and male hormones. These substances easily reach the receiving waters because of their polarity (they are easily soluble) and low degradation rates in conventional treatment plants. Another potentially very important issue is the possibility of the transmission of increased resistance to antibiotics through their uncontrolled release to the environment (Daughton and Ternes 1999). Biological reactors are an excellent environment for the exchange of harmful bacteria.

9.3 REGIONAL PLANNING IN WASTEWATER MANAGEMENT

Regional planning has an important effect on the economics of the wastewater system. Sewage costs are on average 70% of the cost of sewerage plus treatment plant costs in densely populated rural and peri-urban areas in Germany. This figure can well be much higher if circumstances are less favourable. For some years, decentralised on-site treatment has been accepted as a long-term solution in many countries. However, legal requirements are low compared to those for larger WWTPs. It can easily be calculated that on-site plants can contribute far over their population proportion in pollutant loads that are emitted. On the other hand it would be relatively simple to implement new on-site sanitation systems that would fully reuse nutrients.

Proper decisions on where to connect houses to a sewerage system and where to build on-site facilities or small decentralised plants are important. Good regional planning can avoid wasting money and provide highly efficient decentralised treatment and collection systems. Cost calculation procedures should be carried out to include long-term development and balance operation- and investment costs and resulting products (the reuse of water, fertiliser, soil improver). The price of secondary products can be very important in developing and/or water scarce countries where water and industrial fertilisers are not subsidised. Source control sanitation can exceed the performance of the most advanced large plants, often at much lower costs.

One main drawbacks for decentralised technological plants are a lack of maintenance. Legal responsibility and maintenance agreements are essential, and this should be organised in a cost-efficient way. The design of decentralised systems should be in such a way that maintenance and the collection of fertiliser could be combined and take place at regular intervals, such as every 6 or 12 months. Local farmers may be appropriate partners.

9.4 BASIC CONSIDERATIONS FOR THE DESIGN OF SOURCE CONTROL SANITATION AND PROPER WATER MANAGEMENT

The design of source control sanitation aims for a high hygienic standard and full reuse of resources. This is exactly what can be achieved by clever source control. However, the design of the plant must be checked to ensure that these goals can be achieved. Local socio-economic conditions have to be taken very seriously. The background of the new system has to be explained to the users. A fundamental step is the identification of the different characteristics of household wastewater, presented in Table 9.1, which gives a typical range of values.

Table 9.1. Characteristics of the main components of household wastewater

Yearly loads (kg/(P*year)	Greywater 25,000–100,000	Urine ~500	Faeces ~50 (option: add biowaste)
N ~4–5	~3%	~87%	~10%
P ~ 0.75	~10%	~50%	~40%
K ~ 1.8	~34%	~54%	~12%
COD ~ 30	~41%	~12%	~47%
	Treatment → reuse/water cycle	Treatment → fertiliser	Biogas-plant composting → soil conditioner

Table 9.1 suggests the following conclusions:

- Most of the soluble nutrients are found in urine. If urine is separated and converted to agricultural usage, the biggest step towards nutrient reuse and highly efficient water protection will have been taken.
- The health danger of wastewater comes almost exclusively from faecal matter. Separation and low or no dilution paves the way to excellent hygienisation with the end product of an 'organic soil improver'.
- Wastewater that is not mixed with human waste (faeces and urine) is a great resource for high quality reuse of water. Bio-sandfilters and membrane technology are cost-efficient ways of producing secondary water.
- Source control should include evaluating all products that end up in the water. High quality reuse will be far easier when household chemicals are not only degradable but can be mineralised with the technology available. Pipes for drinking water should not emit pollutants (e.g. copper or zinc).
- Rainwater run-off is one of the reasons for building sewerage systems. If decentralised systems are built, rainwater run-off must be taken care of. Economic reasons will often prohibit the construction of sewers for rainwater if decentralised sanitation systems are to be installed. Local infiltration or trenches to surface waters for relatively unpolluted rainwater are often feasible and can be combined with usage. Prevention of pollution includes avoiding copper or zinc gutters or rainwater pipes, as these can cause heavy metal pollution.

At the Global Water Forum in The Hague in October 2000 there were disputes about water scarcity. One delegate from the Centre for Science and

Environment, Delhi, India stated, 'There is no water scarcity, only mismanagement.' The delegate had strong evidence from the incredible success of decentralised rainwater harvesting on a local scale in India. During a devastating drought in Gujarat in 1999 there were many villages that had enough water. These villages had introduced several measures to conserve rainwater, directing it to aquifers with small (only a few metres high) check-dams, and also to wells and cisterns (Manish Tiwari 2000). The introduction of conventional sanitation can well be mismanagement in such a situation, except where the reuse of the mixed wastewater in a combination of irrigation and fertilisation can be carried out year-round. Source control sanitation and greywater reuse can bring the demand for new water down 10% of what is considered efficient today.

9.5 NEW DEVELOPMENT 1: SEPARATION TOILETS AND GRAVITY FLOW

This concept is suitable for single houses and rural settlements. It is based on no-mix toilets (often called separating toilets or sorting toilets). The aim of this is to provide a low-cost and low maintenance system with the potential of full resource recovery. The system collects yellowwater (urine) through a separate pipe to save it in a storage tank until it can be used for agricultural purposes. The storage period should be at least six months.

Brownwater (faeces) is flushed with an appropriate amount of water (normally four to six litres) and is either collected separately or together with greywater and discharged into one chamber of a two-chamber composting tank (with filter-floor or filter bag) (see Figure 9.1) where the solids are pre-composted. After a year of dewatering and composting, the flow is directed to the second chamber while the first one is not added to for one year. This allows further dewatering and composting and makes removal of the faeces from the tank safer.

The products are removed from the composting tank and either used as soil improvers in brown land or composted fully – the compost could be mixed with kitchen or garden waste to decompose completely. Ripe compost is used for soil conditioning and retains or improves soil fertility. The filtrate from the composting tank will be low in nutrients due to the previous separation of urine – dissolved nutrients are mainly found in urine. Therefore, the filtrate can be treated together with the greywater (unless high quality reuse is planned).

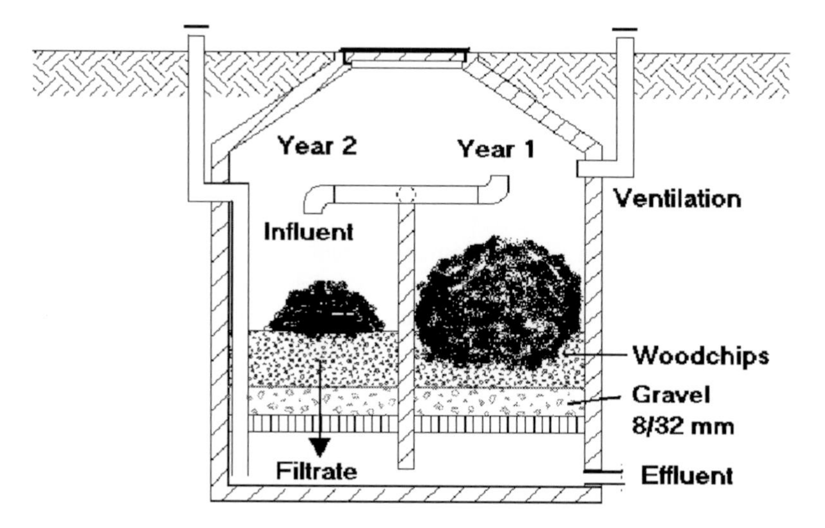

Figure 9.1. A two-chamber composting tank.

Greywater is pre-treated either in the composting tank with the brownwater (avoiding the third pipe, from the house to the tank) or treated separately for quality reuse. The next step can either be the use of a bio-sandfilter (with a vertical intermittent flow) or a combined activated sludge reactor with micro- or nano-filtration. These two technologies form an efficient barrier against pathogens and can achieve high quality effluents with little maintenance. The purified water is discharged to a local receiving tank, infiltrated into the ground or collected for reuse. The constructed wetland requires very little energy but requires about 1 to 2 m² per inhabitant. This concept are shown in Figure 9.2.

The design parameters for the elements of the components of this system can be derived from advanced decentralised technology. Greywater by itself will typically have around half the COD load at two-thirds of the flow. Filtrate from composting chambers will probably not have a big influence except for potential additional pathogen loads. Collection and storage of urine can be carried out in a straightforward way; urine contributes a maximum of 1.5 litres per person per day. Waterless collection is the ultimate goal, although this is not yet fully developed. Flush water must be a small flow, otherwise storage, transport and usage become more difficult. Waterless collection seems to avoid the problems of scaling (where limescale appears on the surface of pipes). The calcium from water contribute to the formation of minerals. Storage tanks must be resistant to chemicals, pipes and tanks must be watertight – small but steady infiltration rates can result in high dilution and more frequent transport requirements. Further experience can be drawn from the pilot projects that are currently being undertaken.

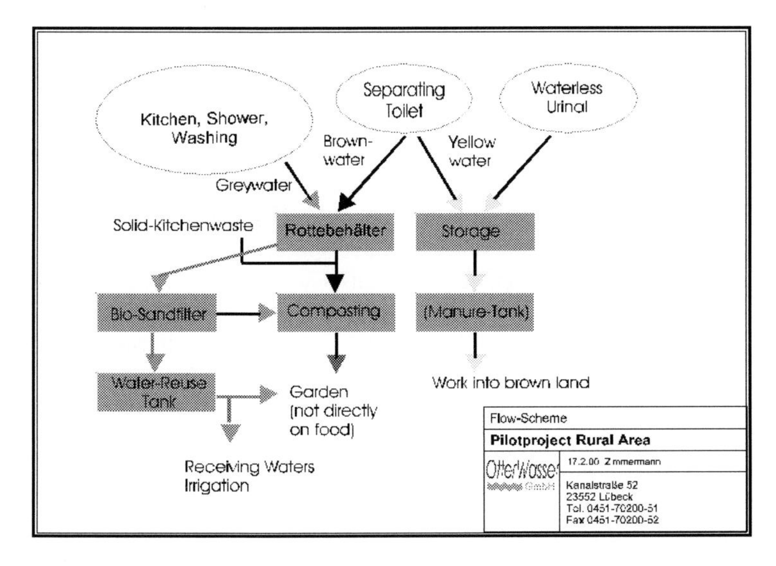

Figure 9.2. Elements of a rural source control sanitation system with a composting chamber (Rottebehälter).

The concept presented here, depending on the boundary conditions, can also be built differently taking into consideration the requirements of regional planning. Using least cost planning, a cost-effective solution for a whole area may be found as well as a gradual introduction. In all cases, however, the ideas behind the concept must be well explained so that inhabitants are motivated to cooperate.

Practical experiments with urine sorting toilets have been carried out mainly in Sweden, which has over 3,000 such toilets installed, thus clearly demonstrating the feasibility of this technology. One drawbacks that has been observed results from using too-small diameters of pipes for urine, that eventually clog from scaling. The final step to waterless collection has not yet been made in Sweden. A German company is currently working on designing a toilet with waterless urine collection. But even with waterless separating toilets, there is one major problem. Men, especially older men, are often reluctant to sit down to urinate. Younger men seem to accept sitting more easily and better understand the positive effect they could have on the local environment by adapting this system. Another solution is development of a waterless urinal. The development of these has had serious problems with the wrong type of cleaning chemicals and faults in construction. New models are available made from ceramics; and a combination with a hydrophobic nano-coating is technically feasible. This type of surface will also be a major step forward for sorting toilets. Another problem with sorting toilets is the disposal of paper that is used after urinating by most women and some men. One solution is a paper bin for this paper, another is to dispose of it into the faecal bowl. If not flushed, there would be no additional water consumption. New solutions to this problem are being sought.

Many composting chambers are being successfully operated in Austria and Germany. Wetlands, constructed with vertical percolation and step feed, are becoming the standard solution, and they have space requirements of less than 3 m^2/person (PE). These space requirements can be smaller for greywater. Small activated sludge plants using membranes for phase separation are becoming increasingly popular and would achieve an even better performance with greywater.

The Technical University of Hamburg (TUHH) and a German company, Otterwasser GmbH in Lübeck, have developed the system described above for on-site treatment at a historic water mill near Burscheid, Cologne, Germany. The system is currently under construction by Lambertsmühle (a private initiative for the restoration of the water mill). This mill is being converted into a museum.

9.6 NEW DEVELOPMENT 2: VACUUM TOILETS AND VACUUM TRANSPORT TO A BIOGAS PLANT

An integrated sanitation concept with vacuum toilets, vacuum sewers and a biogas plant for blackwater is currently being implemented for the new settlement named Flintenbreite within the city of Lübeck in Germany (NN 2000). This area, totalling 3.5 hectares, is not connected to the central sewerage system. The settlement will finally be inhabited by about 350 people and is meant as a pilot project to demonstrate the concept in practice. All components being used in the project have already been in use in different fields of application for many years and are therefore well developed. Vacuum toilets are used in ships, aeroplanes and trains, and some have already been used in apartment buildings for saving water. Conventional vacuum sewerage systems serve hundreds of communities. Anaerobic treatment is in use in agriculture, in industrial wastewater treatment, in biowaste treatment, on many farms and for faeces in tens of thousands of applications in south-east Asia and elsewhere. The system that is being built in Lübeck consists of the following components (Figure 9.3):

- vacuum closets (VC) with collection and anaerobic treatment (along with the co-treatment of organic household waste) in semi-centralised biogas plants, recycling digested anaerobic sludge for use in agriculture with further storage for growth periods. Use of biogas in a heat (for houses) and power generator in addition to natural gas;
- decentralised treatment of grey wastewater in vertical constructed wetlands with interval feeding (very energy efficient); and
- rainwater retention and infiltration in a swale (shallow trench) system.

Heat for the settlement is produced by a combined heat and power generating engine which can be switched over to use biogas when the storage tank is full. This heat is also used to heat the biogas plant. In addition, there is a passive solar system to support heating of the houses and an active solar system for warm water production. Figure 9.3 is not meant to show all the details but gives an idea of the concept, with the collection and treatment of faeces.

At the digestor a vacuum pumping station will be installed. The pumps have an extra unit to be used in the case of failure. The vacuum operates both the vacuum toilets and the vacuum pipes. Pipes have a dimension of 50 millimetres (mm) to allow good transport. They have to lie deep enough to be protected against frost and must be installed with a gradient of around 20 centimetres (cm) every 15 metres to create plugs of the transported matter. Noise is a concern with vacuum toilets but modern units are no louder than flushing toilets, so people will get used to them.

Faeces mixed with shredded biowaste are hygienised by heating the feed to 55°C for 10 hours. The energy is further used by the digestor that is operated at around 37°C with a capacity of 50 m^3. Another concern is the amount of sulphur in the biogas. This can be minimised by controlling the input of oxygen to the digestor or into the gas flow. The biogas plant is meant to be a production unit for liquid fertiliser as well. It is important to consider the pathways pollutants will take, from the beginning. One important source of heavy metals is copper or zinc-plated pipes.

These materials should be avoided and polyethylene pipes used instead. The sludge is not dewatered, to avoid nutrient losses The relatively small amount of water added to the blackwater keeps its volume small enough for transportation. There is a two-week storage period for the collection of the digestor effluent. Biogas will be stored in the same tank within a balloon that gives more flexibility in operation. The fertiliser will be pumped off by a lorry and transported to a farm with a seasonal storage tank, where it will remain for eight months. These tanks are often available or can be built at a low cost. Figure 9.4 shows the building in Flintenbreite where the vacuum pumping station, digestor, heat and power generator and other devices are installed. As well as this equipment, there is a convention room, an office and four flats.

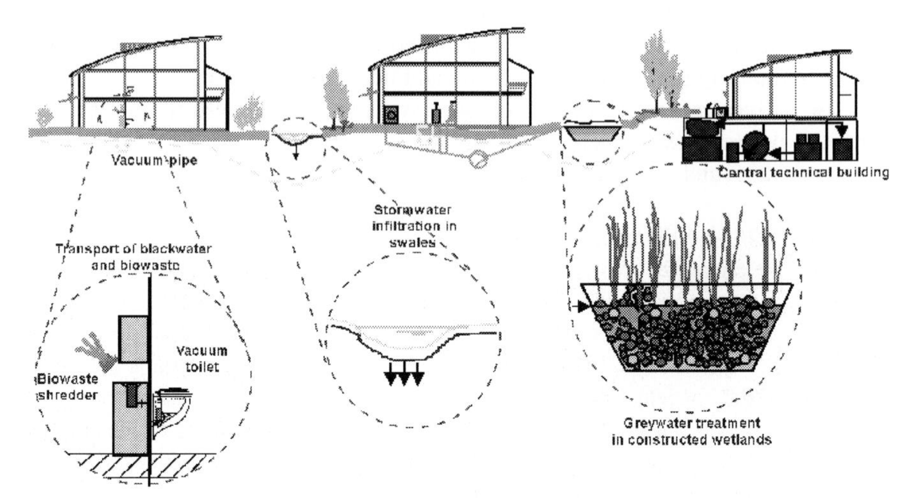

Figure 9.3. A vacuum biogas system, greywater bio-filter and rainwater infiltration.

Decentralised treatment of grey wastewater should be carried out by biofilm processes. Appropriate technologies would be membrane bioreactors or constructed wetlands. These systems both form a barrier against pathogens. Water can be reused in watering gardens or with infiltration to the rainwater system. Greywater is relatively easy to treat because it has a low nutrient content. Several projects carried out on a technical scale have demonstrated the feasibility and good to excellent performance of decentralised greywater treatment (NN 1999). These plants allow reuse of water in toilet flushing, which is not economically feasible in the Lübeck project because of the low water consumption of the vacuum toilets. For Flintenbreite, vertically fed constructed wetlands of 2 m^2 per inhabitant have been constructed. These are relatively cheap to construct and also to operate. There is a primary clarifier as a grit chamber, for solids and for grease control. First measurements of the effluent have shown very low nitrogen concentrations.

The infrastructure for Flintenbreite, including the integrated sanitation concept, has been financed by a German bank and is operated by the private company Infranova. Participating companies, planners and the house- and flat-owners are financially integrated and will have the right to vote on decisions concerning the development. Some of the investment is covered by a connection fee, just as in the traditional systems. Money saved by not having to construct a flushing sewerage system, by a lower freshwater consumption and by the coordinated construction of all pipes and lines (vacuum sewers, local heat and power distribution, water supply, telephone lines) is essential to the economic feasibility of this concept. The fees charged for wastewater and biowaste cover the operation, interest rates on additional investment and rehabilitation of the system. Part of the operating costs must be paid to a part-time operator: this also provides local employment. The company maintains the operation of all the technical structures, including heat and power generation and distribution, active solar systems and an advanced communication system.

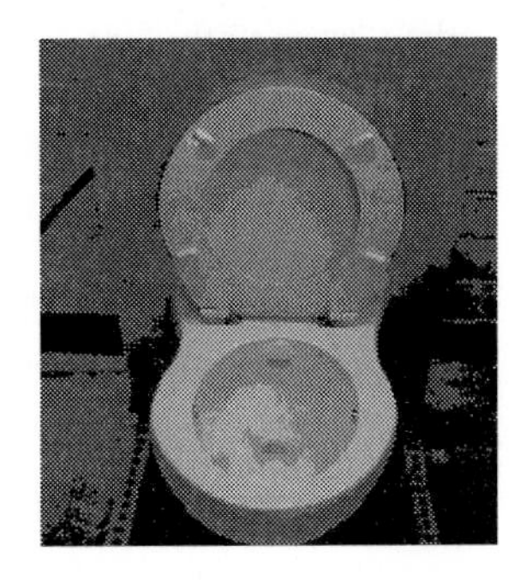

Figure 9.4 (left) The community building in the Flintenbreite settlement; (right) A vacuum toilet that flushes using one litre of water.

A study has been carried out at the Wuppertal Institute in Germany (Reckerzügl and Bringezu 1998) using Material Intensity per Service Unit (MIPS) to compare the material and energy intensity of the structure to a traditional system. Material and energy intensity is half in the decentralised system the amount in a conventional centralised system serving a medium densely populated area (see Table 9.2). In the centralised system, most of the material intensity results from the construction of the sewerage system. The predicted effluent values are based on averages of measurements of greywater. Effluent qualities are shown in comparison to average values of a modern treatment plant with advanced nutrient removal.

Table 9.2 indicates some major advantages for the new system which justify further research. The cumulated reduction of emissions to the sea and savings in energy- and material usage over an average lifetime of 70 years for 350 people would be about 250,000 m^3 of freshwater, 70,000 kilograms (kg) of COD, 1,500 kg of P, 13,000 kg of N, 30,000 kg of K, 5,250,000 kilowatts hours (kWh) of energy and about 56,000 tons of MIPS material usage. The fertiliser produced from this system can also replace fertiliser produced from fossil resources. This can be calculated as another 2,450,000 kWh of energy saved (Boisen 1996). These numbers are important for translation to a large world population and decreasing fossil resources.

The interest in the integrated concept described above has dramatically increased (Otterpohl and Naumann 1993) since the construction of the project in Lübeck. There are other projects where this type of concept will be built. The system in general may well be cheaper than the traditional system. This depends on the local possibilities to infiltrate rainwater, as well as on the size of the area that is served and on the number of inhabitants. An optimum size may be an urban area with 500 to 2000 inhabitants. Smaller units are feasible if the blackwater and biowaste mixture is collected and transported to a larger biogas plant, preferably situated on a farm. Greywater can be treated in an existing wastewater treatment plant if the sewerage system is near by. In some cases this is the most economical way. Nutrient removal can be improved if a certain percentage of the population is served by a separate blackwater treatment.

Source control systems can be considered high efficiency technology. Research on pilot projects will bring a faster pace of development and bring new technologies to all the different social and geographical situations on our crowded planet.

9.7 NEW DEVELOPMENT 3: LOW COST, LOW MAINTENANCE ON-SITE SYSTEMS

There are many ideas and traditional technologies for sustainable sanitation with real source control of human waste (Winblad 1998; Otterpohl *et al.* 1999a). Some are more suitable for rural areas, but there are options for downtown metropolis areas as well. The basic techniques of low-tech collection and treatment (with or without kitchen waste) are:

- dessication (with solar heating, double vault systems); difficult for areas with wet anal cleaning (instead of paper); good with urine collection and reuse;
- composting (often difficult to operate);
- low-diluting toilets with biogas systems; and
- urine collection combined with biogas systems for faeces.

Table 9.2. Estimated emissions, energy consumption and material intensity of the proposed system compared to a traditional system

Advanced traditional sanitation (WC-S-WWTP) concept			New sanitation system		
Emissions			Emissions*		
COD	3.6	kg/(P*a)	COD	0.8	kg/(P*a)
BOD_5	0.4	kg/(P*a)	BOD_5	0.1	kg/(P*a)
Total N	0.73	kg/(P*a)	Total N	0.2	kg/(P*a)
Total P	0.07	kg/(P*a)	Total P	0.01	kg/(P*a)
Total K**	(>1.7	kg/(P*a))	Total K**	(<0.6	kg/(P*a))
Energy			Energy		
Water supply (wide variation)	–25	kWh/(P*a)	water supply (20% water savings)	–20	kWh(P*a)
Wastewater treatment (typical demand)	–85	kWh/(P*a)	Vacuum system	–25	kWh/(P*a)
Consumption			Greywater treatment	–2	kWh/(P*a)
			Transport of sludge (2/month, 50 return)	–20	kWh/(P*a)
Consumption	–110	kWh/P*a)	Consumption	–67	kWh/(P*a)
			Biogas	110	kWh/(P*a)
			Substition of fertilizer	60	kWh/(P*a)
			Win	170	kWh/(P*a)
Total	–110	kWh/(P*a)	Total	103	kWh/(P*a)
Material intensity***	3.6	t/(P*a)	Material intensity***	1.3	t/(P*a)

* measurements of greywater (NN 1999)
** assumption, no data
*** MIPS study (Reckerzügl and Bringezu 1998)

The main problem is to design a toilet system that is comfortable, low-diluting and capable of carrying out the transportation. A promising technique is the NoMix toilet, developed in Sweden. Since most visits to the toilet are to urinate, these systems collect urine using very little water. This allows for simple urine collection or treatment (e.g. drying over a loam wall in hot climates; solar systems to be developed). Urine can be used as a fertiliser directly on brown land or after dilution (with 5 to 10 parts of water) on plants, but not directly on vegetables. Urine should be stored for about six months. Faeces from the NoMix toilet can be sent to biogas plants together with kitchen waste. A source control sanitation system can lead to the proper reuse of fertiliser. At the same time, purified greywater can substitute for freshwater in case of water scarcity. This way systems can be very economic. The system discussed in Section 9.5 could be used in the upper level of low-tech systems in many countries to replace septic tanks with conventional flush toilets.

9.8 NEW DEVELOPMENT 4: UPGRADE EXISTING WASTEWATER INFRASTRUCTURE

Urine collection can convert a conventional sewerage system to one with a very high rate of nutrient reuse and very low nutrient emissions. When most of the urine is kept out of the wastewater treatment plant, nutrient removal becomes obsolete (Larsen and Udert 1999). There are two basic approaches: centralised or decentralised collection. The centralised approach would be to store urine in small tanks and to open them at night when the sewerage system is nearly empty. A remote control system would empty the tanks to create a concentrated flow that can be caught at the treatment plant (Larsen and Gujer 1996). This method is limited to sewerage systems with a good gradient and appropriate retention times; however, it could also be applied to branches of the sewerage system. Decentralised storage and collection is the other possibility.

If all blackwater is collected and treated separately, a conventional sewage system can become a greywater recycling plant and produce secondary water. The conversion could be carried out over decades if necessary. The economic feasibility of such a step has to be well thought-out, because except in very densely populated areas, the rehabilitation of sewage systems requires a high level of investment.

9.9 RISKS, OBSTACLES AND RESTRICTIONS

The first objective for sanitation must be to minimise health risks. New systems should be better than conventional sanitation systems, which are hygienic inside the house but usually not for the receiving waters.

Sanitation is a very sensitive matter with respect to the natural human wish for cleanliness and the taboos surrounding the issue. Failure of new systems can be (and has been in many cases) the consequence of this if this is not considered and included in project development. The issues around new sanitation systems are complex, but they cover an area of basic human need. Keeping food and water cycles separate, returning matter from the land to the land and zero emissions to the water system should all be explained to prospective users of new sanitation systems.

Wastewater infrastructure is usually built to be extremely long-lasting. This permanence of change seems so overpowering to many people that they cannot even imagine different sanitation solutions in the future. We have to consider the lifespan of houses, sewerage systems and treatment facilities in order to avoid financial problems in the future. Change is easier for newly built settlements. The lifespan of a house is far shorter than that of a sewage system. Components of source control sanitation could be installed in flats as they are renovated, and first be connected to the conventional systems. This can be economic with water saving from the beginning. Later, after a group of houses has been converted, separate treatment can be implemented.

9.10 WELCOME TO THE FUTURE!

It is a real challenge to participate in the development of new technology. Professional skills and an open-minded search for solutions are needed in order to find better future sanitation. Open dialogue and exchange of experiences are essential in order to bring the matter forward. There are so many possible options that all social and economic conditions can be met. Creativity is needed to find the appropriate technology and the best way of implementing, operating and financing it. There is an extremely urgent need for new solutions, whether or not this need is ignored by media, politicians and the public. Even though, in many industrialised countries, full conversion will have to be carried out over decades, due to the long-lasting existing sewage infrastructure, these countries have the best resources for research and pilot installations.

9.11 REFERENCES

Agenda 21 (1992) The United Nations Program of Action from Rio. United Nations, New York.

Boisen, T. (1996) Personal communication, TU Denmark, Dept. of Building and Energy.

Daughton, Ch.G. and Ternes, Th.A. (1999) Pharmaceutical and personal care products in the environment: agents of subtle change? *Environmental Health Perspectives* **107**(6), 907.

Harremöes, P. (1997) Integrated water and wastewater management. *Wat. Sci. Tech.* **35**(9), 11–20.

Henze, M. (1997) Waste design for households with respect to water, organics and nutrients. *Wat. Sci. Tech.* **35**(9), 113–120.

Henze, M., Somolyódy, L., Schilling, W. and Tyson, J. (1997) Sustainable sanitation. Selected Papers on the Concept of Sustainability in Sanitation and Wastewater Management. *Wat. Sci. Tech.* **35**(9), 24.

Lange, J. and Otterpohl, R. (1997) Abwasser. Handbuch zu einer zukunftsfähigen Wasserwirtschaft, second edition. Mallbeton Verlag, Pfohren, Germany. (In German.)

Larsen, T. A. and Gujer, W. (1996) Separate management of anthropogenic nutrient solutions (human urine). *Wat. Sci. Tech.* **34**(3–4), 87–94.

Larsen, T. A. and Udert, K.M. (1999) Urinseparierung – ein Konzept zur Schließung der Nährstoffkreisläufe. *Wasser & Boden.* (In German.)

Manish Tiwari, D. (2000) Rainwater harvesting – Standing the test of drought. *Down to Earth* **8**(16), 15 January.

NN (2000) www.flintenbreite.de

Otterpohl, R. and Naumann, J. (1993) Kritische Betrachtung der Wassersituation in Deutschland, in: *Umweltschutz, Wie?,* (ed. K. Gutke), Symposium 'Wieviel Umweltschutz braucht das Trinkwasser?'. Köln, Germany: Kirsten Gutke Verlag, S. 217–233.

Otterpohl, R., Grottker, M. and Lange, J. (1997) Sustainable water and waste management in urban areas. *Wat. Sci. Tech.* **35**(9), 121–133 (Part 1).

Otterpohl, R., Albold, A. and Oldenburg, M. (1999a) Source control in urban sanitation and waste management: Ten options with resource management for different social and geographical conditions. *Wat. Sci. Tech.* **3/4**(2), 153.

Otterpohl, R., Oldenburg, M. and Zimmermann, J. (1999b) Integrierte Konzepte für die Abwasserentsorgung ländlicher Siedlungen. *Wasser & Boden* 51/11, 10.

Pauli, G. (2000) *The Road to Zero Emissions – More Jobs, More Income and No Pollution*, Greenleaf Publishing, Sheffield, UK.

Pimentel, D. (1997) Soil erosion and agricultural productivity: the global population/ food problem. *Gaia* **6**(3), 128.

Reckerzügl, M. and Bringezu, St. (1998) Vergleichende Materialintensitäts-Analyse verschiedener Abwasserbehandlungssysteme, gwf-Wasser/Abwasser, Heft 11/1998.

Winblad, U. (ed.) (1998) *Ecological Sanitation*. Stockholm: SIDA.

10

Potentials of anaerobic treatment of domestic sewage under temperate climate conditions

Youssouf Kalogo and Willy Verstraete

10.1 INTRODUCTION

Since 1881, anaerobic digestion has gradually evolved from taking place in an airtight cesspool and a septic tank to a completely mixed digester and finally to a high rate flow-through waste digester. Interest in anaerobic wastewater treatment was greatly enhanced because of the necessity to decrease the costs of organic waste treatment, together with the rising price of fossil fuels. During its earlier development, the main drawback of anaerobic treatment was the long hydraulic retention time (HRT) needed to achieve an acceptable waste removal efficiency. The HRT was considerably shortened by designing new reactor systems in which a high concentration of anaerobic biomass was

Edited by P. Lens, G. Zeeman and G. Lettinga. ISBN: 1 900222 47 7

retained. In practice, anaerobic digestion has been considered to be feasible only for temperature conditions above 25°C. The anaerobic treatment of domestic sewage at low temperatures still faces major scepticism. Indeed, in the temperate regions of the world, sewage temperatures range from 4–20°C, depending on place and time of sampling, and generally exceed 12°C for only about six months per year (Derycke and Verstraete 1986). Under such moderate climate conditions, wastewater temperatures are therefore considerably lower than the optima for methanogenesis, which are between 35 and 55°C. Currently, treatment of sewage in temperate regions relies primarily on aerobic treatment processes. Aerobic treatment, such as activated sludge or trickling filters, has the disadvantage of high cost, for example, of the order of 50–100 euros per inhabitant equivalent per year (euro $IE^{-1}yr^{-1}$). Treatment costs are halved when anaerobic treatment (that is, upflow anaerobic sludge bed (UASB) reactor technology) is used (Lens and Verstraete 1992). It is therefore important to reconsider the anaerobic treatment of sewage under moderate climate conditions as a prominent alternative technology.

10.2 FUNDAMENTAL ASPECTS OF ANAEROBIC TREATMENT AT LOWER TEMPERATURES

The performance of an anaerobic reactor treating domestic sewage depends strongly on the environmental conditions and the characteristics of the wastewater itself. Anaerobic treatment has also a limited effect on certain pollutants present in domestic sewage such as nutrients (phosphorus and nitrogen) and faecal bacteria. The factors that might affect the process are summarised in Table 10.1 and discussed in the following text.

Table 10.1. Factors that might affect the anaerobic treatment of domestic sewage

FACTOR	EFFECT
Flow and strength variation	Poor effluent quality
Temperature	Slow growth of bacteria
	Low methanogenic activity
	Slow hydrolysis
	Increased gas solubility
	Inhibition by high acetate concentration
Sulphate (SO_4^{2-})	Inhibition of methanogenesis process
	Lower methane production
Suspended solids (SS)	Slow hydrolysis and mass transfer kinetics
	Reduction of specific methanogenic activity
	Disintegration of granules

10.2.1 Flow and strength variation

Domestic sewage is characterised by strong fluctuations in organic matter and flow rate (Figure 10.1). The concentration of organic matter can vary by a factor of 2–10 in a few hours. The flow rate can fluctuate by a factor of 4. This depends mainly on the size of population (the larger the population, the smaller the fluctuation) and the type of sewer. Combined sewers have much higher fluctuations, due to rain and run-off water. Anaerobic bacteria are 'touchy'. They appear to be 'conservatives' and not to have the capacity to rapidly adapt to changing environmental conditions. Therefore, facing strong fluctuation in organic matter and flow rate, these bacteria tend not to work at their real potentials. This generally leads to poor effluent quality. Yet, in practice, operating with a buffer tank can largely quench flow and strength changes.

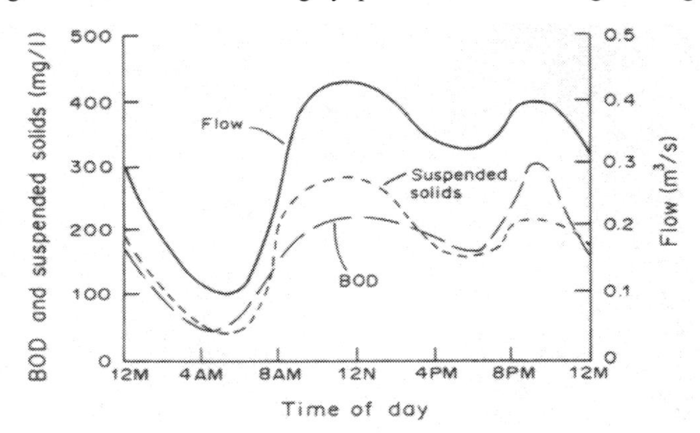

Figure 10.1. Fluctuations in organic matter and flow rate of domestic sewage (after Metcalf and Eddy 1984).

10.2.2 The influence of temperature on the anaerobic digestion process

Anaerobic digestion is considered to be possible under three ranges of temperature labelled psychrophilic, mesophilic and thermophilic. These temperature ranges, roughly represented in Figure 10.2, have no real boundaries. The optimal temperature of psychrophilic microorganisms is around 17°C (Edeline 1997). Low temperature conditions generally lead to a decrease in the maximum specific growth rate. Due to the high doubling time of psychrophilic microorganisms of approximately 35 days, which is 3.5 and 9 times higher than that of mesophilic and thermophilic microorganisms respectively, reactor conversion rates proceed much more slowly at low temperature (Edeline 1997). Except for a few reactions (hydrogenotrophic sulphate reduction (9), hydrogenotrophic methane production (10) and acetate formation from hydrogen and bicarbonate (11)) most of the reactions yield less energy at low temperatures than high temperatures (Table 10.2). However, these changes in free energy are not substantial and do not exclude proper process occurrence.

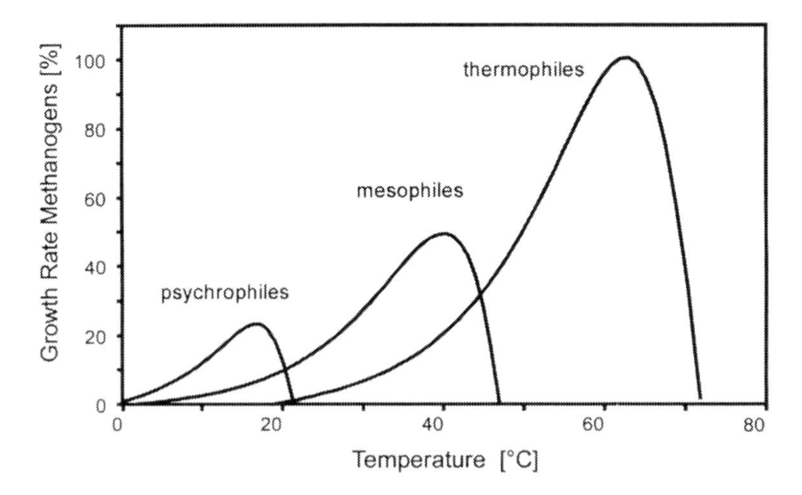

Figure 10.2. Relative growth rate of psychrophilic, mesophilic and thermophilic methanogens (after Wiegel 1990).

Similarly to the specific growth rate, the activity of methanogens is strongly influenced by temperature. This is the case for *Methanothrix* and *Methanosarcina,* two important methanogens in the anaerobic digestion process. These bacteria are mainly responsible for methane production from acetate. The optimal temperature of both bacteria is between 35–40°C (Huser *et al.* 1982; Vogels *et al.* 1988). Figure 10.3 shows that the methanogenic activity of *Methanothrix* (now renamed as *Methanosaeta*) *soengenii* can only be expected if the temperature is a minimum of 10–15°C. Activity at those temperatures will be 10 to 20 times lower than the activity at 35°C. It means that the capacity of an anaerobic reactor seeded with mesophilic biomass will drop sharply during start-up under low temperature conditions. Instead of 10–20, only 0.1–0.2 kg COD will be removed per m^3 reactor per day. Practically, these considerations mean that, to keep the system sufficiently effective, one has to increase the biomass and its retention in the low temperature reactor in order to provide sufficient biocatalyst.

Table 10.2. Stoichiometry and Gibbs free-energy changes of acetatse, propionate, butyrate and hydrogen anaerobic conversion in the presence and absence of sulphate (after Rebac 1998)

	Reactions	ΔG' (kJ/reaction) (37°C)	(10°C)
1	$CH_3CH_2COO^- + 3H_2O \rightarrow CH_3COO^- + HCO_3^- + H^+ + 3\ H_2$	+71.8	+82.4
2	$CH_3CH_2COO^- + 0.75SO_4^{2-} \rightarrow CH_3COO^- + HCO_3^- + 0.75HS^- + 0.25H^+$	–39.4	–35.4
3	$CH_3CH_2COO^- + 1.75SO_4^{2-} \rightarrow 3HCO_3^- + 1.75HS^- + 0.25H^+$	–88.9	–80.7
4	$CH_3CH_2CH_2COO^- + 2H_2O \rightarrow 2CH_3COO^- + H^+ + 2H_2$	+ 44.8	+52.7
5	$CH_3CH_2CH_2COO^- + 0.5SO_4^{2-} \rightarrow 2CH_3COO^- + 0.5HS^- + 0.5H^+$	–29.3	–25.9
6	$CH_3CH_2CH_2COO^- + 2.5SO_4^{2-} \rightarrow 4HCO_3^- + 2.5HS^- + 0.5H^+$	–128.3	–116.4
7	$CH_3COO^- + SO_4^{2-} \rightarrow 2HCO_3^- + HS^-$	–49.5	–45.3
8	$CH_3COO^- + H_2O \rightarrow CH_4 + HCO_3^-$	–32.5	–29.2
9	$4H_2 + SO_4^{2-} + H^+ \rightarrow HS^- + 4H_2O$	–148.2	–157.1
10	$4H_2 + HCO_3^- + H^+ \rightarrow CH_4 + 3H_2O$	–131.3	–140.9
11	$4H_2 + 2HCO_3^- + H^+ \rightarrow CH_3COO^- + 4H_2O$	–98.7	–111.8

Anaerobic treatment at low temperature was reported to be sensitive to high acetate concentrations. The latter inhibited the activity of acetoclastic methanogens (Nozhevnikova *et al.* 1997). However, the relatively low levels of acetate in domestic sewage make it unlikely that inhibition by acetate will take place.

Low temperatures can have a direct impact on the physico-chemical properties occurring in the reactor. As shown in Figure 10.4, the solubility of gaseous compounds present in biogas increases with a decrease in temperature. Therefore, a considerable amount of CH_4, CO_2, H_2 and H_2S will remain in the effluent of reactors operating at low temperatures. As a consequence several problems may occur when operating the reactor. For example in a sludge bed reactor with low biogas production rate, the mixing intensity will decrease. This will result in a poor substrate-biomass contact, but higher upstream velocity can be imposed to compensate this. The pH of the reactor might also decrease with the increase of CO_2 dissolution in water; yet sewage generally has sufficient buffer capacity. Finally, discharged reactor effluent may release considerable amounts of dissolved odorous gases. In that respect, an adequate post-treatment must be conceived.

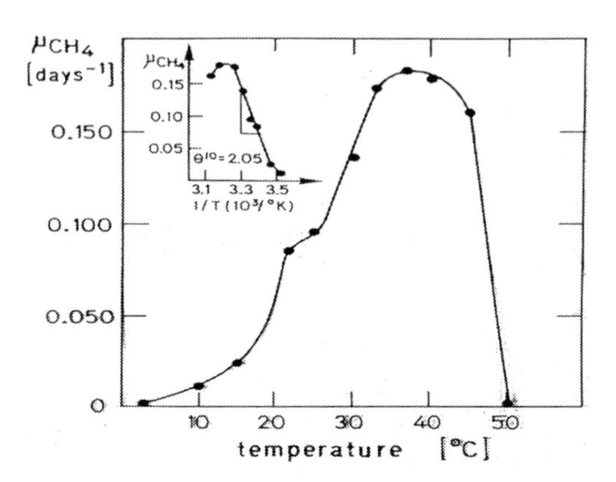

Figure 10.3. Temperature-dependence of methane production from acetate by *Methanothrix soengenii* (after Huser *et al.* 1982).

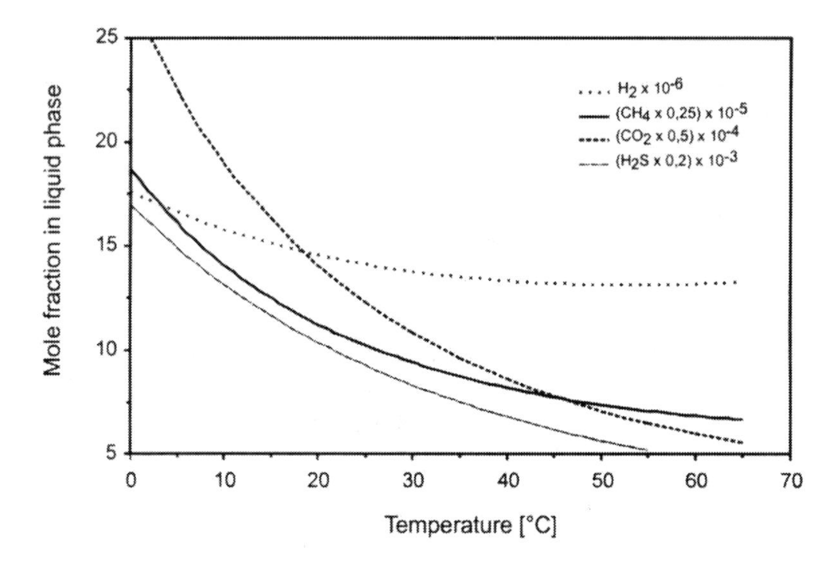

Figure 10.4. Gas solubility in pure water at various temperatures (after Lide 1992).

10.2.3 The influence of sulphate on the anaerobic digestion process

Sulphate-reducing bacteria (SRB) are able to oxidise a part of the COD (mostly via the H_2-intermediary state) present in the wastewater by utilising SO_4^{2-} as an electron acceptor (Widdel 1988). As shown in Table 10.2 and Equation 9, hydrogenosulphate reduction can be favoured at lower temperatures.

SRB can also compete with acetoclastic methanogens for acetate (Visser *et al.* 1992). Yet this conversion is thermodynamically not facilitated at lower temperatures.

At high concentrations of SO_4^{2-}, little CH_4 results since a greater electron flow is directed towards SO_4^{2-} reduction (Harada *et al.* 1994). SRB grow over a wide pH range (5–9), which includes the optimum range (7.0–7.5) for the methanogens. This means that SO_4^{2-} reduction cannot be kept under control in a methanogenic reactor unless SO_4^{2-} is limited. Fortunately, SO_4^{2-} levels are generally low in domestic sewage, for example, 50–200 mg L^{-1} (Yoda *et al.* 1987).

10.2.4 Anaerobic treatment of sewage in relation to its characteristics

As indicated above, the fact that direct anaerobic treatment of domestic sewage is generally not applied is not primarily related to aspects of flow and strength, temperature, free energy change, gas exchange or sulphate concentration. The most important difficulty is the high concentration of SS in sewage. In general, the SS is of the order of 0.3–0.6 g L^{-1} (but can reach values up to 2 g L^{-1}) in municipal sewage so that the soluble COD to volatile suspended solids ratio (CODs/VSS) typically lies around 1 (Mergaert *et al.* 1992). Under low temperature conditions, the SS are hydrolysed very slowly. They tend therefore to accumulate in the reactor, thereby decreasing the reactor volume available for the active biomass sludge and consequently giving rise to low COD conversion efficiencies. A ratio CODs/VSS above 10 is necessary to keep the anaerobic sludge sufficiently active (De Baere and Verstraete 1982). Mathematical modelling has shown that, otherwise, the reactor volume tends to fill up with inactive SS rather than with active microbial biocatalyst (Rozzi and Verstraete 1981). Table 10.3 gives an example of sludge accumulation in a reactor operating on mixtures with different soluble/particulate organics ratios. There is a significant decrease in the proportion of active biomass with the decrease of the ratio of soluble over suspended organics.

Table 10.3. Sludge accumulation in a reactor operating on a mixture of soluble organics (S_S) and suspended (S_P) organics (after Mergaert and Verstraete 1987): HRT = 1 d, SRT = 50 d, biodegradability of particulate organics = 0.4, no wash-out of SS

Influent S_S (kg m^{-3})	Influent S_P (kg m^{-3})	Ratio S_S/S_P	Active biomass (kg m^{-3})	Non-viable SS (kg $m^{-3}r^{-1}$)	% Active on total SS
10	1	10/1	25	45	55
5	1	5/1	12	40	28
1	1	1/1	2.5	35	6

The sewer system itself can be considered as a 'bio-reactor' in which a variety of physical, chemical and microbiological processes take place. For example, most of the rapidly acidified COD (RACOD) present in the sewage is consumed during those processes (Verstraete and Vandevivere 1999). The wastewater arriving at the treatment plant has insufficient RACOD, e.g. only 10–40 mg RACOD out of a total of 210–740 mg COD L^{-1} (Henze *et al.* 2000). As a consequence, the development of acidogenic bacteria, which are intrinsically involved in hydrolysis of SS and play an important role in granular sludge formation (Figure 10.5), is hampered in the anaerobic reactor.

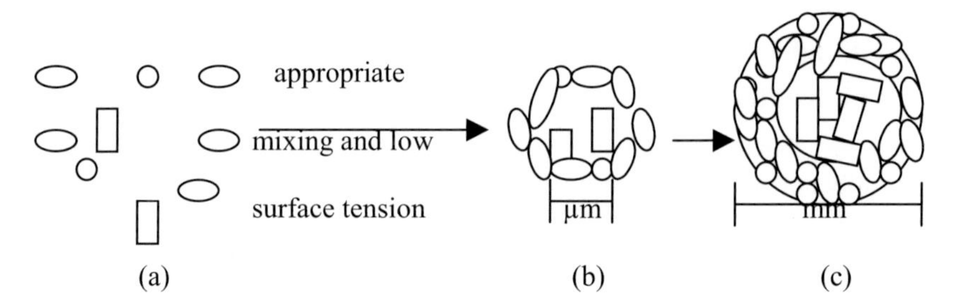

Figure 10.5. Scheme of granule formation according to Thaveesri *et al.* (1995). (a) Free cells of acidifying bacteria (round cells) and methanogenic bacteria (rectangular cells). (b) Acidogens aggregate by means forming exocellular polymers (ECP); they enclose some methanogens. Free cells are washed out. (c) Granule with outer elastic hydrophilic layer formed by ECP-rich acidogens and inner core of hydrophobic methanogens.

Figure 10.6 shows a typical granule grown in a laboratory scale UASB reactor, with cocci bacteria on the surface layer and filamentous cells resembling *Methanosaeta* in the inner part.

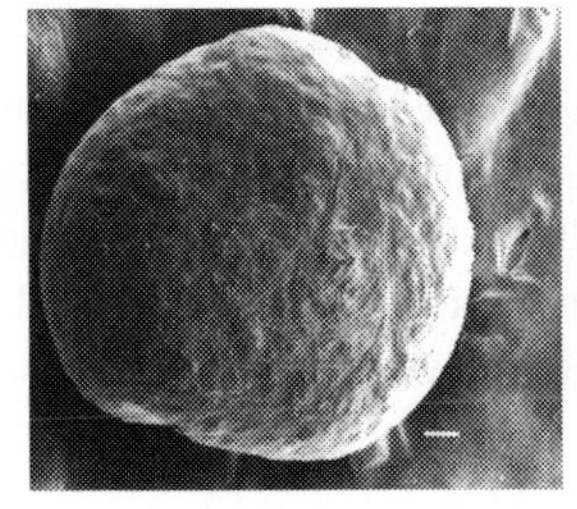

Surface layer: bar = 1 μm Entire granule: bar = 100 μm Inner part: bar = 1 μm

Figure 10.6. Overview of a typical anaerobic granule grown on sewage at 28°C and upflow velocity of 5 m h^{-1} (Photos © LabMET).

In principle, one can obtain granular sludge from an industrial site where granules are readily produced. Such a granular biocatalyst will, when subjected to treated domestic sewage, perform but not tend to grow. For raw domestic sewage, containing a large fraction of SS, granulation will not occur spontaneously in a reactor operating at a low upflow velocity e.g. ≤1 m h^{-1}. In that reactor, flocculent sludge will prevail. Only when a large part of SS has been removed, for instance by primary sedimentation, can granulation take place. Actually, one may prefer to have a reactor with flocculent sludge because the filtration of SS by a flocculent sludge bed might be better than that of a granular sludge bed.

Since hydrolysis is a very slow process at low temperatures, even at long sludge retention times (SRT), very low loading rates should be applied to guarantee methanogenic conditions. This implies that the reactor should operate at long HRTs. When the SRT is known, the corresponding HRT can be calculated from the model of Zeeman and Lettinga (1999) indicated below, provided that the sludge concentration in the reactor (X), the fraction of the influent SS that is removed (R) and the fraction of SS that is hydrolysed (H) are known.

$$\mathrm{HRT} = \left(\frac{\mathrm{C} \times \mathrm{SS}}{\mathrm{X}} \right) \times \mathrm{R} \times (1 - \mathrm{H}) \times \mathrm{SRT}$$

with C = COD concentration in the influent (g COD L^{-1}) and SRT in days.

Consider a sewage with a concentration of 1 g COD L^{-1} of which 65% is suspended and suppose a reactor with a sludge concentration of 15 g VSS L^{-1}. Assuming 175 days SRT as minimum to provide methanogenesis at 15°C, to achieve 75% removal of COD in the form of suspended solids (CODss) with 25% hydrolysis the reactor should operate at the HRT of 4.2 days. From the HRT one can easily calculate the volume (V) of the reactor as follows:

$$V = HRT \times Q$$

with Q = the daily influent flow.

The model of Zeeman and Lettinga (1999) shows also that increasing or decreasing the SS concentration of the influent will change the needed HRT. The model of Rozzi and Verstraete (1981) shows that decreasing the SS concentration (increasing the CODs/VSS ratio) keeps the biomass sufficiently active. Kalogo and Verstraete (2000) observed that increasing the CODs/VSS ratio might allow a decrease in the HRT. Obviously, the overall message is that the key feature in successfully digesting domestic sewage is to observe and if necessary to adjust the amount of soluble COD relative to the amount of particulate COD to around 10.

Due to the slow hydrolysis of SS, anaerobic treatment of domestic sewage at low temperatures has so far not been considered by sanitary engineers as justifiable. This explains why most of the full-scale applications of anaerobic sewage treatment have been related to temperatures above 20°C (Kalogo and Verstraete 1999; Segezzo *et al.* 1998).

10.3 TECHNOLOGICAL ASPECTS OF ANAEROBIC TREATMENT AT LOWER TEMPERATURES

10.3.1 Performance of reactors

Historically, anaerobic sewage treatment at low temperatures started in the Netherlands with UASB reactors. Subsequently, treatments were also examined in other reactors such as the anaerobic filter (AF), the fluidised bed (FB) reactor and the expanded granular sludge bed (EGSB) reactor. The results of these various studies are indicated in Table 10.4. In all these studies, mesophilic sludge (digested sewage sludge or granular sludge) was used as seed material. One can conclude that the results of these tests did not fully encourage the implementation of full-scale installations under temperate climate conditions. These low performances must be evaluated in a context of overall treatment conditions and discharge requirements.

Table 10.4. Performance of anaerobic reactors treating domestic sewage at low temperature.

Reactor type	Volume (L)	T (°C)	Influent concentration ($mg\ L^{-1}$)			Bv (kg COD $m^{-3}\ d^{-1}$)	HRT (h)	Removal efficiency (%)			Authors
			CODt	CODs	SS			CODt	CODs	SS	
UASB	120	7–12	200–1200	100–400	+	+	8–12	65	+	+	Lettinga *et al.* (1981)
UASB	120	12–16	688	+	+	+	24	55-75	+	55–80	Lettinga *et al.* (1983)
AF	160	13–15	467	+	+	1.8	6	35-55	+	+	Derycke and Verstraete (1986)
UASB	110	12–18	465	+	154	+	12-18	65	+	73	Monroy *et al.* (1988)
UASB	20	10–19	900	300	450	1.4–1.7	13-14	35–60	5–26	70–95	De Man *et al.* (1988)
FB	+	10	760	+	+	8.9	1.7–2.3	53–85	+	+	Sanz and Fernandez-Polanco (1990)
EGSB	205	9–11	391	291	+	4.5	2.1	20–48	40	+	Van der Last and Lettinga (1992)
UASB	3.84	13	344	124	82	+	8	59	45	79*	Elmitwalli *et al.* (1999)
UASB	3.84	13	456	112	229	+	8	65	39	88*	Elmitwalli *et al.* (1999)

CODt = total COD; CODs = soluble COD; Bv = volumetric loading rate; + = not indicated; h = hours; T = temperature; * = COD suspended solids (CODss); SS = suspended solids.

10.3.2 Recent developments and strategies

10.3.2.1 Reactor start-up and operation

One of the major current progresses in anaerobic treatment of wastewater at low temperature is the improvement of reactor start-up with mesophilic seed sludge. This was found to be feasible when a high concentration of biomass (30 kg VSS m^{-3}) is introduced in the reactor (Rebac *et al.* 1995). For comparison, this is 10 times higher than the minimum necessary to start a mesophilic reactor (Van Haandel and Lettinga 1994). With such a high concentration, mesophilic sludge exposed to long-term psychrophilic conditions (10–12°C) showed a good activity when the temperature returned to the mesophilic range (Rebac *et al.* 1999). Moreover, the low temperature has no negative effect on the microstructure of granules and seems to favour a layered structure (Gouranga *et al.* 1997). These observations mean that, in temperate regions, the temperature change from winter to summer should not undermine the stability of the reactor. These observations also indicate that seed sludge for inoculation of the reactors operating at low temperature is in principle available since many full-scale mesophilic treatment plants, for example UASB reactors, are operating worldwide. Yet, it must be kept in mind that the granular sludge has a market price of about 1.5 euros per kg VSS. One would have to pay about 45 euros m^{-3} reactor inoculated.

10.3.2.2 Process configurations and efficiencies

Because of the advantage of anaerobic digestion over aerobic treatment (reduction of sludge production, saving of electricity costs), the anaerobic treatment of sewage at low temperature has therefore been reconsidered over the last five years. There is a consensus that significant SS degradation in psychrophilic reactors can be hardly expected, because the hydrolysis rate of SS drops sharply and even may approach zero for various types of solids (Rebac *et al.* 1998; Van der Last and Lettinga 1992; De Man *et al.* 1988). Progress has therefore focused on process configurations using the two-stage reactor concept. In this system, the first reactor with a high HRT is used for the retention and hydrolysis of the SS, while the second reactor with a short HRT is used for methanogenesis. This concept is quite attractive in view of the better performance compared to the single stage presented above (Table 10.5). The major problem with the two-stage reactors is that there is a real need for regular discharge of the excess sludge from the first reactor. Therefore, a third reactor has to be coupled to the system. The latter must stabilise the waste sludge (Figure 10.7).

Two stages of completely mixed sequencing anaerobic bioreactors and clarifiers were studied by Arsov *et al.* (1999) at ambient temperature (25°C). This process was found to substantially accelerate the anaerobic process and therefore merit being investigated under much lower temperature conditions.

Table 10.5. Results of anaerobic treatment of domestic wastewater in two-stage reactors under low temperature conditions compared to one-stage reactor

	Two-stage reactor	Two-stage reactor	One-stage reactor (average)
Parameters	Sayed and Fergala (1995)	Wang *et al.* (1997)	Based on data of Table 10.2
Process configuration	UASB–UASB	HUSB–UASB	
Volume (m^3)	0.042 (0.0046)	200 (120)	
Temperature (°C)	18–20	17	13 ± 3
HRT (h)	8–4 (2)	3 (2)	10 ± 7
Influent characteristics			
CODt (mg L^{-1})	200–700	650	587 ± 299
CODs (mg L^{-1})	+	+	211 ± 126
SS (mg L^{-1})	90–385	217	229 ± 159
Bv (kg COD m^{-3} d^{-1})	1.22–2.75 (1.70–6.20)	5.3 (4.0)	3.7 ± 3.2
Removal efficiency (%)			
CODt	74–82	69	55 ± 17
CODs	73–100	79	31 ± 16
SS	86–93	83	77 ± 13

Data in brackets relate to the second reactor; Bv = volumetric loading rate; + = not indicated; HUSB = hydrolysis upflow sludge blanket.

Another reactor configuration used to treat domestic sewage under low temperature was the anaerobic hybrid (AH) reactor. This system is a combination of an UASB reactor or an EGSB reactor and an AF reactor in one reactor. The reactor bottom is a sludge bed and the top is a filter on which biomass can be attached. An AH reactor investigated by Elmitwalli *et al.* (1999) could remove 92% of CODss and 66% of the total COD at a HRT of 8 hours at a temperature of 13°C. However, the authors indicated that the reactor worked properly on pre-settled sewage. Therefore, pre-treatment for the removal of SS is necessary for treatment at low temperatures.

We believe that in order to guarantee an efficient treatment of domestic sewage under low temperature conditions, at least part of the SS present in the wastewater should be removed before feeding the wastewater to a sludge bed reactor. Thus, Kalogo and Verstraete (1999) have recently suggested a new integrated anaerobic treatment of domestic sewage. This approach combines a UASB reactor and a conventional completely stirred tank reactor (CSTR) for the treatment of wastewater low in SS and the sedimented primary sludge, respectively. The proposed system includes chemical enhanced primary sedimentation (CEPS) to remove the SS. Laboratory experiments (jar and continuous tests) were carried out to investigate the feasibility and efficiency of this new technology. The results indicated that the use of an industrial ($FeCl_3$) or a water extract of *Moringa oleifera* seeds (WEMOS) as a natural coagulant can significantly improve the CODs/VSS ratio (from 1.4 to up to 21) for domestic sewage (Figure 10.8).

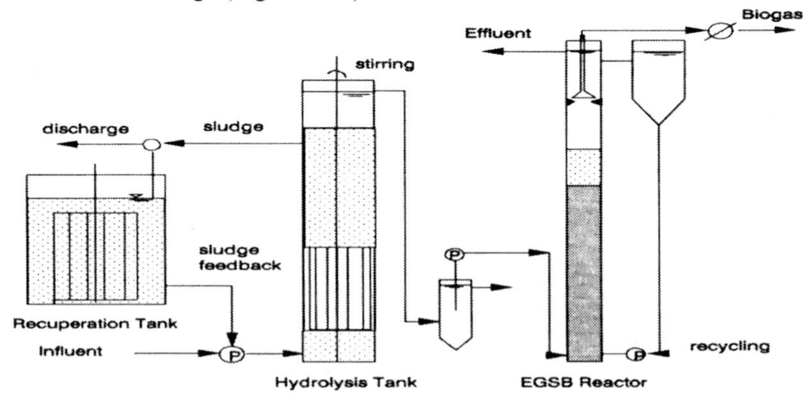

Figure 10.7. Schematic representation of a two-stage reactor (HUSB-EGSB) combined with sludge recuperation reactor (after Wang 1994).

With the natural coagulant, an important part of the increase of the CODs/VSS ratio is due to the increase in CODs by the addition of WEMOS, which is indicated by the increase of CODs and CODt in the supernatant (Table 10.6). This is not the case with $FeCl_3$, which tends to remove part of the CODs (Table 10.7).

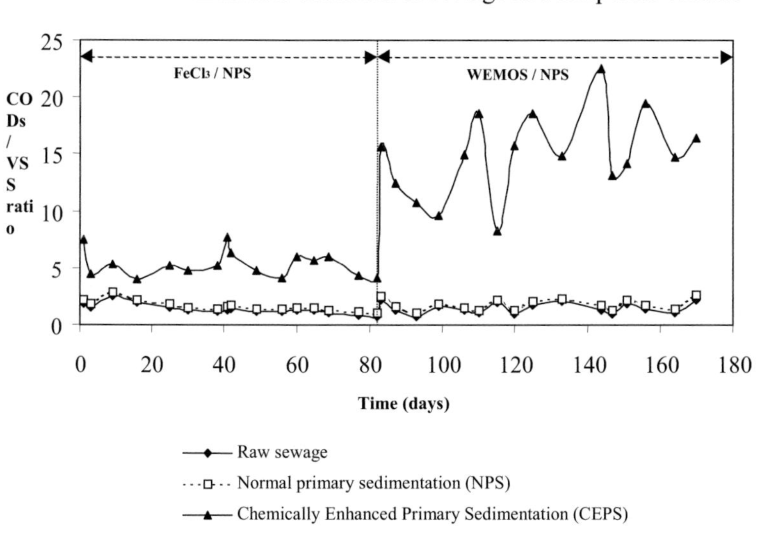

Figure 10.8. Effect of different types of primary treatment on the CODs/VSS ratio of the sewage, continuous test: 70 mg $FeCl_3$ L^{-1} – 24 mL WEMOS (5%, w/v) L^{-1} (after Kalogo and Verstraete 2000).

Table 10.6. Physico-chemical characteristics of the raw wastewater and the supernatant according to the dose of WEMOS (Kalogo and Verstraete 2000)

Dose[a]	SV_{60}[b]	pH	Alka-linity[c]	CODt ($mg\ L^{-1}$)	CODs ($mg\ L^{-1}$)	SS ($mg\ L^{-1}$)	VSS ($mg\ L^{-1}$)	**CODs/ VSS**
*	+	7.6 (0.3)	404 (9)	269 (10)	140 (7)	130 (10)	101 (12)	**1.4 (0.6)**
0	1.2 (0.2)	7.6 (0.1)	392 (6)	195 (12)	133 (9)	93 (11)	74 (12)	**1.8 (0.6)**
0.2	1.2 (0.3)	7.6 (0.2)	390 (8)	194 (10)	144 (9)	90 (8)	72 (9)	**2.0 (1)**
2	1.4 (0.3)	7.4 (0.1)	390 (9)	199 (10)	149 (8)	49 (5)	41 (6)	**3.6 (1.3)**
8	4.3 (0.7)	7.5 (0.3)	385 (7)	209 (13)	184 (7)	28 (5)	23 (4)	**8.0 (2)**
16	9.5 (1.1)	7.6 (0.1)	390 (7)	238 (11)	213 (9)	25 (4)	21 (5)	**10.2 (1.8)**
24	10 (1)	7.6 (0.4)	390 (8)	342 (10)	313 (7)	24 (6)	19 (5)	**16.5 (1.4)**
32	11 (1)	7.5 (0.2)	387 (6)	427 (12)	402 (8)	24 (5)	19 (6)	**21.6 (1.3)**

[a] = $mL\ L^{-1}$; [b] = $mL\ L^{-1}$; [c] = mg $CaCO_3$ L^{-1}; data between brackets correspond to standard deviation; * = related to the raw wastewater; + = not determined; SV_{60} = volume of sludge after settling for one hour.

Table 10.7. Physico-chemical characteristics of the raw wastewater and the supernatant according to the dose of $FeCl_3$ (Kalogo and Verstraete 2000)

Dose[a]	SV_{60}[b]	pH	Alkali-nity[c]	CODt (mg L^{-1})	CODs (mg L^{-1})	SS (mg L^{-1})	VSS (mg L^{-1})	**CODs/VSS**
*	+	7.6 (0.3)	404 (9)	269 (10)	140 (7)	130 (10)	101 (12)	**1.4 (0.6)**
0	1.2 (0.2)	7.6 (0.1)	392 (5)	196 (12)	134 (7)	94 (11)	72 (10)	**1.8 (0.7)**
10	3 (1)	7.6 (0.1)	323 (7)	184 (10)	132 (6)	54 (9)	44 (6)	**3.0 (1.0)**
30	8 (1)	7.3 (0.2)	312 (7)	160 (11)	125 (7)	39 (4)	31 (6)	**4.0 (1.0)**
50	16 (2)	7.1 (0.3)	305 (6)	145 (10)	120 (8)	25 (5)	20 (5)	**6.0 (1.0)**
70	26 (2)	7.1 (0.4)	300 (8)	130 (9)	110 (5)	15 (4)	12 (5)	**9.2 (1.1)**
90	28 (2)	7.1 (0.1)	249 (5)	120 (10)	105 (7)	14 (4)	11 (4)	**9.5 (1.2)**
120	35 (3)	6.9 (0.2)	216 (7)	118 (11)	105 (6)	14 (5)	11 (5)	**9.5 (1.2)**

[a] = mg L^{-1}; [b] = mL L^{-1}; [c] = mg $CaCO_3$ L^{-1}; data between brackets correspond to standard deviation; * = related to the raw wastewater; + = not determined; SV_{60} = volume of sludge after settling for one hour.

After sedimentation of the SS, the supernatant was treated in an UASB reactor. The concentrated sludge from the chemical treatment was alkaline enough to be co-digested with municipal solid waste (MSW) without the supply of external alkalinity (e.g. $NaHCO_3$) to the digester. A flow scheme of the overall concept described is shown in Figure 10.9.

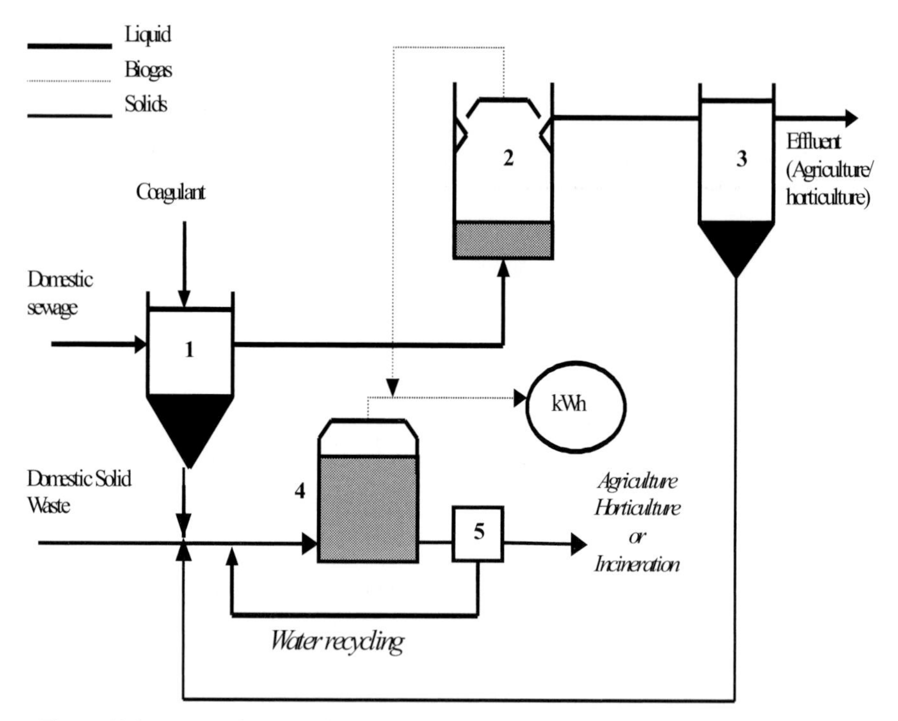

Figure 10.9. Integrated approach to the anaerobic treatment of domestic waste streams for energy recovery and the regeneration of water and bio-solids (carbon and nutrient). 1 = Primary decanter, 2 = UASB reactor, 3 = Post decanter, 4 = CSTR, 5 = Separator.

10.3.3 Outlook

The concept described in Figure 10.9 is fully applicable at 5–10–50 IE. It is clear that in order to make the latter concept an optimal integrated system which would fully agree with the new European legislation (C, N, P removal), the Sharon and the Anammox systems can be incorporated to remove the nitrogen. The removal of P is assured by the CEPS with $FeCl_3$. If necessary an ozonation step, for instance, by installing an ultraviolet (UV) lamp with O_3 emission, can be applied to guarantee a hygienic quality of the final effluent (see Figure 10.10).

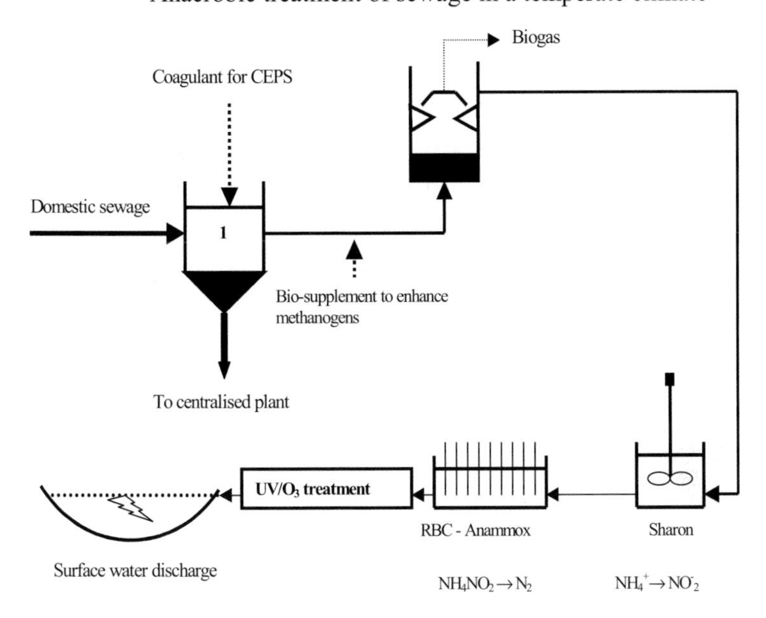

Figure 10.10. The optimal scheme for a decentralised integrated system. 1 = Primary decanter, 2 = UASB reactor, RBC = rotating biological contact reactor, UV/O_3 = ultra violet-ozone generator.

In Figure 10.9, after the primary sedimentation with $FeCl_3$, WEMOS can be dosed directly in the supernatant. Indeed, experimental investigations in our laboratory revealed that besides its coagulating properties, WEMOS contained several nutrients, such as carbohydrates, free amino acids and metal ions. Dosing of 2 mL L^{-1} of a 2.5% (w/v) WEMOS solution was found to enhance the growth and aggregation of anaerobic bacteria.

Due to the slow growth of psychrophilic bacteria, future research should focus on improving the biomass retention in the reactor, for instance, by incorporating a bio-membrane in the reactor. The recent developments in membrane technology make it applicable in anaerobic treatment of wastewater (Visvanathan *et al.* 2000).

10.4COST EVALUATION

10.4.1 Anaerobic versus aerobic treatment

The conventional approach to domestic sewage treatment is very expensive in terms of centralised sewerage (Table 10.8). This approach is also unsustainable in terms of sludge production when activated sludge is used. In Europe, the construction and operation of a domestic sewage treatment plan costs about 33 euros $IE^{-1}yr^{-1}$. With a centralised plant the total cost, including sewer installation and maintenance, is in the order of 100 euros $IE^{-1}yr^{-1}$.

Table 10.8. Running costs per inhabitant and per year for a domestic sewage treatment plant of 55,000 IE. The investment per IE is set at 100 euros (after Rudolph 1999)

INVESTMENT	COST
Sewers	750 EURO IE^{-1}
WWTP	125 EURO IE^{-1}
Total 1	875 EURO IE^{-1}
Annuity[a]	67.5 EURO $IE^{-1}yr^{-1}$
EXPLOITATION	**COST**
Sewers	7.5 EURO $IE^{-1}yr^{-1}$
WWTP	25 EURO $IE^{-1}yr^{-1}$
Total 2	32.5 EURO $IE^{-1}yr^{-1}$
TOTAL[b]	100 EURO $IE^{-1}yr^{-1}$

WWTP = wastewater treatment plant (12.5 euros for sludge disposal included).
[a] = with depreciation over 35 years and interest of 10%.
[b] = wastewater collection and treatment = Annuity + Total 2.

Economic considerations indicate that implementing a decentralised systems will save about two-thirds of the total cost of conventional centralised systems. Moreover, the costs saved can increase further if anaerobic treatment is used instead of activated sludge. Figure 10.11 shows that 1 kg biodegradable COD (CODb) when treated using an aerobic system will consume 1 kWh for aeration. This corresponding cost is 0.1 euro. Subsequently, the process will generate about 0.5 kg cell dry weight (CDW) of sludge. Further treatment in Europe actually costs an average of 0.5 euros. Therefore, full aerobic treatment of sewage will consume about 0.6 euros kg^{-1} CODb. If the same sewage is subjected to anaerobic treatment, one will produce 10 times less sludge and 0.5 m^3 biogas. Of the energy generated in the form of biogas, two-thirds can be used to maintain the temperature of the reactor above 20°C and the remaining one-third can be converted to electricity. The latter will correspond to about 1.5 kWh. Recently, several European Union countries, in order to fulfil the terms of the Kyoto agreements, have been subsidising non-conventional energy at the rate of 0.1 euros kWh^{-1}. Hence, with anaerobic treatment there is a potential for a 'green subsidy' of about 0.15 euros per kg biodegradable COD. Hence the difference in costs (input and output) of aerobic and anaerobic treatment is about 0.75 euros kg^{-1} CODb; a substantial difference.

10.4.2 CEPS anaerobic treatment versus classical anaerobic treatment

This section attempts to provide a cost estimate for a small decentralised low-tech anaerobic treatment plant for 50 IE. Experimental investigations by Kalogo and Verstraete (2000) have shown that the pre-treatment of raw sewage with 70 mg $FeCl_3$ L^{-1} (1 hour detention time) and subsequent treatment in a UASB reactor (2 hours detention time) may remove 74–80% of the total COD. To achieve the same result, a classical one-stage UASB reactor should operate at a temperature $\geq 20°C$ with at least 5 hours detention time. Indeed the model of Van Haandel *et al.* (1996) stipulates that COD removal (%) = $1 - 0.68 \times (HRT)^{-0.68}$. Hence, by coupling the CEPS to the UASB reactor, the volume of the anaerobic reactor (V_{AR}) needed to achieve the same effluent quality can be decreased by a factor of 0.6 (Table 10.9). Taking into account the volume of the primary decanter (V_{PD}) designed with a sedimentation velocity of (V_s) 0.8 m h^{-1} and a retention time of 1 hour, the total volume can decrease by a factor of 0.4. The construction cost of a low-tech UASB reactor is in the range of 200–300 euros per m^3 (Vieira and Sousa 1986; Schellinkhout and Collazos 1992). Consider a high estimate of 300 euros per m^3. The volume decrease by a factor of 0.4 leads to a gain of 120 euros per m^3 of reactor. The primary decanter designed in concrete with equipment for sludge removal at the bottom costs about 120 euros per m^3. Including the reactor construction cost, the primary decanter cost and inoculation, one may save 312 euros for 50 IE (180 L $I.E^{-1}d^{-1}$) (Table 10.9). Since 1 kg of $FeCl_3$ costs 0.46 euros, the 70 mg L^{-1} dosed costs 106 euros per year for 50 IE. This extra cost can be covered in 3 years from the gain on investment cost. Then each IE will have to pay about 2 euros annually for the coagulant. This demonstrates that the CEPS anaerobic treatment constitutes an economically acceptable alternative compared with conventional anaerobic treatment.

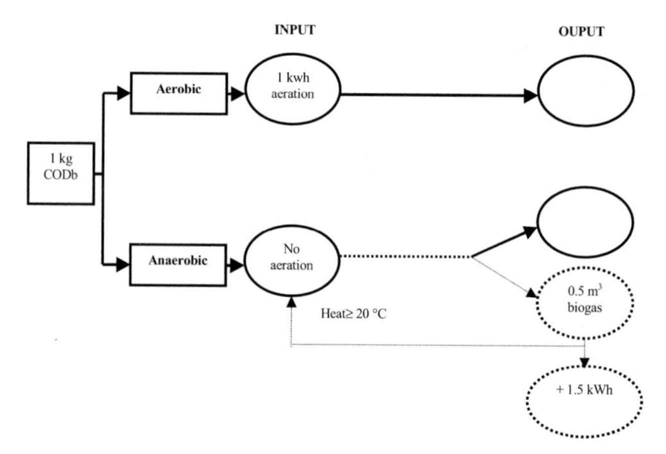

Figure 10.11. A comparison of aerobic and anaerobic treatment of biodegradable organic matter.

Table 10.9. Cost evaluation: conventional anaerobic treatment of domestic sewage versus CEPS anaerobic treatment for 50 IE. UASB is used as the reactor technology

	Conventional anaerobic	CEPS- anaerobic
COD removal efficiency (%)	80	80
Flow (m^3 h^{-1})	0.375	0.375
V_{PD} (m^3)*	–	0.376
V_{AR} (m^3)	1.875	0.735
Total volume (m^3)	1.875	1.111
Cost V_{PD} (euros)	–	43
Cost V_{AR} (euros)	563	221
Inoculum (kg VSS) (inoculation at 1/3 of V_{AR})	18.75	7.35
Cost of inoculation (euros)	28	11
Total investment (euros)	**591**	**279**
Gain (euros)		+ (312)

V_s = depth (m)/HRT (d) = Q (m^3d^{-1})/A (m^2), with Q = wastewater flow and A = surface decanter, * = 2.8% the flow is removed after CEPS treatment.

10.5 ACKNOWLEDGMENTS

This work was partially supported by the government of Côte d'Ivoire through a scholarship. The authors would like to thank Dr G. Zeeman for her constructive suggestions. Biotim's help in providing the investment cost of the primary decanter was also appreciated.

10.6 REFERENCES

Arsov, R., Ribarova, I., Nikolov, N., Mihailov, G., Tolova, Y. and Khoudary, E. (1999) Two-phase anaerobic technology for domestic wastewater treatment at ambient temperature. *Wat. Sci. Tech.* **39**(8), 115–122.

De Baere, L. and Verstraete, W. (1982) Can the recent innovations in anaerobic digestion of wastewater be implemented in anaerobic sludge stabilization? In *Recycling International – Recovery of Energy and Material from Residues and Wastes* (ed. K.J. Tomé-Kozmiensky), Freitag, E. – Verlag Für Umwelttechniek Berlin, pp. 390–394.

Derycke, D. and Verstrate, W. (1986) Anaerobic treatment of domestic wastewater in a lab and pilot scale polyurethane carrier reactor. In *Proceedings of EWPCA Conference on Anaerobic Treatment: a Grown-up Technology*, Schiedam, the Netherlands, 437–450.

Edeline, F. (1997) Epuration biologique des eaux: Theorie et technologie des réacteurs (ed. Lavoisier, Tec & Doc), 4th edn. (In French.)

Elmitwalli, T.A., Zandvoort, M.H., Zeeman, G., Bruning, H. and Lettinga, G. (1999) Low temperature treatment of domestic sewage in upflow anaerobic sludge blanket and anaerobic hybrid reactors. *Wat. Sci. Tech.* **39**(5), 177–186.

Gouranga, C.B., Timothy, G.E. and Dague, R.R. (1997) Structure and methanogenic activity of granules from an ASBR treating dilute wastewater at low temperatures. *Wat. Sci. Tech.* **36**(6–7), 149–156.

Van Haandel, A.C. and Lettinga, G. (1994) *Anaerobic Sewage Treatment. A Practical Guide for Regions with a Hot Climate.* John Wiley & Sons, New York.

Henze, M., Harremoës, P., La Cour Jansen, J. and Arvin, E. (2000) Wastewater treatment. Biological and chemical processes. Springer-Verlag, Germany.

Huser, B.A., Wurhrmann, K. and Zehnder, A.J.B. (1982) Methanothrix soengenii gen. nov. sp. nov., a new acetotrophic non-hydrogen-oxidizing methane bacterium. *Arch. Microbiol.* **132**, 1–9.

Kalogo, Y. and Verstraete, W. (1999) Development of anaerobic sludge bed (ASB) reactor technologies for domestic wastewater treatment: motives and perspectives. *World. J. Microbiol. Biotechnol.* **15**, 523–534.

Kalogo, Y. and Verstraete, W. (2000) Technical feasibility of the treatment of domestic wastewater by a CEPS-UASB system. *Environm. Technol.* **21**, 55–65.

Lens, P.N. and Verstraete W. (1992) Aerobic and anaerobic treatment of municipal wastewater. In *Profiles on Biotechnology* (eds T.G. Villa and J. Abalde), Universidade de Santiago, Spain, pp. 333–356.

Lide, D.R. (1992) *Handbook of Chemistry and Physics*, 73rd edn. CRC Press, Boca Raton, FL.

Lettinga, G., Roersma, R. and Grin, P. (1983) Anaerobic treatment of raw domestic sewage at ambient temperature using a granular bed UASB reactor. *Biotechnol. Bioeng.* **25**, 1701–1723.

Lettinga, G., Roersma, R., Grin, P., De Zeeuw, W., Hulshof Pol, L., Van Velsen, L., Hobman, S. and Zeeman, G. (1981) Anaerobic treatment of sewage and low strength wastewaters. In *Proceedings of the 2nd International Symposium on Anaerobic Digestion* (eds D.E. Hughes, D.A. Stafford, B.I. Wheatley, W. Baader, G. Lettinga, E.J. Nyns, W. Verstraete, and R.L. Wentworth), Elsevier, Amsterdam, pp. 271–291.

De Man, A.W.A., Vanderlast, A.R.M. and Lettinga, G. (1988) The use of EGSB and UASB anaerobic systems for low strength soluble and complex wastewaters at temperatures ranging from 8 to 30°C. In *Proceedings of the 5th International Conference on Anaerobic Digestion* (eds E.R. Hall and P.N. Hobson), Monduzzi S.P.A., Bologna, pp. 197–209.

Mergaert, K. and Verstraete, W. (1987) Microbial parameters and their control in anaerobic digestion. *Microbiol. Sci.* **4**, 348–351.

Mergaert, K., Vanderhaegen, B. and Verstraete, W. (1992) Applicability and trends of anaerobic pre-treatment of municipal wastewater. *Wat. Res.* **26**, 1025–1033.

Metcalf and Eddy Inc. (1984) *Wastewater Engineering: Treatment, Disposal, Reuse.* McGraw Hill, New Delhi, India.

Monroy, O., Noyola, A., Ramirez, F. and Guiot, J.P. (1988) Anaerobic digestion of water hyacinth as a highly efficient treatment process for developing countries. In *Proceedings of the 5th International Conference on Anaerobic Digestion* (eds E.R. Hall and P.N. Hobson), Pergamon Press, London, pp. 347–351.

Nozhevnikova, A.N., Holliger, C., Ammann, A. and Zehnder, A.J.B. (1997) Methanogenesis in sediments from deep lakes at different temperatures (2–70°C). *Wat. Sci. Tech.* **36**(6–7), 57–64.

Rebac, S. (1998) Psychrophilic anaerobic treatment of low strength soluble wastewaters. Ph.D. thesis, Wageningen University, the Netherlands.

Rebac, S., Ruskova, J., Gerbens, S., Van Lier, J.B., Stams, A.J.M. and Lettinga, G. (1995) High-rate anaerobic treatment of wastewater under psychrophilic conditions. *J. Ferment. Bioeng.* **5**, 15–22.

Rebac, S., Van Lier, J.B., Lens, P.N., Van Cappellen, J., Vermeulen, M., Stams, A.J.M., Dekkers, F., Swinkels, K.Th.M. and Lettinga, G. (1998) Psychrophilic (6–15°C) high-rate anaerobic treatment of malting waste water in a two module EGSB system. *Biotechnol. Progress* **14**, 856–864.

Rebac, S., Gerbens, S., Lens, P.N., Van Lier, J.B., Stams, A.J.M. and Lettinga, G. (1999) Kinetic of fatty acid degradation by psychrophilically cultivated anaerobic granular sludge. *Biores. Technol.* **69**, 241–248.

Rozzi, A. and Verstraete W. (1981) Calculation of active biomass and sludge production vs. waste composition in anaerobic contact processes. *Trib. Cebedeau* **455**(34), 421–427.

Rudolph, K.U. (1999) Sewerage charges: a European comparison. *Water Quality International,* March/April, 9.

Sanz, I., and Fernandez-Polanco, F. (1990) Low temperature treatment of municipal sewage in anaerobic fluidized bed reactors. *Wat. Res.* **24**, 463–469.

Sayed, S.K.I.A., and Fergala, M.A.A. (1995). Two stage UASB concept for treatment of domestic sewage including sludge stabilization process. *Wat. Sci. Tech.* **32**(11), 55–63.

Schellinkhout A. and Collazos C.J. (1992) Full-scale application of the UASB technology for sewage treatment. *Wat. Sci. Tech.* **25**(7), 159–166.

Segezzo, L., Zeeman, G., Van Lier J.B., Hamelers, H.V.M. and Lettinga, G. (1998) A review: The anaerobic treatment of sewage in UASB and EGSB reactors. *Biores. Technol.* **65**, 175–190.

Thaveesri, J., Daffonchio, D., Liessens, B., Vandermeren, P. and Verstraete W. (1995) Granulation and sludge bed stability in upflow anaerobic sludge bed reactors in relation to surface thermodynamics. *Appl. Env. Microbiol.* **61**, 3681–3686.

Van der Last, A.R.M. and Lettinga, G. (1992) Anaerobic treatment of domestic sewage under moderate climatic (Dutch) conditions using upflow reactors at increased superficial velocities. *Wat. Sci. Tech.* **25**, 167–178.

Verstraete, W. and Vandevivere, P. (1999) New and broader applications of anaerobic digestion. *Critical Reviews in Environ. Sci. Technol.* **28**, 151–173.

Vieira, S.M.M. and Sousa, M.E. (1986) Development of technology for the use of UASB reactor in domestic sewage treatment. *Wat. Sci. Tech.* **18**(12), 109–121.

Visser, A., Gao, Y. and Lettinga, G. (1992) Anaerobic treatment of synthetic sulfate-containing wastewater under thermophilic conditions. *Wat. Sci. Tech.* **25**(7), 193–202.

Visvanathan, C., Ben Aim, R. and Parameshwaran, K. (2000) Membrane separation biorectors for wastewater treatment. *Critical Reviews in Environ. Sci. Technol.* **30**(1), 1–48.

Vogels, G.D., Keltjens, J. and Van Der Drift, C. (1988) Biochemistry of methane production. In *Biology of Anaerobic Microorganisms* (ed. A.J.B. Zehnder), Willey Editor, New York.

Wang, K. (1994) Integrated anaerobic and aerobic treatment of sewage. Ph.D. thesis. Wageningen University, the Netherlands.

Wang, K., Vander Last, A.R.M. and Lettinga, G. (1997) The hydrolysis upflow sludge bed (HUSB) and the expanded granular sludge blanket (EGSB) reactor process for sewage treatment. In *Proceedings of the 8th International Conference on Anaerobic Digestion*, London, 25–29 May, Pergamon Press, London, **3**, pp. 301–304.

Widdel, F. (1988) Microbiology and ecology of sulfate and sulfur-reducing bacteria. In *Biology of Anaerobic Microorganisms* (ed. A.J.B. Zehnder), Willey Editor, New York, pp. 469–586.

Wiegel, J. (1990) Temperature spans for growth: hypothesis and discussion. *FEMS Microbiol. Rev.* **75**, 155–170.

Yoda, M., Kitagawa, M. and Miyaji, Y. (1987) Long term competition between sulphate reducing bacteria and methane producing bacteria in anaerobic biofilm. *Wat. Res.* **21**, 1547–1556.

Zeeman, G. and Lettinga, G. (1999) The role of anaerobic digestion of domestic sewage in closing the water and nutrient cycle at community level. *Wat. Sci. Tech.* **39**(5), 187–194.

11

Potentials of anaerobic pre-treatment (AnWT) of domestic sewage under tropical conditions

G. Lettinga

11.1 INTRODUCTION

11.1.1 AnWT of domestic sewage

Due to the use of water as the transport medium of human waste, in many parts of the world society is now facing the problem of to treat enormous quantities of (very) low-strength domestic wastewater. Despite the low strength of domestic sewage it is still hazardous to public health, mainly due to the presence of pathogenic organisms. Moreover, in view of the enormous quantities involved, it is also detrimental to the environment. The water transport-based sanitation

Edited by P. Lens, G. Zeeman and G. Lettinga. ISBN: 1 900222 47 7

system has been implemented mainly in the prosperous industrialised world, but increasingly it is also being implemented in less prosperous parts of the world. Regarding the huge investments to be made for installing and maintaining this concept, coming generations will irrevocably have to deal with the 'treatment' problem, wherever it has been implemented.

In the industrialised world, compact advanced aerobic treatment methods are usually applied, which are relatively expensive systems, both in operation and maintenance. In less prosperous countries conventional extensive treatment systems are more popular, mainly because they are considered to be less expensive. However, due to a lack of funds, many countries cannot even afford these systems, and in these cases the sewage collected is being discharged almost untreated.

In view of the well known (Lettinga 1996; van Lier and Lettinga 1999; see also Chapter 10) significant advantages of anaerobic treatment (AnWT) over conventional treatment methods, application of the AnWT method for pre-treatment would represent – when feasible – an extremely attractive alternative. However, as the AnWT method is particularly suited to more concentrated wastewater and at higher ambient temperatures (preferably exceeding 20°C) the crucial question is whether or not the AnWt method is or could be made technically and economically feasible for the pre-treatment of domestic sewage. This obviously would be most easily achievable in tropical regions, where sewage temperatures generally exceed 20°C. The first research studies in the field of AnWT pre-treatment of domestic sewage therefore concentrated on its applicability in tropical regions .

11.1.2 AnWT of sewage under tropical conditions

From available experiences with full scale AnWT installations treating low-strength industrial wastewater in the temperature range 20–40°C, clear evidence has been obtained that AnWT also represents a feasible and attractive solution for the treatment of domestic sewage (van Haandel and Lettinga 1994). Following recent experimental results the modern AnWT concepts with their long sludge hold-up time also offer benefits in warm weather conditions (Jewell 1987; Lettinga *et al.* 1987, 1999; Rebac *et al.* 1997;) and for partially soluble low-strength wastewater such as domestic sewage at temperatures exceeding 20°C.

This section presents the results of a comprehensive and detailed investigation into a large-scale experimental (64 m^3) UASB plant operated over six years from 1983–1989 in Cali, Colombia (Schellinkhout *et al.* 1985, 1989). The results obtained from this study are relevant because they are applicable to a wide range of situations found in tropical regions, and give an insight into:

(1) The proper start-up conditions to be applied.
(2) The preliminary treatment required, i.e. for removing sand and grit.
(3) The treatment efficiency that can be achieved in relation to imposed hydraulics (HLR), the organic loading rates (OLR), and the wastewater characteristics.
(4) The chemical oxygen demand (COD) conversion factors (of removed COD) into sludge COD and methane COD and the sludge characteristics.
(5) Design, construction, operational and maintenance criteria for UASB reactor systems, including the construction materials.
(6) Efficiency and feasibility of various post-treatment methods, including settling, slow sand filtration, trickling filter, anaerobic filter and a small maturation pond.

11.2 TREATMENT OF SEWAGE USING THE UASB CONCEPT

The above-mentioned research in Cali, funded by the Dutch government (DGIS DPO) was carried out by the Agricultural University Wageningen (WAU), Haskoning Consulting Engineers in Nijmegen, the Universidad del Valle (Colombia), and Empresas Municipales de Cali. The research was initiated on the basis of promising results of earlier preliminary pilot plant investigations in this field conducted at the WAU (de Man *et al.* 1996; Grin *et al.* 1983, 1985; Lettinga *et al.* 1983). Along with the results of the Cali project, some relevant experiences with full-scale UASB reactors treating sewage will also be discussed (Schellinkhout and Osario 1992; Draaijer *et al.* 1992).

Figure 11.1 shows a schematic diagram of the 64 m^3 reactor. The influent distribution system of the Cali plant consisted of 16 inlet pipes (fewer could be used) discharging the raw sewage at the bottom of the reactor. The installed flow control box enabled flow adjustment from 0–10 litres per second (l/s) (0–36 m^3/h).

The (mainly) domestic sewage used during the first experimental period of three and a half years was diluted and septic (see Table 11.1). During the last two years of the investigation, sewage from another main collector, conveying mixed domestic and industrial wastewater, was used. The COD concentration of this sewage was higher than in the original sewage (COD total was approximately 400 mg/l, COD fil (paper filtered COD) >150 mg/l); the total suspended solids (TSS) content was similar, but the ash content was lower (30–

35%). Moreover, there were occasional high peaks in the pH of the sewage, due to industrial discharge.

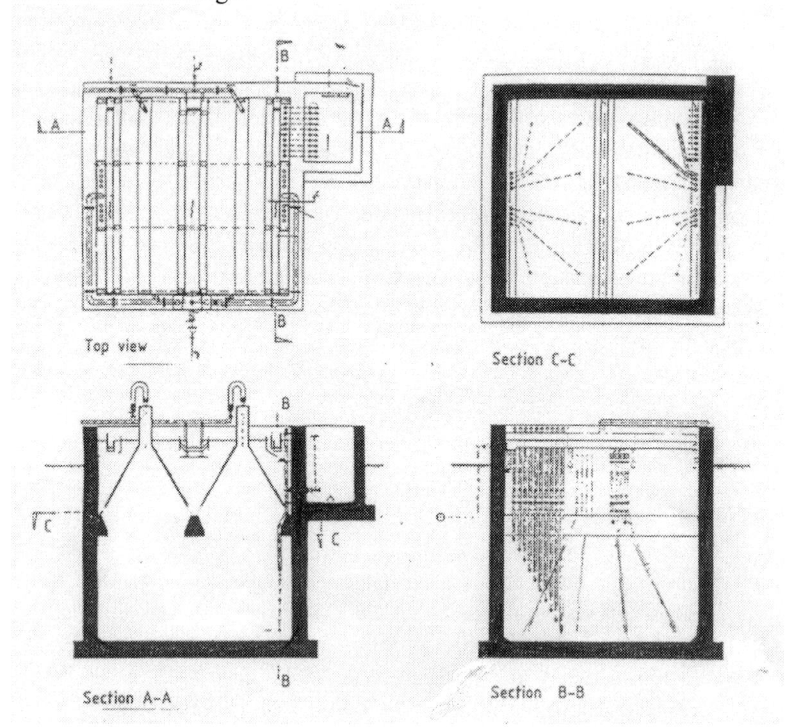

Figure 11.1 Schematic diagram of the 64 m^3 pilot UASB reactor in Cali.

Table 11.1. Main characteristics of sewage from the Cañaverelejo combined collector (the average values as measured over the whole experimental period)

		Period		
		Entire period	Wet season	Dry season
COD total	(g.m^{-3})	267	300	200
COD fil	(g.m^{-3})	112	130	90
BOD	(g.m^{-3})	95	95	95
TSS	(g.m^{-3})	215	189	156
VSS	(g.m^{-3})	108	106	73
Ash content TSS			35–50%	
NTK	(g.m^{-3})	17	18.7	14.3
NH_4^+N (g.m^{-3})	(g.m^{-3})	11	13.6	9.1
Total–PO_4^{3-}	(g.m^{-3})	1.3	0.7	0.8
Temp.	(°C)	25.2	25.0	14.4

11.2.1 The start-up of UASB plants treating raw domestic sewage

For a proper start-up, a well stabilised sludge with a sufficiently high activity and settleability must develop and accumulate in the reactor in the form of a thick sludge bed/blanket. The Cali experiments revealed (later confirmed in various full scale installations (Schellinkhout and Osario 1992; Draaijer *et al.* 1992)) that seeding of the reactor can be delegated to the active biomass present in raw sewage itself. For this purpose (1) the suspended solid fraction of the sewage must accumulate in the reactor, and (2) new bacterial matter should be allowed to grow in the retained sludge.

The performance of the plant in terms of COD removal efficiency can be impressive from start-up, although this mainly results from the removal of suspended settleable matter. The rate of start-up and the treatment performance during the start-up period depend mainly on the imposed HRT (hydraulic retention time) and the sewage characteristics. As was also shown in this experiment, less time is required when a sludge bed is already present in the reactor. During the initial phase, the removal of soluble COD may become temporarily negative, i.e. when hydrolysis of previously accumulated solid substrate ingredients exponentially develops and the total methanogenic activity in the system does not suffice to convert the soluble compounds from the solution. The filling-up time of the reactor with sludge can be estimated from the influent TSS, the imposed HRT and the TSS removal efficiency. For the last factor a value of 70–80% can be set. The total amount of sludge TSS retained in the 64 m^3 Cali UASB reactor was found to amount to 2000 kg.

Start-up can be completed within six months by starting the feeding of the reactor at a relatively long HRT of 12–24 hours during the first week(s), followed by an HRT of 6–12 hours after a few weeks. Since in many regions adequate seed materials for the first start-up are frequently unavailable, it is clear that the experience gained in the Cali plant is of eminent practical importance. The procedures developed in the Cali plant in the meantime have been successfully applied in the start-up of a number of full-scale plants, for example in a 1100 m^3 installation in Kanpur, India (Draaijer *et al.* 1992), a 3300 m^3 UASB plant in Bucaramanga, Colombia (Schellinkhout and Osario 1992) and various other installations (Vieira and Souza 1986).

When the minimum requirements are not met during start-up a balanced digestion process cannot usually be attained. Unfortunately this has been the case in a number of small-scale installations, and this is often due to a complete lack of understanding of the anaerobic digestion process by the operators/contractors of the system. Apart from the treatment being ineffective, neighbourhoods in these cases will also suffer from serious odour nuisance. It should be emphasised that for a smooth start-up it is essential that the process is thoroughly understood.

11.2.2 Operation and performance after start-up

After the start-up has been completed, a certain amount of excess sludge needs to be periodically discharged (for example, once every 4–7 days) in order to prevent the wash-out of sludge by effluent. A decline in the COD_{total} removal efficiency (i.e. based on raw influent COD and raw effluent COD values) can thus be prevented. The maximum achievable efficiency of the system, E^{COD}_{max}, can be derived from the values calculated from the raw influent and filtered effluent samples. Since under tropical conditions the settleability of the major part of the effluent is exceptionally good (SVI, or sludge volume index, = 10–20 ml/g) these E^{COD}_{max}-values can be achieved by using relatively simple means, such as plain post-settling. At temperatures around 25°C and under conditions of continuous operation, E^{COD}_{max} values range from 80–83% at HRT >4 hours and around 73% at a HRT as low as 2.4 hours. The performance of the system also remains excellent under fluctuating flow conditions, with values in the range 78–81%: when applying a day–night regime of HRT = 2.2 hours during the 12 hour day time period and HRT = 6 hours during the night, the average daily HRT = 3.2 hours. Despite the much higher influent COD values during the day (average COD_{tot}: 391 mg/l, average COD_{sol}: 122 mg/l) compared to the night (COD_{tot} average: 183 mg/l and COD_{sol} average 78 mg/l) the E^{COD}_{max} during the day is in the range of 77–93%, averaging out at 82%. The lower efficiencies were found for the lower strength influents during the night, that is, with values ranging from 41–83% (averaging out at 60%). This can be mainly due to the relatively high fraction of soluble COD during the night. Evaluation of the gas production data revealed that a considerable fraction of the COD removed during the day (that is, the entrapped solids) becomes degraded during the night.

With respect to the sludge hold-up, the results of the Cali reactor indicate that approximately 39 kg TSS/m^3 can be retained at HRT = 6 hours, that is, at an average liquid upflow velocity V_f of 0.66 m/hr. When increasing the V_f by a factor of 3, the sludge bed was found to expand at an initial expansion rate of 0.3 m/hr. This might result in heavy sludge wash-out when the high imposed hydraulic load lasted too long and/or the free sludge bed expansion space in the reactor was too small. Following a hydraulic peak load, the bed contracts again. It was found – and this is very important – that expansion proceeds noticeably slower than contraction. As hydraulic loads are generally high during the day and low at night, a sufficient sludge bed expansion space (including the settler volume) should always be available in order to accommodate for sludge bed expansions. The free expansion space at least should be 1.5 metres. This expansion results mainly from the hydraulic peak load imposed. Organic shock loads in the range 1–2.5 kg/m^3 per day resulted in no notable sludge bed expansion.

When a UASB reactor remains unfed for longer periods, the bed will compact within a relatively short period of time to half its height. Experiments conducted in the Cali reactor in which the feed was interrupted for 3 weeks showed that the sludge TSS in the bed reached values exceeding 120 kg/m^3 after 2 days, corresponding to a sludge bed compaction from over 2 m to under 1.4 m after a week. After resuming the operation of the reactor following a 3 week feed interruption at $V_f = 0.66$ m/hr, the bed expanded from 0.8 m to 1.6 m within 2 days, and to 2 m after 7 days. The latter value is distinctly lower than at the beginning of the feed interruption, indicating that the sludge bed expansion proceeds relatively slowly.

BOD treatment results obtained in the Cali plant over the five and a half year experimental period are shown in Figures 11.2–11.3.

Since anaerobic digestion is a mineralisation process, it is clear that little removal of nitrogen (N) and phosphorous (P) compounds can be achieved unless a precipitation reaction occurs. Generally the NH_4^+–N concentration in the liquid phase increases due to the degradation of biodegradable N-compounds. Results obtained in the 64 m^3 UASB plant for N and P removal are summarised in Table 11.2.

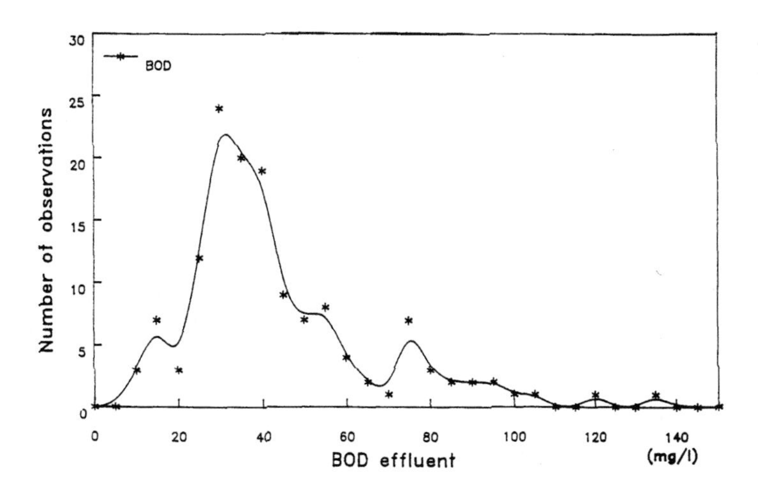

Figure 11.2. Frequency distribution of the weekly averaged total of (raw) effluent BOD values as found over the whole experimental period (also including data obtained during start-up, organic and hydraulic shock loads).

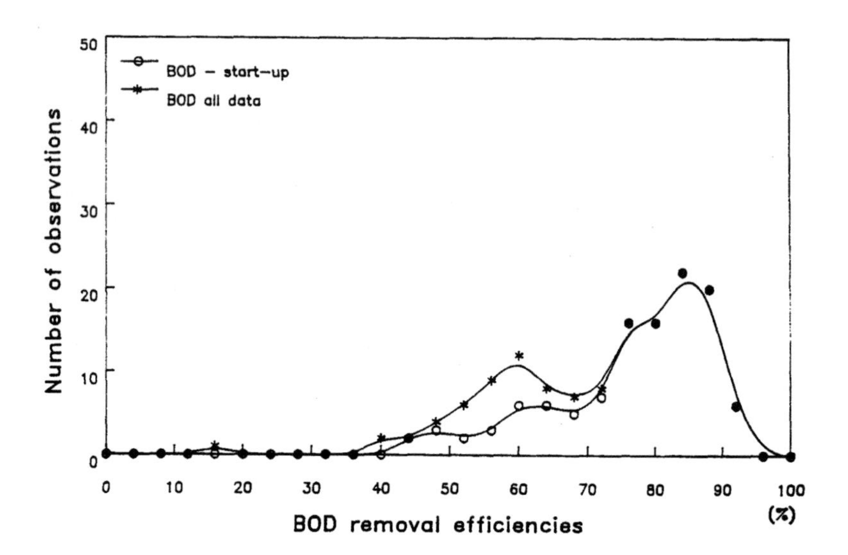

Figure 11.3. Frequency distribution of the calculated weekly averaged BOD efficiencies based on raw effluent and influent values. * = including values found during start-up, (o) = excluding values found during start-up (over the whole experimental period).

Table 11.2. Ammonia-N and phosphate-P concentrations in the influent and effluent solution of the 64 m^3 UASB plant

		NH_4^+–N		P–tot	
		Infl.	Effl.	Infl.	Effl.
Average	Mg/l	10.5	14.9	2.63	1.56
Minimum	Mg/l	4.3	5.2	0.1	0.4
Maximum	Mg/l	20.8	21.8	5.9	3.3

Very similar observations with respect to performance have been obtained in full-scale installations. In the Cali plant, the BOD treatment efficiencies always – except during the start-up phase – exceed the COD treatment efficiency by 2–4%.

11.2.3 Fractional conversion of COD into sludge and into methane

The data obtained in the Cali plant enable a fairly accurate calculation of the COD and TSS conversion into sludge COD and methane COD. The calculated average fractional conversion of COD into methane COD over the whole experimental period amounted to 0.19 Nm^3 CH_4/kg $COD_{removed}$, which corresponds to 0.5 kg CH_4–COD/kg $COD_{removed}$. Consequently approximately 50% of the $COD_{removed}$ will be found in the excess sludge. These figures are likely to be slightly different for other installations in tropical regions, depending on the characteristics of the sewage.

In view of the very low strength character of the sewage, the major part of the produced CH_4 will leave the reactor in dissolved form with the effluent, thus amounting to 56–63% for the Cali plant. If no recovery measures are taken, this methane gas will be lost and will contribute to the greenhouse effect, which would obviously represent a serious disadvantage of anaerobic treatment.

The values found for the excess sludge production in the Cali plant amounted to 0.4–0.6 kg TSS/kg TSS_{in} (= 0.06–0.1 kg TSS/m^3). Based on an average sludge concentration of 31.0–37.5 kg TSS/m^3 (9.4–12.5 kg VSS/m^3) the sludge age was estimated at 35–100 days, depending on the loading rate applied. But in reality the sludge age presumably is significantly higher, because it is likely that the majority of the washed out SS will predominantly consist of the (lighter) TSS ingredients present in the influent. The heavier sludge will remain in the lower part of the sludge bed.

11.2.4 Sludge characteristics

The ash content of the sludge has values in the range 55–65%, which is fairly high, indicating a high sludge stability. Stability digestion tests confirmed this. The specific methane production values found after completion of start-up ranged from 20–50 l CH_4/kg sludge, significantly lower than those found during start-up. Figures found for the sludge stability in other UASB installations treating sewage are very similar.

The assessed values for the maximum specific methanogenic activity (MSA_{max}) amounted to approximately 0.15 kg COD/kg VSS.day at 25°C, which is satisfactory.

The drying characteristics of the sludge in the Cali reactor were exceptionally good, enabling drying bed surface loads of 20 kg TSS/m^2 which resulted in an increase of the TS content from approximately 10% to 35–40% TS within 7 days. Similar values were found in anaerobic sewage treatment plants installed elsewhere.

11.2.5 Maximum loading potential of the reactor, feed inlet system

Using the assessed value for MSA_{max} of 0.15 kg COD/kg VSS.day and an average VSS retention of 10 kg sludge–VSS/m^3, the estimated maximum organic loading rate (OLR) at 25°C is around 1.5 kg $COD_{biodegradable}/m^3$.day. This would correspond to a CH_4 production rate of 1.2–1.6 m^3/hr for a 64 m^3 plant. The maximum gas production rates found in the Cali plant with domestic sewage amounted to 0.8 m^3/hr and 1.22m^3/hr with a feed consisting of sewage supplemented with 0.5 kg COD/m^3 additional *vinasse* (distillery slops), i.e. at an imposed OLR of 2.4 kg COD/m^3.day (soluble COD) and at an HRT of 6 hours. Under these 'high' loading conditions the treatment efficiency of the reactor remained lower than the values found at lower soluble COD loading rates with raw sewage, for example at an HRT <4 hours the soluble COD OLR amounted to a maximum of 1.5 kg COD/m^3.day. This value is the estimated maximum OLR, provided the sludge water contact is satisfactory, which was the case.

With respect to the sludge water contact, the feed inlet distribution system of the reactor is a crucially important factor. For this reason, comprehensive tests were conducted in the Cali plant. The results obtained reveal that a sufficient sludge water contact can be achieved when using only one feed inlet point/4m^2 at an HRT in the range 4–6 hours. This number of feed inlet points has been used with satisfactory performance results at all UASB sewage treatment systems installed so far.

11.2.6 Scum layer formation

Under tropical conditions the extent of scum layer formation at the liquid air interface is of minor importance. Buoying sludge aggregates settle down rapidly after applying minor mixing; that is, rainfall suffices. The stability of scum layer sludge was found to be significantly poorer compared to that present in the sludge bed, i.e. with values of approximately 120 ml CH_4/g VSS and 40 ml CH_4/g VSS respectively. Scum layers accumulating in the gas collector never led to any serious operational problems.

11.3 DISCUSSION AND CONCLUSIONS

From the results obtained in the 64 m^3 UASB plant and in various full-scale plants currently operating, it is obvious that AnWT represents an excellent pre-treatment option for domestic sewage in tropical regions, that is, where sewage temperatures exceed 20°C (Arceivala 1998). Moreover, this method also appears possible at lower temperatures, following on from the results of numerous investigations conducted in this field (de Man *et al.* 1988; Grin *et al.* 1983, 1985; Zeeman and Lettinga 1999; Wang 1994; Metwalli 2000), although – with respect to design and construction – modified systems may need to be applied. The process can be profitably applied at any scale, provided that the system is well designed and properly started up and operated. The first start-up of an one-step UASB plant can be accomplished within a 6–12 week period without using any seeding, at a HRT of approximately 6 hours.

The treatment efficiencies achievable following start-up at HRT 4–6 hours amount to:

COD (total/total)	50–75% (65%)
COD (total/ filtered)	70–90% (80%)
BOD (total/total)	70–90% (80%)
COD (filtered/filtered)	up to 60%
TSS	60–85% (70%)

The above treatment performances will also be achieved under strongly fluctuating flow and influent conditions. By using simple post-treatment systems such as plain settling, a BOD reduction approaching 90% can be achieved. It should be emphasised once more that AnWT is effective in removing organic pollutants only, it virtually comprises a mineralisation process. For removing mineralised ingredients such as ammonia-N, PO_4^{3-} and S^{2-} proper post-treatment systems are available and should be applied, for example (micro-)aerobic methods. The methane production can amount to 0.19 Nm^3/kg $COD_{removed}$ (0.33 CH_4–COD/kg $COD_{removed}$) and the excess sludge production to 0.4–0.6 kg TSS/kg TSS_{in} (= 0.06–0.1 kg TSS/m^3). The sludge retention of the reactor is 31–37.5 kg TSS/m^3 or 9.4–12.5 kg VSS/m^3, resulting in a sludge age of 35–100 days. The sludge is well stabilised, has excellent drying characteristics and its MSA is >0.1 kg COD/kg VSS.day. From the results of the Cali experiments, the required design criteria for one-step UASB reactors for sewage treatment under tropical conditions have been well established. These data in the meantime have been successfully applied in the construction and operation of full-scale UASB installations in India, for example.

Unfortunately, due to opportunism and the negligence of some contractors in some countries, such as Colombia, some poorly designed and/or poorly operated and maintained UASB plants have been installed. This results in a lot of bad publicity and misinformation about AnWT for sewage, which is then immediately used by established sanitary engineering companies, contractors, scientists and so on who want to prevent the implementation of anaerobic treatment systems because this may eventually result in the replacement of conventional systems. Consequently, such poorly performing systems seriously inhibit the much-needed implementation of more sustainable and robust environmental protection concepts. Companies promoting alternative extensive treatment systems, the so-called low-cost extensive systems, which are usually much less sustainable and robust, such as artificial wetlands and lagoons, are not slow to benefit from situations such as this.

11.4 REFERENCES

Arceivala S.J. (1998) Chapter 7 in *Wastewater Treatment for Pollution Control.* Tata McGraw-Hill, New Delhi, India.

De Man, A.W.A., Grin, P.C., Roersma, R., Grolle, K.C.F. and Lettinga, G. (1986) Anaerobic treatment of sewage at low temperatures. *Proc. Anaerobic Treatment: A Grown-up Technology*, Amsterdam, the Netherlands, 451–466.

De Man, A.W.A., Rijs, G.B.J., van Starkenburg, W. and Lettinga, G. (1988) Anaerobic treatment of sewage using a granular sludge bed UASB reactor. *Proc. 5th Int. Symp. On Anaerobic Digestion*, Bologna, Italy, 753–738.

Draaijer H., Maas J.A.W., Schaapman J.E. and Khan A. (1992) Performance of the 5 MLD UASB reactor for sewage treatment at Kanpur, India. *Wat. Sci. Tech.* **25**(7), 123–133.

Grin P., Roersma R.E. and Lettinga, G. (1983) Anaerobic treatment of raw sewage at lower temperatures. In *Proceedings of the European Symposium on Waste Water Treatment*, Noordwijkerhout, the Netherlands, 23–25 November, 335–347.

Grin P., Roersma R. and Lettinga, G. (1985) Anaerobic treatment of raw domestic sewage in an UASB reactor at temperatures from 9–20°C. Proc. Seminar/workshop: Anaerobic treatment of sewage, Amherst, MA, 109–124.

Jewell, W.J. (1987) Anaerobic sewage treatment. *Environ. Sci. Technol.* **21**(1), 14–21

Lettinga, G. (1996) Sustainable integrated biological wastewater treatment. *Wat. Sci. Tech.* **33**(3), 85–98.

Lettinga, G., Roersma, R. and Grin, P. (1983) Anaerobic treatment of domestic sewage using a granular sludge bed UASB reactor. *Bitechnology and Bioengineering* **25**, 1701–1723.

Lettinga, G., Roersma, R. and Grin, P. (1987) Anaerobic treatment as an appropriate technology for developing countries. *Trib. Cebedeau.* **519**(40), 21–32.

Lettinga, G., Rebac, S., Parshina, S., Nozhevnikova, A. and van Lier, J.B. (1999) High rate anaerobic treatment of wastewater at low temperatures. *Applied Environmental Microbiology*, 1696–1702.

Metwalli, T.A. (2000) Anaerobic treatment of domestic sewage at low temperatures. PhD thesis, Wageningen Agricultural University, Wageningen, the Netherlands.

Rebac, S., van Lier, J.B., Jansen, M.C.J., Dekkers, F, Swinkels, K.Th.M. and Lettinga, G. 1997) High rate anaerobic treatment of malting house wastewater in a pilot-scale EGSB system under psychrophilic conditions. *J. Chem. Tech. Biotechnol.* **68**, 135–146.

Schellinkhout, A. and Osario, C.J. (1992) Full-scale application of the UASB technology for sewage treatment. *Wat. Sci. Tech.* **25**(7), 157–166.

Schellinkhout, A., Lettinga, G., Van Velsen, A.F.M., Louwe Kooijmans, J. and Rodriguez, G. (1985) The application of the UASB reactor for the direct anaerobic treatment of domestic wastewater under tropical conditions. *Proceedings of Anaerobic Treatment of Sewage*, Amherst, MA, 27–28 June, 259–276.

Schellinkhout, A., van Velsen, A.F.M., Wildschut, L., Lettinga, G. and Louwe Kooijmans, J. (1989) Anaerobic treatment of domestic wastewater under tropical conditions. Final reports of the first and second phase of experimental studies conducted in Cali, Colombia. Ministry of Foreign Affairs, DPO/OT, February 1985–January 1989.

Van Haandel, A.C. and Lettinga, G. (1994) *Anaerobic Sewage Treatment*. John Wiley, Chichester, UK.

Van Lier, J.B. and Lettinga, G. (1999) Appropriate technologies for effective management of industrial and domestic wastewaters: the decentralised approach. *Wat. Sci. Tech.* **40**(7), 171–183.

Vieira, S. and Souza, M.E. (1986) Development of technology for the use of the UASB reactor in domestic sewage treatment. *Wat. Sci. Tech.* **18**(12), 109–121.

Wang, K.(1994) Integrated Anaerobic and Aerobic Treatment of Sewage. PhD thesis, Wageningen Agricultural University, Wageningen, The Netherlands.

Zeeman, G. and Lettinga, G. (1999) The role of anaerobic digestion of domestic sewage in closing the water and nutrient cycle at community level. *Wat. Sci. Tech.* **39**(5), 187–194.

12

Anaerobic treatment systems for high strength domestic waste (water) streams

G. Zeeman, K. Kujawa-Roeleveld and G. Lettinga

12.1 INTRODUCTION

The application of anaerobic digestion to domestic sewage is becoming more and more popular, especially in tropical countries, where large-scale installations have been installed and are under construction (Hulshoff Pol *et al.* 1997). So far the treatment of domestic sewage is mainly carried out in centralised concepts. A variety of domestic wastewater streams with different qualities and in different quantities is discharged in extended sewer systems in order to be transported over a large distance, frequently along with rainwater. The current sanitation practice leads to the production of very diluted sewage,

Edited by P. Lens, G. Zeeman and G. Lettinga. ISBN: 1 900222 47 7

containing many components, that finally has to be purified. A large fraction of the main components, including organics, nutrients and pathogens, is originally produced in a very small volume, as faeces plus urine. For situations where no sanitation is yet available or when new housing estates or large buildings are going to be built, more sustainable options for collection, transport and treatment can be applied, where concentrated streams are kept separated from diluted streams.

Zeeman *et al.* (2000) showed the potential to separate different waste(water) streams in order to treat them in the most sustainable way with energy production and recovery and the reuse of nutrients as the main objectives. Anaerobic treatment of separately collected domestic sewage in combination with kitchen and food waste (swill) will play a key role in closing the water and nutrient cycle in these sanitation concepts. This chapter will present the different techniques available for the anaerobic treatment of relatively concentrated domestic sewage. The collection system will determine the composition and concentration of the sewage to be treated and the anaerobic technology that can be used. This chapter only considers community on-site treatment of sewage, where the collection of domestic sewage is not combined with the collection of rainwater. Kalogo and Verstraete (2001) present the treatment of total domestic wastewater including rainwater in high rate systems.

12.2 ANAEROBIC WASTEWATER TREATMENT SYSTEMS IN RELATION TO CONCENTRATION

In this chapter six types of domestic waste(waters) are considered, based on different collection systems and different amounts of water for flushing (Table 12.1). The chosen flushing volumes of 1 + 0.5, 4 + 2 and 9 + 6.7 litres for faeces and urine are based on vacuum toilets, low flushing toilets with a booster system in the transport pipe and conventional flushing toilets, respectively (Kujawa-Roeleveld *et al.* 2000).

Anaerobic treatment systems can in general be divided into systems with and without sludge/biomass retention, where:

- the systems without sludge retention are applied for more concentrated waste streams, e.g. wastewaters A and D.
- the systems with sludge retention are applied for more diluted wastewater streams, e.g. wastewaters B, C, E and F.

The applicability of the two types of systems for the above waste(water) streams will now be discussed: that is, treatment systems without biomass retention for waste(waters) A and D and treatment systems with biomass retention for waste(waters) B, C, E and F.

Table 12.1. Different types of domestic waste(waters), based on collection systems and amounts of water used for flushing (Kujawa-Roeleveld *et al.* 2000)

No	Type of waste(water)	Flushing for faeces (l/flushing)	Flushing for urine (l/flushing)	Name of the waste(water)
A	Faeces + urine + kitchen waste	1	0.5	night soil + swill
B	Faeces + urine	4	2	blackwater 1
C	Faeces + urine	9	6.7	blackwater 2
D	Blackwater 1 + kitchen waste	4	2	blackwater 1 + swill
E	Blackwater 1 + greywater	4	2	total domestic wastewater
F	Blackwater 2 + greywater	9	6.7	total domestic wastewater 2

12.2.1 Treatment without biomass retention

12.2.1.1 Composition of wastewaters A and D

Table 12.2 shows the composition and quantity of a mixture of toilet waste(water) and swill for two different flushing scenarios. The values are based on the composition and production of faeces and swill according to Kujawa-Roeleveld (2000) and on urine according to Kujawa-Roeleveld *et al.* (2000). For the production of swill a quantity of 0.4l/p.d is chosen, which is representative of the mean production in a Dutch hospital (Kujawa-Roeleveld *et al.* 2000).

Table 12.2. Theoretical composition of a mixture of faeces + urine and swill for different volumes of flush water (Kujawa-Roeleveld *et al.* 2000)

Parameter	Unit	Value	
Type of wastewater		a	d
Amount of flushing water	(l/p.d)	3.5	14
Protein	gCOD/l	4.90	1.63
Carbohydrate	gCOD/l	11.39	3.77
Lipid	gCOD/l	12.85	4.26
TS	g/l	22.57	7.48
COD*	gCOD/l	29.10	9.65
V_{ww}	l/p.d	5.21	15.71

* Chemical Oxygen Demand

12.2.1.2 Treatment systems

CSTR systems

The CSTR (completely stirred tank reactor) is the most generally applied system for sludge digestion, at mesophilic conditions and retention times between 15 and 30 days (Figure 12.1). The main characteristic of a CSTR system is that its SRT (sludge retention time) is equal to its HRT (hydraulic retention time). In general mesophilic CSTR systems can be applied when the slurry is so concentrated that it will provide enough biogas to produce the energy for heating the system. The higher the concentration, the more surplus energy is produced for other applications. Moreover, reduction in the use of flushing water results in a lower total volume to be treated and therefore in a smaller reactor volume to be installed, providing that the same SRT can be applied for both diluted and concentrated slurry. Moreover, the digested slurry can be more easily applied in agriculture since transport costs will be limited when small volumes of concentrated slurries are produced. So a reduction in the use of water for flushing is a very important aspect in reducing treatment costs.

AC-systems

At low temperature climates, like in Western Europe, it is forbidden to apply fertilisers on the field during the winter period. When the digested slurry will be applied on the fields as a fertiliser, storage of 3-5 months will be necessary to overcome this winter period. In that situation, combined storage and digestion in a fed-batch or accumulation system (AC) at ambient temperatures, is a feasible alternative for a CSTR system (Zeeman, 1991; Zeeman and Lettinga, 1999) for a CSTR system.

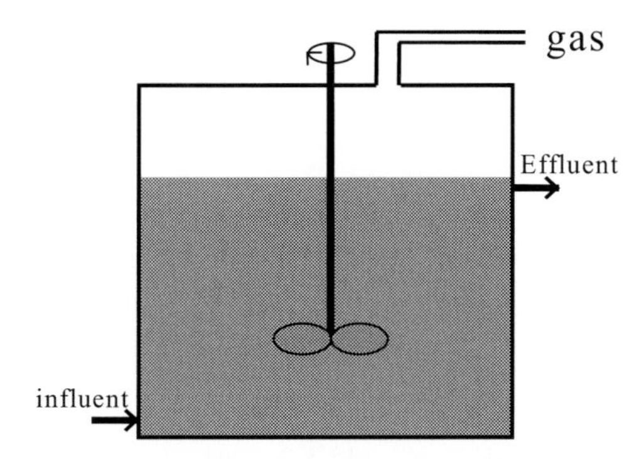

Figure 12.1. Diagram of a CSTR system for the treatment of slurries.

12.2.2 Treatment with biomass retention

For the anaerobic treatment of wastewaters with a large fraction of suspended solids, the hydrolysis of suspended solids is generally the rate-limiting step. Long SRTs are needed to provide a sufficient amount of hydrolysis and methanogenesis. Miron *et al.* (2000) show that an SRT of at least 10 days is necessary to provide methanogenesis at the anaerobic treatment of primary sludge at a process temperature of 25° while a SRT of 15 days is necessary for sufficient hydrolysis and acidification of lipids. For temperatures as low as 15°C, an SRT of at least 75 days has to be provided (non-published results) to achieve methanogenic conditions. To prevent long HRTs and therefore large reactors at the treatment of relatively diluted wastewater streams, the solids must be retained in the system while the liquid needs only a short retention time.

12.2.2.1 Composition of wastewaters B, C, E and F

Table 12.3 presents the concentration and quantity of different types of greywater and blackwater (Kujawa-Roeleveld *et al.* 2000). These results are used to calculate the composition and flow of wastewaters B, C, E and F as presented in Table 12.4.

Table 12.3. Concentration and flow of greywater and blackwater (Kujawa-Roeleveld *et al.* 2000)

Type of wastewater	Volume (l/p/d)	Concentration (g COD/l)	g COD
Kitchen	7.3	1.9	13.87
Laundry	27.6	< 0.>	24.398
Shower/bath	51.5	0.100	5.150
Total	86.4	0.503	43.418
Blackwater 1	4.1 + 11.2 = 15.3	2.6	40.26
Blackwater 2	9.1 + 34.9 = 44	0.9	40.268

Table 12.4. Concentration and flow of wastewaters B, C, E and F

	Units	Value			
Type of wastewater		B	C	e	F
COD	gCOD/l	2.7	0.9	0.823	0.642
V_{ww}^*	l /p.d	15.3	44	101.7	130.4

* Wastewater flow (l /p.d)

12.2.2.2 Treatment systems

The simplest example of such a system, and one that has been applied worldwide for a long time, is the conventional septic tank system. However, the septic tank system can only be applied for low loading rates and only a partial treatment is achieved. These systems are mainly used for house on-site situations. In the 1970s high-rate systems were developed, which are now used more and more for the treatment of domestic sewage.

UASB (upflow anaerobic sludge blanket) systems

In the early 1970s new, high-rate upflow systems were developed, characterised by the achievement of long retention of biomass at short liquid retention time and good contact between biomass and substrate. The most frequently applied system is the UASB (upflow anaerobic sludge blanket) system developed by Lettinga *et al.* (1979). In this chapter calculations for community on-site treatment of wastewaters B, C, E and F are therefore restricted to the application of the UASB system. Figure 12.2 shows a diagram of a UASB reactor. The three-phase separator provides a high retention of sludge in the reactor. The sludge bed in a UASB reactor can either consist of granular or flocculent sludge. Flocculent no granule sludge will develop in the treatment of domestic sewage with a high fraction of suspended solids (SS). Only when two-step systems are applied, where the majority of the suspended solids is removed in the first step (Zeeman *et al.*, 1997; Elmitwally *et al.*, 1999), methanogenic granules can develop in the second step.

Conventional septic tank system

A conventional septic tank in general consists of one or two chambers with a horizontal inlet flow on the top of the reactor. The horizontal influent flow provides that only a part of the suspended solids will be removed by settling. The settled solids will accumulate at the bottom of the reactor. The retention time of the solids depends on the size of the reactor and on the solid concentration that can be provided in the sludge bed. When the reactor is full with sludge, it should be emptied, except for some sludge necessary for inoculation. An important disadvantage of the conventional septic tank system is that no contact is provided between sludge and wastewater, so that no conversion of dissolved components will occur. Moreover the removal of SS is limited to settling.

UASB septic tank system

The UASB septic tank system is a promising alternative to the conventional septic tank (Bogte *et al.* 1993; Lettinga *et al.* 1993). It differs from the conventional system by the upflow mode in which the system is operated, resulting in both improved physical removal of suspended solids and improved biological conversion of dissolved components. The most important difference with the traditional UASB system is that the UASB septic tank system is also designed for the accumulation and stabilisation of sludge. So an UASB septic tank system is a continuous system with respect to the liquid, but a fed-batch or accumulation system with respect to solids. Bogte *et al.* (1993) and Lettinga *et al.* (1993) researched the use of a UASB septic tank system for the on-site treatment of blackwater and total domestic sewage under Dutch and Indonesian ambient conditions. Under low temperature conditions the application of a two step UASB septic tank system could be profitable. Zeeman and Lettinga (1999) present the expected removal efficiencies after treatment of blackwater and greywater plus blackwater at low and tropical temperature conditions.

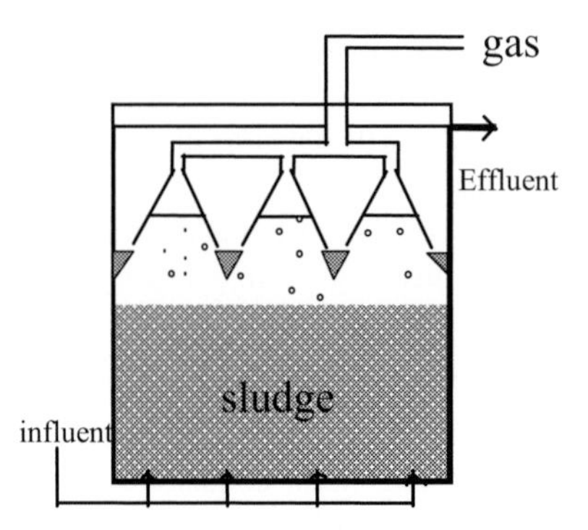

Figure 12.2. Diagram of a UASB system for the treatment of wastewaters.

12.3 MODEL CALCULATIONS

A simplified biological model was used for the model calculations, where acidification and acetogenesis were assumed not to become rate-limiting. Hydrolysis was described by first order kinetics, while methanogenesis was described according to Monod kinetics. The hydrolysis constants were the same as used by Zeeman *et al.* (2000), and are presented in Table 12.5.

The biological model used for the anaerobic digestion of proteins, carbohydrates and lipids to methane gas is schematically presented in Figure 12.3. The biodegradability of all polymers is assumed to be 70%.

Table 12.5 Hydrolysis constants for domestic wastewater components at different temperatures (Zeeman *et al.* 2000).

Component	T (°C)		
	15	20	30
	k_H (d^{-1})		
Lipid	0.010	0.021	0.11
Protein	0.037	0.078	0.40
Carbohydrate	0.044	0.098	0.50

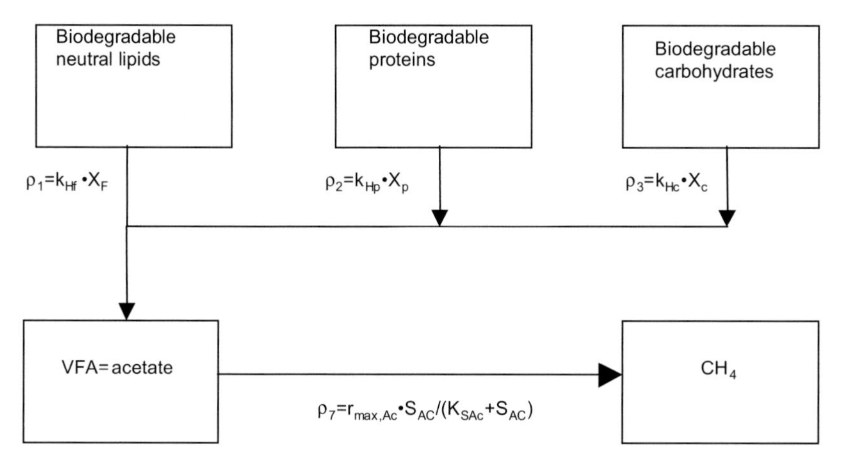

Figure 12.3. Biological model used for anaerobic digestion of proteins, carbohydrates and lipids to methane gas; ρ = conversion rate (g/l.d), k = constant rate (1/d) X = polymer concentration (g/l), F = fats, P = proteins, C = carbohydrates, K_S = affinity coefficient (g/l), S_{Ac} = acetate concentration (g/l).

12.3.1 Model for calculating HRT from the SRT required

For the application of UASB systems, both physical and biological processes will determine the final removal efficiency and conversion of organic compounds to methane gas. For wastewaters with a high fraction of suspended solids, such as sewage, removal of suspended solids occurs by physical processes such as settling, adsorption and entrapment. The subsequent hydrolysis and methanogenesis of the removed solids depends on the process temperature and the prevailing SRT. Zeeman and Lettinga (1999) give a model for the calculation of HRT when a certain SRT is a prerequisite. The model is given below.

The SRT is determined by the amount of sludge that can be retained in the reactor and the daily excess sludge production. The daily excess sludge production is determined by the biomass yield and the removal and conversion of suspended solids. At a certain temperature the SRT will determine whether or not methanogenesis will occur. So when the required SRT is known, the corresponding HRT can be calculated, provided that the sludge concentration in the reactor (X), the fraction of the influent SS that is removed (R) and the fraction of the removed SS that is hydrolysed (H) are known. The HRT of a UASB reactor can be calculated as follows.

$$SRT = X/X_{pr}, \qquad (12.1)$$

where X = sludge concentration in the reactor (g COD/l); 1 g VSS = 1.4 g COD; X_{pr} = sludge production (g COD/l.d)

$$X_{pr} = B_v * SS * R * (1 - H), \qquad (12.2)$$

where B_v = organic loading rate (kg COD/m³.d); SS = $COD_{SS}/COD_{influent}$;

$$HRT = C/B_v, \text{ (days)} \qquad (12.3)$$

where:

$$HRT = (C*SS/X)*R*(1-H)*SRT, \qquad (12.4)$$

where SRT = sludge retention time (days); R = fraction of the COD_{SS} removed; H = fraction of hydrolysed removed solids.

No distinction is made between the fraction of COD_{SS} that is removed but not hydrolysed and the biomass yield.

12.3.2 Energy use for heating of the wastewater

Methane, the end product of mineralisation, has a calorimetric value: 1l of CH_4 (0°C, 760 mm) takes approximately 9.5 kilocalories (kcal). The need to heat the wastewater (Q_H) to 30°C can be calculated according to:

$$Q_H = \phi_V * \tau_F * C_F (30 - t_a)/0.85 \qquad (12.5)$$

where:

Q_H = amount of heat (kcal/d)
ϕ_V = volume flow (l/d)
τ_F = specific weight (≈ 1 for wastewater)

C_F = specific heat (≈ 1 kcal/kg/°C for wastewater)
t_a = temperature of the wastewater
0.85 = efficiency of heat consumption

The heat needed to keep the fermentation space at the right temperature (Q_D) can be calculated as follows:

$$Q_D = F.k.(35-t_{env.})/0.85 \quad (12.6)$$

where:
F = exchange area of the fermentation space (m^2)
k = total heat transfer coefficient ($kcal/m^2/d/°C$)
$t_{env.}$ = temperature of the environment

12.3.3 Results of model calculations and discussion

12.3.3.1 CSTR at 30°C

Based on the models described before, calculations have been made for the conversions taking place in a CSTR at 30°C at two different SRTs, 20 and 30 days. The reactor volume needed per person per day is also calculated. The results are presented in Table 12.6.

Table 12.6. Calculated efficiency of anaerobic mesophilic digestion (30°) of a mixture of night-soil en swill (wastewater a) and a mixture blackwater and swill (wastewater d) in a CSTR system for a SRT of 20 and 30 days

	Wastewater A		Wastewater D	
SRT (d)	20	30	20	30
V_{CSTR} (l/p)	104	156	314	471
VFA production:			2.1	2.3
VFA_{lipid} (gCOD/l)	6.2	6.9	2.1	2.3
VFAcarb (gCOD/l)	7.2	7.5	2.4	2.5
$VFA_{protein}$ (gCOD/l)	3.0	3.2	1.0	1.1
VFA_{tot} (gCOD/l)	16.5	17.5	5.5	5.8
$VFA_{effluent}$ (gCOD/l)	0.36	0.19	0.36	0.19
CH_4 production (l/p.d)	29	32	28	31
Conversion (%) biodegradable COD	79	85	76	83

The results illustrate a methane gas production of ca. 29 l/p.d, at digestion of night soil or blackwater plus swill at 20 days detention time (Table 12.6).

Increasing the SRT to 30 days only results in a 10% increase in gas production. Part of the gas produced is needed to heat the wastewater and the reactor content (Equations 12.5 and 12.6). Applying Equation 12.5 shows that for the treatment of wastewater D (15.7 l blackwater plus swill/day) at 30°C, already 277 kcal are needed for heating up the wastewater from 15 to 30°C, while only 29 × 9.5 = 275.5 kcal are produced as methane gas. Considering that energy is also needed for keeping the reactor content at the right temperature (Equation 12.6), it can be concluded that it is not beneficial to apply a mesophilic CSTR for the treatment of blackwater plus swill, even though the swill considerably increases the COD concentration and therefore the methane production potential.. Moreover, relatively large reactor volumes are needed, viz. 314 l/p. This points to the need to reduce the amount of water used for flushing the toilet. Digestion at low ambient temperatures in either a CSTR or an AC system will lead to even larger reactor volumes to be installed due to the increase in prerequisite SRT at lower temperatures. Therefore only the treatment of the most concentrated domestic wastewater (A) in an AC system at ambient temperatures is calculated. The possibilities of treating blackwaters B and C (without the addition of swill) in a UASB system are discussed later. The sanitation system where very little water is used for flushing the toilet results in the production of a mixture of night soil and swill of only 5.2l/p.d. The latter needs 91.8 kcal/p.d for heating the wastewater from 15 to 30°C. The energy needed for keeping the reactor content at the right temperature depends on the surface of the reactor and the degree of isolation (factor k) as shown in Equation 12.6. For a virtual steel (1.5 cm thick) reactor with a diameter of 10 m, a height of 10 m, a flat roof and a reactor volume (V) of 789 m^3 and isolated with a k value of 0.046 kcal/m^2.h.°C, $Q_{D,total}$ amounts to 179 and 119 kcal/p.d, at ambient temperatures of 15°C and 20°C respectively. Applying heat exchange between influent and effluent can reduce the total heat requirement. Thus an increased surplus energy can be obtained, which can be used in the household.

Another option for reducing the energy needed for heating is to increase the concentration, and therefore reduce the volume to be treated, by separating urine from faeces. Urine contains very little COD and therefore just dilutes the waste. The latter means that instead of a 5.2 litre mixture of night soil and swill, a 1.5 litre mixture of faeces and swill will be produced. This means that the reactor volume will decrease to 29% and the concentration of the influent increases to *ca* 100 g COD/l, resulting in a considerably higher energy surplus. On the other hand, the separation of faeces and urine includes a more complicated collection and transport system. The separate collection of urine and faeces has already been demonstrated. Profits and costs of different systems should be evaluated in order to select the most appropriate collection-, transport- and reactor technology.

12.3.3.2 AC system at 15 and 20°C

The results of the calculations of the treatment of night soil plus swill (wastewater A) are presented in Tables 12.7–10. The results show that a total reactor volume of respectively 584l/p.d and 764 l/p.d is needed at an accumulation period of 100 days at 15 and 20°C respectively. The difference in volume between the two temperatures is caused by the difference in needed inoculum volume, which illustrates that the indoor installation of the reactor, for example in the basement of a building, can considerably reduce the reactor volume needed. The volumes calculated are of course much larger than those calculated for mesophilic CSTR systems. For a CSTR, however, an additional storage of 100 days should be installed when the slurry is to be applied for agriculture. The production of gas is somewhat lower than those in CSTR systems, but on the other hand no gas is needed for heating.

Table 12.7. Calculated inoculum volume and total reactor volume for application of an AC system to treat a mixture of night soil and swill (wastewater A) in an AC system at 20°C for different accumulation periods

Accumulation period (d)	Units	100	180	365
Effective reactor volume	l/p	520	936	1898
Inoculum volume	%	11	7	4
Total reactor volume	l/p	584.3	1006.4	1977.1
Inoculum TS	g/l	33.4	33.4	33.4
Inoculum activity	gCOD/gTS.d	0.03	0.03	0.03

Table 12.8. Calculated efficiency of the anaerobic digestion of a mixture of night soil and swill (wastewater A) in an AC system at 20°C for different accumulation periods

Accumulation period (d)	100	180	365
VFA production:			
VFA_{lipid} (gCOD/l)	7.9	8.4	8.7
VFA_{carboh} (gCOD/l)	6.99	7.43	7.70
$VFA_{protein}$ (gCOD/l)	3.01	3.20	3.31
VFA_{tot} (gCOD/l)	17.9	19.0	19.7
$TS_{hydrolysed}$ (g/l)	11.1	11.7	12.2
$TS_{effluent}$ (gCOD/l)	11.5	10.8	10.4
CH_4 production (l/p.d)	27	29	30

Table 12.9. Calculated inoculum volume and total reactor volume for application of an AC system to treat a mixture of night soil and swill (wastewater A) in an AC system at 15°C at different accumulation periods (days)

Accumulation period (d)		100	180	365
Effluent reactor volume	l/p	520	936	1898
Inoculum volume	%	32	20	11
Total reactor volume	l/p	764	1170	2132
Inoculum TS	g/l	33.4	33.4	33.4
Inoculum activity	gCOD/gTS.d	0.008	0.008	0.008
Max. CH_4 at t=0	gCOD/day	65.4	62.5	62.7

Table 12.10. Calculated efficiency of the anaerobic digestion of a mixture of night soil and swill (wastewater A) in an AC system at 15°C at different accumulation periods

Accumulation period (d)	100	180	365
VFA production:			
VFA_{lipid} (gCOD/l)	6.67	7.67	8.34
VFA_{carboh} (gCOD/l)	5.91	6.80	7.39
$VFA_{protein}$ (gCOD/l)	2.54	2.93	3.18
VFA_{tot} (gCOD/l)	15.12	17.40	18.92
TS_{hydrol} (g/l)	9.37	10.78	11.72
$TS_{effluent}$ (gCOD/l)	13.20	11.79	10.85
CH_4 production (l/p.d)	23	26	29
Conversion (% biodegradable COD)	75	85	93

12.3.3.3 UASB at 10–15°C

To provide methanogenic conditions and both hydrolysis and β-oxidation of lipids at 15°C, a SRT of at least 75 days is necessary. For the treatment of blackwater and total domestic sewage at low temperature conditions, for example, 15°C, a prerequisite SRT of 100 days is assumed. When the sludge bed is assumed to behave like a CSTR and hydrolysis constants are similar to those in Table 12.5, it can be calculated that only *ca.* 30% of the total COD_{SS} retained in the sludge bed will be hydrolysed. Based on these assumptions the HRT to be applied is calculated for the four different wastewaters (Table 12.11).

Table 12.11. Calculated HRT for the treatment of four different domestic wastewaters at 15°C at a prerequisite SRT of 100 days and a sludge concentration in the reactor of 21gCOD/l and an assumed removal of COD_{SS} of 70%

Type of wastewater	B	C	E	F
SRT (days)	100	100	100	100
Wastewater conc. (gCOD/l)	2.7	0.9	0.82	0.64
Conc. SS (gCOD/l)*	2.7	0.9	0.74	0.58
Calculated HRT (days)	6.30	2.10	1.73	1.35
V_{UASB} (l/p)	95	95	176	176
CH_4 production (l/p.d)				
Conversion (%) biodegradable COD	0.4	0.4	0.4	0.4

*The fraction of COD_{SS} in greywater is assumed to be 0.8.

12.3.3.4 UASB at 25–30°C

To provide methanogenic conditions and both hydrolysis and β-oxidation of lipids at 25°C a SRT of at least 15 days is necessary (Miron *et al.* 2000). For the treatment of blackwater and total domestic sewage at tropical conditions, that is, 25–30°C a prerequisite SRT of 30 days is chosen (Table 12.12). It can be calculated that *ca.* 60% of the total COD_{SS} retained in the sludge bed will be hydrolysed when the sludge bed is assumed to behave like a CSTR system and hydrolysis constants are similar as those in Table 12.5 for 30°C. Based on these assumptions the HRT to be applied is calculated for the four different wastewaters (Table 12.12).

The results show an important difference both in the HRT to be applied in a UASB and the methane recovery between a process temperature of 15 and 25°C (Table 12.11 versus 12.12). Treatment in a UASB system under tropical conditions is highly advisable when either blackwater or total domestic wastewater is produced. For low temperature treatment long HRTs are needed. Moreover, low gas recovery is achieved even at the applied low loading rates. Therefore it is recommended to apply a two-step system at low temperatures. The first step basically removes SS, while the second step will provide methane production from dissolved and remaining particulate material. Elmitwalli *et al.*

Table 12.12. Calculated HRT for the treatment of four different domestic wastewaters at 25–30°C at a prerequisite SRT of 30 days and a sludge concentration in the reactor of 21gCOD/l and an assumed removal of COD_{ss} of 70%

Type of wastewater	B	C	E	F
SRT (days)	30	30	30	30
Wastewater conc. (gCOD/l)	2.7	0.9	0.82	0.64
Conc. SS (gCOD/l)*	2.7	0.9	0.74	0.58
Biodegradability SS (%)	70	70	70	70
Calculated HRT (days)	1.08	0.36	0.30	0.23
V_{UASB} (l/p)	16	16	30	30
CH_4 production (l/p.d)	8.6	8.6		
Conversion (%) biodegradable COD	0.85	0.85	0.85	0.85

*The fraction of COD_{SS} in greywater is assumed to be 0.8.

(1999) show that an anaerobic filter (AF) can provide 80% removal of COD_{SS} at an HRT of 4 hours and a temperature as low as 13°C. For the production of methane gas, the sludge produced in the first step can be digested in a CSTR together with swill. Table 12.12 clearly shows the strong effect of the increase in the amount of flushing water on the reactor volume needed per person. The latter stresses the importance of reducing the amount of flushing water and the separation of toilet wastewater and greywater. This separate collection is even more important since the total wastewater contains all the nutrients and pathogens, which were originally produced in a small volume as faeces and urine. After anaerobic digestion of the total wastewater, therefore, a more complex treatment system is needed to subsequently remove these compounds from a large water volume. So 'dilution' by flushing not only affects the anaerobic treatment but also the post-treatment system.

A relatively large difference in reactor volume needed per person per day is shown when comparing digestion of wastewater D (blackwater 1 plus swill) or wastewater A (night soil plus swill) in a CSTR (Table 12.6) with that of wastewater B (blackwater 1) in a UASB system at mesophilic conditions (Table 12.12). This is caused by the addition of swill, and the fact that digested solids are concentrated in the UASB to 21 g $COD/l_{reactor\ vol.}$. The digested slurry after digestion of wastewaters A and D in a CSTR has a concentration of only 12.5 and 4 g COD/l, respectively. One possibility of reducing the volume of night soil produced daily without affecting the gross methane potential is separation of the urine, as mentioned before. Another option is the reduction of the amount of flushing water at the collection of night soil. Vacuum toilets are already available and use flushing water quantities as discussed in this chapter. Zeeman *et al.* (2000) assume a lower water flushing for vacuum toilets, resulting in the production of 2.2 litres mixture of night soil plus swill per person per day with the CSTR volume needed only being 44 litres/p. The development of these toilet systems could even lead to lower water consumption and therefore the need for smaller reactor volumes to be applied.

12.4 REFERENCES

Bogte, J.J., Breure, A.M., van Andel J.G. and Lettinga, G. (1993) Anaerobic treatment of domestic wastewater in small-scale UASB reactors. *Wat. Sci. Tech.* **27**(9), 75–82.

Elmitwalli, T.A., Sklyar, V., Zeeman, G. and Lettinga, G. (1999) Low temperature pre-treatment of domestic sewage in anaerobic hybrid and anaerobic filter reactors. *Proc. 4th IAWQ Conference on Biofilm Reactors*, 17–20 October, New York.

Hulshoff Pol, L., Euler, H, Eitner, A. and Grohanz, T.B.W. (1997) State of the art sector review. Anaerobic trends. *WQI* July/August, 31–33.

Kalogo Y. and Verstraete, W. (2001) Potentials of anaerobic treatment of domestic sewage under temperate climate conditions. Chapter 10 of this volume.

Kujawa-Roeleveld, K. (2001) Types, characteristics and quantities of domestic solid waste. Chapter 5 of this volume.

Kujawa-Roeleveld, K., Zeeman, G. and Lettinga, G. (2000) DESAH in grote gebouwen. EET rapport. In Dutch

Lettinga, G., de Man, A., van der Last, A.R M., Wiegant, W., Knippenberg, K., Frijns, J. and van Buuren, J.C.L. (1993) Anaerobic treatment of domestic sewage and wastewater. *Wat. Sci. Tech.* **27**(9), 67–73.

Lettinga, G., van Velsen, L., de Zeeuw, W. and Hobma, S.W. (1979) The application of anaerobic digestion to ind
ustrial pollution treatment. *Proc. 1st Int. Symp. On Anaerobic Digestion*, 167–186.

Miron, Y., Zeeman, G., van Lier, J.B. and Lettinga, G. (2000) The role of sludge retention time in the hydrolysis and acidification of lipids, proteins, carbohydrates and proteins during digestion of primary sludge in CSTR systems. *Water Research* **34**(5), 1705–1714.

Zeeman, G. (1991) Mesophilic and psychrophilic digestion of liquid manure. Ph.D. thesis, Department of Environmental Technology, Agricultural University, Wageningen, the Netherlands.

Zeeman, G. and Lettinga, G. (1999) The role of anaerobic digestion of domestic sewage in closing the water and nutrient cycle at community level. *Wat. Sci. Tech.* **39**(5), 187–194.

Zeeman, G., Sanders, W. and Lettinga, G. (2000) Feasibility of the on-site treatment of sewage and swill in large buildings. *Wat. Sci. Tech.* **41**(1), 9–16.

Zeeman, G., Sanders, W.T.M., Wang, K.Y. and Lettinga, G. (1997) Anaerobic treatment of complex wastewater and waste activated sludge. Application of an upflow anaerobic solid removal (UASR) reactor for the removal and pre-hydrolysis of suspended COD. *Wat. Sci. Tech.* **35**(10), 121–128.

13

Compact on-site treatment methods for communities – Norwegian experiences

H. Ødegaard

13.1 INTRODUCTION

This chapter discusses compact wastewater treatment units for small communities, i.e. decentralised wastewater systems. The IWA Specialist Group on Small Wastewater Treatment Plants has defined a small treatment plant as being one serving fewer than 2000 persons (pe) or with a daily flow of under 200 m^3/d. We shall use this international definition in this chapter. Even though there is no accepted definition of a decentralised system, *mini treatment* plants (on-site plants) and *small treatment plants* will be distinguished according to the definitions used in Norway.

Edited by P. Lens, G. Zeeman and G. Lettinga. ISBN: 1 900222 47 7

With about 4 million inhabitants and an area of 400,000 km^2, Norway is a sparsely populated country. About 30% of the 450 municipalities have a population of under 3,000. In most municipalities, the population is decentralised in small villages/communities. Even though these small communities have wastewater collection systems leading to small treatment plants, their wastewater systems have all the features of a decentralised system because of the small size. In Norway these plants are called 'small treatment plants' when they serve between 35–500 people, in order to distinguish them from mini treatment plants which serve fewer than 35 people (or 7 houses). Normally, the Norwegian municipalities own the small plants and they fall under another regulatory system in which individual discharge permits are granted by the relevant environmental authorities.

About 25% of the population live in rural areas without any central wastewater collection system, and in these cases on-site wastewater treatment is carried out. According to environmental authority regulations, single houses or groups of up to seven houses can use their own on-site treatment systems. These on-site systems usually consist of a septic tank followed by a flow distributor and a soil infiltration system. In cases where soil infiltration cannot be used due to impermeable soils, compact, pre-fabricated mini treatment plants are often used. They treat all household wastewater (grey- as well as blackwater) and normally use pre-treatment by a septic tank followed by a unit based on biological or chemical processes or combinations of the two. Occasionally, the plants are used for greywater only. Compact, on-site plants that fall under the regulation of scattered dwellings (<7 houses or 35 people) will be referred to here as mini treatment plants. They are normally privately owned. Once the environmental authorities approve a plant, it can be used without an individual discharge permit being needed.

Table 13.1 shows the number of decentralised wastewater treatment plants in Norway in 1996 divided into the three size groups (<35 people, 35–500 people and 500–2000 people) and different type of treatment (biological, chemical or biological/chemical) (Ødegaard 1996). Table 13.1 shows that the majority of the plants are biological/chemical. About 75% of the small plants (35–2000 people) and 70% of the mini plants (<35 people) have phosphorous removal in the form of chemical precipitation (chemical or biological/chemical). This chapter will concentrate on experiences from the two main size groups of plants (mini plants and small plants), and one type of plant from each of these size categories will be dealt with in more detail as examples of technologies used in Norway.

Table 13.1. Number of decentralised wastewater treatment plants in Norway as a function of treatment method and plant size (Ødegaard 1996)

	Total		<35 pe		35–500 pe		500–2000 pe	
Treatment method	No.	%	No.	%	No.	%	No.	%
Biological	1299	30.5	1175	31.3	86	31.2	38	16.8
Chemical	480	11.3	375	10	41	14.8	64	28.3
Bio/chem	2473	58.2	2200	58.7	149	54	124	54.9
TOTAL	4252	100	3750	100	276	100	226	100

Typically the effluent standards for small plants in Norway are given in terms of annual average effluent BOD_7– and total P-concentrations based on 6 flow proportional daily samples per year for plants serving less than 1000 persons or person equivalents (pe) and 12 samples per year for plants serving 1000–2000 pe. The effluent standards vary from one county to another, but are typically as shown in Table 13.2.

Table 13.2. Typical effluent standards for small wastewater treatment plants in Norway

Process category	Tot P (g/m^3)	BOD_7 (g/m^3)
Chemical	0.5–0.6[2]	n/a[1]
Biological	n/a[1]	15–20[2]
Biological/chemical		
Simultaneous precipitation	0.8[2]	15–20[2]
Combined- and post-precipitation	0.4–0.5)	10–15[2]

[1] Standard not given
[2] Average value

13.1.1 Wastewater characteristics

Decentralised wastewater treatment plants normally receive more concentrated wastewater than centralised plants, and variations in composition are greater. The smaller the system, the larger the variation in the composition of wastewater. In addition to the enormous variation in composition of wastewater entering a decentralised plant (from plant to plant as well as over time), there is also an enormous variation in flow. This has to be taken care of by the system, by balancing volumes and/or treatment processes that can tackle these great variations.

In most decentralised treatment plants, pre-treatment takes place in a septic tank in which solids are separated by settling and the settled solids (sludge) are stored in the tank for a longer period (three months to two years) resulting in a certain degree of anaerobic decomposition in the tank. One may say that this results in anaerobically pre-treated water even though the pre-treatment unit is not optimised with respect to solids' hydrolysis and stabilisation. In Norway, there is no experience with specifically designed anaerobic pre-treatment units, for instance based on biofilters or upflow anaerobic sludge blanket (UASB) units. Pre-treatment on anaerobic biofilters is, however, extensively used in on-site treatment plants in Japan (Yang *et al.* 2001). In some cases, though, especially in systems based on activated sludge, no pre-treatment unit is used (see later in this chapter).

In some cases, treatment units do not treat combined sewage, but treat greywater. Greywater arises from all domestic washing operations, such as wastewater from hand basins, kitchen sinks and washing machines, while excluding blackwater (water from toilets). Greywater is generated by the use of soap, shampoo and other toiletries for body cleaning and other such activities. The resulting water quality is by nature very site-specific, varying in strength and composition. It has an organic strength similar to domestic wastewater, but is relatively low in suspended solids and turbidity, indicating that a greater proportion of the waste is dissolved (Jefferson *et al.* 2000). The chemical nature of the organic matter is, however, quite different. The COD/BOD ratio is much greater than is normally reported for sewage. This is coupled with an imbalance of macronutrients such as nitrogen and phosphorus, shown in terms of $COD:NH_3:P$ as 1030:2.7:1 compared to 100:5:1 for domestic wastewater (Jefferson *et al.* 2000). Even though greywater can be treated with the same processes as combined sewage, the special composition requires special attention. Both of the aspects mentioned above reduce, for instance, the efficacy of biological treatment, since organic matter is not the process-limiting component.

13.1.2 Treatment systems

The processes used for treating wastewater (sewage as well as greywater) in compact treatment plants from decentralised wastewater systems, are in principle the same as those used for larger centralised wastewater systems, i.e. physical, chemical and biological processes as well as combinations of these processes. Among the physical processes, sedimentation dominates because of its simplicity. Lamella- or tube settlers are sometimes used for post-settling. The pre-settling step is normally combined with sludge storage, that is, the septic tank or Imhoff tank principle. Of the biological processes, aerobic processes have so far dominated (except for the fact that the septic tank is an anaerobic reactor in addition to being a separation reactor). Of the aerobic biological processes, both activated sludge and biofilm processes (rotating biological contactors (RBCs), trickling filters, submerged biological filters and moving bed biofilm reactors) are in use. Chemical processes are used less than biological processes for small plants. In Norway, however, chemical processes are frequently used, primarily in combination with biological processes but also as the sole treatment. The reason for this is the extensive phosphate removal requirements that exist for most plants in Norway, even for many of the on-site treatment plants.

13.2 NORWEGIAN EXPERIENCE OF COMPACT, PREFABRICATED ON-SITE TREATMENT UNITS (MINI TREATMENT PLANTS)

There has been considerable scepticism towards the use of prefabricated on-site mini-treatment plants in Norway. Earlier, guidelines existed that prohibited the use of such plants and these meant that only plants based on infiltration or constructed sand filters were legal. Many of the houses in Norway are, however, built on rocky ground and there was a pressing need for other solutions for these cases. Since 1985, the authorities have accepted the use of mini treatment plants, provided that the type of plant to be used has been approved.

Six plants have been approved (see Figure 13.1) out of which five are still being used (see Table 13.2). BIOVAC has decided that they will put their emphasis on the fill and draw activated sludge plant (BIOVAC FD), and has withdrawn the flow-through plant (Figure 13.1(a)) from the market. BIOVAC plants dominate the biological and biological/chemical plant market. This plant will be described in more detail later.

Wallax is the only purely chemical plant. It consists of concentric cylinders made out of glass-fibre reinforced plastic, with the pre-settling tank in the outer ring and chemical sludge separation in the central tank. The sludge is taken away by lorry. The Colombio plant is based on pre-precipitation followed by a biofilter, while the Klargester B1 plant is a purely biological plant in two separate stages. The BB plant is also a purely biological plant, but is based on activated sludge. Table 13.3 gives the results of performance tests according to the approval system (Paulsrud and Haraldsen 1993).

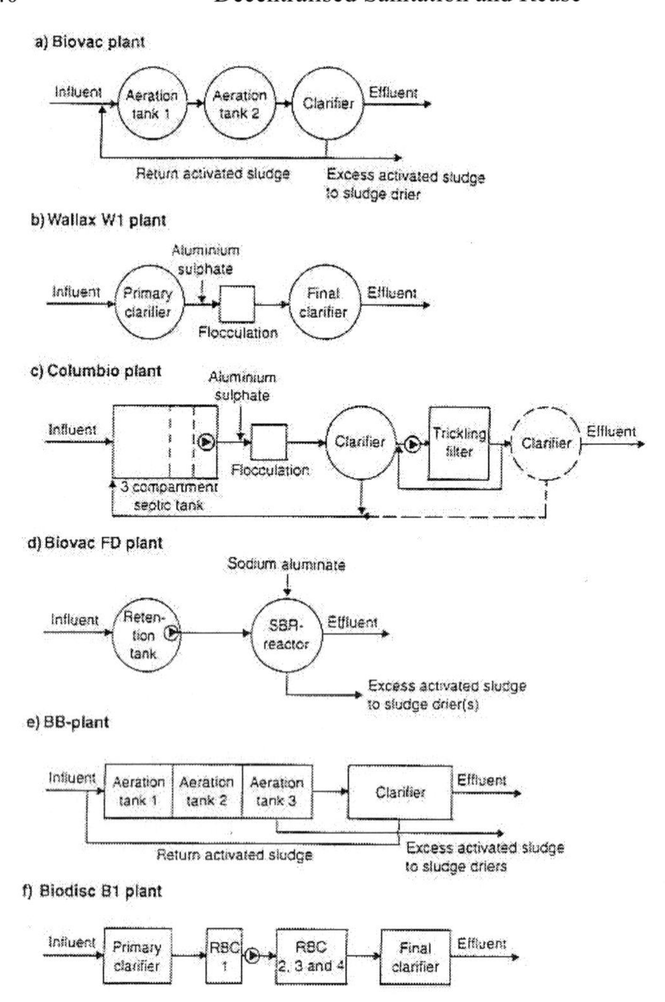

Figure 13.1. Flow sheets for approved mini treatment plants in Norway.

Table 13.3. Average effluent concentrations from different plants in performance test

Plant name	No. of plants	Av..spec. load l/pe·d	BOD_7 mg/l	COD mg/l	Tot P mg/l	SS mg/l
BIOVAC FD	4	124	7	58	0.60	25
Wallax 1 (3)	3	131	n/a	284	0.82	39
Colombio	3	169	20	86	0.37	27
Klargester B1	3	143	15	89	4.5	18
BB-plant	3	139	14	103	8.8	26

13.2.1 Practical experiences

Experience with mini treatment plants in Norway was evaluated in 1994 (Heltveit 1994). All mini treatment plant owners received questionnaires and 65% responded. Control and sampling were carried out at 132 selected plants. However, only grab samples were taken, and the main purpose of the control visit was to evaluate the situation with respect to operation and maintenance.

It was found that 42% of plants were placed in basements, 31% in a small purpose-built house designed for the plant, 10% were in the garage while 13% were dug into the ground without cover. Sixty-one per cent of the plants discharged to a receiving water body (brook, river, lake) while 39% used some kind of soil percolation system downstream from the mini plant. Interestingly, 91% of the plant owners were satisfied with the service they had received from their plant supplier and 94% reported that the supplier had fulfilled their obligations. However, only 50% of plant owners were satisfied with the service they had received from their local authority. Fifty-two per cent reported that that there were no noise or odour problems, while 25% reported odour problems, 19% reported noise problems and 4% reported both. In 53% of plants, assistance had been sought from the plant supplier outside regular service visits (in the five year period) and 67% of the plants had experienced more than one break in operation. These breaks were most often caused by blockages that in most instances had been caused by a failure in electric power supply.

During control visits to the 132 plants, it was evident that plants that were placed in a heated room were better maintained than those placed underground (uncovered) or in unheated rooms. The major disadvantage revealed through the control visits was that many of the plants did not de-sludge frequently enough and contained too much sludge, with the result that sludge escaped in the effluent. Treatment results (as determined via the grab samples) were poorer than expected, mainly due to sludge loss.

13.2.1.1 The BIOVAC fill and draw plant – an example of an on-site mini treatment plant

The BIOVAC FD plant will be discussed here as an example of a mini treatment plant commonly found in Norway. The plant consists of the following main components (see also Figure 13.2):

(1) Reception/flow balancing tank with pump

- The tank does not have any sludge separation function (no septic tank)
- The tank may be located away from the bio-reactor

(2) SBR reactor with control system

- Closed sequencing batch activated sludge reactor with logical process control (PLC)
- Alarm system, logging system for operation- and alarm data
- Control programme adapted to the application (biological, biological/chemical, denitrification and so on)

(3) Sludge dryer with dewatering function

- Dewaters stabilised sludge (aged 20 days) to 15–40% dry solids (DS) by evaporation caused by air suction
- The long storage time (4–6 months) ensures no wet sludge being emptied
- The sludge is removed by BIOVAC as part of their service agreement

(4) Equipment for chemical precipitation

- Mounted directly on SBR – dosage controlled by PLC
- Simultaneous precipitation by sodium aluminate
- Coagulant delivered by BIOVAC as part of their service agreement

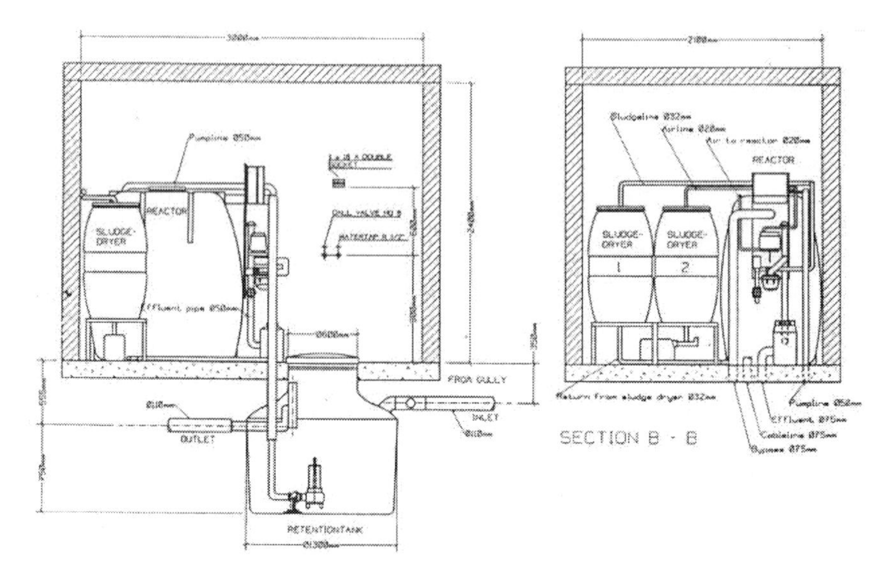

Figure 13.2. The BIOVAC FD single house biological/chemical plant.

In 1996 the cost for a 5 pe biological/chemical plant was 53,300 Norwegian kroner and NOK 48,500 for a purely biological plant while service costs (including chemicals) were NOK 2180 per annum (1780 for a biological plant). The energy consumption is 500 kWh per year.

Table 13.4 gives design and operational specifications for the BIOVAC FD single house (5 pe) plant.

Table 13.4. Design and operation data for a single house (5 pe) BIOVAC FD biol./chem. plant

Input data	Design data	Process values
Design flow: 0.65 m^3/d Max flow: 1.58 m^3/d MLSS: 5000 mgSS/l Reaction time: 180 min Air per aerator: 6 m^3/h	Reactor wet volume: 1.0 m^3 Operation volume: 0.22 m^3 Influent loading (aerated): 6 min Settling time: 90 min Withdraw time: 15 min	Aeration factor: 78% Sludge load: 0.06kg BOD/kgSS*d Sludge production: 0.31 kg SS/d Sludge age (total/aerobic): 16/13 d Air consumption: 68 m^3/d

13.3 NORWEGIAN EXPERIENCE OF COMPACT, SMALL TREATMENT PLANTS (35–2000 PE)

In Norway the majority of small treatment plants (serving 35–2000 people) belong to municipalities. In most counties an 'operation assistance' organisation has been formed to help and support municipalities in operating their wastewater treatment plants. These organisations are often connected to firms with wastewater treatment expertise. Experts from these organisations provide support and help to the operators, collect samples according to the environmental protection authorities' requirements, evaluate the results and present these in annual reports.

Small advanced treatment plants in Norway may also be grouped into three main groups, chemical, biological and biological/chemical plants (see Figure 13.3). There may be many process variations in each of these groups, but there are some dominating subgroups. Among chemical plants, one may differ between primary precipitation plants (without pre-settling tank) and secondary precipitation plants (with pre-settling tank). Among the biological plants, one may differ between activated sludge plants and biofilm plants. RBCs dominate among biofilm plants, but the moving bed biofilm process is becoming increasingly popular for small plants. Among biological/chemical plants, pre-precipitation plants (chemical step ahead of biological step), simultaneous precipitation plants (chemical precipitation takes place in the biological reactor, and is based on activated sludge), combined precipitation plant (chemical precipitation takes place directly after the bioreactor, and is based on a biofilm process, normally RBC) and post-precipitation plants (chemical step after a biological step, normally based on activated sludge) can be distinguished.

The pre-treatment of these plants is normally based on one of the following alternatives: (a) screens and grit removal and possibly conventional pre-settling; (b) overall pre-treatment in a large septic tank/Imhoff tank (combined sludge separation and sludge storage tank); or (c) wastewater shredder base on a grinding action. Sludge handling is normally based on: (a) no treatment other than thickening in the storage volume of the large septic tank; (b) separate thickener/sludge storage (not aerated); and (c) aerated sludge storage/sludge stabilisation tank. A lorry transports the sludge to a larger treatment plant with sludge treatment facilities. Many of the plants in the 500–2000 people range have their own sludge dewatering facilities (mostly centrifuges) in order to cut transportation costs.

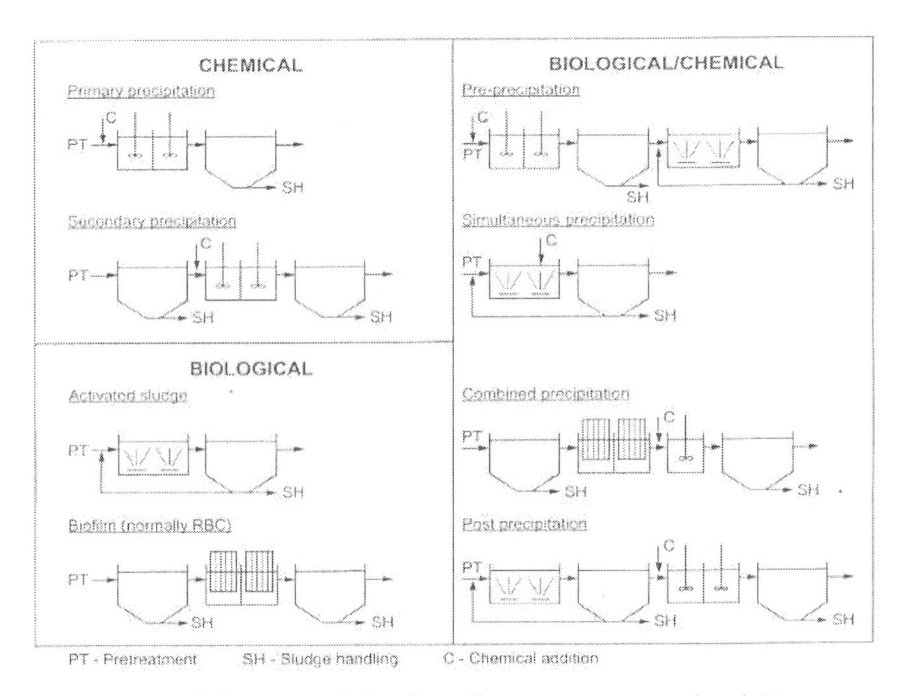

Figure 13.3. Typical treatment solutions for small wastewater treatment plants in Norway.

13.3.1 Performance evaluation investigation

Reports from nine operation assistance organisations were collected, covering influent and effluent data over three years (1994–1996) of 356 plants (Ødegaard and

Skrøvseth 1997). This was about 90% of all the advanced treatment plants in this size category in Norway at this time.

Table 13.5 gives the results of a comparison between the main treatment process categories. It clearly shows that biological/chemical processes are superior to chemical and biological ones with respect to the removal of organic matter. With respect to phosphorous removal, the average chemical process is even better than the average combined biological/chemical one. With respect to suspended solids, the different main process categories are equal. The primary difference between the chemical and biological plants is demonstrated by the greater ability of the chemical ones to remove phosphorus. The average effluent level of the chemical plants is impressive; only 0.42 mg P/l. It is interesting, though, that biological plants remove more than 50% of phosphorous. This is considerably higher than normally experienced in larger centralised biological plants (Henze and Ødegaard 1994). This may be caused by an anaerobic selector in the large septic tank to which the biological sludge is returned, as well as assimilation and the fact that the fraction of particulate phosphorous may be high (Ødegaard 1999).

Table 13.5. Overall treatment results according to parameter and process category

Parameter/process category	$N_{p,tot}$[1]	$N_{p,C}$[2]	$N_{s,C}$[3]	C_{in} (mg/l)	C_{out} (mg/l)	R_1[4] (%)	R_2[5] (%)
COD Chemical	99	98	2811	474±222	108 ±62.0	74.8±10.0	77.2
COD Biological	56	48	730	474±203	88.6±40.9	78.9±8.1	81.3
COD combined	201	194	3593	474±271	60.0±26.2	84.5±7.3	87.3
All	356	340	7134	474±249	78.1±46.8*	81.2±9.6	83.5
BOD Chemical	99	14	316	185±114	48.7±38.9	71.9±16.6	73.7
BOD Biological	56	29	377	222±93	34.7±19.1	81.9±11.5	84.4
BOD combined	201	111	3112	208±136	18.9±21.7	89.3±7.3	90.9
All	356	154	3805	209±127	24.6±25.1	96.2±10.8	88.2
Tot P Chemical	99	99	3557	5.32±2.52	0.42±0.42	90.6±11.0	92.1
Tot P Biol.	56	39	523	6.36±2.75	2.93±1.68	52.9±20.5	53.9
Tot P combined	201	201	6085	6.56±3.28	0.52±0.51	91.1±8.7	92.1
All	356	339	10165	6.17±3.06	0.77±1.07	86.7±16.4	87.5
SS Chemical	99	63	1060	251±167	24.9±18.7	83.4±16.1	90.1
SS Biological	56	34	444	186±119	24.1±13.4	83.3±11.7	87.0
SS combined	201	147	3007	284±191	25.5±29.0	90.7±8.4	91.0
All	356	273	4511	253±174	25.2±24.0	87.4±11.6	90.0

[1] Total number of plants in actual category; [2] Number of plants with parameter analysed; [3] Total number of samples with parameter analysed; [4] Treatment efficiency based on individual samples; [5] Treatment efficiency based on averages of all samples

Similarly, it is interesting to note the relatively high removal efficiency with respect to organic matter in chemical plants. The effluent concentration of COD and BOD are better in the biological plants but not as high as expected. In general, therefore, the difference between the chemical plants and the biological plants was less than expected. Figure 13.4 gives frequency plots for effluent concentrations.

The frequency plots (as well as the standard deviations in Table 13.5) reveal that the process stability with respect to phosphorous is very good in chemical and biological/chemical plants. About 90% of the effluent samples from these plants had under 1 mg P/l. About 80% of the chemical plants had an effluent concentration lower than 0.6 mg P/l while the corresponding figure for the biological/chemical plants was 0.8 mg P/l. The poorer effluent P-concentration in the biological/chemical plant category is primarily caused by a lower process stability among the simultaneous precipitation plants (see Figure 13.7). Among the biological plants, the effluent P-concentration is obviously highly dependent on local conditions.

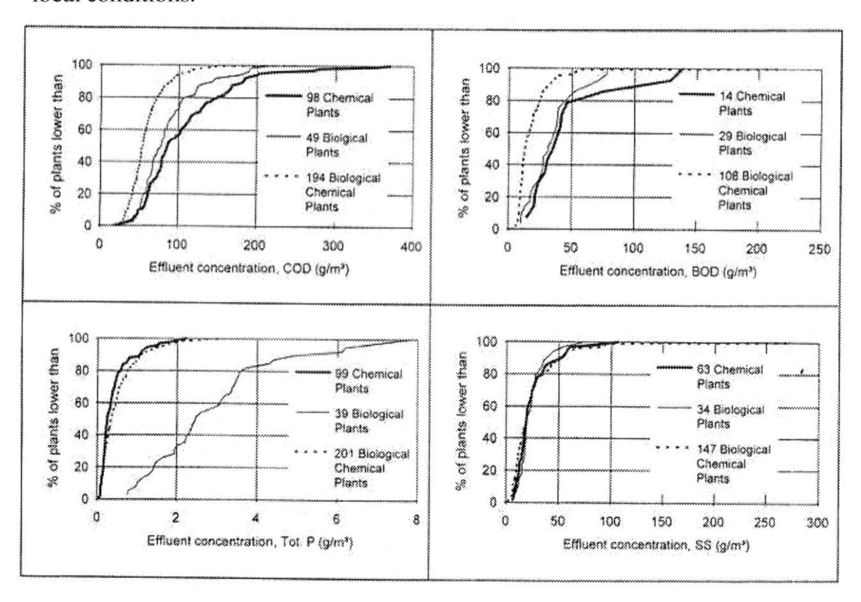

Figure 13.4. Frequency plots of effluent COD, BOD_7, tot P and SS for the main process categories of small wastewater treatment systems.

It is interesting to note that the effluent COD concentration of the median biological respective chemical plant is about the same (70–80 mg COD/l), while at the 80 percentile the biological plants are better – about 100 mg COD/l versus about 150 mg COD/l. This also reflects local conditions. Biological plants are primarily intended to remove organic matter – soluble as well as particulate. Chemical plants are primarily intended for phosphorous and particle removal. If a larger fraction of the organic matter exists in soluble form, biological plants will perform better.

When analysing the effect of plant size, it was found that the results were somewhat better in the 500–2000 pe category than in the <500 pe category among biological and biological/chemical plants, while among the chemical plants the better results were in the <500 pe category. Generally, however, there were small differences between the results of the two plant size categories. This was surprising since one would expect that the bigger the plant, the more stable its operation would be, thus leading to better results. The results indicate that chemical plants are particularly robust when it comes to plant size influence on operation stability.

In the following sections we will analyse more closely the results from each type of process category.

13.3.1.1 Chemical treatment plants

Chemical treatment plants are frequently used in Norway (Ødegaard 1992), first in medium-sized and large plants, but recently increasingly frequently in small plants too, since their operational stability has been found to be very satisfactory. They are normally based on a large septic tank for pre-treatment (secondary precipitation) and the chemically precipitated sludge is pumped back from the final settling tank to the septic tank for storage together with the primary settled sludge. In some cases coagulant addition is carried out directly on the raw sewage and only one unit is used for floc separation and sludge storage (primary precipitation). Coagulants in solution are favoured among the smaller plants and the most frequently used are aluminium sulphate, ferric chloride and pre-polymerised aluminium chloride.

Figure 13.5 gives the frequency plots of the different parameters for the chemical plants. It is demonstrated that in the top 50% of the plants, there is little difference in effluent quality between the two chemical plant categories (primary and secondary precipitation). Among the bottom 50%, however, secondary precipitation plants are clearly better. This is probably a result of better process stability in the secondary precipitation plants because of the balancing effect of the septic tank. The removal of suspended solids (SS) is equally good in the two chemical plant categories.

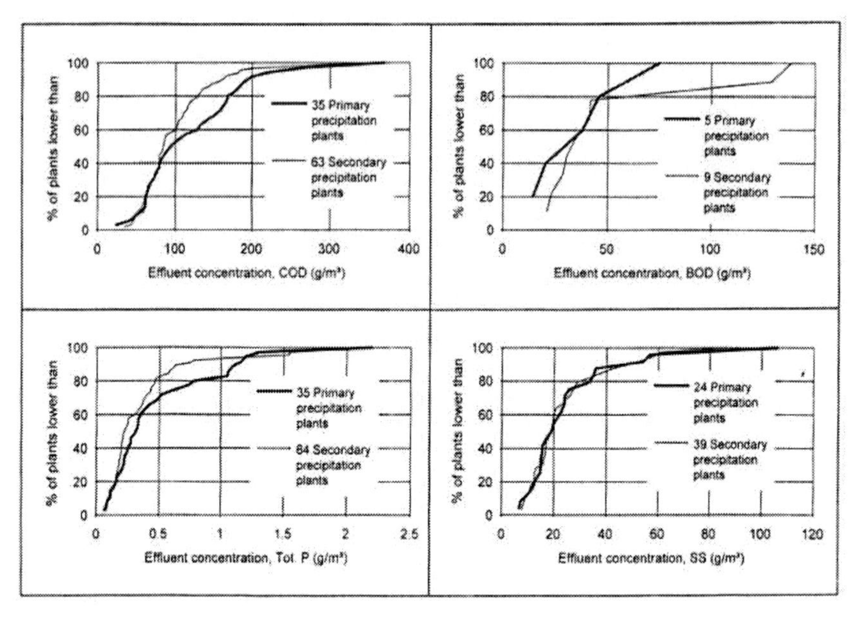

Figure 13.5. Frequency plots of effluent COD, BOD_7, tot P and SS for chemical treatment plants.

13.3.1.2Biological treatment plants

Table 13.5 shows that there are many purely biological plants in Norway. This is because phosphorous removal is normally required. Figure 13.6 shows that activated sludge plants give better results both with respect to COD and SS than the biofilm plants. It can also be seen that the particle separation in a large portion (about 50%) of the biological plants is poor, resulting from sludge loss in the activated sludge plants and pinpoint flocs escaping the separator in the biofilm plant. The effluent phosphorous concentrations in the biofilm plants were significantly poorer than in the activated sludge plants. The low slope of the frequency plot may be taken as an indication of poor process stability among biological plants. The markedly better removal of total P in the activated sludge plants can probably be attributed to a high fraction of particulate P in the wastewater. The activated sludge tank acts as a good flocculator where particles are enmeshed in the flocs and subsequently separated. A similarly high flocculation of particles cannot be expected in the biofilm reactors.

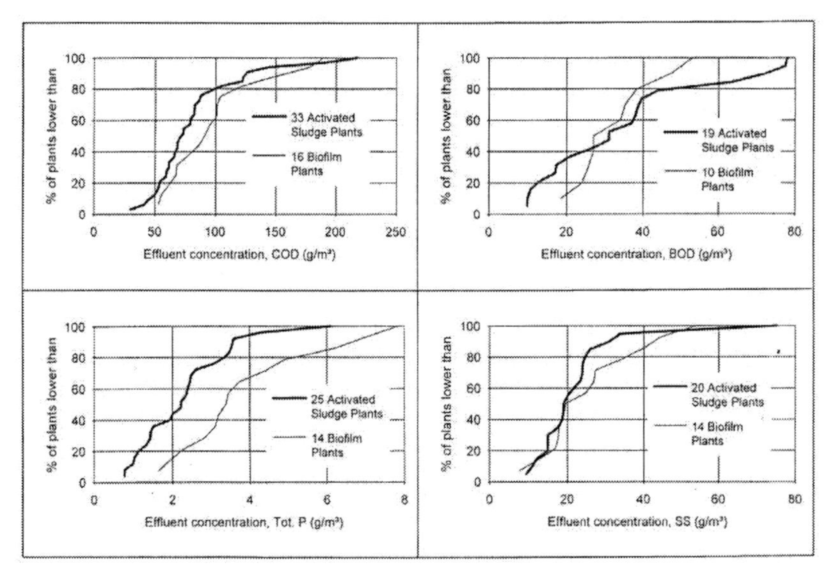

Figure 13.6. Frequency plots of effluent COD, BOD_7, tot P and SS for biological treatment plants.

13.3.1.3 Biological/chemical plants

Figure 13.7 shows the frequency plots for the effluent concentrations of the biological/chemical plants. It demonstrates that the plants in this category all work quite well. The poorest results can clearly be found in the simultaneous precipitation group. This is because activated sludge separation as the last step of the process train often results in poor effluent SS as a consequence of frequent sludge loss.

The combined precipitation plants are almost equally good as the post-precipitation plants with respect to organic matter when top 50% of the plants are considered. But in the bottom 50% of the plants, the combined precipitation plants are somewhat poorer, indicating slightly poorer process stability. In phosphate- and SS removal the post-precipitation plants are clearly best, followed by combined precipitation and simultaneous precipitation plants. The poorer performance of combined precipitation compared to post-precipitation plants may be attributed to less stringent dosage control in the former plants.

The reason for the superiority of the post-precipitation process is the fact that the last chemical step acts as a 'collector' of any sludge lost in previous treatment steps. This ensures a significant improvement in process stability, in accordance with the findings of Storhaug (1990). It should be mentioned that three pre-precipitation plants were also included in the investigation, and that these three plants gave equally good average results as the post-precipitation plants.

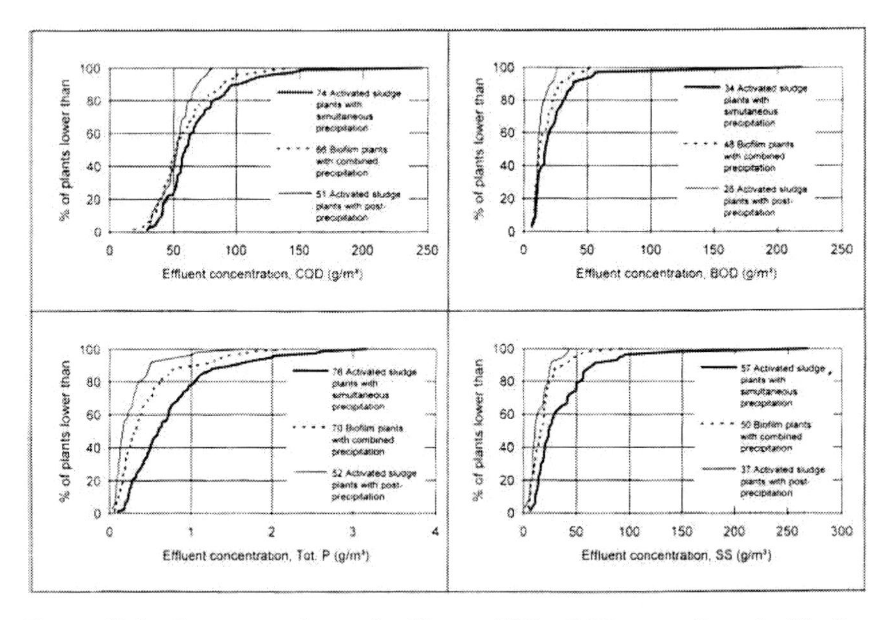

Figure 13.7. Frequency plots of effluent COD, BOD_7, tot P and SS for biological/chemical plants.

3.3.2 Example of a small wastewater treatment plant – the KMT plant

Biofilm reactors have are very popular in Norway for small biological/chemical treatment plants. The reason is that in these, chemical precipitation can take place directly after the biofilm reactor without any intermediate separation step. Most commonly these plants are based on an RBC as the biofilm reactor and a combined primary sludge separation and sludge storage tank as pre-treatment.

A new biofilm reactor, the moving bed biofilm reactor (MBBR) has been developed in Norway and this reactor is also gaining increasing popularity for small plants (Ødegaard *et al.* 1994, 1999a). The idea behind the MBBR was to have a continuously operating, non-cloggable biofilm reactor with no need for back washing, low head-loss and a high specific surface area for biofilm growth. This was achieved by having the biofilm (or biomass) grow on small carrier elements that move along with the water in the reactor. The movement is normally caused by aeration in the aerobic version of the reactor (see Figure 13.8(a)) and by a mechanical stirrer in the anoxic version (Figure 13.8(b)). However, for small plants, the stirrer in the anoxic reactor has been omitted for simplicity reasons and pulse aeration, for a few seconds a few times per day, has been used instead.

The biofilm carriers (Figure 13.8(c)) are made of high-density polyethylene (density 0.95 g/cm^3). The standard carrier (K1) is shaped like a small cylinder with a cross on the inside of the cylinder and 'fins' on the outside. The cylinder is 7 mm long, and 10 mm wide (not including the fins). Lately a larger carrier (K2) of similar shape (length and diameter about 15 mm) has also been introduced, intended for use in plants with coarse inlet sieves and especially for the upgrading of activated sludge plants which have sieves as the only pre-treatment.

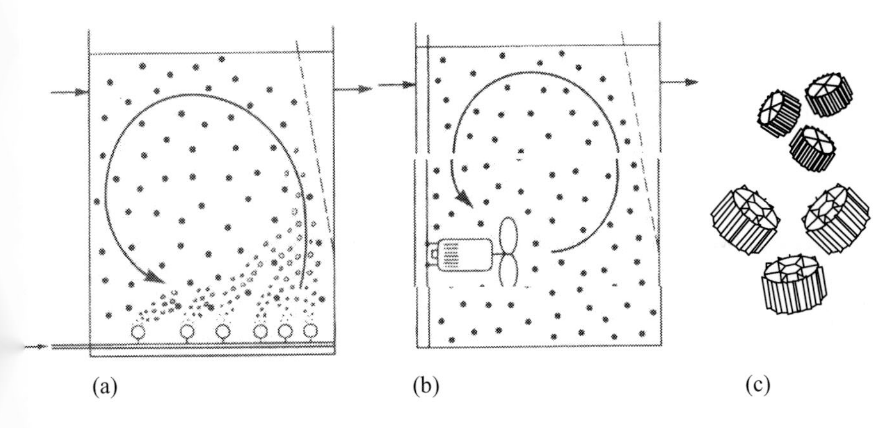

Figure 13.8. The principle of the moving bed biofilm reactor. (a) aerobic, (b) anoxic/anaerobic, (c) carriers (top K1, bottom K2).

One of the important advantages of the moving bed biofilm reactor is that the carrier filling-fraction (the percentage of reactor volume occupied with carriers in an empty tank) can be chosen according to the need for biofilm growtha. The standard filling fraction is 67%, resulting in a total, specific carrier area of 465 m^2/m^3 with the K1 carrier. Since the biomass grows primarily on the inside of the carrier, the effective specific surface area is 335 m^2/m^3 for the K1 carrier and 210 m^2/m^3 for the larger K2 carrier (at 67% filling fraction). It is recommended that filling fractions should be below 70% in order to mix the media properly. The reactor is designed based on a given effective biofilm area load (e.g. kg/m^{2}d) (Ødegaard *et al.* 1999a). If the standard filling fraction results in a growth area larger than needed, because the tank volume is large (for instance in an activated sludge upgrading situation), one may use a lower filling fraction, enough to ensure the necessary effective biofilm growth area defined by the design load.

The typical flow sheet for a small biological/chemical plant based on the KMT MBBR system is shown in Figure 13.9. Whether or not the first compartments of the bioreactor are to be aerated or not depends on the nitrogen removal requirement. There is no effluent standard on nitrogen in small plants in Norway, so small plants are normally not designed for nitrogen removal. However, the moving bed biofilm process has successfully been used for extensive nitrogen removal in larger plants (Ødegaard *et al.* 1994, 1999b).

Table 13.6 gives the average three-year results (1992–1994) for two plants based on the Kaldnes MBBR system (Rusten *et al.* 1997). The Steinsholt plant (plant A) was normally loaded while the Eidsfoss plant (plant B) was very low loaded in terms of organic load (g COD/m^2d).

None of the plants is designed for nitrogen removal, but the Steinsholt plant has a small anoxic reactor ahead of the aerobic ones, resulting in a certain amount of denitrification (42% nitrogen removal). The results are better than those of average small Norwegian biological/chemical treatment plants and the sludge production (gTS/COD_{rem}) is lower. This shows that the plants are operating very well. The reactors of the plants are covered and no smell can be detected. The plants have been proven to be operator-friendly and require little supervision.

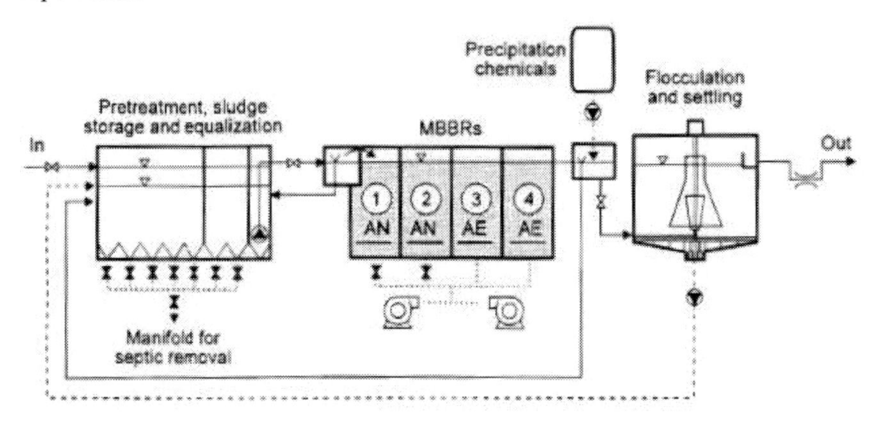

Figure 13.9. Typical flow sheet for biological/chemical plants based on the KMT MBBR system.

Table 13.6. Three-year (1992–1994) average results for two KMT MBBR plants (12 control samples per year)

Plant	Flow (m^3/d) Design/Actual		Persons (pe) Design/Actual		Coagulant dose gAl/m^3 Al/P		Organic load g $COD(BOD_7)$/ m^2d		Sludge production gTS/m^3 $gTS/gCOD_{rem}$	
A	40	40	250	250	14.1	2.5	7.7	4.3	260	0.54
B	160	36	130	1000	15.2	2.2	2.0	0.7	220	0.56

Plant	COD In Out %	BOD_7 In Out %	SS In Out %	Tot P In Out %	Tot N In Out %
A	514 /33/94.0	289/11/ 96.2	220/13/94.1	6.4 /0.17/97.3	50.2/29.2/ 41.8
B	373/32/91.4	126/<10/>92.1	–/10/–	/8.0/0.38/95.2	49.3/38.1/ 22.7

13.4 SUMMARY AND CONCLUSIONS

Experiences from small, compact treatment plants (serving between 35–2000 people) used in decentralised wastewater systems in Norway have been presented in this chapter. The plants are based on chemical, biological and combined biological/chemical processes, and plants were separated into sections to differ between on-site, compact treatment systems (mini treatment plants) for fewer than 7 private houses (<35 pe) and small, centralised systems for villages (35–500 pe and 500–2000 pe).

Regarding the mini-treatment plants, it was found that:

- The treatment results (as determined via the grab samples) were poorer than expected mainly due to sludge loss. Many of the plants did not de-sludge frequently enough, with the result that sludge escaped in the effluent.
- Fifty-two per cent of plant owners reported that that there were no noise or odour problems, while 25% reported odour problems, 19% reported noise problems and 4% reported both.
- Ninety-one per cent of plant owners were satisfied with the service provided to them by their plant supplier and 94% reported that their suppliers had kept their obligations according to their service agreements. Only 50% of plant owners were, however, satisfied with the service provided to them by their local authorities.
- In 53% of plants, assistance was sought from the plant supplier outside regular service visits (over the five-year period) and 67% of plants had experienced more than one break in operation. These breaks was normally

caused by blockages that in most instances were caused by electric power supply failure.
- It was evident that those plants situated in a heated room were better maintained than those placed underground (uncovered) or in unheated rooms.

Regarding the 356 small village treatment plants (serving 35–500 and 500–2000 persons), it was found that:

- Treatment results and process stability were best in the biological/chemical treatment plant category.
- With respect to phosphorous removal chemical plants performed on average better than the biological/chemical plants, but not as well as the post-precipitation plants.
- With respect to organic matter removal, biological plants did not perform as well as biological/chemical plants and not significantly better than chemical plants. Of biological plants, activated sludge plants performed better than the biofilm plants.
- In the biological/chemical treatment plant category, post-precipitation plants (chemical treatment following activated sludge) performed best followed by combined precipitation plants (biofilm reactor followed directly by the chemical step). Simultaneous precipitation plants (chemical addition to an activated sludge plant) gave poorer treatment results and poorer process stability.

In general, greywater may be treated using the same processes and compact plants as pre-settled sewage. However, one has to take into account the special characteristics of greywater (a high soluble COD fraction, low N/COD fraction, and so on) as well as the toxic shock loads. These may result from substances that may enter a greywater source but are not a regular component of it (washing powder, bleach, and so on). Because of this, it may be that physical/chemical processes are more suitable than biological processes if the complete removal of biodegradable matter is not required.

13.5 REFERENCES

Heltveit, S.I. (1994) *Experiences With Mini Treatment Plants.* Report 94:06. Statens Forurensingstilsyn (State Pollution Control Authority of Norway). (In Norwegian.)

Henze, M. and Ødegaard, H. (1994) An analysis of wastewater treatment strategies for Eastern and Central Europe. *Wat. Sci. Tech.* **30**(5), 25–40.

Jefferson, B., Laine, A., Diaper, C., Parsons, S., Stephenson, T. and Judd, S.J. (2000) Water recycling technologies in the UK. *TUWR1 – Proc. 1st Int. Meeting on*

Technologies for Urban Water Recycling, Cranfield University, Cranfield, UK, 19 January.
Ødegaard, H. (1992) Norwegian experiences with chemical treatment of raw wastewater. *Wat. Sci. Tech.* **25**(12), 255–264.
Ødegaard, H. (1999) The influence of wastewater characteristics on choice of wastewater treatment method. *Proc. Nordic Conference on Nitrogen Removal and Biological Phosphate Removal*, Oslo, Norway 2–4.February.
Ødegaard, H. and Skrøvseth, A.F. (1997) An evaluation of performance and process stability of different processes for small wastewater treatment plants. *Wat. Sci. Tech.* **35**(6), 119–127.
Ødegaard, H., Rusten, B. and Westrum, T. (1994) A new moving bed biofilm reactor. *Wat. Sci. Tech.* **29**(10–11), 157–165.
Ødegaard, H., Gisvold, B. and Strickland, J. (1999a) The influence of carrier size and shape in the moving bed biofilm process**.** *Wat. Sci. Tech.* **41**(4–5), 383–391.
Ødegaard, H., Rusten, B. and Siljudalen, J. (1999b) The development of the moving bed biofilm process: from idea to commercial product. *European Water Management* **2**(3), 36–43.
Paulsrud, B. and Haraldsen, S. (1993) Experiences with the Norwegian approval system for small wastewater treatment plants. *Wat. Sci. Tech.* **28**(10), 25–32.
Rusten, B., Kolkinn, O. and Ødegaard, H. (1997) Moving bed biofilm reactors and chemical precipitation for high efficiency treatment of wastewater from small communities. *Wat. Sci. Tech.* **35**(6), 71–79.
Storhaug, R. (1990) Performance stability of small biological chemical treatment plants. *Wat. Sci. Tech.* **22**(3/4), 275–282.
Yang, X.M., Yahashi, T, Kuniyasu, K. and Ohmori, H. (2001) On-site systems for domestic wastewater treatment (Johkasous) and its application in Japan. Chapter 14 in this book.

14

On-site systems for domestic wastewater treatment (johkasous) in Japan

X. M. Yang, T. Yahashi, K. Kuniyasu and H. Ohmori

14.1 INTRODUCTION

There are several systems for wastewater treatment in Japan classified on the basis of the kind of wastewater they treat, facility size and administrative support. They comprise sewerage systems, johkasou systems and night soil treatment facilities, as shown in Figure 14.1.

Johkasou systems, uniquely developed in Japan, are now extensively installed for domestic wastewater treatment throughout the country. The word

© IWA Publishing. Decentralised Sanitation and Reuse: Concepts, Systems and Implementation. Edited by P. Lens, G. Zeeman and G. Lettinga. ISBN: 1 900222 47 7

'johkasou' means here 'a tank for purifying wastewater', or 'an on-site plant for wastewater treatment'. 'Johka' literally means purification and 'sou' means a tank in Japanese. There are two types of johkasou systems: one is 'tandoku-shori johkasou' which is used for flush toilet wastewater treatment only, and the other is 'gappei-shori johkasou' which is used for miscellaneous domestic wastewater treatment.

In Japan, 66% of the population uses gappei-shori johkasous or sewerage systems for miscellaneous domestic wastewater treatment. However, 34% of the population still uses vault toilets or tandoku-shori johkasous for night soil treatment, and grey water is discharged without any treatment in such areas. It is an pressing problem for the government to increase the proportion of people using gappei-shori johkasous or sewerage systems to promote the treatment of grey water and to prevent the pollution of water bodies.

On the other hand, the promotion of the sewerage system has its limitations. The portion of the population served by sewerage system depends on the municipality size as shown in Table 14.1. For municipalities with populations under 50,000 this ratio is only 22%, i.e. one fourth of that for cities with population exceeding one million. The costs of wastewater treatment for small sewerage systems serving under 10,000 inhabitants is 3.6 times higher than that of cities with a population of over one million. The high cost and the economic situation of the people is the reason that municipalities hesitate to construct sewerage systems.

Johkasous are usually installed in rural areas where sewerage systems are not available. There have been more than eight million johkasous installed including tandoku-shori johkasous and gappei-shori johkasous, serving about thirty-six million people. Among the installed johkasous, about 13.5% or one million units are gappei-shori johkasous, serving ten million people.

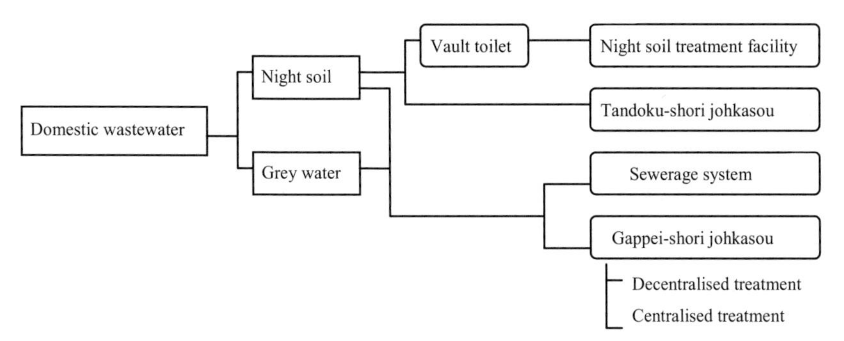

Figure 14.1. Domestic wastewater treatment in Japan (JECES 1998).

Research and development on johkasou technologies has been significant especially in recent years. There are two trends in johkasou technologies: one is to develop johkasous that can remove nitrogen and phosphorous and discharge high quality water. The other one is to develop johkasous that have special functions/benefits for users (Figure 14.2).

This chapter will focus on small-scale johkasous for individual households, or for decentralised treatment. The current situation of johkasou systems in Japan will be explained, and some newly developed unique small-scale johkasous will be presented.

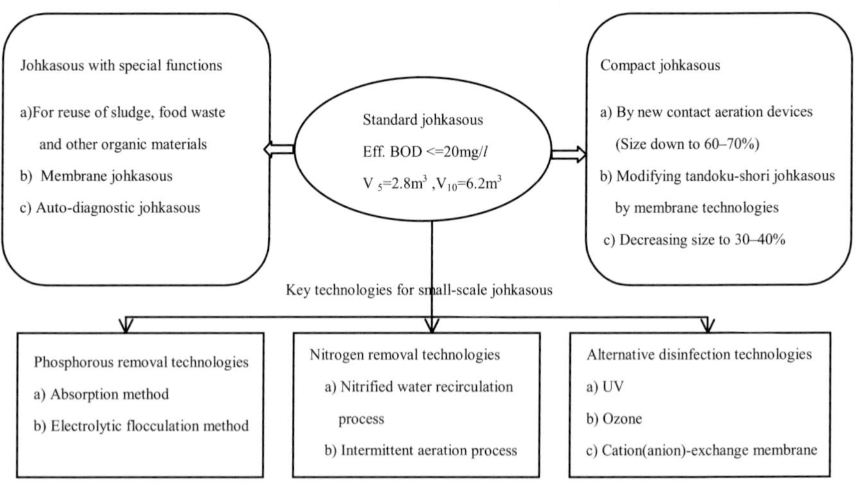

Note: V_5,V_{10} represents the volume of johkasou needed to be used by 5 and 10 people, respectively.

Figure 14.2. Trends in small-scale johkasous.

Table 14.1. Population served by sewerage systems and the cost of wastewater treatment

Scale of city	1,000,000 or more	500,000-1,000,000	300,000-500,000	100,000-300,000	50,000-100,000	50,000 or less	Total/ Average
1. Total population $\times 10^3$	25,030	6620	17,200	26,110	15,820	35,070	125,860
2. Population using sewerage system $\times 10^3$	24,560	4780	11,720	16,700	7710	7640	73,110
(1)/(2) $\times$100, %	98	72	68	64	49	22	58
Number of municipalities	11	10	44	159	228	2781	3233
Cost price, yen/m^3	143	168	168	193	199	278–525	191

14.2 THE STRUCTURE OF JOHKASOUS

14.2.1 Quantities and quality of domestic wastewater

The quantity and quality of domestic wastewater per capita, especially mixed domestic wastewater and pollutant loads, are known to vary with lifestyle, living standard etc. The amount of domestic wastewater ranges from 200 to 250 litres per capita per day in Japan. Pollutant loads in domestic wastewater are said to be 40–50g of BOD, 10–12g of T-N (total nitrogen) and 1.2–1.5g of T-P (total phosphorous) per capita per day.

The amount of wastewater and the pollutant loads of domestic wastewater to be treated by a johkasou are shown in Table 14.2. The differences in the amount of wastewater can partly be due to variations in the number of people present at home during the day. It is worth mentioning that the wastewater from gardening, car washing and so on is not treated in johkasous.

Table 14.2. Amount and pollutant loads of domestic wastewater (per capita per day)

Source of wastewater		Wastewater amount (l)	BOD (g)	T-N (g)	T-P (g)
Flush toilet water	Flushing	50	13		
Miscellaneous domestic wastewater	Cooking	30	18		
	Laundry	40			
	Bathing	50			
	Washing face/hands	20	9		
	Cleaning	10			
Total		200	40	10	1.0

14.2.2 Structure of johkasous

14.2.2.1 Anaerobic filter - contact aeration process

This treatment process is most widely used in small-scale gappei-shori johkasou systems; the required effluent BOD concentrations of this process are less than 20 mg/l. The flowchart of this process is shown in Figure 14.3.

The johkasou consists of an anaerobic filter tank, a contact aeration tank, a sedimentation tank and a disinfection tank. In the anaerobic filter tank and the contact aeration tank, filter media or contact media are filled. An example of this type of johkasou is shown in Figure 14.4. Each tank in this treatment system should have a capacity equal to or exceeding the value calculated by the equations presented in Table 14.3.

The anaerobic filter tank in this treatment system is designed not only for separating/storing solid matter from the influent, but also for reducing the amount of sludge as a result of the anaerobic digestion of organic matter. The anaerobic filter tank of this size is usually divided into two rooms to help the sedimentation of solid matters.

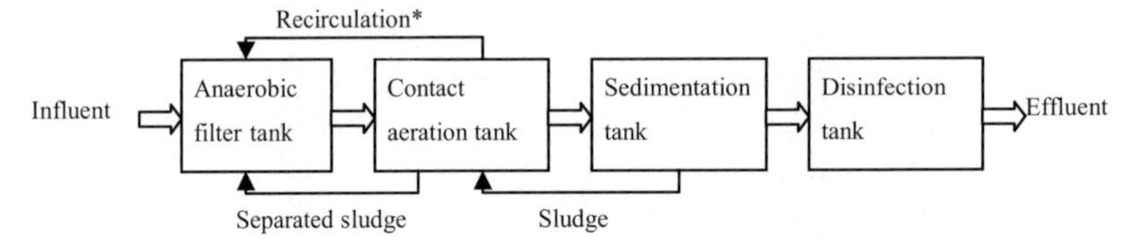

Figure 14.3. Flowchart of anaerobic filter–contact aeration process (a) and denitrification type anaerobic filter–contact aeration process (b). (Note: * for (b) process and part of (a) process.)

Table 14.3. Volume of tanks required in anaerobic filter–contact aeration process

N	Anaerobic filter tank, m^3	Contact aeration tank, m^3	Sedimentation tank, m^3	Disinfection tank, m^3
5 or less	1.5	1.0	0.3	
6–10	1.5+(n–5)×0.4	1.0+(n–5)×0.2	0.3+(n–5)×0.08	0.2×n×1/96
11–50	3.5+(n–10)×0.2	2.0+(n–10)×0.16	0.7+(n–10)×0.04	

Note: n denotes the number of users of each design

14.2.2.2 Denitrification type anaerobic filter–contact aeration process

Johkasous of this treatment process are designed for both BOD and nitrogen removal with effluent BOD and T-N concentrations less than 20 mg/l. The flowchart of this process is shown in Figure 14.3. Each tank in this treatment system should have a capacity equal to or exceeding the value calculated by the equations presented in Table 14.4.

To ensure a smooth nitrification, the design volume of the contact aeration tank is bigger and the aeration intensity is higher than that in the anaerobic filter–contact aeration process, respectively. The denitrification is realised by recirculating aerobically treated wastewater from the contact aeration tank to the anaerobic filter tank.

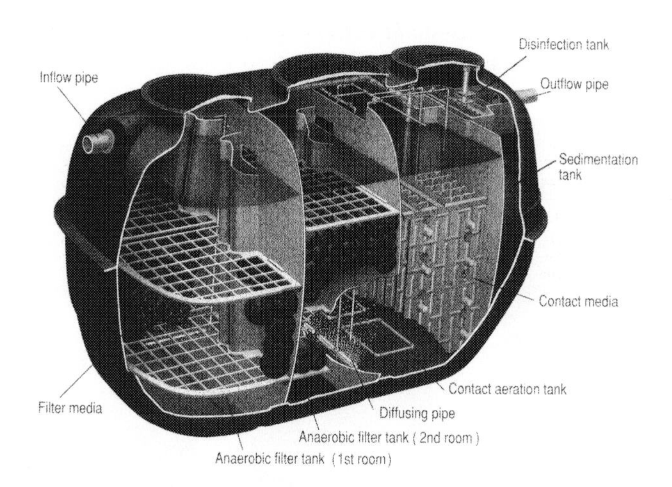

Figure 14.4. An example of a small-scale johkasou.

Table 14.4. Volume of tanks required for the denitrification anaerobic filter–contact aeration process

N	Anaerobic filter tank, m^3	Contact aeration tank, m^3	Sedimentation tank, m^3	Disinfection tank, m^3
5 or less	2.5	1.5	0.3	
6–10	2.5+(n–5) × 0.5	1.5+(n–5) × 0.3	0.3+(n–5) × 0.08	0.2 × n × 1/96
11–50	5.0+(n–10) × 0.3	3.0+(n–10) × 0.2	0.7+(n–10) × 0.04	

Note: n denotes the number of users of each design

14.2.3 Filter and contact media

Several kinds of filter and contact media used in the anaerobic filter tank or in the contact aeration tank, are shown in Figure 14.5. Most are made of plastics.

The configuration and shape of the media in the anaerobic filter tank are determined according to the purpose of this tank. If the tank is designed for capturing the suspended solids and storing them in the upper space of the tank as scum, media with high capture ability, such as *Balllike* media shown in Figure 14.4, can be used. If the tank is just used for rectifying influent, settling suspended solids and storing them in the bottom of the tank, *Plate/Netlike-cylinder* media, also shown in Figure 14.4, can be adopted.

Media in the contact aeration tank are selected on the balance of the high biological capacity and easy backwashing of the media. There are some newly developed media, such as small ceramic porous balls and small cubic sponges, for new treatment processes.

Figure 14.5. Filter and contact media used in johkasous.

14.3 TREATMENT PERFORMANCE OF SMALL-SCALE JOHKASOUS

14.3.1 Treatment performance of johkasous for BOD removal

Effluent BOD concentrations of johkasous of the anaerobic filter–contact aeration process are shown in Figure 14.6. It is seen that 79% of the effluent BOD concentrations were smaller than 20 mg/l, and 55% of those were smaller than 10 mg/*l*. The average effluent BOD concentration is 14.3 mg/*l* and the standard deviation is 14.2 mg/*l*. As the effluent BOD criterion for design is 20 mg/*l*, the treatment performance of the johkasous was satisfactory.

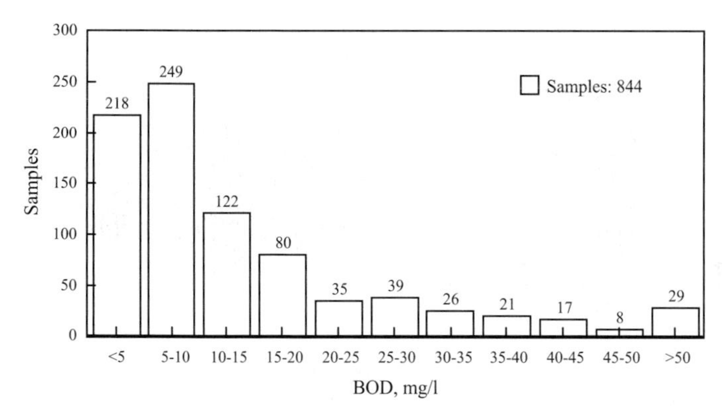

Figure 14.6. Distribution of effluent BOD concentration of anaerobic filter–contact aeration johkasous (Yang *et al.* 1996).

Effluent quality is usually affected by two factors. One is the pollutant load, which could be represented by the ratio of the number of users to the number of users the system was designed for (R_u). Another factor depends on the way the johkasou is operated, with or without water recirculation from the contact aeration tank to the anaerobic filter tank. The average effluent BOD concentrations rearranged by the two factors above are shown in Table 14.5. It was found that the average of the effluent BOD concentration increased with the increase of R_u, and that the effluent BOD concentrations were lower when the johkasou was operated with recirculation, as compared to the case without recirculation.

Table 14.5. Average of effluent BOD of anaerobic filter–contact aeration johkasous unit:mg/l (values in brackets indicate the number of samples)

LR_u	0.25	0.25–0.50	0.50–0.75	0.75	Sum
With recirculation	7.8 (22)	11.2 (143)	14.4 (136)	19.6 (54)	13.5 (355)
Without recirculation	10.1 (41)	13.5 (188)	16.0 (179)	18.1 (81)	14.9 (489)

The recirculation from the contact aeration tank to the anaerobic filter tank plays an important role not only in preventing the washout of suspended solids, but also in the promotion of denitrification. Table 14.6 shows the effect of recirculation on the effluent quality. It can be seen that the average effluent water quality was improved when the johkasou was operated with recirculation, compared to without recirculation. For instance, the average effluent T-N concentration decreased from 30 mg/l without recirculation to 14 mg/l with recirculation. Most johkasous of this treatment process can be expected to remove nitrogen, and this led to the development of small-scale johkasous for nitrogen removal.

Table 14.6. Average of effluent water quality

	Without recirculation				With recirculation			
	SS	BOD	T-N	T-P	SS	BOD	T-N	T-P
Number of samples	28	36	36	28	22	111	111	22
Average (mg/l)	11	18	30	4.3	9	16	14	3.2
Minimum (mg/l)	0.5	<3	8.6	1.5	1	<3	3.3	1.6
Maximum (mg/l)	44	51	74	7.6	36	41	33	5.1

14.3.2 Treatment performance of johkasous for both BOD and nitrogen removal

14.3.2.1 Johkasous for experiments

Three kinds of johkasous (below referred to as types A, B and C) were used for investigation, i.e. a johkasou with an anaerobic filter tank for primary treatment followed by either a contact aeration tank, a biofilm filtration tank or a moving bed biofilm tank for secondary treatment.

The filter plastic in the anaerobic filter tank filled up the tank volume up to about 60%. The biofilm filtration tank was filled with porous ceramic balls of 6–9 mm in diameter and with a specific gravity ranging from 1.05–1.2, or with cubic plastics medium of 1 cm^3 and specific gravity of 1.0. In the moving bed biofilm tank, cubic polyurethane media of 10 mm in length and specific gravity of 1.0 were filled up to 50% of the volume of the tank.

These johkasous have the following common characteristics: (a) an equalisation device and a recirculation device; (b) when the influent flows in abruptly, it is equalised by raising the water level in the anaerobic filter tank.

For the biofilm filtration tank, a washing device is added to the system, which is used for automatic backwashing for ten minutes each day. The flowchart and volume of each tank is shown in Figure 14.7 and Table 14.7.

14.3.2.2 Maintenance

Maintenance was carried out every three months for each johkasou. During maintenance, johkasou operators remove slime from the pipes, checked the situation of sludge accumulation in the anaerobic filter tank, determined the timing of desludging, adjusted the air flow rate for aeration and adjusted the equalisation and recirculation devices. In the biofilm filtration tank, johkasou operators examined the sludge attached to the media and adjusted the time required for backwashing.

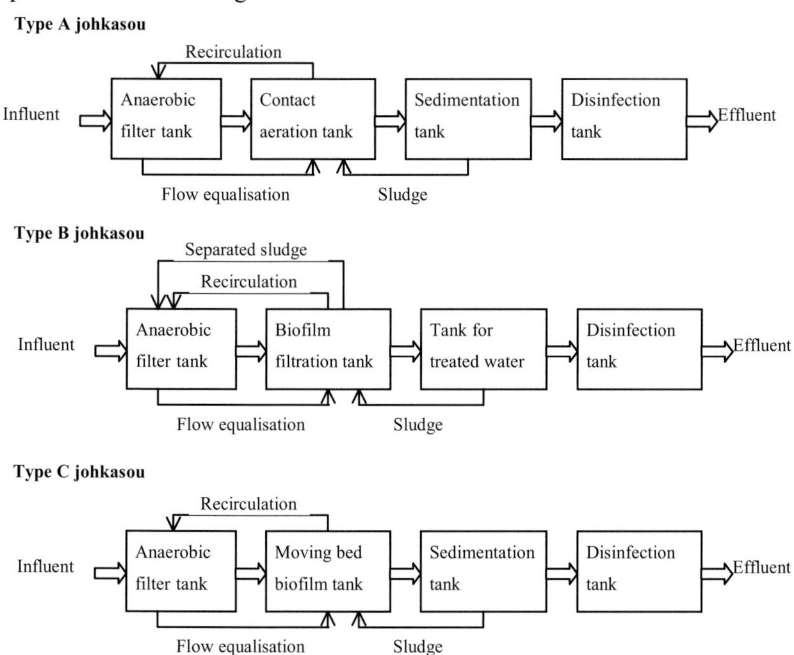

Figure 14.7. Flowchart showing different types of johkasous.

Table 14.7. Volume of each tank of johkasous

		Type A			Type B			Type C		
No of users for design		5	7	10	5	7	10	5	7	10
Inflow rate for design (m^3/day)			1.0			1.4			2.0	
Required quality of treated water		BOD and T-N equal to or smaller than 20 mg/l								
Effective volume (m^3)	Anaerobic filter tank	1.90	2.32	3.54	1.72	2.24	3.22	1.54	2.19	3.23
	Contact aeration tank	1.13	1.41	2.10						
	Biofilm aeration tank				0.31	0.44	0.59			
	Moving bed biofilm tank							1.10	1.57	2.04
	Sedimentation tank	0.48	0.48	0.71				0.34	0.50	0.75
	Tank for treated water				0.45	0.70	0.91			
	Disinfection tank				0.03					
	Total	3.55	4.25	6.37	2.50	3.41	4.76	3.00	4.29	6.05
Blower no. 1	Power, W	60	95	125	50	51	71	80	95	118
	Flow rate, l/min	73	102	132	50	60	71	80	100	120
Blower no. 2	Power, W		30		51	71	105		26	
	Flow rate, l/min		31		60	84	84		28	

Since inflow conditions in johkasous vary, one should be careful when determining the amount of equalisation water and recirculation water and air flow rate. Effluent BOD concentration could be estimated from the operation conditions and the transparency of the effluent. To examine the concentration of nitrogen in the effluent, it is necessary to measure NH_3-N and NO_x-N using portable instruments.

14.3.2.3 Results and discussion

The distribution of BOD and T-N of johkasous is summarised in Table 14.8. About 64% of type A johkasous met the required water quality. For type B and type C johkasous, this increased to more than 80%. Eighty-two per cent of effluent BOD concentrations were lower than 20 mg/l for type A, 94% for type B and 98% for type C. Meanwhile, percentages of effluent T-N concentrations below 20 mg/l were 75% for type A, 84% for type B and 85% for type C.

Type B johkasous showed better performance than type A. This can be due to differences in filter media in the aeration tank. Filter media in the biofilm filtration tank of type B johkasous are very small and have a high specific surface area, meaning that the capacity of biological treatment and ability to capture suspended solids is enhanced. However, type B johkasous are very complicated to maintain and operate. It is essential that type B johkasous are operated using backwashing according to the amount of biofilm attached to the filter media.

The type C johkasou showed a treatment performance as good as that of type B johkasous. Although type C johkasous have the advantage of being free of backwashing, the air flow rate should be adjusted according to the size and weight of media in the moving bed biofilm tank. Therefore, the maintenance and operation of type C johkasou is slightly more complicated than that of type A johkasous.

Table 14.8. Distribution of effluent BOD and T-N

	Type A	Type B	Type C
Number of samples	233	150	46
Number of johkasous investigated	126	95	29
BOD and T-N ≤20 mg/l (%)	64	81	83
BOD ≤20 mg/l (%) T-N >20 mg/l (%)	18	13	15
BOD ≤20 mg/l (%) T-N ≥20 mg/l (%)	11	3	2
BOD and T-N >20 mg/l (%)	7	3	–

14.4 TREATMENT PERFORMANCE OF SMALL-SCALE JOHKASOUS USING A WASTE DISPOSAL UNIT

Four small-scale anaerobic filter–contact aeration process johkasous were selected to investigate the effect of using a waste disposal unit (WDU) on treatment performance (Yang *et al.* 1994).

Field surveys were carried out every two months for the four johkasous. The discharge from a WDU was collected after operation, and a composite sample was taken. Influent and wastewater in each tank were sampled hourly over a period of 24 hours, and then blended for water quality analysis. The thickness of scum layers and settled sludge in each tank, especially in the anaerobic filter tank, were measured. Specifications of the WDU used in the experiment are shown in Table 14.9.

The influent characteristics are shown in Table 14.10. Both blackwater and greywater, including discharge from a WDU, were introduced into the johkasous. Although the average daily flow was under or near the designed loading for all four johkasous, the situations were very different. In johkasou 2, peak flow, peak coefficient (the ratio of the maximum hourly flow rate vs. the average flow rate in 24 hours) and BOD load greatly exceeded those of the designed loading. This means that this johkasou was heavily overloaded. Johkasous 3 and 4 showed a slight overloading, and the pollutant load of johkasou 1 was below the designed loading.

Table 14.9. Waste disposal unit specifications in the experiment

Weight	4.5 kg
Dimensions	H 360mm × L 275mm
Power	3/4 HP
Rotation rate	2700 rpm
Voltage	AC 100 V, single phase, 50/60 Hz

Table 14.10. Influent characteristics

	Johkasou				Planned
	No.1	No.2	No.3	No.4	loading
Average daily flow (m^3/day)	0.93	1.21	0.95	1.05	1.2
Peak flow in one hour (m^3/h)	0.24	0.42	0.28	0.24	0.30
Peak coefficient	6.5	8.8	7.1	5.6	6.0
SS (g/day)	87	221	145	205	240
BOD (g/day)	203	393	306	345	240
T-N (g/day)	22	35	31	26	*
T-P (g/day)	2.1	3.4	2.9	3.2	*

* No regulation

The BOD concentrations of the influent and effluent are shown in Figure 14.8. The BOD concentrations of the influent showed a maximum value of 508 mg/l. However, it was found that BOD concentrations of effluent from the

anaerobic filter tank decreased to about one-third of the influent values, and thus the pollutant load on the contact aeration tank was greatly reduced. This implies that most of the solid matter in the influent was removed from the wastewater in the anaerobic filter tank and was captured in the tank as scum or sludge. Further, it appears that BOD concentration of the effluent from the anaerobic filter tank is heavily affected by the high inflow rate as represented by the peak coefficient; this resulted in turbulence in the tank and brought scum/sludge into the contact aeration tank. A scum layer of more than 30cm in the anaerobic filter tank characterised all johkasous except johkasou 2. The difference between the scum deposited in the anaerobic filter tank in No.4 and the other johkasous might result from the different inflow rates.

The measured average effluent BOD concentrations, except those of johkasou 2, were near to or under 20 mg/*l*. The high concentrations of effluent BOD discharged from johkasou 2 could be attributed to the fact that the average detention time of the treated water in the sedimentation tank was too short to effectively settle the suspended solids, due to the high hydraulic shock load, i.e. the high peak coefficient. Therefore, it is extremely important when examining the performance of a johkasou to first assess whether or not the inflow rate and peak coefficient are within the overload range. The high influent BOD load due to use of a WDU can be reduced by filter media in the anaerobic filter tank and exerts little effect on effluent BOD.

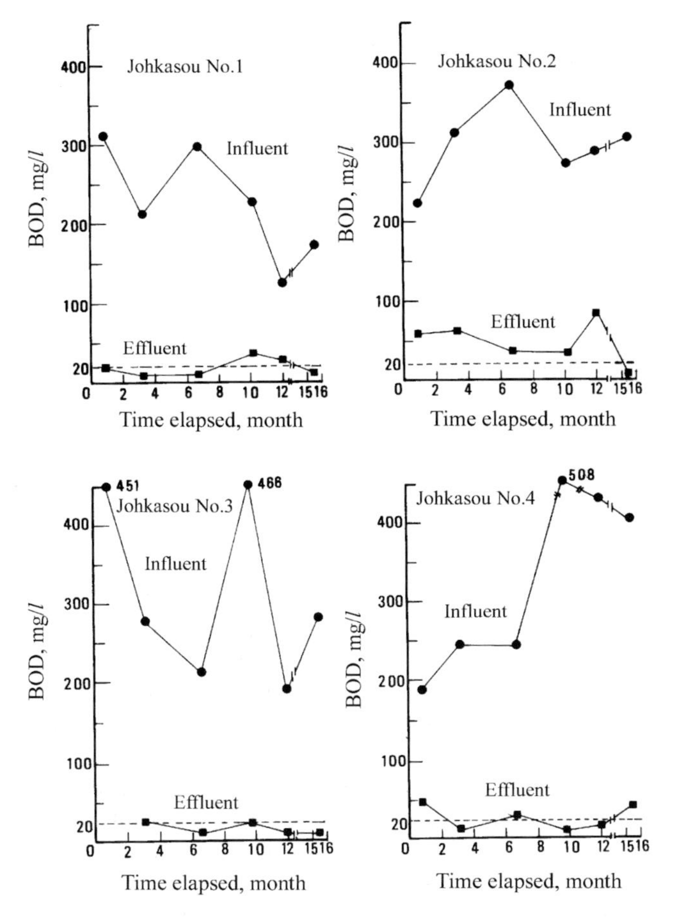

Figure 14.8. BOD concentrations in influent and effluent.

The treatment performance of the johkasous is shown in Table 14.11. The average BOD concentration of the effluent from johkasous 1, 3 and 4 are 20 mg/l, 14 mg/l and 25 mg/l, respectively, thus near to the required value of 20 mg/l. The average BOD concentration of effluent from johkasou 2 showed a relatively high value of 52 mg/l, which could be attributed to the high flow rate. The situation for SS was very similar to that of BOD. All johkasous except No. 2 showed good treatment performance, and demonstrated that small-scale johkasous are capable of treating domestic wastewater including discharge from WDUs, provided that the flow rate and the pollutant loads are within the design values.

Table 14.11. Treatment performance of the johkasous

	Johkasou No.1			Johkasou No.2			Johkasou No.3			Johkasou No.4		
	Inf.	Eff.	Rem	Inf.	Eff.	Rem	Inf.	Eff.	Rem.	Inf.	Eff.	Rem
Parameter	mg/l		%	mg/l		%	mg/l			mg/l		%
SS	94	14	85	183	38	79	154	8	95	207	20	90
BOD	212	20	91	288	52	82	307	14	95	333	25	92
T-N	24	31	–	30	34	–	33	20	–	24	18	–
T-P	2.3	4.5	–	2.8	3.0	–	3.1	3.6	–	3.1	3.4	–

14.5 NEWLY DEVELOPED JOHKASOUS WITH MEMBRANE SEPARATION

14.5.1 Advantages of membrane johkasous

The treatment performance of johkasous is usually evaluated on the basis of effluent BOD values. However, it is difficult to prevent washout of fine suspended solids from the sedimentation tank of a johkasou, similar to conventional biological wastewater treatment processes.

A johkasou system with membrane separation technologies (to be referred to as membrane johkasous in this chapter) was developed and applied in advanced wastewater treatment to remove both BOD and nitrogen. Membrane johkasous are expected to have the following advantages, in addition to the production of an effluent with a low BOD concentration:

- Treatment performance seems to be unaffected by settling characteristics of activated sludge, as membrane modules are immersed in the bioreactor and treated wastewater is sucked directly from membrane modules.
- Nitrogen removal will be promoted by intermittent aeration or recirculation of nitrified water to an anaerobically operated tank.
- Operation and maintenance will become simpler. Operation and maintenance of the membrane johkasous will focus on ensuring steady

physical solids separation by membranes, rather than controlling biochemical reaction, as in standard johkasous.

- Treated wastewater contains very little suspended solids and little micro-organisms, thus has a low health risk. Therefore, it may be used for gardening, car washing, toilet flushing, fire fighting and so on.

A newly developed membrane johkasou, based on application of the intermittent activated sludge process and using plate and frame membrane (PFM) modules, has been investigated (Ohmori *et al.* 1999). The membrane modules used in this johkasou are prepared from polyethylene with a pore size of 0.3 μm. An example of the membrane johkasou is shown in Figure 14.9.

The membrane johkasou consists of a sedimentation/separation tank, an intermittent aeration tank with PFM modules and a disinfection tank. The experimental conditions applied are summarised in Table 14.12, and the flowchart is shown in Figure 14.10. Treated wastewater was sucked from the system using a siphon system.

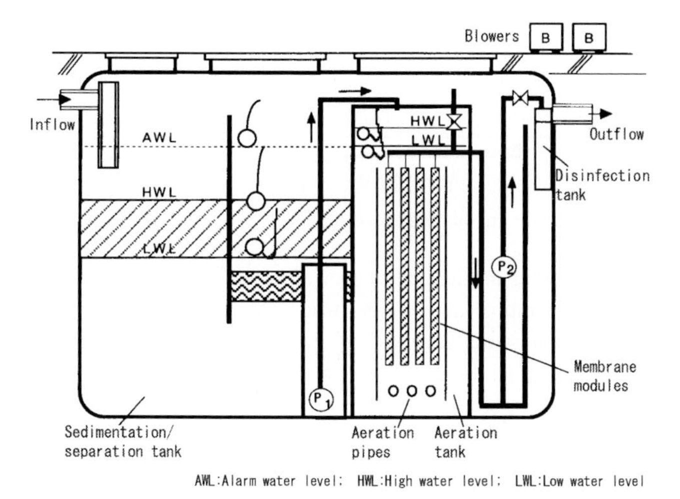

Figure 14.9. Diagram of the membrane johkasou.

Table 14.12. Experimental conditions

Number of users for design		5 persons
Number of users		5 persons
Inflow rate for design		1.25 m^3/day
Treated wastewater quality for design		BOD<=5 mg/*l*, T-N<=10 mg/*l*
Water permeability for design (max.)		3.2 m^3/day
Effective	Sedimentation/separation tank	1.926 m^3
Volume	Intermittent aeration tank	0.536 m^3
	Disinfection tank	0.026 m^3
Membrane	PFM module	8 sheets
Module	Surface area	6.4 m^2
	Permeation flux	0.45 m^3/m^2 day

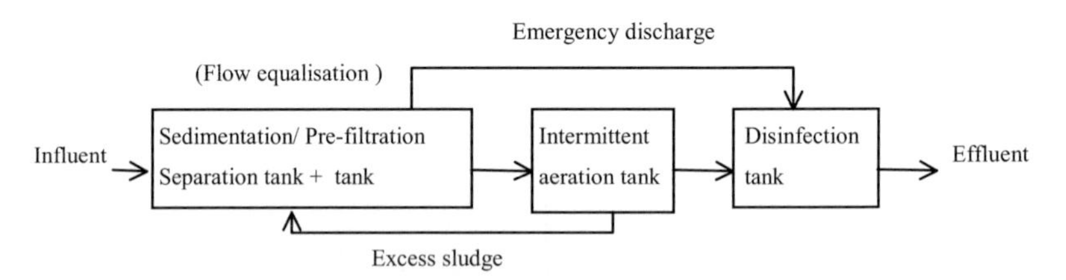

Figure 14.10. Flowchart of the membrane johkasou.

During the experiment, the treated wastewater quality, mixed liquor suspended solids (MLSS) concentration and water head loss were measured periodically. Maintenance operations were carried out every three months, and membrane cleaning (by sodium hypochlorite solution and desludging/adjusting the activated sludge concentration) every six months.

14.5.2 Treatment performance of membrane johkasous

The average amount of wastewater flowing in the johkasou amounted to 850 l/day, and the peak coefficients were usually around 5.0 and 7.4 at a maximum. During the experiment, the aeration condition was changed only twice from intermittent aeration to continuous aeration by the automatic control system to cope with high flow rates, but no overflow was observed even at the maximum peak coefficient of 7.4. The variations in MLSS concentration and water head loss are shown in Figure 14.11.

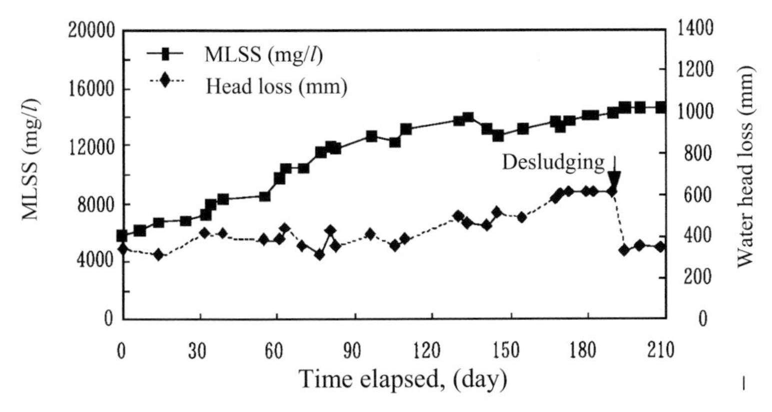

Figure 14.11. Variation in MLSS concentration and head loss.

Activated sludge in the aeration tank was automatically returned to the sedimentation/separation tank in order to adjust the MLSS concentration three months after the start of the operation. From the fourth month, the MLSS concentration in the aeration tank was kept in the range of 13,000–14,000 mg/l. The average sludge yield was found to be 0.28 kg-SS/kg-BOD. The low sludge yield was realised by the high MLSS concentration and the operational condition of endogenous phase of activated sludge.

The johkasou functioned steadily without any problems. It took six months before the head loss had increased to 288 mm, implying that membrane cleaning by spiral flow in the aeration tank was effective.

The variations in BOD and T-N concentration of treated water are shown, together with those in other water samples, in Figures 14.12 and 14.13.

The average BOD concentration of the treated wastewater was 2.3 mg/l with variations from 1.0–3.9 mg/l, which is well below the design value of 5 mg/l. The results show that the johkasou exerts a high and steady ability to remove organic materials. During the experiment, the water temperature in the aeration tank was between 12.7–28.9 °C.

The average T-N concentration of the treated wastewater was 7.9 mg/l and ranged between 4.7–9.9 mg/l, which is also well below the designed value of 10 mg/l. Even on the day that the water temperature dropped to 12.7 °C, the T-N concentration of treated water was found to be 8.5mg/l. These results show that the membrane johkasou can remove nitrogen with a T-N concentration below 10 mg/l, as long as water temperature is kept higher than 13° C.

Since the treated water was separated from activated sludge through membrane modules, the SS concentration of the treated water was usually below 5 mg/l. The transparency of treated wastewater showed values greater than 100 cm, and total coliform density was below 100 cells/ml.

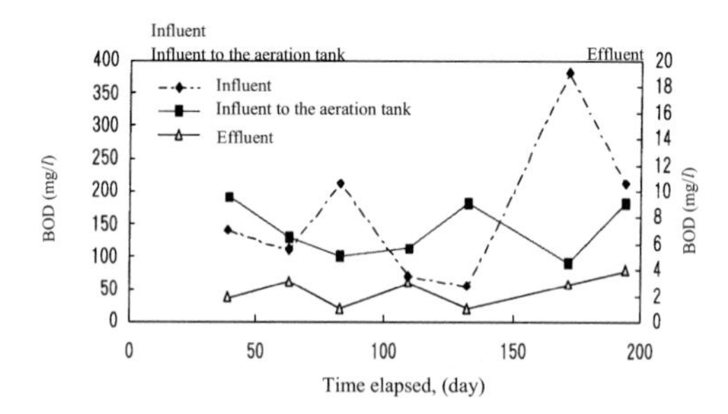

Figure 14.12. Variations in BOD concentrations.

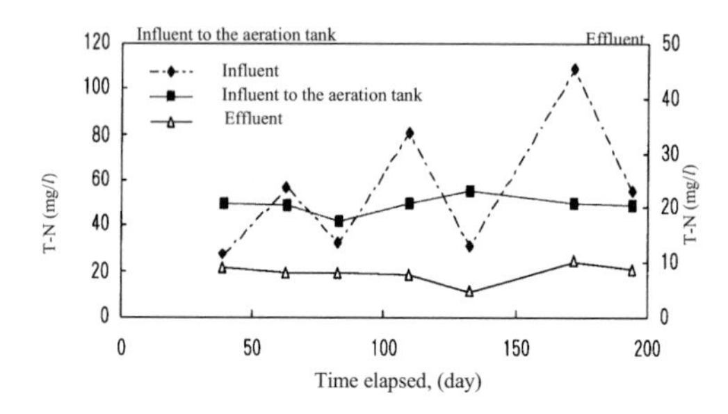

Figure 14.13. Variations in T-N concentrations.

The excess sludge settled in the sedimentation/separation tank was 6.5 dry-kg after six months of operation. The settled excess sludge consisted of 76% organic matter and 24% inorganic matter. It was pumped out from the system within 20 minutes using a vacuum car, showing that desludging can be done as in normal johkasous.

14.5.3 Effects of membrane cleaning on treatment performance

In order to recognise the effects of membrane cleaning on treatment performance, two experiments on membrane cleaning were carried out by the method shown in Table 14.13. In the first experiment, after membrane cleaning and neutralisation, effluent was pumped and sampled for measuring residual chlorine and pH. In the second experiment, after membrane cleaning and neutralisation, the concentration of NO_x-N and NH_3-N in the water in the aeration tank was measured continuously. The measurement results are shown in Figures 14.14 and 14.15.

The residual chlorine was almost zero and the pH decreased from 7.5 to 7.2 in one hour after membrane cleaning, though the cleaning wastewater was turbid temporally at the end of cleaning process. This showed that neutralisation of cleaning wastewater is effective.

In the experiment monitoring NO_x-N and NH_3-N, the NO_x-N concentration first decreased from 15 mg/l to 9 mg/l due to the dilution effect by cleaning wastewater, and then returned to its initial value within five hours. There was no variation in NH_3-N concentration during the 24 hour survey, showing that membrane cleaning exerted no effects on nitrification.

Table 14.13. Membrane cleaning and neutralisation

Membrane cleaning
Reagent: Sodium hypochlorite solution of 5,000 mg-Cl/l, 15l
Procedure: Injecting the solution into the membrane modules at a rate of 1.0 l/min during cleaning, the aeration tank was aerated normally
Neutralisation of cleaning wastewater
Reagent: 0.5 % sodium thiosulfate solution, 5*l*
Procedure: Injecting the solution into the membrane modules at a rate of 1.0 l/min during neutralisation, the aeration tank was aerated normally

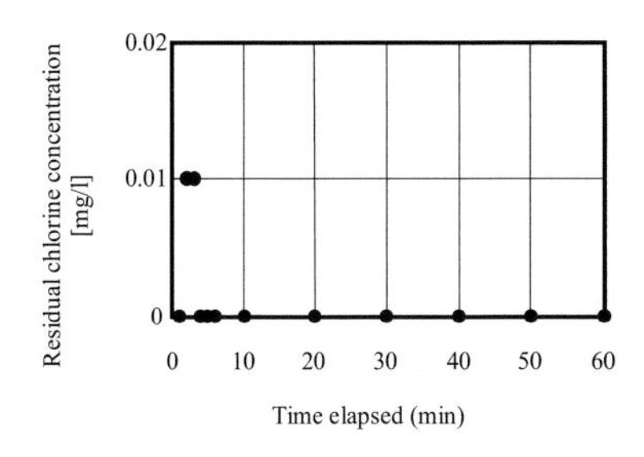

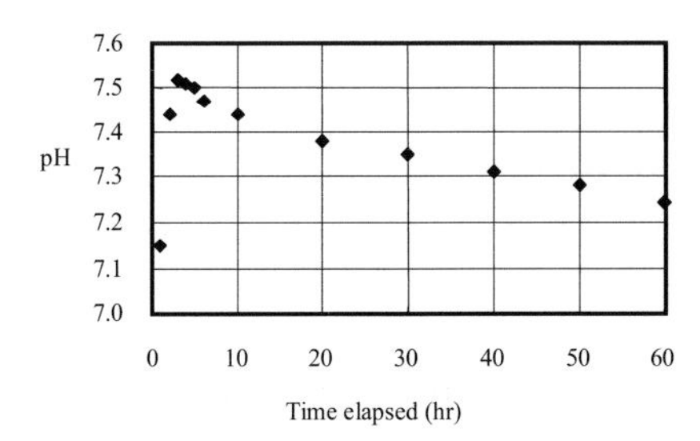

Figure 14.14. Variations of residual chlorine and pH following membrane cleaning.

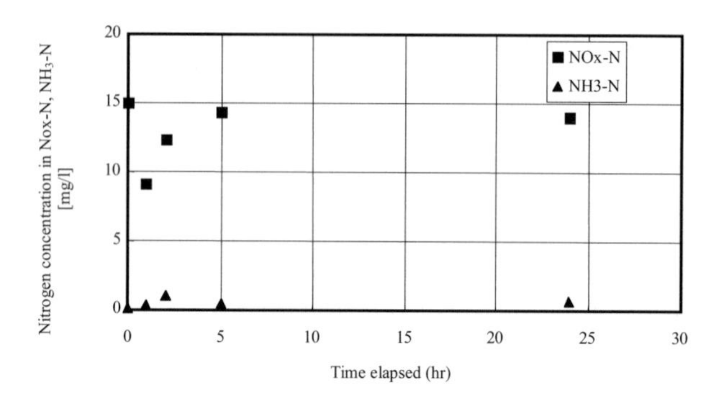

Figure 14.15. Variations of NO_X-N and NH_3-N concentrations.

14.6 MAINTENANCE AND DESLUDGING OF SMALL SCALE JOHKASOUS

14.6.1 Obligations on johkasou users

Small-scale johkasou users must use johkasous correctly and must maintain and desludge them according to local regulations. Johkasou users must also submit to an annual water quality examination. This examination should be conducted by the inspecting agency specified by the regional governor.

Since johkasou users do not always have expert knowledge concerning maintenance and desludging, those jobs are usually entrusted to johkasou maintenance vendors and johkasou desludging vendors. The general procedure for the maintenance and desludging of johkasou systems and water quality examination is summarised in Figure 14.16.

14.6.2 Maintenance

Maintenance and operation are necessary to adjust the johkasou or repair it as required. More specifically, these operations include monitoring the operating condition of each tank and auxiliary accessories and the water quality of effluent, in order to discover faults or defects and take action to repair thems.

The procedures to be followed and the minimum frequency of maintenance are stipulated according to the treatment process and size of the johkasou system. For a gappei-shori johkasou for individual households based on the anaerobic filter–contact aeration process, for example, the procedures for maintenance, which are to be followed at least three times a year, are stipulated as follows:

For the anaerobic filter tank, the transparency, effluent pH and sludge accumulation should be measured. When the amount of settled sludge is approaching the storage capacity limit, the johkasou user must contact the johkasou desludging contractor and request desludging.

For johkasous with flow equalisation and recirculation devices, the amount of water required for flow equalisation and recirculation must be adjusted as necessary, and pipes must be cleaned to remove slime.

For the contact aeration tank, the water temperature, transparency, pH and dissolved oxygen in the tank should all be measured, and the amount and colour of the biomass attached to the contact media should be checked. When biomass forms a thick layer, it should be separated by backwashing. The separated sludge is transferred to the anaerobic filter tank.

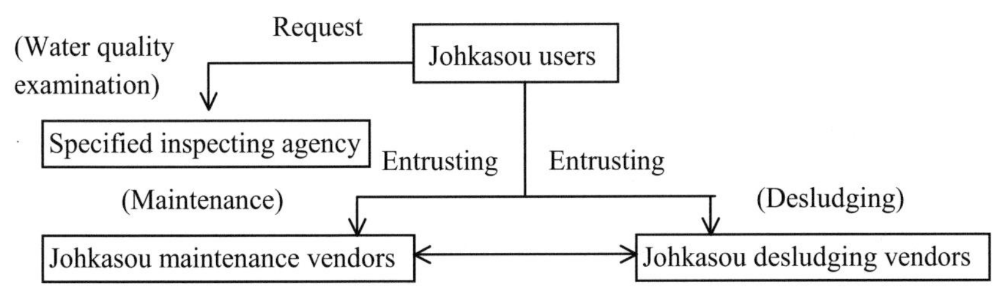

Figure 14.16. Organisation of maintenance, desludging and water quality examination (Ohmori 1996).

The transparency and pH of the effluent sedimentation tank should be measured. Also, sludge accumulation has to be recorded. If there is scum or deposited sludge in the tank, it should be transferred to the anaerobic filter tank.

For the disinfection tank, check on the residual amount of chlorine disinfectant in the column and measure the residual chlorine in the effluent. Replace the column with chlorine disinfectant if necessary.

For the blower, check, clean and repair each component according to its maintenance specifications.

14.6.3 Desludging

Desludging involves draining the deposited sludge and scum from johkasous and conditioning sludge in the tanks. The primary purpose of desludging is to avoid its washout.

For small-scale gappei-shori johkasous for individual households based on the anaerobic filter–contact aeration process, desludging should be done at least once a year according to the following procedure:

Use a vacuum car capable of carrying 2 to 4 metric tons of sludge.

In the first room of the anaerobic filter tank, drain all scum and deposited sludge, wash the cover of the filter media with tap water, then drain all the water from the tank.

In the second room of the anaerobic filter tank, contact aeration tank, sedimentation tank and disinfection tank, the necessary amount of water to be drained is variable, depending on the condition of sludge accumulation in each tank. Fill the johkasou with tap water to the designated level. Carry the drained sludge to the sludge treatment facility (for example, night soil treatment facilities) for sanitary treatment.

In the night soil treatment facilities, sludge from johkasous is mixed with night soil from vault toilets and is introduced to anaerobic treatment processes and/or aerobic treatment processes. In some night soil treatment facilities, sludge is used to produce biogas by anaerobic digestion and generating electricity; some compost plants use dewatered sludge. Thus sludge from johkasous and vault toilets was treated with material recovery and energy recovery in the centralised treatment plants.

14.6.4 Water quality examination

Small-scale johkasous for individual households are subject to an annual water quality examination by specified inspecting agencies. The examination consists of three parts: (1) visual inspection, (2) water quality examination, (3) inspection of maintenance and desludging records.

Visual inspection involves inspecting the condition of devices, the water flow, the johkasou, any odour problems, the condition of disinfection and so on.

Effluent pH, dissolved oxygen, transparency, residual chlorine and BOD should all be measured. If the johkasou is designed for nitrogen removal, total nitrogen of effluent will also be measured.

Maintenance records and desludging records will be inspected to discover whether or not the johkasou was maintained and desludged appropriately.

The results of visual inspection, water quality examination and inspection of the records will be used to judge treatment performance of the johkasou. If any malfunction of the johkasou examined is found, the inspector will advise the johkasou user on how to improve the operation of the johkasou.

14.6.5 The energy consumption of johkasous

It is essential for a johkasou to be constantly supplied with electric power. Blowers used in johkasous play an important role in supplying air not only for aeration, but also for flow equalisation and recirculation. The energy cost of a johkasou in operation depends on the power of the blowers.

For small-scale johkasous, the power of blowers varies with the type and the capacity of johkasous. Some examples of the energy consumption of different type johkasous are shown in Table 14.14.

Table 14.14. The energy consumption of johkasous

Type/capacity	Power of blowers (W)	Energy consumption (kW/Year)
Anaerobic filter–contact aeration process*		
For 5 users	58	508
For 10 users	112	981
Type A johkasou in Figure 14.6**		
For 5 users	90	788
For 10 users	155	1358
Type B johkasou in Figure 14.6**		
For 5 users	91	797
For 10 users	176	1542
Type C johkasou in Figure 14.6**		
For 5 users	106	929
For 10 users	144	1261
Membrane johkasou**		
For 5 users	158	1384
For 10 users	320	2803

Note: * BOD removal only; **Both BOD and N removal

14.7 FURTHER DEVELOPMENTS IN JOHKASOU TECHNOLOGY

Research and development on small-scale johkasou technology has been significant in Japan. There are several trends in the research and development, as shown in Figure 14.2 earlier in this chapter.

Some key technologies developed in wastewater treatment facilities have been applied to small-scale johkasous. For example, an absorption method and an electrolytic flocculation method have been introduced at the experimental level in small-scale johkasous for phosphorus removal.

Membrane modules for modifying tandoku-shori johkasous to gappei-shori johkasous have been developed and will come into use in the near future. Using these membrane modules, tandoku-shori johkasous will be easily changed to gappei-shori johkasous with a comparatively small volume.

Johkasous with special functions, such as johkasous for WDU use and membrane johkasous for water reuse, have been developed and are now on the market. Johkasous for WDU use will not only offer new benefits to the johkasou user, but also are expected to promote the reuse of organic matter. Effluent from membrane johkasous may be used for gardening, car washing, toilet flushing, fire fighting and so on. Membrane johkasous are very useful in such areas where water supply is limited and water reuse is essential.

Compact johkasous for BOD removal, which have 60–70% of the volume of standard johkasous, are now available, so that people who previously did not have enough room for a standard johkasous can now fit a compact model.

Small-scale johkasous for both BOD and nitrogen removal are becoming more common. This kind of johkasou with advanced treatment performance is mainly applied to such areas where effluent is discharged to closed or semi-closed water bodies.

14.8 REFERENCES

Japanese Education Centre for Environmental Sanitation (JECES) (1998). Johkasou systems for the treatment of domestic wastewater.

Ohmori, H. (1996) Maintenance and management of johkasou systems. *Proceedings of Japan–China Symposium on Environmental Science*, Tokyo, 2224 November.

Ohmori, H., Yahashi, T., Furukawa, Y., Kawamura, K. and Yamamoto, Y. (2000) Treatment performance of newly developed johkasous with membrane separation. *Wat. Sci. Tech.* **41**(10/11), 197–204.

Yang, X.M., Kuniyasu, K. and Ohmori, H. (1996) An investigation on treatment performance of small-scale gappei-shori johkasous. Proceedings of Japan–China Symposium on Environmental Science, Tokyo, 22–24 November, 215–218.

Yang, X.M., Watanabe, T., Kuniyasu, K. and Ohmori, H. (1994) The influence of discharge from food waste disposers on johkasou performance. *Proceedings of International Conference on Asian Water Technology*, Singapore, 22–24 November.

15

Extensive water-based post-treatment systems for anaerobically pre-treated sewage

Nigel Horan

There are many important factors to consider when selecting the most appropriate sewage treatment option for a given wastewater in a given location. When considering any of the wide range of process options, the single most important factor is whether the options under consideration are able to meet the effluent quality required. After all, the sole point of installing a wastewater treatment system is to achieve improvements in the health of the population and to protect the aquatic environment from pollution. The required improvements are in turn dictated by the needs and resources of the community, which should be reflected in the severity of effluent standards imposed. In practice, however, important factors such as land restrictions or site geology may mean that the most suitable option, technically, is not feasible in engineering terms.

Edited by P. Lens, G. Zeeman and G. Lettinga. ISBN: 1 900222 47 7

Modern, well-designed sewage treatment facilities can remove a wide range of contaminants from both domestic and industrial wastewaters. These contaminants typically include:

(1) Organic material, which is routinely expressed as a Biochemical Oxygen Demand (BOD) or Chemical Oxygen Demand (COD)
(2) Suspended solids
(3) Ammonia
(4) Nitrate
(5) Phosphate
(6) Heavy metals
(7) Faecal coliforms
(8) Helminths

This ability is reflected in the standards with which the plant must comply and such standards would always include 1 and 2. Additionally, where the effluent is required for reuse in irrigation of edible crops, 6, 7 and 8 would be included and where the effluent discharges to a watercourse likely to suffer from eutrophication, 3, 4 and 5 would also be included. Unfortunately, no single unit process is able to remove all the above eight parameters efficiently. It is usual practice therefore to assemble a treatment train of the most appropriate unit options able to treat each of these parameters. The job of the process engineer is to ensure that the most cost-effective and appropriate treatment train is assembled, which meets the requirements for effluent quality and yet which is compatible with the engineering limitations imposed by the site. A bewildering range of options is available to achieve this (Table 15.1).

Although each option in Table 15.1 demonstrates good contaminant removal, each has a number of advantages and disadvantages associated with them. The three major criteria which must be balanced are: cost, land requirement and sludge production. As a general rule, as a process becomes more complex, the amount of land it requires (generally referred to as the site footprint) reduces and the effluent quality it can attain will improve. However, this is achieved at greater cost and with an increased sludge production. These observations have led to the increasing use of centralised urban sanitation systems in order to demonstrate cost savings as a result of scale, and to minimise the number of sludge reception and treatment sites.

Conversely, as a treatment system is simplified its site footprint increases (often significantly) and it cannot achieve the same consistently high effluent quality. However, the amount of sludge it produces is generally much lower, and often such a high effluent quality is not necessary in view of the fate of the final effluent (Johnstone and Horan 1994). Such simple systems are ideal as a basis for a decentralised sanitation system. This chapter considers the operation and treatment efficiency of one particularly attractive option; the waste stabilisation pond system (Arthur 1983).

Table 15.1. Unit processes suitable for removal of a range of sewage contaminants

Contaminant	Unit process
Organic material	Anaerobic pre-treatment (e.g. UASB), facultative ponds[1], trickling filter, activated sludge, land treatment, chemical addition, membrane bioreactor
Suspended solids	Screening and comminution, septic tanks, anaerobic ponds[1], primary sedimentation, lamella settlers, dissolved air flotation
Ammonia	Trickling filters, activated sludge, ion exchange, air stripping
Nitrogen (total and oxidised)	Activated sludge
Phosphate	Chemical addition, activated sludge
Excreted pathogens	Maturation ponds[1], chlorination, ozonation, membrane bioreactor, UV

[1] the components of a waste stabilisation pond (WSP) system

It is apparent from Table 15.1 that a treatment train comprising of anaerobic, facultative and maturation pond will be able to remove all the contaminants, with the exception of nitrogen-based pollutants and phosphate (in other words the nutrients). Within the European Community (EC), the need for nutrient removal is restricted to sites discharging into sensitive watercourses. Consequently, pond systems could potentially find widespread application. For instance, waste stabilisation ponds (WSPs) are widely used in France where there are over 2500 such systems (Racault *et al.* 1995).

15.1 EFFLUENT STANDARDS

Wastewater treatment systems are designed to ensure improvements in environmental quality (including public health) and effluent standards should be set to balance what is required in order to achieve these improvements, and what a region can afford economically (but see Johnstone and Horan 1994). The EC Urban Wastewater Treatment Directive (EC 1991) and the EC Bathing Water Directive (EC 1994), provide good example, of effluent standards designed to protect, respectively, the aquatic environment and bathing water quality (Table 15.2) which are realistic and with adequate time provisions for compliance.

Table 15.2. The standards prescribed by the Urban Wastewater Treatment Directive (UWWTD) and EC Bathing Water Quality Directive

Parameter	Percentage reduction	Mandatory Limit	Reference
BOD_5 (mg/l)	70–90	25	EC 1991
COD (mg/l)	75	125	EC 1991
SS (mg/l)	35	35	EC 1991
Escherichia coli/100 ml	–	2000	EC 1994

However, in many arid and semi-arid areas of the world, large-scale reuse of sewage effluents is necessary because of the water shortages that result from increasing populations and agricultural demand. In such cases standards are required which focus more on the protection of public health. The health risks associated with human waste reuse have been widely examined over the past twenty years and many epidemiological studies have shown demonstrable health effects from wastewater reuse (for example, Shuval *et al.* 1986).

On the basis of these studies, the relative health risks associated with reuse of untreated wastewater have been quantified in terms of the excess infection caused by different classes of pathogens. This scheme has been used as a basis for the provision of firm guidelines aimed at minimising the health risks associated with reuse of wastewaters. It has resulted in the recommendation of standards for the microbiological quality of treated wastewaters which are both technically feasible to achieve and also, on the basis of the best epidemiological evidence to date, will minimise associated health risks to an acceptable level (see Table 15.3). However, there is still debate as to whether certain standards are unnecessarily strict and might thus limit the full potential of wastewater reuse (Shuval 1996).

The standards for wastewater reuse are based solely on the removal of intestinal nematodes and faecal coliforms and, in the case of the latter, a 5 or 6 log removal is required. A waste stabilisation pond is the only form of wastewater treatment that can guarantee such high removal rates without resorting to disinfection of effluents. In hot climates these standards are readily achievable by a series of five ponds, each with a retention time of five days; this will also produce an effluent with a low BOD and nutrient concentration which is suitable for unrestricted irrigation.

Table 15.3. Microbiological quality guidelines for treated wastewater used for irrigation (summarised from WHO 1989)

Reuse conditions	Exposed group	Intestinal nematodes[a] (arithmetic mean no. of eggs per litre)	Faecal coliforms (geometric mean no. per 100 ml)
Unrestricted irrigation (crops likely to be eaten uncooked, sports fields, public parks)	Workers, consumers, public	<1	<1000[b]
Restricted irrigation (cereal crops, industrial crops, fodder crops, pasture and trees[c])	Workers	<1	No guideline required

[a] *Ascaris lumbricoides*, *Trichuris trichura* and human hookworms.
[b] A more stringent guideline (<200 faecal coliforms per 100 ml) is appropriate for public lawns, such as hotel lawns, with which the public may come into direct contact.
[c] In the case of fruit trees, irrigation should cease two weeks before fruit is picked, and no fruit should be picked off the ground. Sprinkler irrigation should not be used.

15.2 EFFLUENT QUALITY FROM ANAEROBIC PRE-TREATMENT

Previous authors have considered the application of anaerobic technology as a pre-treatment option for domestic wastewaters (Zeeman *et al.* 2001, Kalogo and Verstraete 2001). In addition, many authors have undertaken field studies to assess the application of anaerobic technology as a pre-treatment option. Chernicharo and Machado (1998) demonstrated a BOD removal of 87% in an upward flow anaerobic sludge blanket (UASB) reactor with a hydraulic retention time (HRT) of 6 hours. This UASB was able to produce an effluent quality of 55 mg BOD/l, which was then further treated in an anaerobic filter. Removal of suspended solids was even better and an effluent quality of 24 mg/l was achieved from an influent containing 456 mg suspended solids/l. Similar results were also obtained by de Sousa and Foresti (1996) who achieved an effluent BOD of 36 mg/l from an influent of 257 mg/l with an HRT (defined as the volume of reactor divided by the average daily flow) of only 4 hours. It should be noted, however, that the sewage was a synthetic one and a good performance could be expected. Additional removal of 66% in a sequencing batch reactor provided a final effluent BOD of only 6 mg/l.

However, it is not only purely domestic wastewaters that are amenable to anaerobic pre-treatment. Dean and Horan (1995) treated a mixed domestic and industrial wastewater containing predominantly textile effluent in Mauritius. Using a UASB with an HRT of 10h they achieved a soluble COD removal of 78% from an initial COD of around 750 mg/l. Austermann-Haun *et al.* (1998) demonstrated the feasibility of anaerobic pre-treatment of a brewery wastewater. The influent had a BOD of 1496 mg/l and this was treated in a UASB with an HRT of 8.3h to achieve a removal of 94% and an effluent quality of 95 mg BOD/l. Successive treatment in a nutrient removal activated sludge plant (with an F/M of 0.068/d) and sand filter resulted in a final effluent quality of 3.6 mg/ BOD and 1.8 mg SS/l.

It is apparent from the above that in the absence of specific performance data, a conservative approach for domestic sewage is to assume a removal of 80% of the influent BOD for a UASB with an HRT >6 hours. An alternative is to use the empirical expression (Equation 15.1) proposed by van Haandel and Lettinga (1994) for temperatures in excess of 20 °C:

$$E = 1 - 0.68(HRT)^{-0.68} \tag{15.1}$$

where HRT is in hours. This gives a value for the COD removal efficiency of 70% at an HRT of 6 h. For the majority of wastewaters, this means that a post-treatment option is required to treat a wastewater which has a BOD and COD <100 mg/l. In addition it will also be necessary for the post-treatment option to remove excreted pathogens and nutrients, if this is a requirement of the effluent discharge standards.

Conventional treatment systems that could be exploited for post-treatment require a constant input of mechanical energy together with regular maintenance if they are to achieve consistent BOD and nutrient removal. In many situations it is not practical or desirable to provide either the energy or the maintenance requirements. Under these circumstances extensive, water-based treatment systems provide cheaper alternatives to conventional processes. There is a hierarchy of such processes, which at their simplest would comprise a series of shallow excavations where the only input of energy derives from sunlight. These systems are generally termed waste stabilisation ponds (WSPs) and are effective for removal of BOD and excreted pathogens, but have only a limited nutrient removal capacity. The addition of fixed or floating aerators to a pond will uprate it to an aerated lagoon and increase the rate of BOD removal. However, pathogen removal is lost and nutrient removal is still limited. Provision of a quiescent area within the lagoon to permit solids settlement, together with the addition of a pump to recirculate and/or waste the settled solids, will uprate an aerated lagoon to a simple activated sludge system. This should achieve up to

95% removal of organic material and also permit nutrient removal. Again, however, pathogen removal is lost. The following sections will describe the application of each of these options for the post-treatment of anaerobically pre-treated effluents with a particular emphasis on waste stabilisation pond systems.

15.3 WASTE STABILISATION PONDS

A waste stabilisation pond is a shallow excavation which receives a continuous flow of wastewater (Figure 15.1). A number of these are required, generally arranged in series such that successive ponds receive their flow from the previous pond. The most common arrangement of ponds is to have an anaerobic pond together with a facultative pond arranged in series and followed by a number of maturation ponds, although other arrangements are possible (Figure 15.2).

Figure 15.1(a). Typical decentralised treatment system featuring a five-pond series comprising anaerobic, facultative and three maturation ponds. (b) Facultative waste stabilisation pond illustrating their simple construction (Photographs courtesy of D.D. Mara).

Since the normal configuration for a pond train involves anaerobic pre-treatment by an anaerobic pond, their conventional design criterion requires very little modification for a decentralised sanitation system with anaerobic pre-treatment by, for instance, an UASB reactor. The degree of treatment achieved across the complete treatment system will be a function of the number of ponds in the series and the retention time of the wastewater in each pond (Table 15.4).

Although the most important role of waste stabilisation ponds is in the removal of pathogenic microorganisms, they are still capable of producing an effluent that is low in organic carbon. The oxygen necessary to satisfy the carbonaceous and nitrogenous oxygen requirements is provided by algal photosynthesis, and the rate at which this oxygen is produced will determine the rate at which aerobic, heterotrophic oxidation takes place. In the absence of algal-generated oxygen, BOD removal will still take place as a result of sedimentation, anaerobic metabolism and aerobic respiration driven by surface aeration.

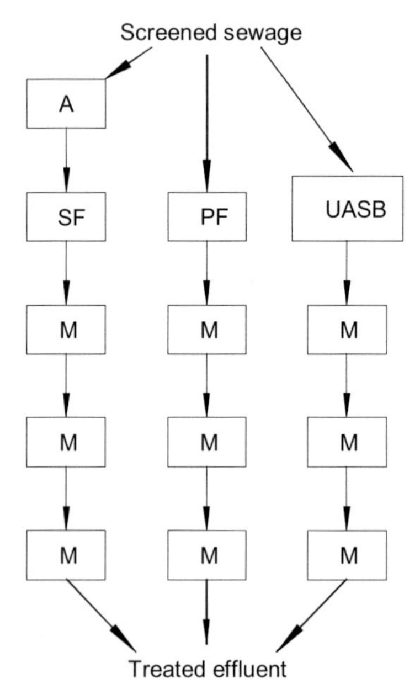

Figure 15.2. Potential arrangements of a pond treatment train (A = anaerobic pond; PF = primary facultative; SF = secondary facultative; M = maturation).

Since pond systems rely entirely on natural processes, a complex biological ecosystem develops within the pond. Consequently, despite the fact that they are the simplest form of wastewater treatment system, they are the most poorly understood in terms of the chemical and biochemical reactions that take place within them. As a result of this, models for the design of waste stabilisation ponds tend to be purely empirical.

Table 15.4. The principal functions of the main pond types and their typical performance and operating data (from Horan 1990)

Pond type	Depth (m)	Retention time (d)	Major role	Typical removal efficiencies	
				Type	%
Anaerobic	2–5	3–5	Sedimentation of solids, BOD removal, stabilisation of influent, removal of helminths	BOD SS FC Helminth	40–60 50–70 1 log 70
Facultative	1–2	4–6	BOD removal	BOD SS FC	50–70 Increases 1 log
Maturation (for a typical system comprising of three ponds)	1–2	12–18	Pathogen removal, nutrient removal	BOD SS FC Helminth	50–70 20–40 4 log 100

15.4 FACULTATIVE PONDS

15.4.1 Principles

A facultative pond is the first pond to receive anaerobically pre-treated effluent and, strictly speaking, it is referred to as a secondary facultative pond (a primary facultative pond is the one that receives raw wastewater). In many circumstances only anaerobic pre-treatment and facultative ponds will be required: for example, prior to restricted crop irrigation and fishpond fertilisation (Mara and Pearson 1997).

The major role of facultative ponds is for BOD removal. The presence of both aerobic and anaerobic (strictly speaking, anoxic) environments within the same pond means that both anaerobic metabolism associated with anaerobic environments and oxidative metabolism associated with an aerobic environment will occur. Unlike conventional processes, the oxygen required to satisfy the carbonaceous and nitrogenous oxygen demand is supplied by biological means and not mechanically. This is achieved by constructing shallow ponds (1–2 m deep) which maximise capture of solar radiation and encourage the growth of photoautotrophic organisms (principally algae). The algae utilise the carbon dioxide evolved in bacterial heterotrophic metabolism to provide them with a source of carbon, using energy from photosynthesis. This results in the production of gaseous oxygen. This relationship between the algae and bacteria can be considered as a form of symbiosis, and their interactions have been summarised in Figure 15.3.

In order to maximise the growth of algae, the loading to a facultative pond must be controlled so that the oxygen demand of the influent wastewater does not exceed the rate at which oxygen can be supplied by photosynthesis. Because of the requirements of algae for light, the rate of oxygen production and thus its concentration in the pond will vary both diurnally and with pond depth. As the pond depth increases, light penetration decreases and thus less oxygen is available. The lower half of the pond is therefore generally devoid of oxygen and hence anaerobic metabolism occurs.

During daylight the majority of the BOD is removed by facultative aerobic bacteria at the aerobic surface layer. As the intensity of the incident sunlight decreases, photosynthetic activity declines until at low light levels the algae switch from photosynthesis to respiration. During the hours of darkness, as the residual oxygen is utilised, the pond will slowly become anaerobic. Anaerobic metabolism in the pond sediment serves to degrade sedimented sludge, and thus increases the times between desludging. Typically the desludging period for facultative ponds is between 5 and 10 years (Saqqar and Pescod 1995).

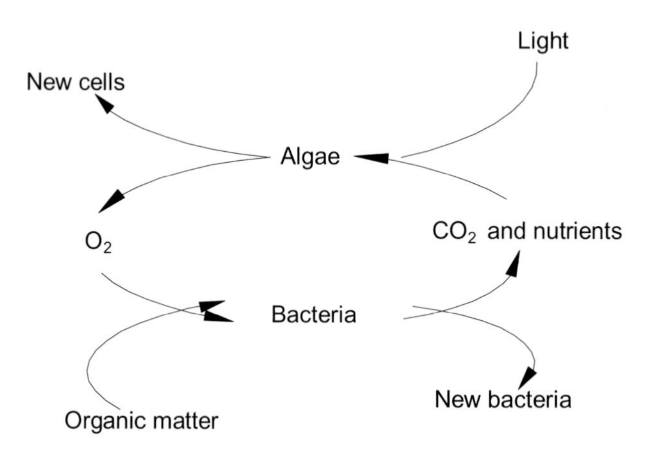

Figure 15.3. Symbiotic relationship between the bacteria and algae in a facultative pond system.

15.4.2 Design

Early models for the removal of BOD in primary facultative ponds assumed that they behaved as completely mixed reactors, with BOD following first-order removal kinetics (for example, Marais and Shaw 1961). It is more common, however, to adopt an empirical approach to design based on the permissible surface BOD loading ($\lambda_{s(max)}$ in kg/ha d). This is defined as:

$$\lambda_{s(max)} = 10LiQ/A_f \tag{15.2}$$

where A_f is the required pond area (m^2), L_I is the influent BOD (mg/l) and Q is the design flow rate (m^3/d).

The value for the areal loading increases with increasing temperature and the maximum loading that can be applied is given by Equation 15.3 (McGarry and Pescod 1970):

$$\lambda_{s(max)} = 60.3\ (1.099)^T \tag{15.3}$$

However it is not usual to design at the maximum loading as it allows little scope for failure or expansion. A more appropriate equation is that of Mara and Pearson (1997) which applies at temperatures >10°C (Equation 15.4). Below this temperature the maximum loading is 100 kg/ha d.

$$\lambda_{s(max)} = 350(1.07 - 0.002T)^{T-25} \tag{15.4}$$

where T is the temperature of the coldest month in °C.

The required pond area for a given areal BOD loading (calculated as the mid-depth area, A_f) is then calculated from Equation 15.2. The mean hydraulic retention time (t_f) of the facultative pond can now be determined from Equation 15.5 since:

$$t_f = \frac{A_f D_f}{Q} = \frac{10 L_i D_f}{\lambda_s} \tag{15.5}$$

where D_f is the mean depth of the facultative pond in metres and is typically around 1.5 m. A minimum value for the retention time is 4 days at temperatures above 20°C and 5 days below this temperature. This will ensure that the algae are able to multiply without cell washout and it also helps to minimise short-circuiting within the pond.

15.4.3 Performance

Many waste stabilisation pond systems do not include an anaerobic pond and thus the facultative pond is a primary pond. Dixo *et al.* (1995) have carried out a comprehensive study into the removal of a range of parameters following treatment in a UASB pre-treatment system. They monitored the effluent quality for a range of parameters in a treatment train that comprised a full-scale UASB (160 m^3) with a retention time of 0.29 days, followed by one facultative and three maturation ponds on a pilot-scale. The ponds were each of 5 d retention time giving a total retention time of 20 days. This system was capable of achieving the WHO Guidelines for unrestricted irrigation. Inspection of Table 15.5 suggests that the effluent would comply with the EC UWWTD (see Table 15.2).

Table 15.5. Performance of an anaerobic pre-treatment system (UASB) followed by four stabilisation ponds with a total retention time of 20 days

Parameter	Influent	Effluent from				
		UASB	F1	M1	M2	M3
BOD (mg/l)	771	83	54	47	44	43
Suspended solids (mg/l)	681	66	63	86	75	72
pH (units)	7.6	7.1	7.6	8.3	8.8	9.1
Faecal coliforms (/100 ml)	2.8×10^7	9.2×10^6	6.0×10^5	2.8×10^4	3.7×10^3	6.1×10^2
Nematode eggs (/L)	16774	1740	3.5	ND	ND	ND

15.5 MATURATION PONDS

15.5.1 Principles

Maturation ponds are also referred to as tertiary lagoons or occasionally as aerobic ponds. They may receive the effluent from a facultative pond or, depending on the effluent quality required, may receive effluent directly from the pre-treatment process. Their major role is the removal of pathogenic microorganisms such as viruses, bacteria and helminths. They are not suitable for BOD removal and thus the influent to a maturation pond should meet the discharge consent for this parameter. Indeed the BOD can increase across a series of maturation ponds as a result of additional algal growth. For this reason some regulators (EC 1991) allow analysis from maturation ponds to be carried out on filtered samples and will permit unfiltered suspended solids of 150 mg/l.

Pathogen removal in a maturation pond is achieved by providing a retention time long enough to reduce the numbers to the required level. The number and size of maturation ponds will therefore depend on the standard of effluent that is required and this is normally expressed as faecal coliforms/100 ml. Rational design of maturation ponds is hindered by a lack of information on the mechanism or mechanisms of pathogen removal. There is general agreement that the excreted eggs of helminths such as *Ascaris, Trichuris* and *Taenia* are removed by sedimentation, due to their large size (from 20 to 70μm). Protozoal cysts of organisms such as *Giardia* and *Entamoeba* behave in a similar way, although they are generally much smaller (14 μm long and 8 μm broad in the case of *Giardia*), and thus require longer retention times. Removal will take place across the pond series and complete removal can be expected in ponds with overall retention times of 11 days or more. Although they are removed from the pond effluent, they are not necessarily inactivated, and can remain viable in the sludge layer for several years (Mara and Pearson 1997). This is an important consideration when designing a desludging regime.

Over 100 different viruses are excreted in faeces by man, and very little is known about their survival in ponds. Since most viruses carry a strong negative charge, it is assumed that the major mechanism for their removal is adsorption to particulate material, followed by sedimentation. The limited data available suggest that ponds with overall retention times of greater than 30 days should achieve at least a 4 log reduction in enteroviruses and a 3 log reduction in numbers of rotaviruses. In a similar way to the helminths, removal from the effluent does not mean inactivation, and viruses may be capable of remaining viable in the sludge layer for long periods.

The faecal coliform bacteria have been universally adopted as indicators of excreted pathogen removal, and consequently much of the work into elucidation of removal mechanisms has been carried out using these bacteria. The rate of die-off of faecal coliforms in maturation ponds is much faster than in other pond types and increases with increasing temperature. A number of mechanisms have been suggested for pathogen removal in maturation ponds of which the most important ones are nutrient starvation, enhanced pH, high dissolved oxygen concentration, lethal UV irradiation from sunlight and protozoal predation (see Curtis *et al.* 1992; Saqqar and Pescod 1991; Davies-Colley *et al.* 1999). All of the above mechanisms are closely interlinked; for instance increased sunlight causes increased UV irradiation and stimulating algal photosynthesis leads to an increased pH and dissolved oxygen concentration. It is difficult, therefore, to delineate and quantify the contribution of each potential removal mechanism, thus this approach has not yet featured in pond design.

In addition to reduced pathogens, the effluent from maturation ponds is also lower in suspended solids. The suspended solids content of a facultative pond effluent is composed mainly of algae. These are predominantly motile flagellates which form dense bands to effectively utilise the incident sunlight. These dense bands can often appear in the effluent and thus give very a high suspended solids content. In maturation ponds these are replaced by non-motile algae which outgrow the flagellates due to the increased light penetration. Since these algae are non-motile, they are distributed evenly throughout the pond and their concentration in the effluent is reduced, as dense bands are not encountered.

As well as their role in pathogen removal by pH elevation, algae play a further role in maturation ponds by removing nitrogen and phosphorus. The major mechanism of phosphorus removal is sedimentation as organic phosphate, associated with the algal cell. In addition, at elevated pH values, phosphate becomes insoluble and chemical precipitation occurs. Many algae, such as the *Cyanophytae*, are able to store phosphate in the form of granules, and algae also represent the largest fraction of bound organic phosphate in the pond. A similar removal mechanism appears to operate for nitrogen in maturation ponds. Soluble nitrogen is almost exclusively in the form of ammonia, and this form is taken up by algae in preference to nitrate nitrogen. This nitrogen will enter the pond sediments as organically bound nitrogen when the algae die and settle, and although a fraction of it is biodegradable, up to 60% remains undegraded in the pond sediment.

15.5.2 Design

The removal of faecal coliform bacteria from any pond follows a first-order removal kinetics and if complete mixing is assumed then:

$$N_e = \frac{N_i}{1 + K_b t} \tag{11.6}$$

where N_e is the number of faecal coliforms in 100 ml of effluent, N_i is the number of faecal coliforms in 100 ml of influent, K_b is the first-order rate constant for faecal coliform removal (d^{-1}) and t is the *r*etention time in the pond (d). For a number of ponds (*N*) in series, Equation 11.6 takes the form:

$$N_e = \frac{N_i}{(1 = K_b t_a)(1 + K_b t_f)(1 + K_b t_m)^n} \tag{11.7}$$

where t_a ,t_f, t_m are the retention times of the anaerobic pre-treatment system, facultative and maturation ponds in the series. *n* represents the number of maturation ponds required. The first-order removal constant K_b is a lumped parameter which takes into account all the factors which affect pathogen removal. It is particularly sensitive to temperature and this effect has been modelled empirically as:

$$K_{b(T)} = 2.6(1.9)^{T-20} \qquad (11.8)$$

This equation suggest that for every 1°C rise in temperature, the value of K_b increases by 19%. As more quantitative information becomes available on the removal of pathogens, then more deterministic equations for K_b can be formulated.

In order to use Equation 11.7 to design maturation ponds, a value for both the number of maturation ponds (n), and their retention times (t_m) is required. This is generally carried out by iteratively increasing n, in order to find the combination of n and t_m which has the least areal land requirements. This can easily be done with a simulation program on a personal computer. This occurs with a maximum value of n and consequently a minimum value of t_m, with the following boundary conditions:

(1) The minimum acceptable value of t_m is three days. Below this value the danger of hydraulic short-circuiting becomes too great.
(2) The value for t_m should not be higher than that for t_f.
(3) The areal BOD loading on the first maturation pond should not exceed the areal BOD loading on the facultative pond, assuming a BOD removal of 70% in the anaerobic pre-treatment system (pond or UASB) and the facultative pond.

The value of N_i should ideally be obtained from analysis of the wastewater for which the pond is intended to treat. If this is not available at the design stage, then a value of 10^7 faecal coliforms/100 ml can be used. This is quite conservative since the effluent from a UASB contains in the range 10^6 to 10^7 faecal coliforms/100 ml (van Haandel and Lettinga 1994).

Table 15.6. Removal of microorganisms from maturation ponds worldwide (adapted from Maynard *et al.* 1999)

Country	Retention time (d)	FC removal (%)	Helminth removal (%)
Tanzania	8	90.8	–
Kenya	3	89.5	–
	27.6	–	99.96–100
Morocco	–	73–95	–
	7	99.6	100
	7.5	74.4–84.3	–
Tunisia	–	99.97	–
S. Africa	3	99.99	99.9999
Israel	–	90	–
India	–	–	38.5–100
Thailand	20	88	–
New Zealand	–	92–99.9	–
Australia	16	98.8–99.96	–
	3	–	100
France	40–70	99.95	–
Portugal	–	96.5	–
Cayman Islands	3	66–79	–
Peru	5.5	–	100
Brazil	15	83.5–95.3	–
	9.8	–	87–100

The value selected for Ne will depend upon the intended use of the pond effluent. In view of the high effluent quality that a pond is capable of producing, the effluent is frequently exploited for reuse, either in agriculture or aquaculture. Typical guideline values for the microbiological quality of treated pond effluents intended for agricultural reuse are given in Table 15.4. Thus if a value for N_e of 10^3 faecal coliforms/100 ml is selected, then the effluent from the pond should be suitable for unrestricted irrigation, which includes irrigation of edible crops. Van Haandel and Lettinga (1994) found that in north-east Brazil this microbiological quality could be achieved with a UASB at a retention time of 7.2 hours followed by 3 maturation ponds each of 2.5 days retention. The overall value of K_b for this series was 4.3/d.

15.5.3 Performance

The performance of maturation ponds worldwide has recently been comprehensively reviewed by Maynard *et al.* (1999). This review raised many questions. The authors concluded that little information is available as to the removal mechanisms for BOD, suspended solids and heavy metals. In addition, there is conflicting evidence for mechanisms to describe bacterial removal. They provided a very useful summary of worldwide performance of maturation ponds in terms of microbiological removal (Table 15.6), which clearly demonstrates the wide variability in faecal coliform removal over a range of 73% to 99.99% at similar hydraulic retention times.

15.6 OTHER CONSIDERATIONS IN POND DESIGN

15.6.1 Mixing and short-circuiting

The mixing of the pond contents is an important mechanism which helps to convey oxygen produced at the pond surface to the lower layers. In addition, it contributes oxygen in its own right by the process of reaeration. This mixing also helps to reduce thermal stratification which aids in dispersing bacteria and algae throughout the pond, thus producing an effluent of a more consistent quality. In order to facilitate wind-inducing mixing, the longest dimension of the pond should lie in the direction of the prevailing wind. Since stratification is more pronounced during the summer months, the direction of the prevailing wind during this period should be selected, if it is seasonally variable. Care must be taken in site selection, however, to ensure that ponds are sited downwind of the community, in case problems arise through pond odours, although this should not occur with a well designed system (Mara and Pearson 1997). Short-circuiting in ponds is a common problem as it reduces the pond volume available for wastewater treatment. These effects may be minimised by careful consideration of pond geometry, and also by sensible placement of inlet and outlet structures. Facultative and maturation ponds should be designed to approximate plug-flow mixing as far as possible and thus high length to breadth ratios are employed. Site constraints frequently mean that the required length to breadth of up to 20:1 cannot be achieved. In such situations, baffles placed within the pond will help to ensure plug-flow. These ponds should lie parallel to the direction of the prevailing winds, so badly placed inlet and outlet structures could favour transport of inlet wastewater directly to the outlet by surface wind action. *In situ* studies have shown that a bacteriophage tracer introduced to the inlet of a secondary facultative pond with a retention time of 10 days appears in the effluent within 6 hours (Horan and Naylor 1988).

Siting of the inlet pipe such that the wastewater flows against the prevailing wind will effectively prevent wind-induced short-circuiting, In addition, elevated inlet pipes will produce better mixing and dispersion of the influent due to the turbulence they create. They have the added advantage that samples of pond influent can easily be taken. In large ponds the influent is often split into a number of inlets. Outlet pipes are generally (but not always) sited at the opposite end of the pond to the inlet and should always be fitted with scum guards to prevent the discharge of accumulated scum into the next pond. The level at which the pond contents are withdrawn is controlled by the depth of the scum guard, the height of the pond is controlled by the depth of the scum guard and the height of the outlet pipe within the scum guard. In facultative ponds the guard should extend below the level of the algal band (usually no more than 60 cm below the surface) to ensure that the effluent does not have an excessive algal content. It is often convenient to fit outlet devices with facilities for varying the level of the pond, to permit essential maintenance such as desludging and repairs to the base and embankment.

15.7 AERATED LAGOONS

15.7.1 Principles

An aerated lagoon resembles a waste stabilisation pond in that it is a shallow basin between 2 and 5 m deep, with a larger surface area, which receives a continuous flow of wastewater. It differs, however, in that the oxygen for BOD removal is provided not by algal photosynthesis but by mechanical aeration. It is difficult to see the attractions in aerated lagoons. They have the disadvantages of pond systems in that they need a large land area, together with the higher cost and mechanical problems associated with activated sludge processes. The rationale for their adoption is that in most temperate climates the reduced levels of sunlight mean that algal photosynthesis is also much reduced, and thus longer retention times (and therefore a larger land requirement) are required in order to ensure acceptable BOD removal. It was thought that providing supplementary oxygen using mechanical aerators could reduce this land requirement. A similar philosophy is often followed for the uprating of overloaded facultative ponds in hot climates. Unfortunately, in ponds where mechanical aerators have been installed, there is a drastic change in the ecology of the lagoon, leading to a complete disappearance of the algae and their replacement by a mixed heterotrophic bacterial community, which grows in the form of flocs that resemble activated sludge flocs. Thus, ecologically, aerated lagoons most resemble an activated sludge process operated without cell recycle. It is apparent, therefore, that instead of supplying supplementary oxygen, aerators must supply the total oxygen demand of the wastewater. In addition, as a result of floc formation, the microorganisms will agglomerate and settle under quiescent conditions; aerators must also provide adequate energy for mixing, in order to keep the solids in suspension and disperses the dissolved oxygen throughout the basin.

Power inputs to aerated lagoons are not as high as for activated sludge, so larger solid particles are able to settle out. When lagoons are operated in climates where the bottom sludge temperature exceeds 20°C, this sludge is removed by digestion. If the temperature falls below this, then solids accumulate faster than they can be digested, with a resultant sludge build-up. Under certain climatic conditions, the digestion rate during the summer months is adequate to reduce sludge accumulated during winter.

15.7.2 Design

Aerated lagoons are not completely mixed, but due to their large area and shallow depth, together with the high degree of mixing provided by the aerators, their flow regime closely approximates a completely mixed reactor. It is common practice, therefore, to use a completely mixed model and first-order removal kinetics (Crites and Tchobanoglous 1998). Thus for *n* equal-sized lagoons the required hydraulic retention time (t) is given by:

$$t = \frac{(n/k)}{[(S_o/S_e)^{1/n} - 1]} \qquad (15.9)$$

where k is the BOD rate removal constant (/d), S_o is the influent BOD concentration (mg/l) and S_e is the effluent BOD concentration (mg/l).

A value of k of 0.276 /d is recommended at 20°C and this is corrected for temperature using Equation 15.10.

$$k_T = k_{20}\,(1.035)^{T-20} \qquad (15.10)$$

The total volume of lagoon required decreases rapidly as the number of lagoons in series increases, up to a limit of four lagoons.

15.7.3 Aerators

Aerated lagoons generally employ mechanical surface aerators, which are either floating or fixed. Occasionally, in deeper lagoons, submerged turbines must be provided in order to provide adequate mixing. Concrete pads are usually placed under the mixers in order to prevent scour of the lagoon bottom. These are not required, however, when the lagoon has been completely lined. Aerators are selected based on the amount of oxygen needed to satisfy the oxygen demand of the influent. One equation widely used in the UK estimates the total oxygen demand (TOD) of a wastewater and takes the form:

$$TOD\,(kg/d) = 0.0864\,Q[0.75\,(S_o - S_e) + \frac{5.25\ x\ 10^{-4}\ XV}{Q} \qquad (11.11)$$

where Q is the influent flow rate (l/sec) and X is the reactor mixed liquor solids (mg/l). This value for oxygen demand into wastewater must be converted into a standardised value into clean water in order to select the appropriate size of the aerator. When an aerated lagoon is the system of choice to post-treat a pre-treated wastewater, the value for the TOD is likely to be very small. This is due to the low BOD of the influent wastewater and the low operating MLSS of an aerated lagoon, which is generally <500 mg/l. In addition to the aeration requirements, the installed aerator power must also be adequate to mix the lagoon contents. This can be calculated from Equation 15.12 and it is important that the larger figure from Equations 15.11 and 15.12 is used for aeration selection.

$$p(\mathrm{W/m^3}) = 0.004X + 5 \text{ for values of } X < 2000 \text{ mg/l} \qquad (15.12)$$

where X is the concentration of suspended solids in the lagoon. As the concentration of suspended solids in aerobic lagoons is typically in the range 200–300 mg/l, a power requirement for mixing of ~6 W/m^3 is generally required.

15.7.4 Effluent treatment

The effluent from an aerated lagoon requires some form of settlement stage before it is fit to discharge to a watercourse. Although this can be a conventional sedimentation tank, it is more usual to discharge into one or more maturation ponds, dependent upon the final effluent quality required. A single maturation pond will generally provide adequate solids removal. Where pathogen removal is necessary, three ponds are needed in order to render the effluent fit for unrestricted irrigation.

15.8 REFERENCES

Arthur, J..P. (1983) Notes on the design and operation of waste stabilization ponds in warm climates of developing countries. Technical Paper No 7. The World Bank, Washington DC.

Austermann-Haun, U., Lange, R., Seyfried, C-F. and Rosenwinkel, K-H. (1998) Upgrading an anaerobic/aerobic wastewater treatment plant. *Wat. Sci. Tech.* **37**, 243–250.

Chernicharo, C.A.L. and Machado, R.M. (1998) Feasibility of the UASB/AF system for domestic sewage treatment in developing countries. *Wat. Sci. Tech.* **38**, 325–332.

Crites, R. and Tchobanoglous, G. (1998) *Small and Decentralized Wastewater Management Systems*. McGraw-Hill, Singapore.

Curtis, T.P., Mara, D.D. and Silva, S.A. (1992) Influence of pH, oxygen and humic substances on ability of sunlight to damage faecal coliforms in waste stabilization ponds. *Applied and Environmental Microbiology* **58**, 1335–1343.

Davies-Colley, R.J., Donnison, A.M., Speed, D.J., Ross, C.M. and Nagels, J.W. (1999) Inactivation of faecal indicator microorganisms in waste stabilization ponds: interactions of environmental factors with sunlight. *Water Research* **33,** 1220–1230.

Dean, C. and Horan, N.J. (1995) Applications of UASB technology in Mauritius. Research Monographs in Tropical Public Health Engineering, No. 7. Department of Civil Engineering, University of Leeds, UK.

de Sousa, J.T. and Foresti, E. (1996) Domestic sewage treatment in an upflow anaerobic sludge blanket-sequencing batch reactor system. *Wat. Sci. Tech.* **33,** 73–84.

Dixo, H.G., Gambrill, M., Catunda, P.F.C. and van Haandel, A.C. (1995) Pathogen removal in stabilization ponds treating UASB effluents. *Wat. Sci. Tech.* **31**(12), 275–285.

EC (1991) Council Directive of 21 May 1991 concerning urban waste water treatment (91/271/EEC). *Official Journal of the European Communities* L268/1–14 (24 September).

EC (1994) Council Directive of 29 March 1994 concerning the quality of bathing water (94/C 112/03). *Official Journal of the European Communities* 94/0006 (22 April).

Horan, N.J. (1990) *Biological Wastewater Treatment: Theory and Operation*. John Wiley & Sons, Chichester.

Horan, N.J. and Naylor, P.J. (1988) The potential of bacteriophage to act as tracers of water movement. *Proceedings of 2nd IAWPRC Asian Conference on Water Pollution Control,* 699–705.

Johnstone, D.W.M. and Horan, N.J (1994) Standards, costs and benefits: an international perspective. *Journal of the Institution of Water and Environmental Management* **8**, 450–458.

Kalogo, Y. and Verstraete, W. (2001) Potentials of anaerobic treatment of domestic sewage under temperate climate conditions. Chapter 10 in this volume.

Mara, D.D. and Pearson, H.W. (1997) Waste Stabilization Ponds: Design Manual for Mediterranean Europe. Copenhagen, Denmark: World Health Organization Regional Office for Europe.

Marais, GvR. and Shaw, V.A. (1961) Rational theory for the design of waste stabilization ponds in South Africa. *Transactions of the South African Institute of Civil Engineers* **3,** 205.

Maynard, H.E., Ouki, S.K. and Williams, S.C. (1999) Tertiary lagoons: a review of removal mechanisms and performance. *Water Research* **33**, 1–14.

McGarry, M.G. and Pescod, M.B. (1970) Stabilization pond design criteria for tropical Asia. In *Proceedings of the Second International Symposium on Waste Treatment Lagoons (*ed R.E. Mckinney), pp. 114-132. Laurence, KS: University of Kansas..

Racault, Y., Boutin, C. and Seguin, A. (1995). Waste stabilization ponds in France: a report on fifteen years experience. *Wat. Sci. Tech.* **31**(12), 91–101.

Saqqar, M.M. and Pescod, M.B. (1991) Microbiological performance of multi-stage stabilization ponds for effluent reuse in agriculture. *Wat. Sci. Tech.* **23**(7/9), 1517–1524.

Saqqar, M.M. and Pescod, M.B. (1995) Modelling sludge accumulation in anaerobic wastewater stabilization ponds. *Wat. Sci. Tech.* **31**(12), 185–190.

Shuval H. (1996) Do some current health guidelines needlessly limit wastewater recycling in agriculture? A risk assessment/cost-benefit approach. Paper presented at the World Bank Meeting on Recycling Waste for Agriculture: The Rural–Urban Connection. 23–24 September, Washington DC.

Shuval, H.I., Adin, A., Fattal, B., Rawitz, E. and Yetutiel, P. (1986) Wastewater irrigation in developing countries. Technical Paper No. 51. The World Bank, Washington DC.

Van Haandel, A.C. and Lettinga, G. (1994) Anaerobic Sewage Treatment – A Practical Guide for Regions with a Hot Climate. Wiley, Chichester, UK.

WHO 1989 Health Guidelines for the Use of Wasteater in Agriculture and Aquaculture. Technical Report Series No. 778. Geneva: World Health Organisation.

Zeeman, G., Kujawa-Roeleveld, K. and Lettinga, G. (2001) Anaerobic treatment systems for high-strength domestic waste(water) streams. Chapter 12 in this volume.

16

Management of wastewater by natural treatment systems with emphasis on land-based systems

A.N. Angelakis

16.1 INTRODUCTION

In the natural environment, physical, chemical and biological processes occur when water, soil, plants, microorganisms, and the atmosphere interact. Land application of wastewater was the first natural technology to be reported (Angelakis and Spyridakis 1996). However, land treatment systems were developed in the nineteenth century and then were forgotten about until the 1960s. In the United States there were about 400 natural treatment systems for wastewater (NTSW) in the 1970s and today it is estimated that there are more than 2000 (Reed *et al.* 1995).

Edited by P. Lens, G. Zeeman and G. Lettinga. ISBN: 1 900222 47 7

Natural treatment systems are designed to take advantage of these processes to provide wastewater treatment. Natural systems consider the processes that depend primarily on their natural components to achieve the intended efficiency in wastewater treatment. The processes involved in natural systems include many of those used in mechanical or in-plant treatment systems (sedimentation, filtration, gas transfer, adsorption, ion exchange, chemical precipitation, chemical oxidation and reduction, and biological conversion and degradation) as well as others unique to natural systems such as photosynthesis, photooxidation, and plant uptake. In natural systems, the processes occur at 'natural' rates and tend to occur simultaneously in a single 'ecosystem reactor', as opposed to mechanical systems in which processes occur sequentially in separate reactors or tanks at accelerated rates as a result of energy input (Metcalf and Eddy 1991).

Most natural treatment systems are environment-friendly and economical for many rural locations. They have evolved into significant alternatives for centralized wastewater management. These natural systems, when site conditions are favorable, could usually be constructed and operated for less cost and with less energy than the more popular and more conventional mechanical technologies. In addition, reuse of treated water is possible with all natural systems (Angelakis 1997).

Most of the early wastewater treatment facilities in the United States and Canada were land treatment systems. In the United States, federal legislation, commencing with the Clean Water Act of 1972 (PL 92-500), proposed a 'zero discharge' goal and encouraged a reuse and recovery philosophy. Land application of pre-treatment wastewater is the only economical way to achieve all these goals, and so the concept was reborn (Crites *et al.* 2000).

The natural treatment systems are: (1) the soil-based or land-treatment systems (slow rate, rapid infiltration, and overland flow), (2) the aquatic-based systems (constructed and natural wetlands) and aquatic plant treatment systems, (3) the various configurations of ponds and (4) various types of on-site systems (Angelakis and Tchobanoglous 1995; Metcalf and Eddy 1991). In this chapter, emphasis is given to land-based treatment systems. Topics to be examined include (1) the need for NTSW, (2) the basic characteristics of NTSW, (3) some basic aspects of on-site systems, (4) the slow-rate systems, (5) the rapid infiltration systems, (6) the overland-flow systems, (7) the floating aquatic systems, (8) the constructed wetlands, (9) ponds, and (10) combination systems. In addition, the treatment performance of various NTSWs is presented.

16.2 THE NEED FOR NATURAL TREATMENT SYSTEMS

All over the world wastewater technology has improved substantially during the last few decades. However, in several parts of the world, decisions on the more efficient use of the appropriate technology are often made not on the basis of their general benefit but are dictated by various other factors, which often result in wrong directions and decisions and in high cost installation.

The high cost (for construction, maintenance, and operation) of the most conventional treatment processes has brought about economic pressures to society even in developed countries, and has forced engineers to search for creative, cost-effective, and environmentally sound ways to control water pollution. One technical approach is to construct artificial ecosystems as a functional part of wastewater treatment. In such cases, wastewater could be treated and reused successfully as water and nutrient resources in agriculture, silviculture, and landscape irrigation practice, such as parks, schoolyards, and green belts (Angelakis 1997).

At present, it is recognized that even in the developed world, it may never be possible to implement complete sewerage systems covering 100% of the population, mainly due to economic reasons. Given that complete sewerage system and wastewater treatment will not be possible in the foreseeable future, and increasing demands are being made on fresh water supplies, it is clear that decentralized wastewater management, considering NTSW, is of great importance in developing long-term strategies for the management of our environment worldwide (Angelakis 1997).

NTSW such as ponds are always a strong contender for the best treatment option, particularly in warm climates (Arthur 1983). However, there is reason to suggest that ponds, even in the temperature climate of the UK, could be suitable (Mara 1997).

In many developing countries large quantities of wastewater, domestic and industrial alike, are discharged into stream beds without any treatment. Aquatic pollution and other adverse environmental impact is consequently very high. People often respond to this by saying, ‘We knew this would happen, but we don’t have enough funds to construct a modern treatment plant in order to avoid these adverse effects.’ Perhaps a treatment plant is constructed after this has happened. One may think that this is a panacea providing a solution to almost all problems. However, this deals with only part of the problem. There are very expensive conventional wastewater treatment facilities that are currently facing problems, even in developed countries. In Greece for example, such plants cost from US$15-30/pe, but in many cases they are not properly operated, for many reasons, such as (a) insufficient funds for operating and maintaining the plant,

(b) suitable personnel are not available, and (c) poor construction of the plant causes it to fail.

By the end of the last century, almost two-thirds of the 250 wastewater treatment plants (WWTPs) in Greece (mainly activated sludge and extended aerated systems) were facing problems, and about 50 such plants had gone out of operation (Tsagarakis *et al.* 1998).

Presently, in many cases, raw wastewater is discharged and the aquatic pollution is being continued. Thus, NTSW especially the stabilization ponds have recently been in consideration, although some years ago, these processes seemed to be unrealistic, due to the increased area required for NTSW particularly in developed countries. However, this is only true for the coastal tourist area where land is very expensive and the land availability is relatively low. Assuming for example, a very conservative design allowance of 5 m^2 of land/person for a typically constructed wetland, it should use only 0.05% of the total area of the country, that is, 4500 ha, to permanently treat the wastewater of the country. Natural treatment systems for wastewater must be feasible for at least small communities. These are of low cost in construction, operation and maintenance, low energy-consuming, and environment-friendly.

Another example is copied from Mara (1997), who visited a small-activated sludge plant in northern Africa. It produced a good effluent but consumed 50% of the town council's electricity budget. On top of this, because the plant did not have the capacity to treat all the city's effluent, over 80% of it was directed untreated straight to the river. Therefore, the plant treated less than 20% of the flow at an extremely high cost and the river was grossly polluted.

Too many wastewater treatment engineers seem to be seduced by electro-mechanical equipment. There are many examples of this. Such an example comes from a small town in Latin America; some twenty years ago, of an aerated lagoon operating very well and serving a town of around 50,000 people (Mara 1997). Recently, it was reported that the aerator had fallen off its shaft, and the pond became a waste stabilization one, where fishes were swimming around. The plant was located in an area of scrub as far as the eye could see. Why weren't ponds built in the first place? Presumably they were not sophisticated enough.

With respect to wastewater treatment, it is now possible to produce any required treated effluent quality. While the cost of this may be high with some technologies, developments are proceeding so rapidly that it is fair to say that treatment costs will be comparable to those of centralized, conventional facilities or even lower, especially when the costs of wastewater collection and/or transportation are also considered. Tchobanoglous (1999) introduced the multiple quality concept for the treatment of wastewater where different levels of treatment are used regarding the disposal site and/or reuse practice planned (Figure 16.1). In such a scheme NTSWs play an important role.

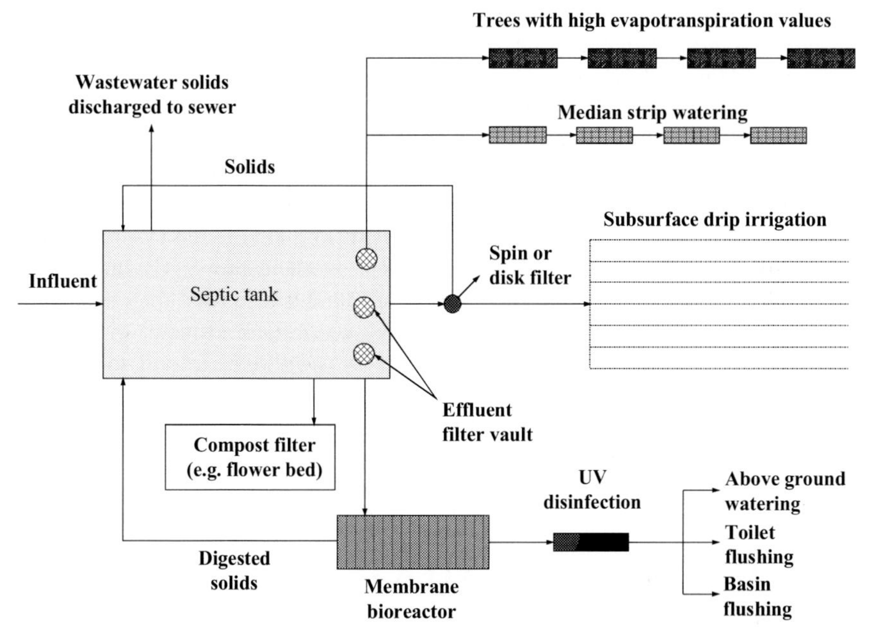

Figure 16.1. Multiple quality concept treatment and reuse scheme for an apartment building (Tchobanoglous 1999).

16.3 CHARACTERISTICS AND OBJECTIVES OF NATURAL TREATMENT SYSTEMS

The use of land-based natural treatment systems in the United States dates from the 1881. As in Europe, sewage farming (the older term used in the early literature) became relatively common as a first attempt to control water pollution. In the first half of the twentieth century, these systems were generally replaced either by in-plant treatment systems or by managed farms. These newer land-treatment systems tended to predominate in the western United States, where the resource value of wastewater was an added advantage (Metcalf and Eddy 1991).

The physical features, design objectives, and treatment capabilities of the various types of natural systems are described and compared in this section of the chapter. Comparisons of major site characteristics, typical design features, and the expected quality of the treated wastewater from the principal types of natural systems are presented in Tables 16.1 and 16.2 (Angelakis 1994).

All forms of natural treatment systems are preceded by some form of mechanical pre-treatment. For wastewater, a minimum of fine screening or primary sedimentation is necessary to remove gross solids that can clog distribution systems. The need to provide pre-application treatment beyond a minimum level will depend on the system objectives and regulatory requirements. The capacity of all natural systems to treat wastewater sludge is finite, and systems must be designed and managed to function properly within that capacity.

The major types of natural treatment systems are land-based systems. Land treatment processes include slow-rate (SR), overland-flow (OF) and rapid infiltration (RI). In addition to these processes, land is also used for various on-site soil absorption systems designed to treat septic tank effluent. In general, land treatment is the controlled application of wastewater to soil to achieve treatment of constituents in the wastewater through natural processes. A comparison of expected effluent quality of treated water from SR, RI, and OF systems is given in Table 16.1. More detail on each type of system is discussed in subsequent sections.

16.4 ON-SITE WASTEWATER TREATMENT SYSTEMS

There are over 20 million soil absorption systems used for on-site wastewater management in the United States alone. These systems generally operate with minimal or no operational attention for many years and can be permanent management systems if they are properly sited and designed.

Typical on-site treatment and disposal system for individual homes consists of a septic tank and a gravity, subsurface soil absorption system. Alternative treatment systems may include grease traps for industrial, commercial, or restaurant wastewater, Imhoff tanks for multi-family systems or clusters of homes, and additional treatment beyond primary (septic tank) treatment, such as sand filters and aerobic treatment units. In most cases, septic tanks (of various sizes) are the preferred initial treatment unit (Tchobanoglous and Angelakis 1996).

Final treatment and disposal of the effluent from a septic tank or other treatment units are currently accomplished; most commonly, by means of subsurface soil absorption. An absorption system, commonly known as a leachfield, consists of a series of trenches (0.9–1.5 m) filled with a porous medium such as gravel. The porous medium is used (1) to maintain the structure of the disposal field trenches; (2) to provide partial treatment of effluent; (3) to distribute the effluent to the infiltrative soil surfacesl and (4) if the trenches are not full, to provide temporary storage capacity during peak flows (Crites and Tchobanoglous 1998).

Effluent from the septic tank is applied to the disposal field by intermittent gravity flow, or by periodic dosing using a pump or a dosing siphon. Unfortunately, conventional trench designs fail to take maximum advantage of the treatment capabilities of the soil because they are typically located below the region of maximum bacterial activity. New trench designs now involve the use of very shallow trenches with no porous medium (see Figure 16.2). The use of shallow trenches enhances the treatment of the effluent with respect to the removal of BOD, SS, phosphorus, and nitrogen. It is interesting to note that the use of such shallow trenches was recommended as early as 1915 in an early Public Health Bulletin (Lumsden *et al.* 1915).

Table 16.1. Comparison of expected effluent quality of treated water from slow-rate, rapid infiltration, and overland-flow natural treatment systems (Metcalf and Eddy 1991)

Constituent	Value, mg/l: Slow rate[a]		Rapid infiltration[b]		Overland-flow[c]	
	Av.	Max.	Av.	Max	Av.	Max.
BOD	<2	<5	2	<5	10	<15
Suspended solids	<1	<5	2	<5	15	<25
Ammonia nitrogen as N	<0.5	<2	0.5	<2	1	<3
Total nitrogen as N	3	<8	10	<20	5	<8
Total phosphorus as P	<0.1	<0.3	1	<5	4	<6

[a] Percolation of primary or secondary effluent through 1.5 m of soil.
[b] Percolation of primary or secondary effluent through 4.5 m of soil.
[c] Run-off of continued municipal wastewater over about 45 m of slope.

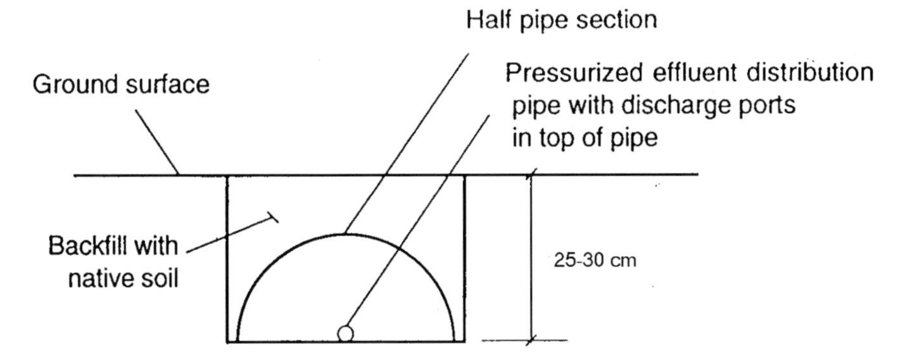

Figure 16.2. A typical shallow unfilled leachfield.

16.5 SLOW-RATE TREATMENT

16.5.1 General characteristics

Slow-rate (SR) treatment (the predominant natural treatment process currently in use) involves the application of wastewater to vegetated soil to provide treatment and to meet the growth needs of the vegetation. The applied water either is consumed through evapotranspiration or percolates vertically and horizontally through the soil profile (see Figure 16.3). Any surface run-off is usually collected and reapplied to the system. Treatment occurs as the applied water percolates through the soil profile. In most cases, the percolate will enter the underlying groundwater, but in some cases, the percolate may be intercepted by natural surface waters or recovered by means of drains or recovery wells. The rate at which water is applied to the land per unit area (hydraulic loading rate) and the selection and management of the vegetation are functions of the design objectives of the system and the site conditions. SR systems are often classified as type 1 or type 2 depending on their design objectives.

A SR system is considered to be type 1 when its principal objective is wastewater treatment and the hydraulic loading rate is not controlled by the water requirements of the vegetation but by the limitations imposed by the design parameters soil permeability or constituent loading. Type 2 systems, designed with the objective of water reuse through crop production or landscape irrigation, are often referred to as wastewater irrigation or crop irrigation systems. Wastewater can be applied to crops or vegetation (including forest land) by a variety of sprinkling methods (see Figure 16.3) or by surface techniques such as graded border and furrow irrigation (see Figure 16.4).

Intermittent application cycles, typically every four to ten days, are used to maintain predominantly aerobic conditions in the soil profile. The relatively low application rates combined with the presence of vegetation and the active soil ecosystem provide slow-rate systems with the highest treatment potential of the natural treatment systems (see Table 16.1). A list of large municipal SR systems is presented in Table 16.3.

16.5.2 Removal mechanisms

SR systems are very effective in removing BOD, nitrogen, phosphorus, metal, trace elements and microorganisms.

BOD: On the basis of successful operations where high-strength food-processing wastewater is applied, BOD loading rates of up to 500 kg/ha·d can be achieved. Most municipal SR systems are loaded at less than 11 kg/ha·d of BOD, which is an order of magnitude lower than the capacity of the soil. As

Table 16.2. Comparison of site characteristics and design features of alternative natural treatment systems (Metcalf and Eddy Inc. 1991 and Crites *et al.* 2000)

Feature	Slow-rate (type 1)	Slow-rate (type 2)	Rapid infiltration	Overland- flow	Wetland application	Floating aquatic plant
Grade	20% cultivated and 40% uncultivated sites		Not critical	2–8%	0.2%	0.4–0.5%
Soil permeability	Moderate		Rapid	Slow to none	Moderate	Not applicable
Groundwater depth (m)	>1.5		1 during application 1.5–3.1 during drying	Not critical	Not critical	Not critical
Climate	Winter storage in cold climates		Not critical	Not critical	Depends on vegetation	Depends on vegetation
Application techniques	Sprinkler or surface[a]		Usually surface	Sprinkler or surface	Sprinkler or surface	Surface
BOD loading rates (kg/ha.d)	3–11		45–180	5.5–22.5	<70	20–500
Annual hydraulic rate (m/yr)	1.7–6.1	0.61–2.04	6.1–91.5	7.3–56.7	5.5–18.3	5.5–18.3
Area required [ha/(10^3 m^3/d)[b]]	6.0–21.4	18.2–58.8	0.4–6.0	0.65–4.80	1.9–6.6	1.9–6.6
Minimum pre-treatment provided	Primary sedimentation[c]		Primary sedimentation	Screening	Primary sedimentation	Primary sedimentation
Disposition of applied wastewater	Evapotranspiration and percolation		Mainly percolation	Surface run-off and evaporation with some percolation	Evapotranspiration, percolation, and run-off	Some evapotranspirati on
Need for vegetation	Required		Optional[d]	Required[e]	Required	Required

[a] Includes furrow and graded border.
[b] Field area in ha not including buffer area, roads, or ditches.
[c] Depends on the use of the effluent and the type of crop.
[d] Grass is sometimes used.
[e] Water-tolerant grasses are used.

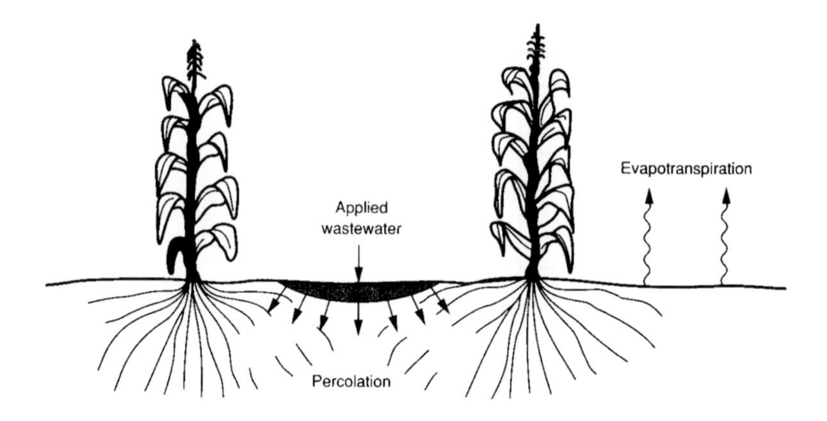

Figure 16.3. Slow-rate treatment system hydraulic pathway.

Figure 16.4. Surface distribution using the furrow method (Angelakis and Tchobanoglous 1995).

presented in Table 16.4, BOD removal in SR systems regularly exceeds 98% (Crites and Tchobanoglous 1998).

Table 16.3. Selected municipal slow-rate land treatment systems in the US and Canada (Reed *et al.* 1995)

Location	Flow (m^3/d)	System area (ha)	Application method
Bakersfield, CA	73,600	2060	Ridge and furrow and border strip surface application
Clayton County, GA	75,950	960	Solid-set sprinklers
Lubbock, TX	62,500	2000	Center-pivot sprinklers
Mitchell, SD	9300	520	Center-pivot sprinklers
Muskegon County, MI	110,400	2160	Center-pivot sprinklers
Petaluma, CA	20,000	220	Travelling-gun sprinklers
Vernon, BC, Canada	10,300	591	Travelling-gun and side-roll sprinklers

TSS: Filtration through the soil is responsible for TSS removals of 99% or more. Percolate SS values of 1 mg/l or less are usually achieved if the loading is below 27.5 kg SS/ha·d.

Nitrogen removal: Design for nitrogen is accomplished using a nitrogen balance that matches the expected removal plus a percolate nitrate nitrogen of less than 10 mg/l (drinking water standard) at the project boundary against the nitrogen loading. Nitrogen removal in SR systems is presented in Table 16.5.

Table 16.4. BOD removal for SR systems in the US (Crites and Tchobanoglous 1998)

Location	Flow	BOD loading,	Applied BOD	Percolate BOD	Removal
	(m^3/d)	(kg/ha·d)	(mg/l)	(mg/l)	(%)
Dickinson, North Dakota	0.038	3.3	42	<1	>98
Hanover, New Hampshire	Pilot				
(a) Primary effluent		13.2	101	1.4	98.6
(b) Secondary effluent		4.4	36	1.2	96.7
Roswell, New Mexico	0.015	3.3	43	<1	>98
San Angelo, Texas	0.020	12.1	119	1.0	99.1
Yarmouth, Massachusetts	Pilot	11.0	85	<2	>98

Nitrogen uptake by crops is not constant. It depends on the crop yield and the nitrogen content of the harvested crop. Nitrogen uptake rate in forage, field, and forest crops (Angelakis 1994; Angelakis and Tchobanoglous 1995; US EPA 1981) are presented in the system design section of this chapter. Biological nitrogen removal in SR systems occurs by nitrification/denitrification. The loss to denitrification depends on the BOD:N ratio and the soil temperature, pH, and soil water content. Intermittent application, which is characteristic of SR

Table 16.5. Nitrogen removal for agricultural crop SR systems in the US (Crites and Tchobanoglous 1998)

Location	Crop	Total N applied (mg/l)	Percolate total N (mg/l)	Removal (%)
Dickinson, North Dakota	Brome grass	11.8	3.9	67
Hanover, New Hampshire				
(a) Primary effluent	Reed canary grass	28.0	9.5	66
(b) Secondary effluent		26.9	7.3	73
Pleasanton, California	Pasture	27.6	2.5	91
Roswell, New Mexico	Corn	66.2	10.7	84
San Angelo, Texas	Sorghum, oats, grass	35.4	6.1	83
Yarmouth, Massachusetts	Reed canary grass	30.8	1.8	94

systems, also serves to enhance nitrification followed by denitrification. The typical loss to denitrification is 25% of the applied nitrogen. If the BOD:N ratio is high (20:1 or 40:1), denitrification can be responsible for 80% or more of nitrogen loss (Crites and Tchobanoglous 1998).

Ammonia volatilization losses of 10% can be expected if the soil pH is above 7.8 and the cation exchange capacity is low (low absorption of ammonium by the soil).

Nitrogen can also be stored in the soil. In arid regions and for sites initially low in organic matter, the nitrogen applied can be stored at rates up to 220 kg/ha.yr. For soils that are rich in organic matter (>70%), the nitrogen storage will be insignificant (Crites and Tchobanoglous 1998).

Phosphorus: Phosphorus adsorption occurs rapidly in soils, with chemical precipitation being a slower process. Acidic conditions favor complexes with aluminum and iron. Alkaline soil conditions favor phosphate complexes with calcium. Phosphorus removal at various SR sites is presented in Table 16.6.

Table 16.6. Phosphorus removal for SR systems in the US (Crites and Tchobanoglous 1998)

Location	Total applied phosphate (mg/l as P)	Percolate total phosphate (mg/l as P)	Removal (%)
Camarillo, CA	11.8	0.2	98.3
Dickinson, ND	6.9	0.05	99.3
Hanover, NH	7.1	0.03	95.8
Roswell, NM	7.95	0.39	95.1
Tallahassee, FL	10.5	0.1	99.0
Helen, GA	13.1	0.22	98.3
State College, PA	5.6	0.08	98.6
Clayton County, GA	4.9	0.02	99.6
West Dover, VT	4.2	0.4	90.5
Wolfeboro, NH	3.3	0.02	99.4

Phosphorus removal can be estimated conservatively by Equation 16.1 (Reed and Crites 1984):

$$P_x = Pe^{-kt} \quad (16.1)$$

where P_x = total phosphorus at a distance x along the flow path, mg/l; P = total phosphorus in the applied wastewater, mg/l; k = 0.048 d^{-1}; t = detention time; d = $x(0.40)/k_x S$; x = distance along flow path, m; k_x = hydraulic conductivity in soil in direction x, m/d; S = hydraulic gradient; S = 1 for vertical flow.

Metals: Some trace elements can be toxic to plants and/or consumers of plants. In most cases, maintenance of soil pH at or above 6.5 will retain trace elements as unavailable insoluble compounds. If the soil pH falls below 6.5, the metals tend to become more soluble and are therefore able to leach deeper into the soil or to the groundwater (Crites and Tchobanoglous 1998).

Trace organics: Trace organics are generally not of concern for small systems, since SR is an effective process for the removal of trace organics (Reed *et al.* 1995; Crites and Tchobanoglous 1998).

Microorganisms: The SR process is effective in removing bacteria and viruses from wastewater. Because of this effective removal and the practice of disinfection prior to irrigation, removal of microorganisms is not a limiting factor in the design of SR systems (Crites and Tchobanoglous 1998).

16.6 RAPID INFILTRATION

16.6.1 General characteristics

The use of treated wastewaters for groundwater recharge leads to two main benefits: (a) an improvement in the quality of water and its protection from pollution; and (b) control of groundwater depletion in over-exploited aquifers. An additional, but no less important, advantage of this practice is its low cost in comparison to other commonly used technologies. Groundwater recharge can be operated by injection wells or by infiltration basins. The first technique is much more expensive than the second one, since before injection into the aquifer, the wastewater must be carefully pre-treated (filtration, the removal of organic material, and disinfection), especially if the water is intended for potable use (Romano and Angelakis 1998). The groundwater recharged by infiltration basins, commonly called soil-aquifer-treatment (SAT) systems, is economically more feasible because it requires less intensive pre-treatment.

Groundwater aquifers provide natural mechanisms for the treatment, storage, and subsurface transmission of reclaimed water. By the application of the SAT technique (or geopurification), an improvement occurs in the water quality as the water moves down the vadose zone of the aquifer and then laterally through the aquifer to the collection system (pumped wells, gravity subsurface drains, and so on) (Bouwer 1993). For suitable soils, the filtration process essentially removes all solids, BOD, and microorganisms, most of the meta-phosphates, and a considerable amount of nitrogen. The nitrogen and microorganisms removal processes are renewable and can continue indefinitely, while other constituents, such as metals and phosphates, can accumulate in the soil and in the vadose zone. However, the rate of accumulation is slow and it may take decades or even centuries before the soil porosity, the hydraulic conductivity and consequently the infiltration capacity of the unsaturated zone is affected. For these reasons, SAT systems usually have a long life span (Bouwer 1991).

Infiltration basins are intermittently flooded and periodically cleaned. Infiltration rates are typically of the order of a few decimeters per day during flooding but, due to regular drying, long-term average infiltration rates are more of the order of 100–400 m/yr. At these rates, one ha of infiltration basin can infiltrate 1.0–4.0 mm^3/yr (Bouwer 2000). Typical SAT and recovery systems are shown in Figure 16.5.

Soil and aquifer characteristics are the main factors influencing pathogen survival. Therefore, the selection of sites for SAT systems should consider some requirements for suitable soils and aquifers. Recharge systems require permeable soil for adequate infiltration rates, vadose zones without confining layers or other problems, such as contaminated zones or undesirable chemicals that can be leached out. The aquifers must be unconfined and have good quality groundwater. An ideal surface soil is uniform and coarse enough to give high infiltration rates but fine enough to give good filtration treatment. Permeability values of the order of 25 mm/h or more are necessary for rapid infiltration, therefore, sandy loams and loamy or fine sands and gravel are preferred Very coarse sand and gravel are not suitable because they allow wastewater to pass too rapidly through the first part of the soil, where the major biological and chemical action takes place. A uniform soil with a thickness of more than 3 m is preferred (Reed *et al.* 1995).

SAT systems are generally designed and managed so that all the water that infiltrates as sewage effluent can be recovered with wells, drains, or via seepage into surface water. The design of the average annual hydraulic loading rate, L, based on permeability is determined by Equation (16.2):

$$L = aNI \qquad (16.2)$$

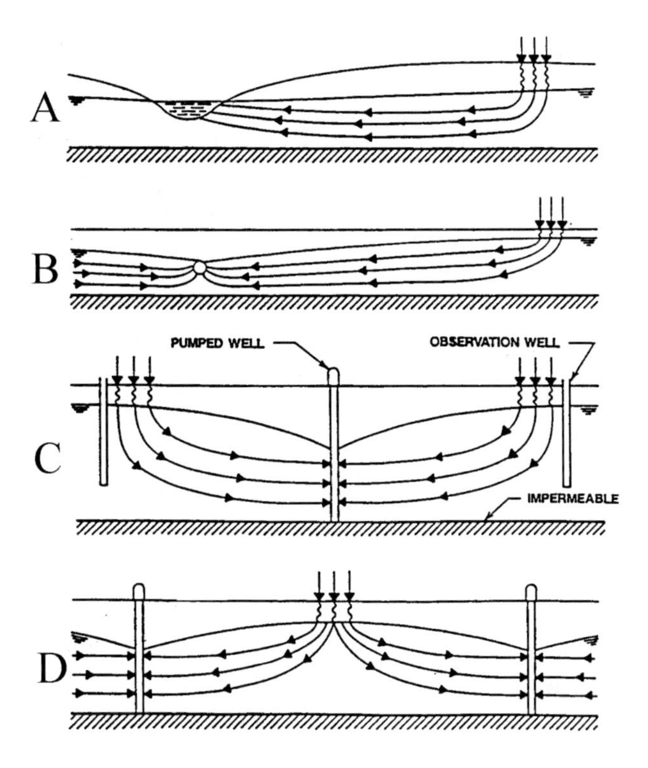

Figure 16.5(A) Schematic of a soil–aquifer–treatment system with natural drainage of renovated water into stream, lake, or low area; (B) collection of renovated water by subsurface drain; (C) infiltration areas in two parallel rows and line of wells midway between; and (D) infiltration areas in the centre surrounded by a circle of wells (Source: Bouwer 1991).

where a = application factor based on field measurements (usually ranges from 0.002 to 0.015); N = number of operation days per year (d/yr); and I = infiltration rate in mm/h.

16.6.2 Removal mechanisms

In general, key removal mechanisms include volatilization, chemical, and/or biological conversion reactions, precipitation (metals, P), and sorption (adsorption and precipitation reactions), within the soil matrix. The removal of most chemical and microbiological constituents occurs in the top two centimeters of the vadose zone. Biological breakdown of organics, nitrification, and denitrification occur in both the unsaturated and saturated zones. Pathogen removal depends on the soil and aquifer conditions (Romano and Angelakis 1998). Examples of selected SAT projects are shown in Table 16.7 (National Research Council 1994). Quality parameters from the Arizona project, using mildly chlorinated secondary effluent from an activated sludge plant as it entered into the infiltration basin, and afterwards it had been pumped from a well located in the centre of the infiltration basin of the SAT system, are shown in Table 16.8. The performance of the Dan Region wastewater reclamation project in the Metropolitan Tel Aviv area before recharge and after SAT is shown in Table 16.9 (National Research Council 1994).

Table 16.7. Selected SAT systems

System/region	Treatment of wastewater applied	Method of application	Volume applied (mm^3/yr)
Water Factory 21, Orange County, CA	Advanced	23 wells	35
Montebello Forebay, CA	Tertiary	Basins in 3 sites	35–40
Phoenix, AZ (two projects)	Secondary	Basins	10–15
El Paso, TX	Advanced	10 wells	14
Long Island, NY	Stormwater run-off	Basins	85
Orlando, FL	50% stormwater run-off 45% wetland overflow 5% untreated wastewater	310 drainage wells	35
Dan Region, Israel	Secondary	Basins in 2 sites	80

BOD: Organic material is degraded primarily by aerobic microorganisms in the soil profile. When the BOD loading is high, bacteria multiply rapidly and form slime layers, which can eventually clog the soil pores and reduce, not only the infiltration rate, but also the rate of soil recreation during drying. By-products from the activity of anaerobic bacteria tend to accelerate the rate of soil clogging. Thus, excessive BOD loading will result in system failure. Design of the BOD loading should be in the range of 45–180 kg/ha·d (Metcalf and Eddy 1991).

Nitrogen: In a SAT system, the principal nitrogen removal mechanism is denitrification. The maximum amount of nitrogen that can be effectively denitrified in such a system, under optimum operating conditions (ΔN), may be estimated by Equation 16.3:

$$\Delta N = \frac{TOC - 5}{2} \tag{16.3}$$

where TOC = total organic carbon.

Table 16.8. Quality effluent parameters obtained from the salt river floodplain west of Phoenix (Arizona) SAT system (Source: Bouwer 1993)

Parameters	Secondary effluent (mg/l)	Recovery well samples (mg/l)
Total dissolved solids	750	790
Suspended solids	11	1
Ammonium nitrogen	16	0.1
Nitrate nitrogen	0.5	5.3
Organic nitrogen	1.5	0.1
Phosphate phosphorus	5.5	0.4
Fluoride	1.2	0.7
Boron	0.6	0.6
Biochemical oxygen demand	12	0
Total organic carbon	12	1.9
Zinc	0.19	0.03
Copper	0.12	0.016
Cadmium	0.008	0.007
Lead	0.082	0.066
Faecal coliforms per 100 ml.	3500	0.3
Viruses, PFU/100 l	2118	0

Metals: Removal of trace elements (principally metals) occurs mainly through sorption. Metals are retained in the soil profile. The retention capacity for most metals in soils is generally very high, especially at pH values above 6.5. Under low pH and anaerobic conditions, some metals are more soluble and can be released into the soil solution. Removal efficiencies for most metals generally range between 85–95% (Metcalf and Eddy 1991).

Trace organic compounds: Trace organic compounds are removed from wastewater through volatilization and absorption followed by biological breakdown. In general, SAT systems are capable of removing large fractions of trace organic compounds.

Microorganisms: Removal mechanisms for microorganisms (bacteria, protozoa, and helminths) include die-off, straining, entrapment, predation, desiccation, and adsorption. Viruses are removed almost exclusively by adsorption and subsequent die-off. Relatively long travel distances through the soil profile are required for the complete removal of microorganisms in a SAT system, with a distance depending on the soil permeability and hydraulic loading rate (Gerba and Goyal 1985).

Table 16.9. Performance of Dan Region project: basic wastewater parameters (averages for 1990)

Parameters	Units	Before SAT	After SAT	Percentage Removal
SS	mg/l	17	0	100
BOD_5	mg/l	19.9	<0.5	>98
BOD filtered	mg/l	3.1	<0.5	>84
COD	mg/l	69	12.5	82
COD filtered	mg/l	46	12.5	73
Total organic carbon	mg/l	20	3.3	84
Dissolved organic carbon	mg/l	13	3.3	75
UV 254 absorbance	$cm^{-1} \times 10^3$	298	64	79
$K\mu nO_4$ as O_2	mg/l	14.1	2.3	84
$KMnO_4$ filtered as O_2	mg/l	12.6	2.3	82
Detergents	mg/l	0.5	0.078	84
Phenols	μg/L	8	<2	>75
Ammonia, as N	mg/l	7.56	<0.05	99
Kjeldahl nitrogen	mg/l	11.5	0.56	95
Kjeldahl nitrogen filtered	mg/l	10.2	0.56	95
Nitrate	mg/l	2.97	7.17	
Nitrate	mg/l	1.24	0.10	92
Nitrogen	mg/l	15.7	7.83	50
Nitrogen filtered	mg/l	14.4	7.83	46
Phosphorus calcium	mg/l	3.4	0.02	99
Alkalinity, as calcium carbonate	mg/l	306	300	–
PH	–	7.7	7.9	–

16.7 OVERLAND FLOW

16.7.1 General characteristics

In overland flow (OF) systems, pre-treated wastewater is distributed across the upper portions of carefully graded, vegetated slopes and allowed to flow over the surfaces of the slope to run-off collection ditches at the bottom of the slopes. A process schematic is shown in Figure 16.6. Overland flow is normally used at sites with relatively impermeable surface soils or subsurface layers, although the process has been adapted to a wide range of soil permeabilities because the soil surface tends to seal over time. Percolation through the soil profile is, therefore, a minor hydraulic pathway, and most of the applied water is collected as surface run-off. A portion of the applied water will be lost by evapotranspiration. The percentage of the applied water lost varies with the time of the year and local climate. Systems are operated by using alternating application and drying periods, with the lengths of the periods depending on the treatment objectives. Distribution of wastewater may be accomplished by means of high-pressure sprinklers, low-pressure sprays, or surface methods such as gated pipe (Figure 16.6).

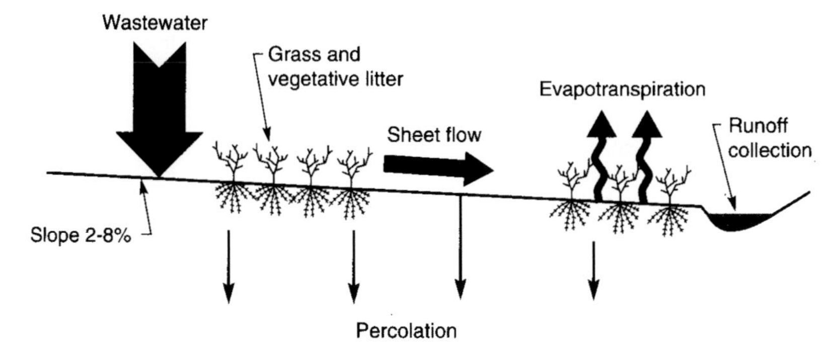

Figure 16.6. Overland-flow process schematic (Crites *et al.* 2000).

16.7.2 Removal mechanisms

OF systems are effective at removing BOD, TSS, nitrogen and trace organic compounds. However, they are less effective at removing phosphorus, heavy metals, and microorganisms.

BOD: The BOD loading rate typically ranges from 5.5–22.5 kg/ha.d. Biological oxidation accounts for the 90–95% removal of BOD normally found in OF systems. BOD removals from four overland-flow systems are presented in Table 16.10 along with the application rate and slope length.

TSS: Overland flow is effective in removing most suspended solids, with effluent TSS levels commonly between 10–15 mg/l. Algae are not removed effectively in most OF systems because many algal types are buoyant and resist removal by filtration or sedimentation.

Nitrogen: The removal of nitrogen by OF systems depends on nitrification/denitrification and the crop uptake of nitrogen. The removal of nitrogen in several OF systems is presented in Table 16.11. Up to 90% removal of ammonia was reported at 0.10 $m^3/m \cdot h$ in the OF system in Davis, California, where oxidation lagoon effluent was applied. At Sacramento County (also in California), secondary effluent was nitrified at an application rate of 0.54 $m^3/m \cdot h$. Ammonia concentrations were reduced from 14 to 0.5 mg/l (Crites and Tchobanoglous 1998).

Table 16.10. BOD removal for overland-flow systems (Crites and Tchobanoglous 1998)

Location	Wastewater type	Application rate (m^2/m.h)	Slope length (m)	BOD (mg/l) Influent	BOD (mg/l) Effluent
Ada, OH, USA	Raw wastewater	0.08	37	150	8
	Primary effluent	0.11	37	70	8
	Secondary effluent	0.22	37	18	5
Easley,	Raw wastewater	0.24	55	200	23
SC, USA	Pond effluent	0.26	46	28	15
Hanover, NH	Primary effluent	0.14	51	72	9
USA	Secondary effluent	0.08	31	45	5
Melbourne, Australia	Primary effluent	0.27	252	507	12

Phosphorus: Phosphorus removal in OF is limited to about 40–50% because of the lack of soil–wastewater contact. If needed, phosphorus removal can be enhanced by the addition of chemicals such as alum or ferric chloride.

Heavy metals: Heavy metals are removed in OF by the same general mechanisms as phosphorus absorption and chemical precipitation. Heavy metal removal will vary with the constituent metal from about 50-80% (Crites and Tchobanoglous 1998).

Trace organic compounds: Trace organic compounds are removed in OF systems by a combination of volatilization, absorption, photodecomposition, and biological degradation.

Microorganisms: Overland flow is not very effective in removing microorganisms. Fecal coliforms are reduced by about 90% when raw or primary effluent is applied.

16.8 FLOATING AQUATIC PLANTS

In floating aquatic plant systems, the plants are floating species such as water hyacinth and duckweed (see Figure 16.7). Water depths are typically deeper than wetlands systems, ranging from 0.5–1.8 m. Supplementary aeration has been used with floating plant systems to increase treatment capacity and to maintain aerobic conditions necessary for the biological control of mosquitoes.

Table 16.11. Nitrogen removal in overland-flow systems (Crites and Tchobanoglous 1998)

Parameter	Ada, Oklahoma	Hanover, New Hampshire	Utica, Mississippi
Type of wastewater	Screened raw wastewater	Primary effluent	Pond effluent
Application rate (m^2/m.h)	0.08	0.13	0.07
BOD:N ratio	6.3	2.3	1.1
Total nitrogen (kg/ha.yr)			
Applied	1177	935	649
Removed	1078	869	490–589
Crop uptake	110	209	242
Nitrification-denitrification	968	660	248–358
Removal, mass basis (%)	92	94	75-90
Total nitrogen (mg/l)			
Applied	23.6	36.6	20.5
Run-off	2.2	5.4	4.3-7.5
Removal, concentration basis (%)	91	85	63-79

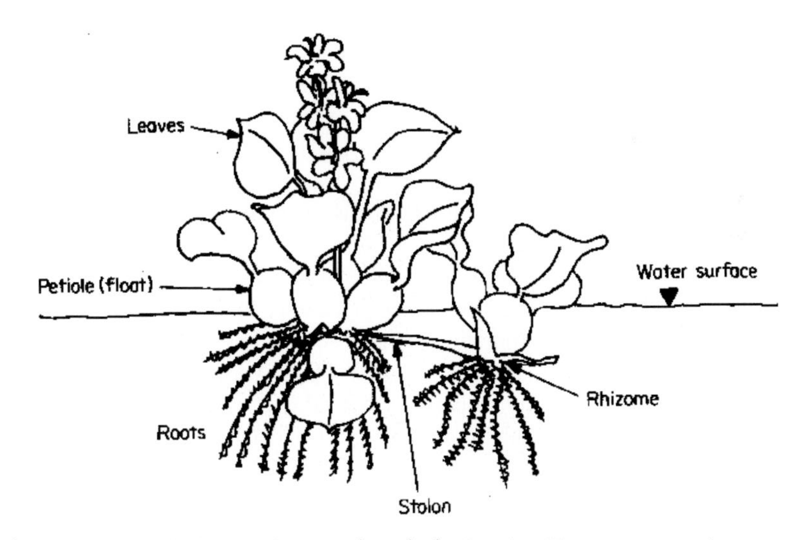

Figure 16.7. Morphology of the water hyacinth plant (*Eichhornia crassipes*).

Floating aquatic treatment systems have been used for a variety of treatment purposes including secondary treatment, advanced secondary treatment, and nutrient removal. Floating aquatic treatment is defined as the use of aquatic plants or animals as a component in a WWTS. The two principal types of floating aquatic plant systems are the water hyacinth (*Eichhornia crassipes*) and the duckweed (*Lemna spp.*) systems. The water hyacinth systems involve floating or suspended plants with relatively long roots of 0.6–1.2 m. The root structures serve as a medium for the attached growth of bacteria. The duckweed systems on the other hand, have short roots, usually under 10 mm, and therefore, functions as a surface shading systems. The principal difference in design and physical requirements between floating aquatic and wetlands systems are associated with the plants.

The two principal types of water hyacinth wastewater treatment systems are aerobic non-aerated and aerobic aerated systems. In the latter, supplemental air is provided and the operating depth is usually greater (1.2–1.4 m). With duckweed the depth can be 1.5–2.5 m. Water hyacinth plants grow in the temperature range of 20–30 ^{o}C. However, water temperatures as low as 10 ^{o}C can be tolerated, so long as the air temperature does not drop below 5–10 ^{o}C. Duckweed plants can survive in water temperatures up to 5 ^{o}C.

For floating aquatic treatment systems a slight sloping level and uniform topography is preferred. However, ponds and channels may be constructed on steeper sloping or uneven sites.

Pre-treatment requirements for water hyacinth systems, especially aerobic aerated systems, are typically those of primary treatment. However, duckweed systems have been designed primarily to upgrade facultative pond effluents. High removals of BOD and TSS have been reported in properly designed floating aquatic systems, especially those employing water hyacinths (Tchobanoglous 1999). However, lower efficiencies for nutrients, metals, and pathogens removal have been demonstrated.

The range of loading rates for BOD for floating aquatic systems is from 100–500 kg/ha.d to 22–28 kg/ha.d for water hyacinths and duckweed systems, respectively. The Lemna Corporation, which sells proprietary floating plastic barriers and harvesting equipment, suggests that wastewater entering the duckweed portion of the facility be partially treated to a BOD level of 60 mg/l or less by facultative ponds, aerated ponds, or mechanical treatment plants (Crites and Tchobanoglous 1998). To achieve a 20 mg BOD/l in the effluent, Lemna suggests a target influent of 40 mg/l, 20 d detention time and a pond sizing of 12.8 m^2/m^3.d, based on a minimum pond depth of 1.5 m. To achieve a final BOD of 10 mg/l, a target influent BOD of 30 mg/l, a 28 day hydraulic detention time and a pond sizing of 18.4 m^2/m^3.d is necessary (Tchobanoglous 1999).

Floating aquatic plants shield the water from sunlight and reduce the growth of algae. Floating plants have also been shown to reduce BOD, nitrogen, metals, and trace organic compounds. Water hyacinth systems are an emerging technology that is being developed in large-scale pilot systems, such as at San Diego, California, and full scale systems at Austin and San Benito, Texas, in Orlando, Florida and in other places. Annual hydraulic loadings and specific area requirements for floating plant systems are similar to those of wetland systems (Table 16.1).

16.9 WETLAND SYSTEMS

16.9.1 General characteristics

In general, wetlands are defined as lands where the water surface is near the soil surface for long enough each year to maintain saturated soil conditions, along with related vegetation. Wetlands are inundated land areas with water depths typically less than 0.6 m that support the growth of emergent plants (Stowell *et al.* 1981) such as cattail (*Typha* spp.), yellow flag (*Iris pseudacorus*), bulrush (*Scirpus* spp.), arrowhead (*Sapittaria* spp.), reeds (*Phragmites australis*), sedges (*Carex* spp.), and rushes (*Juncus* spp.). The vegetation, rhizomes, and roots provide surfaces for the attachment of bacteria films, aid in the filtration and adsorption of wastewater constituents, transfer oxygen into the water column, and control the growth of algae by restricting the penetration of sunlight. All three types (natural, mitigation and enhancement, and constructed) wetlands have been used for wastewater treatment. However, constructed wetlands are used most often. The use of natural wetlands is generally limited to the polishing or further treatment of secondary or advanced treated effluent.

The principal types of constructed wetlands systems used are (a) free water surface (FWS), and (b) subsurface water flow (SWF). Schematic representations of FWS and SWF systems are shown in Figures 16.8 and 16.9, respectively. Usually, depths of 0.1–0.45 and 0.45–1.0 m are used for FWS and SWF systems, respectively. SWF systems have the advantages of smaller land area requirements and avoidance of odor and mosquito problems compared to the FWS systems. Disadvantages of SWF systems are their increased cost due to the gravel media and the potential for clogging of the media. Based on the existing conditions in Yemen and the experience gained using stabilization ponds for several years, FWS systems appear to be most favorable.

Vegetation used in both systems are similar to those referred before. In addition to them, arrowarum (*Peltandra* spp.) and pickerel weed (*Pontederia* spp.) have been used in constructed wetlands. Common reed (*Phragmites australis*) which is very well adapted and grows in many areas of the world can be used in constructed wetlands for wastewater treatment. These species have the following typical characteristics: temperature 12–23 °C (10–30 °C is required for seed germination), very rapid growth via rhizomes (lateral spread about 1 m/yr), annual yield about 40 metric tonnes/ha of dry matter, pH ranges from 2 to 8, maximum salinity tolerance 45 g/l, and low food value for most birds and animals.

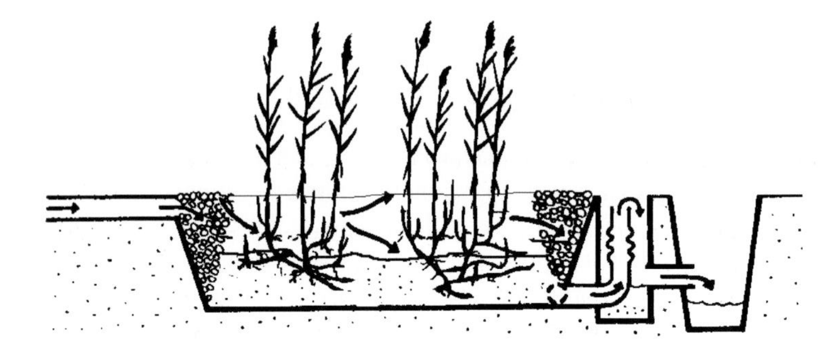

Figure 16.8. Schematic representation of an FWS system.

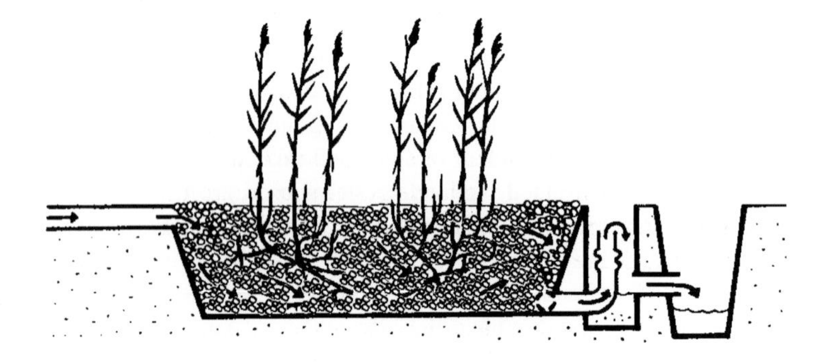

Figure 16.9. Schematic representation of an SWF system.

Pre-treatment for wetlands usually consists of settling (septic tanks or Imholf tanks), screening with a rotary disk filter, a micro screen or a micro strainer, or stabilization ponds. Pre-treatment for SWF wetlands typically consists of primary treatment. BOD loading rates generally need to be kept below 110 kg/ ha.d. In SWF systems the major source of oxygen to the root zone is vegetation. Site features for potential wetlands are similar to those for wastewater treatment ponds. Slopes of 1% or slightly greater, particularly for SWF systems, are most favorable.

The removal of the constituents and the transformation mechanisms are biological (attached bacteria), physical and chemical. In SWF systems, treatment mechanisms occur as wastewater flows through the porous media. The principal removal mechanisms include biological conversion, physical filtration and sedimentation, and chemical precipitation and adsorption. High removal efficiencies of BOD and TSS, along with significant removal efficiencies of nitrogen, metals, trace organics, and pathogens have been reported (Tchobanoglous 1999). Removal mechanisms are dependent on detention time, media characteristics, loading rates, temperature, and management practices.

Free water surface systems may also be designed with the objective of creating new wildlife habitats or enhancing nearby existing natural wetlands. Such systems normally include a combination of vegetated and open water areas and land islands with appropriate vegetation to provide waterfowl with breeding habitats. Subsurface flow systems are designed with the objective of secondary or advanced treatment. These systems have also been called 'root zone' or 'rock-reed filters' and consist of channels or trenches with relatively impermeable bottoms filled with sand or rock media to support emergent vegetation (see Figure 16.9).

16.9.2 Constituent removal and transformation mechanisms

High removals of BOD and TSS can be expected from FWS wetlands, along with significant removals of nitrogen, metals, trace organics, and pathogens. The degree of removal is usually dependent on detention time and temperature. The removal mechanisms for FWS constructed wetlands are described below.

BOD: Soluble and particulate BOD is removed by different mechanisms in FWS constructed wetlands. Soluble BOD is removed by biological activity and adsorption on the plant and detritus surfaces and in the water column. The low velocities and emergent plants facilitate flocculation/sedimentation and entrapment of the particulate BOD. Organic solids, removed by sedimentation and filtration (as discussed below), will exert an oxygen demand, as does the decaying vegetation. As a result, the influent BOD is removed rapidly down the wetland cell. The observed BOD in the wetland will also reflect the detrital and benthic demand, which leads to a 'background' concentration.

TSS: The principal removal mechanisms for TSS are flocculation and sedimentation in the bulk liquid, and filtration (mechanical straining, chance contact, impaction, and interception) in the interstices of the detritus. Most of the settleable solids are removed within 15–30 m of the inlet. Optimal removal of TSS requires a full stand of vegetation to facilitate sedimentation and filtration and to avoid regrowth of algae. Algal solids may take 6–10 days of detention time for removal.

16.9.3 Process design considerations

The principal process design criteria for FWS constructed wetlands are detention time, organic loading rate, required surface area, and water depth. Hydraulic loading rate is a common basis of comparison, either in L/d or m^3/ha.d, but both rates are calculated from the area and the flow. Other design considerations include aspect (length to width) ratio, hydraulic considerations, thermal considerations, and vegetation harvesting. Typical process design criteria are presented in Table 16.12.

Table 16.12. Typical design criteria and expected effluent quality for constructed wetlands (both FWS and SWF)

Item/design parameter		FWS wetlands	SWF wetlands
	Unit	Value	Value
Detention time	d	2–5(BOD) 7–14(N)	3–4(BOD) 6–10(N)
BOD loading rate	kg/ha·d	<70	<70
Water depth	m	0.09–0.60	0.30–0.75
Medium depth	m		0.50–0.8
Hydraulic loading rate	$m^3/m^2 \cdot d$	0.014–0.047	0.014–0.047
Specific area	$ha(10^3 m^3/d)$	2.14–7.15	2.14–7.15
Aspect ratio		2:1 to 4:1	
Mosquito control		Required	Not needed
Harvesting interval	yr	3–5	Not needed
Expected effluent quality [a]			
BOD_5	mg/l	<20	<20
TSS	mg/l	<20	<20
TN	mg/l	<10	<10
TP	mg/l	<5	<5

[a] Expected effluent quality based on a BOD loading equal to or less than 112 kg/ha·d and typical settled municipal wastewater.

Nitrogen: Nitrogen removal in constructed wetlands is accomplished by nitrification and denitrification. Plant uptake accounts for only about 10% of nitrogen removal. Nitrification and denitrification are microbial reactions that depend on temperature and detention time. Nitrifying organisms require oxygen and an adequate surface area to grow on and, therefore, are not present in significant numbers either in heavily loaded systems (BOD loading >112 kg/ha.d) or in newly constructed systems with incomplete plant cover. Based on field experience with FWS systems, one to two growing seasons may be needed to develop sufficient vegetation to support microbial nitrification. Denitrification requires adequate organic matter (plant litter or straw) to convert nitrate to nitrogen gas. The reducing conditions in mature FWS constructed wetlands resulting from flooding are conducive to denitrification. If nitrified wastewater is applied to a FWS wetland, the nitrate will be denitrified within a few days of detention.

Phosphorus: The principal removal mechanisms for phosphorus in FWS systems are adsorption, chemical precipitation, and plant uptake. Plant uptake of inorganic phosphorus is rapid; however, as plants die, they release phosphorus, so that long-term removal is low. Phosphorus removal depends on soil interaction and detention time. In systems with zero discharge or very long detention times, phosphorus will be retained in the soil or root zone. In flow-through wetlands with detention times between 5 and 10 days, phosphorus removal will seldom exceed 1 to 3 mg/l. Depending on environmental conditions within the wetland, phosphorus, as well as some other constituents, can be released during certain times of the year, usually in response to changed conditions within the system such as a change in the oxidation-reduction potential (ORP).

Metals: Heavy metal removal is expected to be very similar to that of phosphorus removal although limited data are available on actual removal mechanisms. The removal mechanisms include adsorption, sedimentation, chemical precipitation, and plant uptake. As with phosphorus, metals can be released during certain times of the year, usually in response to changes in the ORP within the system.

Trace organics: Limited data are available on removal of trace organics. Removals of 88–99% have been reported (Reed *et al.* 1995). Removal mechanisms include volatilization, adsorption, and biodegradation.

Microorganisms: Pathogenic bacteria and viruses are removed in constructed wetlands by adsorption, sedimentation, predation, and die-off from exposure to sunlight (UV) and unfavorable temperatures.

16.10 PONDS

The most prevalent natural system is the wastewater treatment pond. Stabilization ponds have been employed for the treatment of municipal wastewater for over 3000 years. Currently, large numbers of pond systems are used throughout the world. Over 7000 pond systems are used in the United States for the treatment of municipal and industrial wastewater under a wide range of weather conditions. These pond systems are used alone or in combination with other wastewater treatment processes.

As in any other system, if stabilization ponds are overloaded and/or untreated industrial wastewater are mixed with municipal wastewater, as in many developing countries (Egypt, Yemen, Jordan, and others), then this may cause problems related to negative public health and environmental effects. Stabilization ponds require minimal operational and maintenance skills and energy. When ponds in series are operated properly, a relatively high quality effluent results with few settleable solids, a safe level of pathogenic microorganisms, no helminths, and high nutrient content. This is particularly true in arid and semi-arid climates. Multi-level pond systems with appropriate design characteristics including a total hydraulic detention time of ≥15 days and with proper operation and maintenance might have removal efficiencies of 4–6 $\log_{10}$ of total coliforms and helminths from raw wastewater, and in warm climatic conditions fecal coliforms may be reduced to about 1000 counts/100 mL (Al-Layla 1992, 1997). However, stabilization ponds with aquatic or constructed wetland systems have been used successfully for advanced treatment of effluent from stabilization ponds. Such systems also appear to be cost-effective and environmentally sound. Wastewater treatment ponds can be classified based on the depth and biological reactions that occur in the pond. Under this classification there are four main types of ponds: (a) aerobic, (b) facultative, (c) aerated, and (d) anaerobic. Aerobic ponds are relatively shallow with usual depths ranging from 0.3 to 0.6 m. Detention is provided by algae during photosynthesis and by wind-aided surface re-aeration. These ponds are often mixed by recirculation to maintain dissolved oxygen throughout their entire depth. Aerobic ponds are typically limited to warm, sunny climates and are used relatively infrequently in the United States (Angelakis 1997).

Facultative ponds are the most common type of treatment pond. Facultative ponds, also referred to as oxidation ponds, are usually 1.5–2.5 m deep with detention times ranging from 25–180 days. The surface layers of the ponds are aerobic with an anaerobic layer near the pond bottom. Oxygen is supplied by surface re-aeration and by photosynthetic algae. The major problem with facultative ponds is the production of algae that remain in the effluent, which often causes suspended solids in the effluent to exceed discharge requirements.

Aerated ponds or lagoons can be either partially mixed or completely mixed. Oxygen is usually supplied by mechanical floating aerators and sometimes by diffused aeration. Aerated ponds are usually 2–6 m deep with detention times ranging from 7–20 days. Aerated ponds can accept higher BOD loadings than facultative ponds, are less susceptible to odors, and usually require less land.

Anaerobic ponds are loaded heavily with organics and do not have an aerobic zone. Anaerobic ponds are 2.5–5 m deep and have detention times of 20–50 days. The biological activity is usually low, compared to a mixed anaerobic digester (Angelakis 1997). Anaerobic ponds have been used for strong industrial wastewaters and for rural communities with a significant organic load from industries such as food processing. They are not widely used for municipal wastewater in the United States.

Ponds can also be categorized by the duration and frequency of their effluent discharges, as follows: (a) total containment ponds, (b) controlled discharge ponds, (c) hydrograph controlled release ponds, and (d) continuous discharge ponds.

The total containment pond or evaporation pond is only used in climates where evaporation exceeds the precipitation on an annual basis. Controlled discharge ponds discharge only once or twice a year when stream conditions are satisfactory. The hydrograph controlled release (HCR) pond is a variation of the controlled discharge pond in which the discharge rate is matched to the stream flow rate. As with the controlled discharge ponds, the HCR pond only discharges when the stream flow is above an acceptable minimum value. In the continuous discharge pond, the effluent is discharged at the same rate (less evaporation and seepage losses) as the influent wastewater flow.

16.11 COMBINATION SYSTEMS

There are a number of systems that combine natural treatment processes. In areas where effluent quality must be very high, or where a high degree of treatment reliability must be maintained, combinations of natural treatment processes may be desirable.

For example, either a SR, a SAT, or a wetlands treatment system could follow an OF system and would result in better overall treatment than the OF system alone. These combinations could be used to improve levels of BOD, SS, nitrogen, and phosphorus removal. Similarly, OF could be used prior to SAT to reduce nitrogen levels to acceptable levels. This combination was demonstrated successfully in a pilot study at Ada, Oklahoma (USA), using screened raw wastewater for the OF portion. SAT treatment may also precede SR land treatment. In this combination, water quality following a SAT is expected to be high enough that even the most restrictive requirements regarding the use of renovated water on food crops can be met. Also, the ground water aquifer can be used to store treated water to correspond with crop irrigation schedules.

In addition, combination systems are applied for advanced or tertiary treatment for removal of metals, ammonia, pathogens, and toxicity from the treated wastewater. Such systems should be facultative ponds followed by constructed wetlands or aquatic floating system. Crites and Tchobanoglous (1998) indicated that wetlands at Arcata, California can be used for treatment, habitat enhancement, and educational benefits. A constructed wetland is used to treat 8700 m^3/d of effluent from a facultative pond. The effluent from the wetland is then discharged into 12.5 ha of enhanced wetland (marshes). Aquatic and wetland systems can be used in combination, usually in series, to achieve specific water quality objectives. For example, duckweed or hyacinth systems could be used prior to a constructed wetland to minimize algae concentrations. A combined aerated aquatic treatment system with a constructed wetland has been studied at Harwich, Massachusetts (Nolte and Associates 1989).

Acknowledgements

The author is grateful to Dr N. Paranichianakis for his valuable assistance on collecting information and Mr N. Papadoyannakis for his assistance in typing this chapter.

16.12 REFERENCES

Al-Layla, A.A. (1992) Global aspects of water resources scarcity and solutions. In *Proc. of Nat. Seminar on Wastewater Reuse,* A.A. Al-Layla, Eng. Sana'a, Univ. Sana'a, Yemen.

Al-Layla, A.A. (1997) Major aspects of wastewater reuse. Unpublished data.

Angelakis, A.N. (1994) Natural treatment systems of municipal wastewaters. *Technica Chronica* **14**(4), 125–140.

Angelakis, A.N. (1997) *Wastewater Reclamation and Reuse for Urban Areas of Yemen.* Food and Agriculture Organization of UN, Rome, Italy, p. 144.

Angelakis, A.N. and Spyridakis, S.V. (1996) The status of water resources in Minoan times – A preliminary study. In *Diachronic Climatic Impacts on Water Resources with Emphasis on the Mediterranean Region* (eds. A.N. Angelakis and A. Issar), Springer-Verlag, Heidelberg.

Angelakis, A.N. and Tchobanoglous, G. (1995) *Municipal Wastewaters: Natural Treatment Systems, Reclamation and Reuse and Disposal of Effluents.* Crete University Press. (In Greek.)

Arthur, J.P. (1983) Notes on the design and operation of waste stabilization ponds in warm climates of developing countries. Technical Paper No. 7, The World Bank, Washington DC.

Bouwer, H. (1991) Role of Groundwater Recharge in Treatment and Storage of Wastewater for Reuse. *Wat. Sci. Tech.* **24 (**9): 295–302.

Bouwer, H. (1993) From sewage farm to zero discharge. *European Water Pollution Control* **3**, 1.

Bouwer, H. (2000) Unpublished data. Agricultural Research Service, US Water Conservation Laboratory, Phoenix, Arizona.

Crites, R. and Tchobanoglous, G. (1998) *Small and Decentralized Wastewater Management Systems*, WCB and McGraw Hill, New York.

Crites, R.W., Reed, S.C. and Bastian, R.K. (2000*) Land Treatment Systems for Municipal and Industrial Wastes*, McGraw Hill., New York.

Gerba, C.P. and Goyal, S.M. (1985) Pathogen removal from wastewater during groundwater recharge. In: *Artificial Recharge of Groundwater* (ed. T. Asano), Butterworth Publishers, Stoneham, MA, USA. **9:** 283-317.

Lumsden, L.L., Stiles, C.W. and Freeman, A.W. (1915) Safe Disposal of Human Excreta at Unsewered Homes. Public Health Bulletin No. 68, United States Public Health Service, Government Printing Office, Washington, DC.

Mara, D.D. (1997) Ponds' wasted opportunity. *Water Quality International* Sept. 1, IAWQ, London, UK.

Metcalf and Eddy, Inc (1991) *Wastewater Engineering: Treatment, Disposal and Reuse*, 3rd edn, McGraw Hill, New York.

National Research Council (1994) *Groundwater Recharge Using Waters of Impaired Quality*. National Academy of Science, National Academy Press, Washington DC.

Nolte and Associates (1989) *Harwich septage treatment pilot study: evaluation of technology for a solar aquatic treatment system*. Prepared for Ecological Engineering Associates, Marion, MA, USA.

Reed, S.C. and Crites, R.W. (1984) Handbook on Land Treatment Systems for Industrial and Municipal Wastes, Noyes Data, Park Ridge, NJ, USA.

Reed, S.C., Crites, R.W. and Middlebrooks, E.J. (1995) *Natural Systems for Waste Management and Treatment,* 2nd edn, McGraw Hill, New York.

Romano, P. and Angelakis, A.N. (1998) Groundwater recharge with reclaimed wastewater effluents. In *Wastewater Management in Yemen* (eds A.N. Angelakis and S. Thirugnanasambanthar), pp. 143–154, FAO, Rome.

Stowell, R., Ludwig, R., Colt, J. and Tchobanoglous, G. (1981) Concepts in aquatic treatment system design, *Journal of Environmental Engineering Division*, Proceedings ASCE, vol. 107, no EE5.

Tchobanoglous, G. (1999) Wastewater: an undervalued water source for sustainable development. In *Proc. Environmental Technology for the 21st Century*, Heleco 99, Thessaloniki, 3–6 June, **1**, 3–9.

Tchobanoglous, G. and Angelakis, A.N. (1996) Technologies for wastewater treatment appropriate for reuse: potential for applications in Greece. *Wat. Sci. Tech.* **30**(10–11), 17–26.

Tsagarakis, K.P., Mara, D.D., Horan, N.J. and Angelakis, A.N. (1998) Evaluation of reuse and disposal sites of effluent from municipal wastewater treatment plants in Greece. *Proc. of Inter. Conf. on Advanced Treatment, Recycling and Reuse*, Milan, 14–16 September, **II**, 867–870.

US EPA (1981) *Process Design Manual for Land Treatment of Municipal Wastewater*, EPA 625/1-81-013, Cincinnati, Ohio.

17

Treatment methods for grey water

B. Jefferson, S. Judd and C. Diaper

17.1 INTRODUCTION

Over the coming decades, the management of water resources will become one of the most important issues across the industrialised world. Key drivers are region-specific but relate largely to climatic and regional factors. In certain countries around the world these are compounded by a shift towards hotter, drier climates. For example, the annual rainfall in London is now similar to that in Istanbul and considerably less than perceivably warmer regions such as Lisbon and Texas (Figure 17.1).

Recycling can make a contribution to the sustainability of available water resources. Central to the approach of water recycling is the concept of the utility of water whereby water is used of a quality commensurate with its application (Thomas and Judd 2000). This then permits the exploitation of large water resources that are not necessarily of the highest purity. Grey water recycling is one aspect of this and relates principally to the reuse of lower polluted water generated within buildings for uses such as toilet flushing or irrigation. This

Edited by P. Lens, G. Zeeman and G. Lettinga. ISBN: 1 900222 47 7

chapter will discuss the technologies that are currently used throughout the world for the treatment of grey water, and includes the latest understanding of some of the problems faced when selecting technologies for grey water treatment.

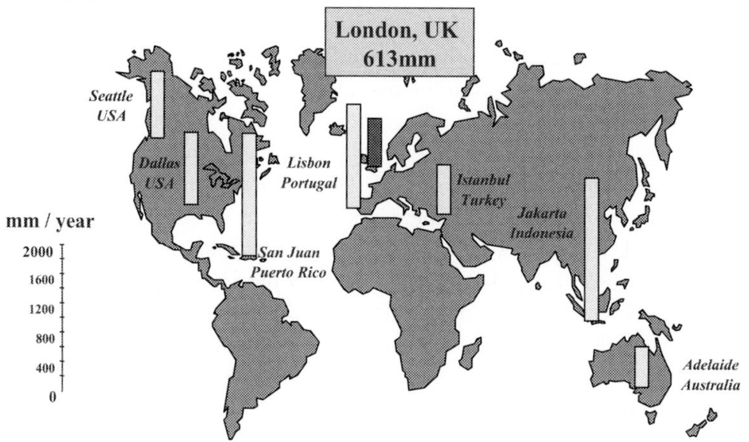

Figure 17.1. Typical rainfall patterns around the world.

17.2 CHARACTERISATION OF GREY WATER

Grey water arises from domestic washing operations. Sources include waste from hand basins, kitchen sinks and washing machines, but specifically exclude black water from toilets, bidets and urinals. A further subdivision is often made restricting grey water sources to hand basins, showers and baths thus removing sources of grease or other highly polluting substances. Grey water is generated by the application of soap or soap products to the human body and so characteristics are a function of, amongst other things, geographical location, demographics and level of occupancy.

17.2.1 Flow

In the domestic environment the amount of grey water produced by a household, about 30% of the total water use, approximately equals the volume of water required for toilet flushing. This balance is very different for buildings of different occupancy, for example in an office building the amount of grey water produced is considerably less than the requirement for toilet and urinal flushing, 27% and 60% respectively (Griggs *et al.* 1998).

Notwithstanding the apparent balance between grey water produced and WC flush water required, the dynamics of the situation are not ideal. In the home, grey water is generated over short time periods at times slightly offset from toilet flushing, whereas toilet flushing takes place more consistently through the day. This generally results in a deficit in water during the afternoon and late evening. This situation may be rectified by buffering using appropriately-sized storage tanks, but this increases both the overall system size and capital cost, as well as complicating installation procedures. Detailed examination of the benefits of storage reveal that 1 m^3 of storage is suitable for a wide range of occupancy scales (Figure 17.2, Dixon *et al.* 1999). Increasing storage capacity above this provides only marginal increases in water savings and increases problems associated with grey water degradation and disinfection reliability.

Generally the requirement for irrigation water in the UK is low, only 3% of the total water used within the home. In the US the requirement is much higher and accounts for nearly 60% of the total water used, thus the feasibility of grey water use for irrigation is very location-specific. Utilisation of grey water for irrigation is also dependent on climatic conditions. Higher rainfall implies reduced demand for a grey water source. However, there are few countries where rainfall is equally distributed throughout the year and the possibility of utilising grey water during times of low or no rain is a viable option, provided storage is available. Also, in situations where landscaping is used for aesthetic improvement of the area, i.e. hotel gardens, utilisation of grey water is a valid option since plants can be viewed as non-vital users.

Figure 17.2. Relationship between storage size and water saving efficiency (PI_E) for residences of varying occupancy

17.2.2 Composition

Grey water can have an organic strength similar to domestic wastewater but is relatively low in solids suggesting that a greater fraction of the organic load is dissolved (Table 17.1). Although the concentration can be similar to domestic wastewater the composition is not. The COD:BOD ratio can be as high as 4:1 with a corresponding deficiency in macro nutrients such as nitrogen and phosphorus. Grey water exhibits significant variations in composition; within a specific sample group, within an individual showering or bathing operation and also between reported schemes (Table 17.1). The variation between the schemes reflects differences in washing habits both in terms of product type and concentration used by an individual. The relatively small scale of the majority of grey water schemes means that the variations seen from an individual can have a pronounced impact on the overall characteristics of the grey water to be treated. The minimum limit to these variations is shown within the grey water produced by an individual. In trials conducted on the authors' bath water, in which both the type and concentration were strictly controlled, the variation in quality was measured at 17% relative standard deviation (RSD) for BOD and up to 100% RSD for total coliform concentration.

Although storage is a key facet of any recycling scheme, little has been reported on the effects of storage on grey water quality. A study conducted at Cranfield University in the UK revealed a rapid decline in organic strength with both real and synthetic grey water under quiescent and agitated conditions (Figure 17.3). The degradation was shown to follow a first-order decay over an initial 7 day period irrespective of storage conditions, with rate constants varying between 0.011 day^{-1} (quiescent) and 0.622 day^{-1} (agitated). Further, the COD:BOD ratio of the water decreased, indicating that the waste was becoming more biodegradable. These results correlate well with data reported from operational schemes (Jefferson *et al* 2000a). The concentration of coliforms was shown to increase by 3 log over an initial 5 day period and then remained stable for a further 15 days before numbers started to decrease. However, although there is an increase in indicator species this does not imply an increase in pathogenic micro-organisms and hence risk (Lee 1999). Thus, a major factor affecting the characteristics of grey water between different recycling schemes is the residence time of the grey water in the collection network, which can range from minutes to days. In one specific example in an accommodation block of a university the collection network was significantly altered to reduce the residence time of the system. A significant increase in organic strength was observed almost immediately (Ward 2000).

A major issue of concern is the comparison of real and synthetic grey water sources. Standard recipes are available in the literature (Holden and Ward 1999; Laine *et al.* 1999) based on combinations of soap, shampoo, oil, hair and effluent from a sewage works. These recipes can be tailored to obtain any strength of grey water required.

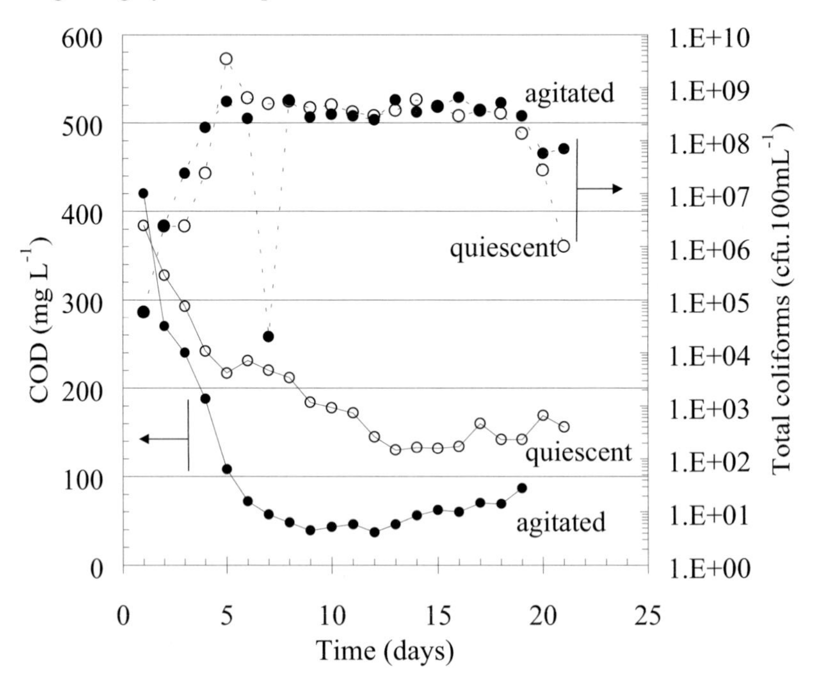

Figure 17.3. Impact of storage on water.

However, the nature of the synthetic grey water is very different to real grey water matrices in terms of concentrations of macro and micro nutrients (Jefferson *et al.* 2000b). Nutrient balancing in both real and synthetic sources revealed an increase in the oxygen uptake rate and the COD removal in comparison to non-nutrient balanced samples. However, the magnitude of the changes and the effects of micro nutrients were very different, indicating a difference in the biological degradation pathways for the different grey waters at a fundamental level. This can have a pronounced effect when assessing the suitability of a new technology for grey water recycling. The most appropriate approach to technology assessment is to conduct trials on real grey water sources but introduce higher levels of dissolved and suspended organic matter, as provided by a grey water analogue, both gradually and as shock loads. This then allows the trial to be conducted at a range of grey water strengths and should improve understanding of how the technology will operate in real situations.

17.3 WATER RECYCLING STANDARDS

The acceptability of recycled water for any particular end use is dependent on its physical, chemical and microbiological quality. Development of a clear strategy towards domestic water recycling has to a certain degree been hampered by a lack of water quality standards that address these needs. No standards exist in the UK and apart from a proposed criterion by the Building Services Information and Research Association (Mustow *et al.* 1997) and a guidance note on reclaimed water systems (WRAS 1999) no standards have been developed. Table 17.2 shows the appropriate standards for water recycling in other parts of the world where grey water reuse is practised or where standards have been suggested as appropriate for grey water applications.

There is some commonality in the water quality determinants used to define bacteriological quality, biodegradability, clarity and acidity, though the actual absolute permitted levels vary considerably. Two groups of standards can be identified from two different ideologies. The first of these is based on the quality of the grey water being commensurate with its application, and in such cases the standards are similar to those for bathing water since the level of risk to the user is about the same. In these more pragmatic approaches the main water quality criteria relates to coliform concentration which is set in the order of a few thousand colonies per 100 millilitres (ml). The alternative approach is more conservative and considers grey water treatment in a similar manner to that of municipal or industrial effluent. In these cases the standards include terms for BOD_5, turbidity as well as more restrictive levels of coliforms. In countries such as the US and Japan where grey water recycling has been practised for some time, the more conservative approach is taken.

Public health protection is a key consideration and as such all water recycling standards include parameters relating to the potential for disease transmission. Indicator organisms are generally preferred due to their ease of measurement and the familiarity with their use in the water industry. However, there is no consensus about which organisms should be used and how standards based on them should be interpreted with respect to recycling (Asano 1998). One issue is

Table 17.1. Characterisation of grey water

	Shower	Bath	Hand basin	HoR 1[A] (Holden and Ward 1999)	HoR 2[B] (Holden and Ward 1999)	Single house[C]	USA[D]	Sweden[E]	Australia[F]
BOD_5	146	129	155	33	96	38	162	196	159
(ppm)	(55)	(57)	(49)		(103)	(38)			(69)
COD	420	367	587	40	168	163	366	–	–
(ppm)	(245)	(246)	(379)		(91)	(107)			
Turbidity	84.8	59.8	164	20	57	38	–	–	113
(NTU)	(70.5)	(43)	(171)		(138)	(47)			(55)
TC	6800	6350	9420	–	5200000	0–>2419	24000000	3600000	–
(cfu/100ml)	(9740)	(9710)	(10100)		(3600000)				
E.coli	1490	82.7	10	–	–	ND	–	–	–
(cfu/100ml)	(4940)	(120)	(8750)						
FS	2050	40.1	1710	–	479	ND	1400000	880000	–
(cfu/100ml)	(4440)	(48.6)	(5510)		(859)[a]				
TN	8.7	6.6	10.4	–	–	13		6.5	11.6
(ppm)	(4.8)	(3.4)	(4.80)			(6)			(10.2)
PO_4^{3-}	0.3	0.4	0.4	0.4[b]	2.4	1.13		7.8	–
(ppm)	(0.1)	(0.4)	(0.3)		(0.7)	(1.03)			
PH	7.52	7.57	7.32	–	7.7	7.6	6.8	–	7.3
	(0.28)	(0.29)	(0.27)		(0.4)	(0.35)			(0.6)

Table 17.1: Numbers in brackets relate to one standard deviation around the mean, HoR = university hall of residence, TC = total coliforms, FS = faecal streptococci, TN = total nitrogen, A = Holden and Ward (1998), B = Surendran and Wheatley (1998), C = Sayers (1998), D = Brandes (1978), E = Olsson *et al.* (1968), F = Christova-Boal *et al.* (1996).

Table 17.2. Summary of water quality standards and criteria suitable for domestic water recycling (from Jefferson 2000b)

	Total coliforms count/100 ml	Faecal Coliforms	BOD_5 (mg/l)	Turbidity (NTU)	Cl_2 (mg/l)	PH
Bathing water Standards*	10000 (m) 500 (g)	2000 (m) 100 (g)				6–9
USA, NSF		<240	45	90		
USA EPA	Non detectable		10	2	1	6–9
Australia	<1	<4	20	2		
UK (BSIRA)	Non detectable					
Japan	<10	<10	10	5		6–9
WHO*	1000 (m) 200 (g)					
Germany (g)	100	500	20	1–2		6–9

*bathing water standards suggested as appropriate for domestic water recycling; (g) = guideline, (m) = mandatory

that no correlation exists between the concentration of indicator organisms and actual pathogens; a positive coliform concentration only demonstrates a potential pathway for disease transmission rather than any actual risk of illness. At one stage the EPA suggested that a level of 200 counts per 100 ml of faecal coliforms could be adopted from bathing water regulations. However, the added gastrointestinal illness rate at this level has been calculated at 19 illnesses per 1000 swimmers which is now being considered too high by some of the regulatory bodies in the US (Asano 1998). In addition, some virus strains are more resistant to disinfection than the indicator species. This has led to a conservative approach in some areas with a non-detectable level of indicator species being introduced. Combining technology and quality requirements improves the applicability of surrogate measures such as total coliforms and is now forming the basis of reuse standards proposed by the EPA (Asano 1998). In the case of the standards relevant to domestic reuse the technologies stated are secondary treatment followed by filtration and disinfection.

17.4 TECHNOLOGIES

Currently a plethora of technologies are being developed and installed around the world varying greatly both in complexity and performance. Technologies developed range from simple systems for single house applications to very advanced treatment trains for large-scale reuse.

17.4.1 Single house systems

The most common systems currently employed in the UK for grey water recycling are based at the single house scale with in excess of 20 reported schemes currently operating (Diaper *et al.* 2000a). A number of companies market products but all tend to be based on the same generic processes. These include two different stages (Figure 17.4): a coarse filtration stage for the removal of large particles such as hair and skin which could adversely affect the aesthetic quality of the water, and a disinfection process, usually either chlorine or bromine based on slow release solid blocks. Investigations into suitable filter types in Australia have shown the benefits of disposable filters made either from nylon or geotextiles (Christova-Boal *et al.* 1996). UV irradiation has also been tested in place of disinfection, although its effectiveness is limited by the transmission of water at 254 mm which usually means that filtration alone is insufficient for UV applications (Kreysig 1996).

These systems are designed to meet less stringent standards, akin to those for bathing water. They rely on a short residence time and the nature of the grey water remains unchanged from the point of production, and the effluent organic load and turbidity are ostensibly the same as for the raw grey water. The high organic load and turbidity may lead to problems during disinfection since both can limit the effectiveness of the disinfection stage for two main reasons.

First, grey water contains flocculent particles above 40 μm in diameter, and these are known to reduce the chemical disinfection capability because the disinfectant cannot diffuse far enough into the flocs to kill all microbial species. Second, organic matter in the water imparts a disinfectant demand. Trials have shown that this can constitute up to 99% of the total chlorine demand and can cause coliform breakthrough since all the disinfectant is utilised to combine with organic contaminants. Also, in the case of chlorine, disinfectant by-products such as chloramines and trihalomethanes are generated which do have some disinfectant capability but which can adversely affect human health.

The water-saving potential of these single house systems is highly dependent on the reliability of system components and the householder's awareness of the status of the system. The key components of the system are the filter and the pump; if there is a blockage in the filter or the pump fails, the water-saving efficiency of the system is reduced to zero.

In order to ensure a continual supply of water for toilet flushing all systems contain a potable water back-up device. In many systems there are control measures which ensure that if the disinfectant runs out then grey water is no longer processed and potable water is fed to the system. This ensures that non-disinfected grey water does not enter the system. Similarly, when there is not enough grey water to supply WC flushing demand (this could be due to filter blockage) the system will switch to potable supply. This means the system requires regular maintenance checks in order to ensure that maximum water savings are achieved. In many of the trials on these systems householders have been unaware that the system has failed and requires maintenance (Diaper *et al.* 2000b). Thus, relaying the status of the system to the householder or training the householder to carry out regular maintenance procedures is the key to ensuring maximum water savings.

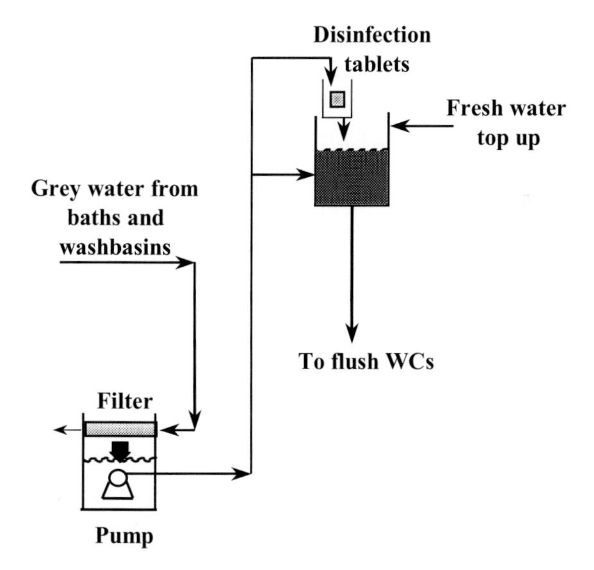

Figure 17.4. An example of a single house grey water recycling system.

A full 30% water saving has been achieved with many of systems under trial (Diaper *et al.* 2000b; Sayers 1998). However, the economic benefits are minimal and are directly related to the occupancy level of the house, higher occupancy implying to increased financial benefits and shorter payback periods. For single occupancy households the payback period for a system with a representative annual water saving of 19.3% is in excess of 50 years for the highest possible UK water charge (Sayers 1998).

17.4.2 Physical systems

Physical systems such as depth filters and/or membranes have been developed because they produce a higher quality effluent than the single house systems described above. Two distinct types of operation have been reported in the literature, purely physical processes such as sand filters (Costner 1990) or membranes (Holden and Ward 1999). The alternative is the use of a membrane downstream of a biological unit, in which case the process acts as a polishing stage (Shin *et al.* 1998).

Both slow and rapid gravity depth filters have been tested for grey water treatment. The low tech sand filter is usually constructed from a 250 litre drum containing layers of media, coarse gravel at the base and sand at the top. The level of treatment and capacity of the system will depend on the properties of the media used and, to some extent, climatic conditions, i.e. increased loading rates can be achieved in hotter climates or with coarser sand. Maintenance of such systems includes occasional backwashing to decompact and regenerate the media bed, and removal of the top layer of sand. In order to provide continuous treatment a duplex system is often used, with one filter backwashing and drying out while the other is online (der Ryn 1995). Typical reported removal efficiencies for physical systems alone are 63% BOD_5 and 28% NTU for a depth filter and 86% BOD and 99% NTU for a membrane (Holden and Ward 1999). Membranes produce the highest level of removal as the membrane acts as a direct sieve whereby all particles and molecules above the pore size of the membrane are removed.

The project water quality from a membrane is generally high but such processes are known to suffer from operational and economic constraints (Stephenson *et al.* 2000). The key factor limiting the economic viability of membrane systems is the fouling of the membrane surface by pollutant species. This increases the hydraulic resistance of the membrane, thereby commensurately increasing the energy demanded for membrane permeation and/or decreasing the permeate flux. Fouling can be suppressed by operation at a lower membrane flux or can be substantially removed by cleaning (Stephenson *et al.* 2000), the former requiring larger membrane areas to process the same volume of grey water. Both factors increase the overall process cost and cleaning also imparts an undesirable chemical load on the waste stream. Fouling can be such that the flux declines by up to 90% after just one hour of operation (Gander *et al.* 2000).

Problems of poor treated water quality have been reported with membrane systems treating grey water, as well as difficulties in effectively cleaning the membrane. The residence time of the system has been identified as having a major effect on fouling propensity and system performance. When stored for extended periods of time, grey water can become anaerobic, resulting in the generation of organic components which are less readily rejected by the membrane (Holden and Ward 1999). Membrane trials have shown that purely physical systems do not reject all coliforms from the waste stream. This has been explained in terms of protein migration through the membrane pores which aids the transport of coliform species (Judd and Till 2000).

The alternative membrane process is to operate membranes as a downstream process after a biological stage. In these situations the membrane acts as a polishing process and thus the pore size of the membrane is normally much smaller than those reported above. Examples of this technique include a membrane in series with a sequencing batch reactor (SBR) in Korea (Shin *et al.* 1998) and downstream from a biological aerated filter (BAF) at the Millennium Dome in London (Smith *et al.* 2000). The rejection characteristics of a range of pore size membranes have been investigated for the treatment of grey water after a BAF (Smith *et al.* 2000). The soluble BOD concentration in the treated stream decreased from 8.3 $mg.L^{-1}$ with an ultrafiltration membrane to 2.4 $mg.L^{-1}$ with a nanofiltration membrane.

17.4.3 Biological treatment options

Biological treatment options are required to remove organic contamination. A number of different biological systems have been trialled or installed around the world ranging from rotating biological contactors (RBC) in Germany (Clarke *et al.* 1998) to membrane bioreactors (MBR) in Japan (Huitorel 2000) (Figure 17.5). These are especially important for systems that include large distribution networks such as hotels and office blocks where organic levels in the effluent will cause problems with aesthetic quality and the potential for regrowth in pipes. This is reflected in the development of such technologies in Japan where grey water recycling is commonly adopted in such buildings. In fact, by 1983, 100 grey water treatment systems had already been installed in Japan, 25% of which were based on MBR technology (Huitorel 2000).

Biological treatment alone is not usually sufficient to produce an effluent suitable for reuse. In all cases, the biological reaction must be accompanied by a physical process to retain active biomass and prevent the passage of solids into the effluent. Technologies that encompass several of these requirements offer an attractive alternative to some of the more traditional approaches. This provides one of the reasons for the increased interest and substantial market penetration of membrane bioreactors (MBR). The MBR is the amalgamation of a suspended growth reactor and membrane filtration device into a single unit process. The membrane can be configured external to, as in side stream operation, or immersed in the bioreactor. The process represents an intensification of traditional biological processes with the added advantage that the membrane retains particles including bacteria and viruses. Effluent from MBR treating grey water is typically solid-free, low in organic pollution and contains non-detectable levels of coliforms (Stephenson *et al.* 2000).

Intensive trials on the application of biological processes for the treatment of grey water have been carried out (Stephenson and Judd 2000). The processes tested were an MBR, BAF and a membrane aeration bioreactor (MABR). The latter process utilises a membrane to supply pure oxygen to a biological process allowing high oxygen mass transfer rates (Stephenson *et al.* 2000). The technology is gaining application in high strength waste industries where an oxygen limitation exists (Brindle 1998). The technology is unsuitable for grey water, however, since the high surfactant levels within the water prevent the reliable formation of a biofilm on the membrane surface.

The BAF and, in particular, the MBR processes were generally able to effectively treat the organic and physical pollution (Figure 17.6). The main difference in performance was most apparent from the microbial quality (as total coliforms, *E.Coli* or faecal *Streptococci* concentration). The difference can be attributed to the fact that the BAF is essentially a depth filter whereas the MBRs are barrier processes. Hence, whilst the MBRs were able to conform to all existing water recycling standards around the world, the BAF was unable to attain the microbial quality of any of the standards listed in Table 17.2.

The role of the membrane and the bioreactor were further investigated by operating the membrane with and without a connection to a bioreactor. Membrane filtration, evaluated by operating the side stream unit without connection to a bioreactor, gave reduced organic and microbial removal and, most notably, incomplete coliform rejection. As stated above, this is attributed to protein adsorption on the membrane surface during experimentation, which may have provided a transport pathway for micro-organisms. An advantage of the MBR is that they retain bacteria, since protein adsorption into the flocs and subsequent digestion eliminates this pathway. Fouling problems are also decreased: direct membrane filtration is known to lead to irreversible fouling, demanding vigorous and ultimately high-cost cleaning regimes.

Whilst MBR membrane configuration has little impact on pollutant removal and rejection, hydraulic performance is markedly influenced. Submerged

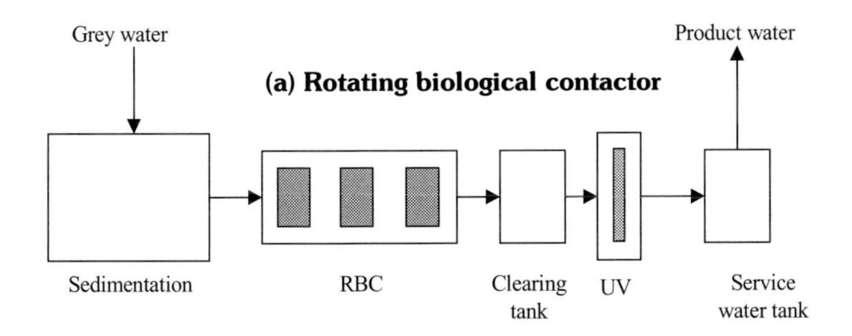

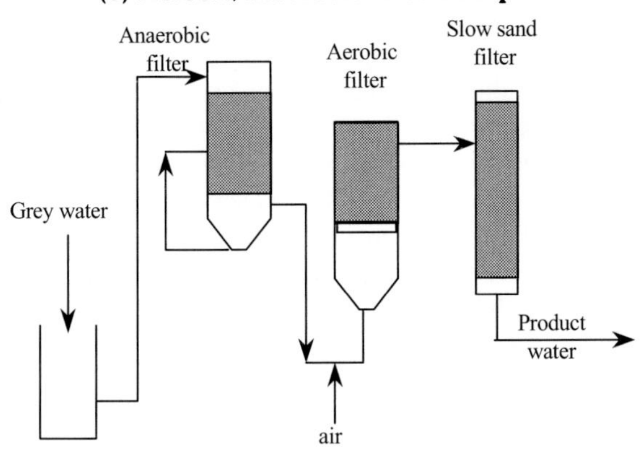

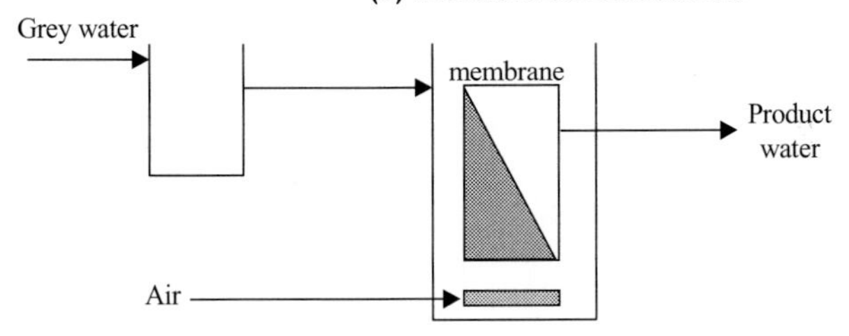

Figure 17.5. Process flow diagrams for biological treatment processes. (a) after Nolde 1999; (b) after Surendran and Wheatley 1998; (c) after Jefferson *et al.* 2000a.

systems produce low but stable fluxes due to low operational pressures. Side stream MBRs operate at higher pressures and cross flow velocities and yield higher initial fluxes. However, the flux tends to decline rapidly, and backflushing and chemical cleaning is regularly required to recover the high flux (LeClech *et al.* 2000). Energy demand calculations reveal that submerged configurations require 100 times less energy than the side stream for operation, but require 3–4 times more membrane area.

A major barrier to the uptake of biological processes, especially on a smaller scale, is their susceptibility to toxic shock loads, such as the spiking of grey water sources by substances that are not traditional components of it, for example bleach. A recent public survey suggests that bactericidal agents such as bleach and bathroom cleaners are the most likely substances to be spiked into grey water (Laine *et al.* 2000). Food, alcohol and washing powder have also been identified as likely spike substances. Critical concentrations of a range of possible substances have been determined in terms of the concentration at which inhibition of the biological reaction begins. The most important substance is bleach with inhibition beginning at concentrations as low as 1.4 $mL.L^{-1}$ with very substantial inhibition occurring at 3 $mL.L^{-1}$. The effect of such chemicals is to provide a limitation on the size at which biological treatment systems can be reliably used for grey water treatment. Plant sizes serving below 10–20 people will have a higher risk of failure due to uncontrolled introduction of onerous substances. However, the exact size is dependent upon the buffering capacity of the storage system as well as the risk of discharge of onerous substances.

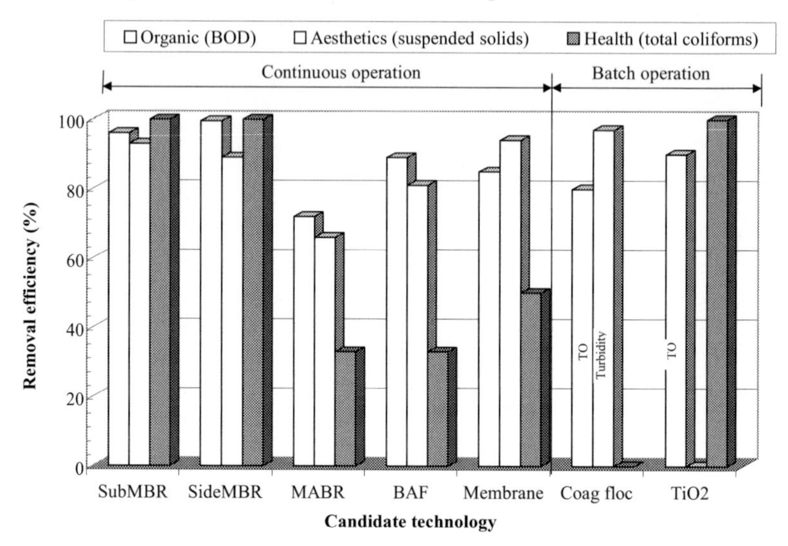

Figure 17.6. Removal efficiencies of the candidate technologies in grey water. SubBMR = Submerged membrane bioreactor; Side BMR = Side stream membrane bioreactor; MABR = Membrane aeration bioreactor; BAF = Biological aerated filter; Coagfloc = Coagulation flocculation; TiO_2 = Titanium dioxide catalysed.

17.4.4 Natural treatment systems

Many natural systems exist for the treatment of grey water, both bespoke and package plant. Some are based on simple infiltration techniques whereby the primary aim is to allow the grey water to percolate into the soil. This will provide some treatment before the waste stream infiltrates groundwater sources, the level of treatment depending on soil conditions and location of the infiltration site. Other natural treatment systems are based on the approach that the nutrients in the waste stream can be utilised to provide plant growth.

The process operations used do not differ from those natural systems utilised for sewage effluent treatment but, since grey water contains a lower concentration of contaminants, the maintenance frequency is often reduced. In addition, for systems where final disinfection is necessary i.e. where direct human contact is highly likely, the disinfection dose is reduced due to the reduction in contaminant concentrations. Reduced disinfection dosage will reduce operational costs and also reduce possible detrimental environmental effects.

The simplest method for reuse of grey water is referred to in the literature as the 'Mexican Drain' (der Ryn 1995). This is a basic system where grey water is collected and then fed directly to plants via a bucket or hose and no treatment occurs (Figure 17.7). Care should be exercised in ensuring that the application area is changed every few days and ideally application should be rotated with rainwater. These steps are necessary since irrigation with untreated grey water can lead to accumulation of sodium in the soil. Untreated grey water is usually alkali in nature and so irrigation of acid-loving plants such as azaleas, rhododendrons and citrus fruits should be avoided.

Despite the seemingly innocuous nature of grey water, care should be taken in selecting irrigation methods and plants types for irrigation. Grey water may contain high concentrations of micro-organisms and application methods producing aerosols should be avoided. Also, it is questionable whether grey water should be used for the irrigation of edible plants such as leafy vegetables and root crops.

An advancement on the simple bucket method referred to above is the use of a septic tank. This method allows settling of solid material and biological degradation of some grey water components. Since the contaminant concentration in grey water is considerably less than in domestic sewage the septic tank will require less frequent emptying. The outlet from the septic tank can be further treated by many other methods. Usually the effluent will feed into a system of leach lines or a drainfield, the selection and design of which will depend on soil properties and location. Design parameters will correlate with those for septic tanks treating domestic sewage. With soils of low porosity or areas with high water tables, above-ground infiltration mounds can be constructed. The waste stream fed to the mound is dispersed by evaporation to the air as well as by percolation into the soil.

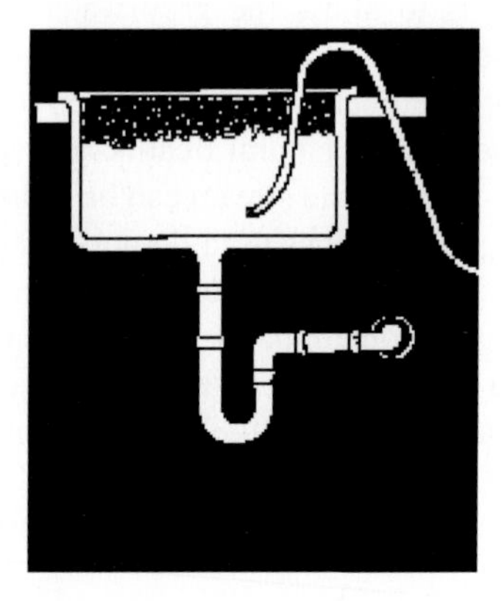

Figure 17.7. Mexican drains (Mann and Williamson 1996).

Septic tanks can be combined with other treatment processes to improve the quality of the effluent. For example, an anaerobic filter at the outlet of the septic tank will reduce possible solids carry-over and alleviate possible clogging of drainage field pipes. Alternatively, rather than allowing the outlet from septic tanks to percolate into the soil, the stream can be used to promote plant growth. The effluent stream can be fed to soil beds, either in greenhouses or outdoors, containing evergreen plants or trees (silviculture).

A final natural method of grey water treatment can be broadly termed aquaculture, whereby the nutrients in the grey water stream are utilised to grow water-loving plants. Many wetland plants have the ability to take up excess nutrients and store them, allowing the plants to continue to grow when these nutrients are limited, as can happen with grey water. These treatment methods can vary from pond systems to engineered gardens. Pond systems can be integrated into landscaping. One such plant has been operating at a technical university in Sweden for over three years and provides good bacterial and nutrient removal (Günther 2000). Pond systems can be used to grow algae that can then be fed to fish or molluscs. Algae will also shift pH to more alkaline conditions favouring ammonia volatisation and nitrification, as well as bacterial kill (Mars *et al.* 1999).

In engineered gardens, suitable plants are grown in an evapotranspiration bed lined with an impervious membrane often all enclosed in a greenhouse (Del Porto 1999). The capital costs of such systems will be higher but there is more control over the plant types and the structure and characteristics of the growth medium. The plants can be used for landscaping, or harvested and used as animal fodder or construction materials, or can be composted for land disposal. Plants utilised in these systems include bamboo, water hyacinth, *Phragmites* spp. (reeds), *Strezlitzia reginae* (Bird of Paradise), *Gracillima variegata* (ornamental corn) (Del Porto 1999), *Triglochin hueglii* and *Schoenplectus validus* (Mars *et al.* 1999).

When using grey water for irrigation and crop production, care should be taken in the use of soaps, cleansers and other household chemicals because certain components of these products have been shown to reduce plant yield (Garland *et al.* 2000). Control of these grey water contaminants at source will ensure minimal detrimental effects on plant growth.

The performance of natural and low-tech treatment systems does not vary greatly for different scales and methods of treatment (Table 17.3).

Table 17.3. Performance of selected natural treatment processes

Treatment method	Volume m^3/day	BOD or COD removal (%)	Bacterial reduction log number	Reference
Three chamber sedimentation – horizontal flow reed bed – sandfilter – pond	Average 10.7	BOD >97*	6	Fittschen and Niemczynowicz 1997
Two chamber Sedimentation – sand filter – horizontal flow reed bed**	Average 0.4	BOD 95.8 COD 91.4	Not reported	Schönborn *et al.* 1997
Lime/gravel filter - three ponds – sandfilter	~1.2	BOD 100*	2–3	Günther 2000

*BOD_7
** Feed of grey water plus urine

17.5 REFERENCES

Asano, T. (1998) *Wastewater Reclamation and Reuse.* Technomic, Pennsylvania, US.

Brandes, M. (1978) Characteristics of effluent from gray and black water septic tanks. *JWPCF* 50, 53–63.

Brindle, K. (1998) Membrane aeration bioreactors for the treatment for high oxygen demanding wastewater. Ph.D. thesis, Cranfield University, UK.

Christova-Boal, D., Eden, R.E. and McFarlane, S. (1996) An investigation into greywater reuse for urban residential properties. *Desalination* **106**, 391–397.

Clarke, J., Holden, B. and Ward, M. (1998) *Grey Water Recycling in the Hotel Industry.* Anglian Water Report, Anglian Water, Peterborough, UK.

Costner, P. (1990) We all live downstream: a guide to waste treatment that stops water pollution. The National Water Centre, Waterworks Publishing Company, Eureka Springs, Arkansas, US.

Del Porto, D. (1999) Zero-effluent-discharge systems prevent pollution: conserving, separating and using up effluents on site. In: *Proceedings of International Ecological Engineering Conference*. Norway, 6 June 1999.

Diaper, C., Jefferson, B. and Parsons, S.P. (2000a) Water recycling technologies in the UK. *J. CIWEM.*

Diaper, C., Dixon, A., Butler, D., Fewkes, A., Parsons, S. A., Strathern, M., Stephenson, T. and Strutt, J. (2000b) Small scale water recycling systems – risk assessment and modelling. In: *1st World Congress of the International Water Association*, Paris, 3–7 July, p. 9.

Dixon, A., Butler, D and Feweks, A. (1999) Water saving potential of domestic water reuse systems using greywater and rainwater in combination. *Wat. Sci. Tech.* **39**(5), 25–32.

Fittschen, I. and Niemczynowicz, J. (1997) Experiences with dry sanitation and grey water treatment in the eco-village Toarp, Sweden. *Wat. Sci. Tech.* **35**(9), 161–170.

Gander M., Jefferson B. and Judd S. (2000). MBRs for use in small wastewater treatment plants. *Wat. Sci. Technol* **41**(1), 205-211.

Garland, J.L., Levine, L.H., Yorio, N.C., Adams, J.L. and Cook, K.L. (2000) Gray water processing in recirculating hydroponic systems: phytotoxicity, surfactant degradation and bacterial dynamics. *Wat. Res.* **34**, 3075–3086.

Griggs, J.C., Shouler, M.C. and Hall, J. (1998) Water conservation and the built environment. In *Proc Conf 21AD: Water*. Linacre College, Oxford, 24–25 April.

Günther, F. (2000) Wastewater treatment by grey water separation: Outline for a biologically based grey water purification plant in Sweden. *Ecological Engineering* **15,** 139–146.

Holden, B. and Ward, M. (1999) An overview of domestic and commercial re-use of water. *Proc. IQPC Conf. on Water Recycling and Effluent Reuse*. Copthorne Effingham Park, London, UK.

Huitorel, L. (2000) The treatment and recycling of waste water using an activated sludge bioreactor coupled with an ultrafiltration module. In *TUWR1 – 1st Intl. Mtg. on Technologies for Urban Water Recycling*, Cranfield, UK, 19 January.

Jefferson, B., Laine, A., Parsons, S., Stephenson, T. and Judd, S. (2000a) Technologies for domestic wastewater recycling. *Urban Wat.* **1**(4), 285–292.

Jefferson, B., Laine, A., Stephenson, T. and Judd, S. (2000b) The characterisation of grey water and its impact on the design of urban water technologies. *Wat. Res.* (in preparation).

Judd, S.J. and Till, S.W. (2000) Bacteria breakthrough in crossflow microfiltration of sewage. *Desalination* **127**, 251–260.

Kreysig, D. (1996) Greywater recycling: treatment techniques and cost savings. *World Water and environmental engineering*, **32**, 18-19.

Laine, A.T., Jefferson, B., Judd, S. and Stephenson, T. (1999) Membrane bioreactors and their role in wastewater reuse. *Wat. Sci. Tech.* **41**(1), 197–204.

Laine, A.T., Jefferson, B., Judd, S. and Stephenson, T. (2000) Water recycling from grey to black water: the process engineering approach to water resource problems. *Proc.*

Research 2000 – Stretching the Boundaries of Chemical Engineering, Bath, 6–7 January.

LeClech, P., Jefferson, B., Laine, A., Smith, S. and Judd, S. (2000) The influence of membrane configuration on the efficacy of membrane bioreactors for domestic waste water recycling. In *WEFTEC 2000 – Proc. of the 73rd Annual Conf. and Expo. On Water Quality and Wastewater Treatment*, 14–18 October, Anaheim, p. 20.

Lee, J. (1999) Public health laboratory service. Personal communication.

Mann, H.T. and Williamson, D. (1996) *Water Treatment and Sanitation*. Intermediate Technology Publications, Nottingham, UK.

Mars, R., Mathew, K and Ho, G. (1999) The role of the submergent macrophyte *Triglochin hueglii* in domestic grey water treatment. *Ecological Engineering* **12**, 57–66.

Mustow, S., Grey, R., Smerdon, T., Pinney, C. and Waggett, R. (1997) Implications of using recycled grey water and stored rainwater in the UK. Report No. 13034/1, BSRIA, Bracknell, March.

Nolde, E. (1999) Grey water reuse system for toilet flushing in multi-storey buildings – over ten years experience in Berlin. *Urban Wat.*, **1**, 275–284.

Olsson, E., Karlgren, L. and Tullander, V. (1968) Household wastewater. National Swedish Institute for Building Research, Stockholm, Report No. 24.

der Ryn, S.V. (1995) The toilet papers. Recycling waste and conserving water. Ecological Design Press, Sausalito, California.

Sayers, D. (1998). A study of domestic greywater recycling. Interim report, National Water Demand Management Centre, Environment agency, Worthing, UK.

Schönborn, A., Züst, B. and Underwood, E. (1997) Long term performance of the sand-plant-filter Schattweid (Switzerland). *Wat. Sci. Tech.* **35**(5), 307–314.

Shin, H-S., Lee, S-M., Seo, I-S., Kim, G-O., Lim, K-H. and Song J-S. (1998) Pilot scale SBR and MF operation for the removal of organic and nitrogen compounds from grey water. *Wat. Sci. Tech.* **38**(6), 79–88.

Smith, A., Khow, J. and Hills, S. (2000) Water reuse at the UK's Millennium Dome. *Membrane Technology* **118**, 5–8.

Stephenson, T. and Judd, S. (2000) In-building wastewater treatment for water recycling. EPSRC final report, number GR/K84967.

Stephenson, T, Judd, S., Jefferson, B. and Brindle, K. (2000) Membrane bioreactors for wastewater treatment. IWA publishing, London.

Surendran, S. and Wheatley, A.D. (1998) Grey water reclamination for non-potable reuse. *J. CIWEM* **12**, 406–413.

Thomas, D.N. and Judd, S.J. (2000) Entropy in water management. *CIWEM J,.* in press.

Ward, M. (2000) Anglian Water Services. *Personal communication.*

Water Regulations Advisory Service (1999) IGN 9-02-04 Reclaimed Water Systems: Information about installing, modifying or maintaining reclaimed water systems. Water Regulations Advisory Scheme, London.

18

Aspects of groundwater recharge using grey wastewater

A. Ledin, E. Eriksson and M. Henze

18.1 INTRODUCTION

Mismanagement and pollution of the environment has been in public focus since the 1960s, when it became clear that the basic principle for urban water use, 'to use and contaminate water', caused serious damage to receiving waters around cities and settlements. Effects observed, such as oxygen depletion of surface waters, microbial pollution and even epidemics, smell and extensive fish death led to the technical development of wastewater treatment plants under municipal auspices. These centralised urban sanitation systems, used for treating wastewater produced in households, can be very effective, but can also be expensive and resource consuming. This is one reason for the growing demand for integrated decentralised sanitary systems providing opportunities to save and

Edited by P. Lens, G. Zeeman and G. Lettinga. ISBN: 1 900222 47 7

reuse wastewater. Another reason to look for alternative ways of handling wastewater is water shortage, which is a problem in large parts of the world. One way to reduce the need for freshwater is to reuse wastewater, after some decentralised, often low-tech treatment or without any treatment at all.

Among the alternatives for treatment of wastewater, infiltration into the soil has been suggested. The objectives stated for the implementation of infiltration varies greatly, but covers mainly two types of 'uses' for the infiltrated water:

- As a pre-treatment step, where the water after infiltration is applied for irrigation.
- As a method for recharge of the water and thereby making a shortcut in the hydrological cycle.

The focus is today on the possibility for reusing/infiltrating so-called 'diluted', 'light' or 'grey' wastewater. All three terms refer to wastewater produced in households, office buildings and schools as well as some types of industries, where there is no contribution from toilets or heavily polluted process water. Grey wastewater is wastewater from baths, showers, hand basins, washing machines and dishwashers, laundries and kitchen sinks. This type of wastewater, referred to in this chapter as grey wastewater, has been estimated to account for about 73% of the volume of combined residential sewage (Hansen and Kjellerup 1994).

In general terms, grey wastewater has lower concentrations of organic matter, some nutrients (e.g. nitrogen) and microorganisms than combined wastewater. However, the concentration of phosphorus, heavy metals and xenobiotic organic pollutants are around the same levels. The main sources for these pollutants are chemical products such as laundry detergents, soap, shampoo, toothpaste and solvents. The soaps are alkali salts of long-chained fatty acids, while the detergents consist of surfactants as well as a number of other chemicals to improve their function, such as builders, bleaches, enzymes etc. Microorganisms may also be introduced into grey wastewater by hand-washing after using the toilet or changing nappies, baths, washing babies and small children, and from uncooked food products in the kitchen.

The major problem related to infiltration of grey wastewater is the risk of contamination of the soil and receiving waters (mainly groundwater). Due to the relatively high content of different types of pollutants, there is an obvious risk of contamination from both chemical compounds and microorganisms. The objective of this chapter is to evaluate this risk. In order to do this, it is necessary to have a good knowledge of the water to be infiltrated as well as knowledge of the soil characteristics and the processes determining the fate of the pollutants in soil and water. We have therefore chosen to focus on current knowledge of the characteristics of grey wastewater, and to briefly summarise the main processes determining the behaviour of pollutants in soil, using knowledge gained from wastewater infiltration plants.

18.2 CHARACTERSISTICS OF GREY WASTEWATER

The chemical content of grey wastewater depends on the source of the water; whether it is households or commercial laundries, and from what installation the water is drawn kitchen sink, bathroom, hand basin or laundry wash. Furthermore, lifestyle, customs and use of chemical products will be of importance. Biological growth within the transport system is another source of microorganisms and chemical substances. During storage and transportation of grey wastewater, biological growth may lead to increased concentrations of microorganisms including faecal coliforms. This may also cause new organic and inorganic compounds to be produced: metabolites from partly degraded chemicals present in the grey wastewater. The presence of nutrients such as phosphate, ammonium/nitrate and organic matter will promote this microbial growth. Chemical reactions could also take place during storage and transportation of grey wastewater, and thereby cause changes in the chemical composition of the water.

It is clear that there are a large number of compounds and microorganisms that could potentially be present in grey wastewater. It is therefore necessary to identify which microorganisms are actually present, in order to obtain an accurate picture of the wastewater, to be used to evaluate the risk of contamination for soil and receiving water. This characterisation has to include physical parameters, chemical compounds and microorganisms. Information needed can be obtained by combining the available data with a survey of chemical compounds and microorganisms that could theoretically be present. The content of chemicals can be based on the 'declaration of contents' present on the packages of chemical products as well as on industrial production statistics (Eriksson *et al.* 2001).

18.2.1 Physical parameters

Relevant physical parameters are temperature, colour, turbidity and content of suspended solids. Observed maximum and minimum values for these parameters in different types of grey wastewater are given in Table 18.1. Grey wastewater temperatures are often higher than the temperature of the water supply (18–38 °C) due to hot tap water used for personal hygiene and laundry.

High temperatures may be unfavourable since they favour microbial growth and could, in supersaturated waters, induce the precipitation of calcite, for example.

Table 18.1. Characteristics of different types of grey wastewaters

	Laundry	Bathroom	Kitchen sink
Physical properties	in mg/L	in mg/L	in mg/L
Colour (Pt/Co units)	50–70[A]	60–100[A]	
Suspended solids	79–280[ACG]	48–120[AG]	134–1300[FG]
TDS		126–175[E]	
Turbidity, NTU	14–296[ABC]	20–370[ABE]	
Temperature in °C	28–32°C (83–90 °F)	18–38[D]	
Chemical properties	in mg/L	in mg/L	in mg/L
pH	9.3–10[A]	5–8.1[ABDE]	6.3–7.4[F]
Electrical conductivity	190–1400[A]	82–20000[AD]	
Alkalinity	83–200 as ($CaCO_3$)[A]	24–136 (as $CaCO_3$)[AE]	20.0–340.0[F]
Hardness		18–52 (as $CaCO_3$)[E] 112–	
BOD_5	48–380[AC]	76–200[A]	
BOD_7	150[G]	170[G]	387–1000[G]
COD	375[G]	280[G] up to 8000 COD_{Cr}	26–1600[FG]
TOC	100–280[C]	15–225[E]	
Dissolved oxygen		0.4–4.6[D]	2.2–5.8[F]
Sulfate		12–40[B]	
Chloride (as Cl)	9.0–88[A]	3.1–18[AB]	
Oil and grease	8.0–35[A]	37–78[A]	
Nutrients			
Ammonia (NH_3–N)	<0.1–3.47[ABCG]	<0.1–25[ABDG]	0.2–23.0[FG]
Nitrate and nitrite* as N	0.10–0.31[A]	<0.05–0.20[A]	
Nitrate (NO_3–N)	0.4–0.6[C]	0–4.9[B]	
Phosphorus as PO_4	4.0–15[C]	4–35B[D]	0.4–4.7[F]
Nitrogen as total	1.0–40[A]	4.6–20[A]	15.4–42.8[F]
Tot–N	6–21[CG]	0.6–7.3[BG]	13–60[G]
Tot–P	0.062–57[ACG]	0.11–2.2[AG]	3.1–10[G]
Ground elements	in µg/l	in µg/L	in µg/L
Aluminium (Al)	<0.1–21[A]	<1.0[A]–1.7[G]	0.67–1.8[G]
Barium (Ba)	0.019[G]	0.032[G]	0.018–0.028[G]
Boron (B)	<0.1–0.5[A]	<0.1[A]	
Calcium (Ca)	3.9–14[AG]	3.5–21[AG]	13–30[G]
Magnesium (Mg)	1.1–3.1[AG]	1.4–6.6[AG]	3.3–7.3[G]
Potassium (K)	1.1–17[AG]	1.5–6.6[AG]	19–59[G]
Selenium (Se)	<0.001[A]	<0.001[A]	
Silicon (Si)	3.8–49[A]	3.2–4.1[A]	
Sodium (Na)	44–480[AG]	7.4–21[AG]	29–180[G]
Sulphur (S)	9.5–40[A]	0.14–3.3[AG]	0.12[G]

Heavy metals			
Arsenic (As)	0.001–<0.038[AG]	0.001[A]–<0.038[G]	<0.038[G]
Cadmium (Cd)	<0.01–<0.038[AG]	<0.01[AG]	<0.007[G]
Chromium (Cr)	<0.025[G]	0.036[G]	<0.025–0.072[G]
Cobalt (Co)	<0.012[G]	<0.012[G]	<0.013[G]
Copper (Cu)	<0.05–0.27[AG]	0.06–0.12[AG]	0.068–0.26[G]
Iron (Fe)	0.29–1.0[AG]	0.34–1.4[AG]	0.6–1.2[G]
Lead (Pb)	<0.063[G]	<0.063[G]	<0.062–0.14[G]
Manganese (Mn)	0.029[G]	0.061[G]	0.031–0.075[G]
Mercury (Hg)	0.0029[G]	<0.0003[G]	<0.0003–0.00047[G]
Nickel (Ni)	<0.025[G]	<0.025[G]	<0.025[G]
Silver (Ag)	0.002[G]	<0.002[G]	<0.002–0.013[G]
Zinc (Zn)	0.09–0.44[AG]	0.01–6.3[AG]	0.0007–1.8[G]
Xenobiotic organic compounds			
Detergents		Identified[D]	
Long chained fatty acids		Identified[E]	
Microbiological parameters			
Campylobacter spp.	n.d[A]	n.d[A]	
Candida albicans		n.d[E]	
Colifager PFU/mL	102 × 103[G]	388 × 103[G]	<3[G]
Cryptosporidia	n.d[A]	n.dA	
*Eschericia coli**	8.3x106[G]	3.2 × 107[G]	1.3 × 105–2.5×108[G]
Faecal coliforms*	9–1.6 × 104[ABC]	1–8 × 106[ABC]	
Faecal streptococci*	23 - 1.3 × 106[ABCG]	1–5.4 × 106[ACG]	5.15×103–5.5×108[G]
Giardia	n.d[A]	n.d[A]	
Heterotrophic bacteria*		up to 1.8 × 106[D]	
Pseudomonas aeruginosa		n.d[E]	
Salmonella spp.	n.dA	n.d[A]	
*Staphylococcus aureus***		1–5 × 105[E]	
Thermotolerant coli*	8.4 × 106 G	up to 8.9 × 106[DG]	0.2×106–3.75×108[G]
Total coliform*	56–8.9 × 105 ABC	70–2.8 × 107[ABCE]	
Total bacterial population (cfu/100 mL)		300–6.4 × 108[EB]	

= per 100 mL; ** = per mL.
[A]Christova-Boal *et al.* (1996); [B]Rose *et al.* (1991); [C]Siegrist *et al.* (1976); [D]Santala *et al.* (1998); [E]Burrows *et al.* (1991); [F]Shin *et al.* (1998); [G]Hargelius *et al.* (1995).
Identified: only qualititative analyses, no quantifications were performed

Measurements of turbidity and suspended solids give some information about the content of particles and colloids that could induce clogging of the soil pores, and thereby influence the permeability of the soil. The highest concentrations of particles and colloids are generally found in grey wastewater from kitchen sinks and washing machines (see Table 18.1). Although the amount of solids is generally lower than in combined wastewater, the risk of practical problems related to clogging should not be neglected. This is because the combination of colloids and surfactants (from detergents) could cause a stabilisation of the colloidal phase, due to sorption of the surfactants on the colloid surfaces. This prevention from agglomeration of the colloidal matter will reduce the efficiency of a pre-treatment step including settling of solid matter before the infiltration. However, this stabilisation does not mean that the colloids will not induce clogging of the soil matrix.

18.2.2 Chemical parameters

The chemical parameters of interest include a number of different sub-groups. Measured values of the general hydrochemical parameters, such as pH, alkalinity, hardness and electrical conductivity are given in Table 18.1. The table also covers standard wastewater parameters such as BOD, COD and oxygen content as well as the concentration of nutrients (N and P). Some data are available in the literature on heavy metals in grey wastewater, but information concerning the content of the heterogeneous group of xenobiotic organic compounds (XOCs) is thus far unavailable.

The pH of combined wastewater is generally between 7.5–8; indicating a small difference in pH between grey wastewater and combined wastewater. However, grey wastewater originating from laundry has found to be alkaline (pH range 9.3–10), which could influence the fate of chemical compounds, as well as the biological activity in the soil .

The effects of the infiltration of grey wastewater on soil pH and buffering capacity will be determined by the alkalinity, hardness and pH of the infiltrating water. However, the effect observed will also be influenced by the natural buffering capacity of the soil. The properties of the soil, regarding, for example, sorption capacity of pollutants, will change as a result of the infiltration. In addition, measurements of alkalinity and hardness will also provide information on the risk of clogging the soil. These parameters are largely determined by the quality of the drinking water, while the influence of chemicals added during the use of the water is generally limited.

The concentration of the wastewater components is a function of the amount of pollutants discharged during water usage and consumption (that is, water volume) related to this (Henze and Ledin 2001). Thus the contribution of traditional wastewater parameters (BOD and COD) to the quality of grey wastewater is less than for combined wastewater. COD values in the range 210–740 mg/l and BOD values in the range 150–530 mg/l have been reported for household wastewater (Henze *et al.* 2000; Henze and Ledin 2001), which is higher than values of grey wastewater (Table 18.1). There are two main reasons for these differences: faecal matter and toilet paper are not present in grey wastewater, and the water consumed in the production of grey wastewater is, in general, proportionally higher than for the production of combined wastewater.

The concentration of nitrogen is also lower in grey wastewater (in the range 0.6–60 mg/l) compared to household wastewater (in the range 20–80 mg/l; Henze *et al.* 2000; Henze and Ledin 2000) since no urine is present in grey wastewater. The total concentration of phosphorus varies considerably, because contributions from washing detergents, the primary source for P in grey wastewater, differs depending on the product used. The relatively low concentrations of N and P indicate that the risk of eutrophication of receiving waters due to the infiltration of grey wastewater is less than for combined wastewater. But it should also be kept in mind that the concentrations of these nutrients will have impact on the microbiological activity of the soil and will thereby contribute to the microbiological removal of organic matter from the infiltrating water in the soil.

The chemical composition of the suspended solids (organic or inorganic; mineral composition) plays a significant role with respect to the clogging potential. Material that can be volatilised due to transformation processes such as biological degradation or photo-oxidation, or that which can be dissolved due to biological and chemical reactions, can under appropriate conditions be removed from the soil matrix, thereby reducing the risk of clogging.

Plastic and metal piping both release compounds, such as XOCs and heavy metals, to the supply water and to the grey wastewater. The concentration of metals and some of the other elements observed (Table 18.1) will therefore depend on the contribution from three sources:

(1) Chemical products, resulting from water use
(2) The type of pipes used for transportation
(3) The quality of the water supply when it leaves the water works

Laundry wastewater was found to contain elevated amounts of sodium, compared to other types of grey wastewater. This may be due to the use of sodium as counter ion to several anionic surfactants used in powder laundry detergent or the use of sodium chloride in ion exchanges. Only relatively small quantities of heavy metals have been reported in grey wastewater, with one exception. Christova-Boal *et al.* (1996) found notably high levels of zinc in laundry wastewater (0.09–0.34 mg/l) and in bathroom wastewater (0.2–6.3 mg/l), compared to the concentrations found by other authors (which were in the range 0.01–0.44 mg/l). One reason for these relatively high values in bathroom wastewater could be the use of chlorine tablets, which had been used for disinfecting the grey wastewater before reuse for toilet flushing. The tablets were acidic and that probably caused leaching of zinc from the plumbing. Infiltration of grey wastewater containing heavy metals constitutes a potential risk for contamination of both the soil and the receiving waters, since most heavy metals are toxic to plants, animals and humans.

Only two studies, reporting on the presence of XOCs in grey wastewater have been found in the literature (Table 18.1). Santala *et al.* (1998) used a screening method with GC-MS and showed that the majority of the XOCs consisted of detergents. The other study, also describing the results from a GC-MS screening of shower wastewater, revealed that the even-numbered long chain fatty acids of C_{10} to C_{18} originating from soap were present (Burrows *et al.* 1991). These very limited results concerning the presence of XOCs in grey wastewater are not representative of the XOCs that could potentially be present. Thousands of different compounds have been mentioned in the literature and for combined wastewater at least 500 different XOCs have been identified and quantified (Eriksson *et al.* 2001).

One way of selecting the compounds that should be included in a monitoring programme or a risk evaluation study could be based on the data that are available on the statistics for the production of chemical household products. The ‘large volume chemicals’ i.e. substances that are produced and consumed in the highest quantities could be expected to give the highest concentrations and thereby cause the greatest effects when introduced into the environment. However, micropollutants should not be neglected since they could also be expected to cause damage, although their concentrations are low compared to the large volume chemicals.

Another feasible way to identify potentially relevant XOCs is to use the product information available on most common chemical household products. Eriksson *et al.* (2001) showed that at least 900 different substances or groups of substances could be present in grey wastewater, due to the use of chemical household products. The study was largely based on the information available in the declaration of contents present on the different types of common household products, covering products from shampoos and toothpaste to washing powders. Eriksson *et al.* (2001) divided the XOCs into eleven different groups according to their purpose. The major compounds in the list were surfactants used in detergents, dishwashing liquids and hygiene products i.e. non-ionic, anionic and amphoteric surfactants. Other large groups were solvents and preservatives. Solvents are added in order to improve the solubility of organic compounds such as fragrances in household chemicals, which is necessary since the main solvent is water. Preservatives are added to the vast majority of household chemicals to prevent microbiological growth in the product, as well as to act as disinfectants when they are used.

In order to identify those XOCs that could constitute a risk for pollution of soil and receiving waters, it is not enough to find those compounds that potentially could be present. It is also necessary to include some information about their effects in the environment. Eriksson *et al.* (2001) applied a method that usually is employed in environmental risk assessment of new chemical compounds. The evaluation carried out was based on the classification of the XOCs with respect to toxicity, bioaccumulation and biodegradation (see van Leeuwen and Hermens 1995). The compounds were divided into eight different groups according to this classification, and indicate how environmentally hazardous the compounds are estimated to be.

Out of the approximately 900 substances identified as being potentially present in household chemicals, 10% were categorised as priority pollutants i.e. were placed in the first three groups with the highest environmental impact. Among these were anionic, nonionic, cationic and amphoteric surfactants, including compounds or groups of compounds such as linear alkylbenzene sulphonates (LAS), nonylphenol- and other alkylphenol-ethoxylates. Different preservatives and softeners were also among the prioritised compounds. The softeners are mainly esters of phthalic acid, e.g. di-ethyl-hexyl phthalate (DEHP). It should, however, be mentioned that this type of classification is hampered by the limited information available about toxicity, bioaccumulation and biodegradation for a large number of the relevant compounds. The number of compounds listed in the three priority groups is not final and could increase dramatically if more information becomes available.

18.2.3 Microorganisms

The major risk associated with the presence of microorganisms in grey wastewater is the risk of infection due to direct contact with the water during the infiltration, where the application on/in the soil is the most critical moment. There is also a potential risk of the contamination of soil used for gardening or agriculture or receiving waters used as drinking water supplies. There is generally very little known about the presence of microorganisms in grey wastewater (Table 18.1). Four types of pathogens may be present: viruses, bacteria, protozoa and intestinal parasites. It can, however, be expected, when evaluating microbiological parameters, that microbial populations of faecal origin in grey wastewater cause the major health risk.

Studies have shown that grey wastewater originating from bathrooms contained up to 3.2×10^7/100 mL of *E.coli*, 1–8 $\times 10^6$/100 mL faecal coliforms and 7–2.8 $\times 10^7$/100 mL total coliforms. These concentrations were slightly higher than those measured in grey wastewater from laundry (Table 18.1). Burrows *et al.* (1991) analysed grey wastewater from showers in some US military facilities. In that study *Candida albicans*, *Pseudomonas aeruginosa* and *Staphylococcus aureus* were included, since those microorganisms are commonly found in the human mouths, noses and throats. Neither *C. albicans* nor *P. aeruginosa* were observed. In another study, *Campylobacter spp.*, *Cryptosporidia*, *Giardia* and *Salmonella spp.* were tested for; however, none was detected (Christova-Boal *et al.* 1996).

Kitchen wastewater may contain several types of microorganisms due to contamination from uncooked food, such as raw meat. No analyses of faecal coliforms or total coliforms in grey wastewater from kitchen sinks have been found in the literature. *E.coli* has been observed in the range of 0.1×10^6 to 2.5×10^8 /100 mL, while a thermotolerant coli has been found in the range of 0.2×10^6 to 3.8×10^8 /100 mL (Hargelius *et al.* 1995).

18.3 THE FATE OF POLLUTANTS IN SOIL AND WATER

18.3.1 Main processes

Evaluating the risk of soil and receiving water pollution due to the infiltration of grey wastewater requires knowledge of the fate of the chemical compounds and microorganisms in soil and water, including the residence times and the transfer factors to the adjacent compartments (air, water and sediments). This in turn requires an understanding of the reactions and transport mechanisms to which pollutants are subjected. The main processes determining the fate of pollutants in soils and waters are sorption, volatilisation and degradation (Connell 1997).

Sorption is a general term that refers to the binding of a pollutant to a solid constituent present either in the soil matrix or suspended in water. There are numerous physical-chemical mechanisms determining the sorption behaviour of a pollutant, and the detailed description is beyond the scope of this chapter (see Stumm and Morgan 1996).

Assuming that the equilibrium distribution of a contaminant between the solid and aqueous phases can be described by a linear isotherm, the distribution is given by:

$$C_s = K_d \cdot C_w \qquad (18.1)$$

where C_w and C_s are the contaminant concentrations in the water and solid phase, respectively, and K_d is the partition coefficient. Values for K_d are generally determined experimentally. For hydrophobic XOCs they can be estimated from the fraction of organic carbon in the solid phase (f_{oc}) and the organic carbon–water partition coefficient (K_{oc}) for the XOC, since sorption behaviour is largely determined by hydrophobic interactions between the contaminant and organic matter in the solid phase. This is done according to:

$$K_d = f_{oc} \cdot K_{oc} \qquad (18.2)$$

Values for K_{oc} can be found in the literature, or can be estimated from experimentally determined expressions for the relationship between K_{oc} and the octanol–water partition coefficient (K_{ow}), which in turn can be found in the literature for a large number of compounds (see, for example, Fisk 1995).

Hydrochemical conditions such as pH and ionic strength will largely determine the sorption behaviour of heavy metals and hydrophilic XOCs. A relationship between the content of organic matter in the solid phase and the sorption behaviour of this type of pollutants has been recognised, but it is not as unambiguous as for the hydrophobic XOCs. It has also been shown that the contents of complexing ligands and competing cations, as well as the mineralogical composition of the solid phase, have an influence on the sorption of heavy metals (see Ledin 1993).

K_d-values can be used to estimate the relative mobility of pollutants in both soils and waters. In soils this transport in general is referred to as *leaching*. It should be noted that sorption of a pollutant to a solid phase does not necessarily slow down its mobility. If the solid phase is mobile, which could be the case for small particles (colloids) in both soils and waters, it will follow the flow of water. These colloidal particles can become immobile due to attachment to the soil matrix (in soils) or undergo settling (in waters) and thereby act as a sink for the sorbed contaminant. Figure 18.1(a) and (b) illustrate some of the processes and hydrochemical parameters that are usually included in the term ‘sorption’.

XOCs that can be volatilised can transfer into the gas state from a solid or a liquid state and migrate considerable distances as vapours through a soil. *Volatilisation* results in a decrease in the concentration of volatile XOCs from both soils and waters, and a large part of the pollutant mass can be transferred to the atmosphere. Equilibrium partitioning between air and soil and water, respectively, is represented by Henry’s law:

$$K_H = C_w / p_i \qquad (18.3)$$

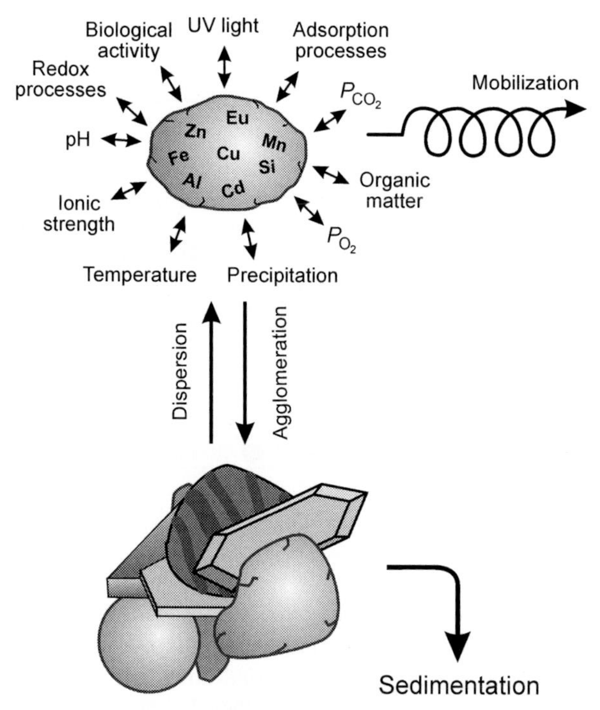

Figure 18.1(a). Some processes affecting the distribution of pollutants between mobile and stationary phases in water (from Ledin 1993).

where K_H is the Henry law constant, p_i is the partial pressure of compound i in air, and C_w is the concentration of the compound in water. Values for K_H can generally be found in the literature, and can be used for evaluating the potential for removal of a specific compound from the contaminated soil or water.

The term *degradation* refers to the transformation of a compound into other compounds, which under optimised conditions will lead to complete mineralisation of the compound. Degradation could occur either abiotically (chemically and photochemically) or biotically, where the latter occurs with the involvement of microorganisms. Abiotic processes that may be important in soil and water include hydrolysis, phytolysis and reduction/oxidation processes. Microbially mediated degradation, which can also be called biodegradation, is frequently the most significant degradation mechanism for XOCs in soil.

The degradation rate of pollutants in soils and waters is determined by several factors, where the properties of the contaminant (chemical structure, concentration, and so on) are among the most important. Environmental factors

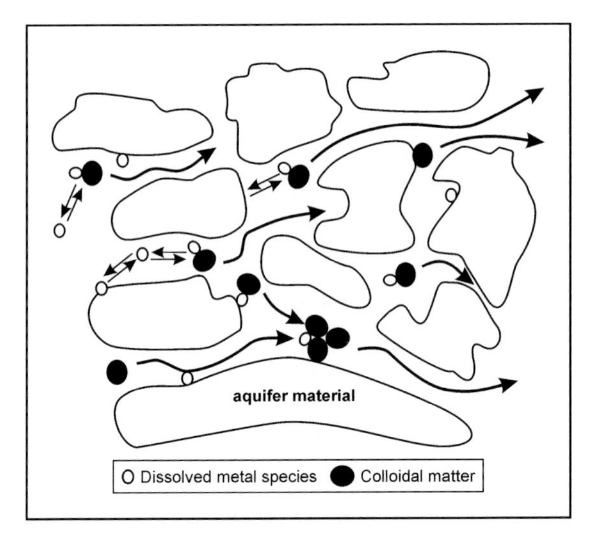

Figure 18.1(b). Some processes affecting the distribution of pollutants between mobile and stationary phases in soil (modified from Buffle and Leppard 1995).

such as temperature, soil moisture content, pH, redox conditions and the presence of nutrients are also important, as well as the microbial density and capacity to degrade the pollutants. Redox conditions will determine the type of biodegradation that occurs, which will either be aerobic degradation with oxygen as the electron acceptor, or anaerobic with, for example, nitrate or sulphate as electron acceptors. A period of adaptation, known as a lag phase, is commonly observed during microbiological degradation. However, when microorganisms at a given location, such as an infiltration plant, are repeatedly exposed to an organic contaminant, adaptation can lead to a phenomenon known as enhanced biodegradation. Standardised methods for estimating biodegradation of XOCs in different environments (soil, groundwater, surface water; aerobic and anaerobic conditions) are under development (EU 1995). The results obtained could be used to evaluate the relative degradability of pollutants, for instance during infiltration.

Microorganisms themselves will be eliminated through numerous interacting reactions including physical, chemical and biological processes. However, physical removal (filtration) is in most cases the dominating process.

Numerous simulation models exist that describe the transport of substances in soil and water. The models range from simple one-dimensional models to much more complicated three-dimensional flow models, including non-equilibrium sorption phenomena. The main components in all models are, however, a description of the hydrodynamics/flow conditions for water, transport mechanisms and conversion processes for substances, where the number and complexity of the equations describing these processes depend on the type of model. To the best of our knowledge, no one has successfully applied simulation models to evaluate the risk of contamination of soil and receiving waters with respect to heavy metals and XOCs due to the infiltration of grey wastewater.

18.3.2 Experience obtained from wastewater infiltration plants

Infiltration through soil is one of the oldest techniques for purifying combined wastewater. Several studies have shown that this is a simple and effective technique. The microorganisms that are attached to the soil particles use the wastewater as a nutrient, and thereby form the basis for the treatment (see Droste 1997). Generally work has concentrated on the capacity to remove suspended solids, BOD, COD, nutrients and microorganisms, while studies focusing on the removal of heavy metals and XOCs are more scarce.

The type of soil (e.g. sandy, loamy or clay) involved will influence the fate of the contaminants to a great extent. However, the following will focus on mineral soils, since these soil types are the most common ones.

Carré and Dufils (1991) evaluated the efficiency of an infiltration basin, which was aimed on treating combined wastewater. The study showed that the water transferred a part of the contamination into the unconfined groundwater, which lay at a depth of 6 m below the bottom of the infiltration basins. Elevated concentrations of a unspecified anionic detergent and boron were observed in the groundwater, while the removal of total coliforms, faecal coliforms and faecal streptococci was satisfactory.

Fittschen and Niemczynowicz (1997) tested a facility including a three-chamber sedimentation tank followed by a root-zone facility, a sand filter and finally an artificial pond for treatment of mixed grey and yellow (urine) wastewater. This is one of the few studies focusing on infiltration as a method for treatment of grey wastewater. Among potential pollutants present in the grey wastewater only the efficiency in removing some heavy metals (Cu, Pb and Zn) were discussed. The results showed reductions in the concentrations of both Cu and Zn after each step in the treatment, where a concentration below the detection limit (Cu 0.02 mg/l and Zn 0.005 mg/l) was obtained after the sand filter. The concentration of Pb was below the detection limit (0.05 mg/l) after the sedimentation tank. The study also illustrated that infiltration through a sand filter reduced the concentrations of thermostable coliform bacteria to satisfactory concentrations.

18.3.3 Experiences obtained from other infiltration systems

It can be concluded that experiences discussed in the literature about the infiltration of wastewater and, in particular, the infiltration of grey wastewater with respect to the behaviour of the pollutants is scarce. Additional knowledge could be obtained from similar 'systems' such as infiltration from rivers to groundwater, infiltration of road run-off and percolation of rainwater through landfills.

Ding *et al.* (1999) studied the transport of some selected parameters in the Santa Ana River and the underlying aquifer, which is recharged by the river. The river contained a high proportion of tertiary treated wastewater effluents, and the authors selected the following four specific wastewater indicator compounds to be included in their monitoring programme: ethylene diaminetetraacetic acid (EDTA), nitrilotriacetic acid (NTA), a naphthalene dicarboxylate (NDC) isomer and alkylphenol polyethoxy carboxylates (APECs). Both NTA and the APECs appeared to be attenuated significantly during the infiltration of polluted river water and during the groundwater transport. The attenuation of NTA and APECs could be explained by biotransformation of the compounds during the infiltration. The less biodegradable compounds, EDTA and NDC were, however, detected in wells 1.8 and 2.7 km down-gradient with little apparent attenuation.

The behaviour of various persistent metabolites derived from the group of non-ionic surfactants nonylphenol polyethoxylates (NPEOs) has been studied during the infiltration of river water to groundwater at two Swiss field sites (Ahel *et al.* 1996). The study showed that nonylphenol (NP), nonylphenol monoethoxylate (NP1EO) and nonylphenol diethoxylate (NP2EO) were removed during the infiltration to groundwater, while the reduction in the concentrations of the metabolites from the degradation of NPEOs; nonylphenoxy acetic acid (NP1EC) and nonylphenoxy(ethoxy) acetic acid (NP2EC) were limited. The data suggested that low temperatures, which previal in some areas in winter, significantly reduced the elimination efficiency of NP, while the behaviour of NP1EO and NP2EO were less affected by temperature.

Infiltration of road run-off could be seen as an analogue to the infiltration of grey wastewater. Pollutants focused on have mainly been heavy metals and polyaromatic hydrocarbons (PAHs), and studies have indicated that the potential for contamination of groundwater is limited (see Mikkelsen *et al.* 1996). The attenuation observed in the concentrations of both heavy metals and PAHs with depth in the soil profile is believed to be a result of sorption to the soil matrix.

Leaching of highly polluted water from landfills to soil and groundwater could be seen as another parallel to infiltration of grey wastewater. The major difference is that the wastes are deposited on the soil and the infiltration of polluted water is due to percolation of rainwater through the wastes. Current knowledge of the behaviour of some pollutants in landfill leachate plumes has recently been summarised by Christensen *et al.* (2001). The study states that heavy metals do not seem to constitute a significant risk for pollution of receiving waters, partly because the heavy metal concentrations in the leachate often are low, and partly because of the strong attenuation by sorption and precipitation. The latter process is mainly due to the anaerobic conditions in the leachate-affected aquifers studied, and is therefore probably irrelevant in infiltration plants. Several XOCs have been found to be degradable in the leachate-contaminated groundwater, although there are anaerobic conditions. It should be kept in mind that most studies have based their conclusions on the observation that the compounds have disappeared, but this does not necessarily mean that the compound is fully degraded. It could, as in the example with NPEOs and NPECs, be a result of converting the compound into a metabolite. These metabolites are generally more hydrophilic than the original compounds and could therefore be more mobile in soil.

18.4 REFERENCES

Ahel, M., Schaffer, C. and Giger, W. (1996) Behaviour of alkylphenol polyethoxylate surfactants in the aquatic environment – III. Occurrence and elimination of their persistent metabolites during infiltration of river water to groundwater. *Wat. Res.* **31**(1) 37–46.

Buffle, J. and Leppard, G. (1995) Characterization of aquatic colloids and macromolecules. 1. Structure and behaviour of colloidal material. *Environ. Sci. Tech.* **29**(9), 2169–2184.

Burrows, W.D., Schmidt, M.O., Carnevale, R.M. and Schaub, S.A. (1991) Non-potable reuse: Development of health criteria and technologies for shower water recycle. *Wat. Sci. Tech.* **24** (9), 81–88.

Carré, J. and Dufils, J (1991) Wastewater treatment by infiltration basins: usefulness and limits – Sewage plant in Creances (France). *Wat. Sci. Tech.* **24** (9), 287–293.

Christensen, T.H., Kjeldsen, P., Bjerg, P.L., Jensen D.L., Christensen, J.B., Baun, A., Albrechtsen, H.-J. and Gorm, H. (2001) Biogeochemistry of landfill leachate plumes. *Applied Geochemistry* (in press).

Christova-Boal, D., Eden, R.E. and McFarlane, S. (1996) An investigation into greywater reuse for urban residential properties. *Desalination* **106**, 391–397.

Connell, D.W. (1997) *Basic Concepts of Environmental Chemistry*. CRC Press, Boca Raton, FL.

Ding, W.H., Wu, J., Semadeni, M. and Reinhard, M. (1999) Occurrence and behaviour of wastewater indicators in the Santa Ana River and the underlying aquifers. *Chemosphere*, **39**(11), 1781–1794.

Droste, R.L. (1997) *Theory and Practice of Water and Wastewater Treatment*. John Wiley & Sons, New York.

Eriksson, E., Auffarth, K., Henze, M. and Ledin, A. (2001) Characteristics of grey wastewater (submitted for publication).

EU (1995) Technical Guidance Documents in Support of the Commission Regulation (EC) No. 1488/94 on Risk Assessment for Existing Substances in Accordance with Council Regulation (ECC) No. 793/93, Brussels.

Fisk, P.R. (1995) Estimation of physicochemical properties: theoretical and experimental approached. In *Environmental Behaviour of Agrochemicals* (ed. T.R. Roberts and P.C. Kearney), Vol 9 of *Progress in Pesticide Biochemistry and Toxicology*, John Wiley & Sons, Chichesters.

Fittschen, I. and Niemczynowicz, J. (1997) Experiences with dry sanitation and greywater treatment in the eco-village Toarp, Sweden. *Wat. Sci. Tech.* **35**(9), 161–170.

Hansen, A.M. and Kjellerup, M. (1994) Vandbesparende foranstaltninger. Teknisk Forlag. (In Danish.)

Hargelius, K., Holmstrand, O. and Karlsson, L. (1995) Hushållsspillvatten. Framtagande av nya schablonvärden för BDT-vatten. In: Vad innehåller avlopp från hushåll? Näring och metaller i urin och fekalier samt i disk-, tvätt-, bad- & duschvatten, Naturvårdsverket, Stockholm. (In Swedish.)

Henze, M. and Ledin, A. (2001) Types, characteristics and quantities of combined domestic wastewaters. Chapter 4 in this book.

Henze, M., Harremoës, P., la Cour Jensen, J. and Arvin, E. (2000) Wastewater Treatment, Biological and Chemical, 3rd edition, Springer-Verlag, Berlin.

Ledin, A (1993) Colloidal Carrier Substances. Properties and impact on trace metals distribution in natural waters. (Diss). Linköping Studies in Art and Science, Linköping University, Sweden.

Mikkelsen, P., Häflinger, M., Ochs, M., Tjell, J.C., Jacobsen, P. and Boller, M. (1996) Experimental assessment of soil and groundwater contamination from two old infiltration systems for road run-off in Switzerland. *Sci. Tot. Environ.* **189/190**, 341–347.

Rose, J.B., Sun, G., Gerba, C.P. and Sinclair, N.A. (1991) Microbial quality and persistence of enteric pathogens in greywater from various household sources. *Wat.Res.* **25**(1), 37–42.

Santala, E., Uotila, J., Zaitsev, G., Alasiurua, R., Tikka, R. and Tengvall, J. (1998) Microbiological greywater treatment and recycling in an apartment building. In conference proceedings: Advanced Wastewater Treatment, Recycling and Reuse, 14–16 September, Milan, 319–324.

Shin, H.-S., Lee, S.-M., Seo, I.-S. Kim, G.-O., Lim, K.-H. and Song, J.-S. (1998) Pilot-scale SBR and MF operation for the removal of organic and nitrogen compounds from greywater. *Wat.Sci.Tech.* **38**(6), 79–88.

Siegrist, R., Witt, M. and Boyle, W.C. (1976) Characteristics of rural household wastewater. *Journal of the Environmental Engineering Division* **102**(EE3), 533–548.

Stumm, W. and Morgan J.J. (1996) *Aquatic Chemistry*, 3rd edition. John Wiley & Sons, New York.

van Leeuwen, C.J. and Hermens J.L.M (ed.) (1995) *Risk Assessment of Chemicals: An Introduction*. Kluwer Academic Publishers, Dordrecht.

19

Potentials of water reuse in houses and other buildings

P.M.J. Terpstra

19.1 INTRODUCTION

The primary motive for the study of alternative water systems is the search for a sustainable society. This search implies mankind's struggle to plan our society in such a way that future generations can be guaranteed a sustainable environment. Brundtland's definition reads: 'meeting the needs and aspirations of the present generation without compromising the ability of future generations to meet their needs' (Brundtland 1987).

Natural resources and the environment should be available for future generations. The general principle of sustainability is put into practice to counteract three effects: running out of resources; pollution of ecosystems; and disrupting natural systems by the destruction of land and biodiversity.

Edited by P. Lens, G. Zeeman and G. Lettinga. ISBN: 1 900222 47 7

Present use of water by our society is not sustainable in the sense that we take more clean water from the ecosystem than is naturally replenished and, moreover, discharge too much polluted water into the ecosystem. Part of this problem is caused by domestic water use (Hedberg 1995).

In this chapter various options for domestic water use systems are presented, which could lead to a more efficient and sustainable use of water for domestic and institutional purposes in urban areas, in a future sustainable society. These systems can be regarded as alternatives to present water supply and use, and are designed for a hypothetical sustainable society in which the pollution issue in general has been solved. In such a society rainwater would be pure and of drinking water quality (Meijer and van Leeuwen 1996).

Water use systems should enable an appreciable reduction in the use of domestic (drinking) water and domestic waste water which enables an effective purification and reuse of waste products. To fit in with a sustainable living environment, these systems should meet the following requirements (Worp and Don 1996):

- preservation of the water quality focused on reuse and recycling
- no accumulation of contaminants (in the soil, surface water and ground water)
- no damage to ecosystems by drought
- the water supply should satisfy the user's minimum requirements as to quantity and quality
- there should be no substantial shift of environmental impact from the water system to other areas of the ecosystem (e.g. energy consumption, CO2 production and use of raw materials)
- The above criteria have been used as the framework to define alternative options to the present water supply system

19.2 HOUSEHOLD WATER USE

In a household, water is used for a number of processes including washing clothes, preparing food, heating, personal hygiene, cleaning, and so on. Purposes and applications can be split into four main categories: personal hygiene, toilet flushing, consumption and cleaning. Domestic water use in the Netherlands increased over the years to 1992, when it stabilised. In 1996, the average amount of drinking water per member of a household was 134 litres per day (NIPO 1996), of which about 5% is rinse water and water for watering gardens (Witteveen and Bos 1994). Table 19.1 shows domestic water use in the Netherlands in 1992 split up according to the various uses.

Table 19.1. Household water in 1992 (NIPO 1992)

Type of use	Per event (l)	Use frequency (event d^{-1})	Degree of market penetration [1] (%)	Use ($l\,p^{-1}d^{-1}$)
Personal hygiene				
Shower	63.5	0.63	99	39.5
Bath	120	0.17	39	8.0
Washbasin	4	0.97	95	3.7
Toilet flushing	7.2	5.94	100	42.7
Food preparation				2.6
Cleaning				
Washing, hand	40	0.06	100	2.5
Washing, machine	100	0.25	94	23.2
Washing up, hand	11.2	0.78	100	8.8
Washing up, machine	25	0.22	13	0.7
Other				3.3
Total				**135**

[1] Percentage of households that own the appliance or equipment.

Water use does not only occur by virtue of the water flowing through the household system via the water mains. Water use is also connected with the use of goods and appliances that do not directly use water, and even with energy. Indirect water use is also an essential part of domestic water use. Water is used in the production of almost all household articles, foodstuffs, equipment and also for the generation of electricity. However, even though this indirect water consumption is an essential part of domestic water consumption, the impact of these indirect water flows is beyond the scope of this chapter.

19.2.1 Sources at household level

The current most important source of household water is the drinking water that is supplied via the water mains. It is expected that in the future recycled water will also be supplied via a different water supply system. It is most likely that production of this water will be small-scale and of a quality not fit for drinking. In addition to mains water, rainwater can also be used for household purposes. Assuming an average rainfall of 700 mm/year, a roof surface of 60 m^2 and a recuperation of 75 % (a reasonable estimation) some 31.5 m^3 per year could be collected. This would mean 36 litres per person per day ($l\ p^{-1}d^{-1}$) for an average household of 2.4 persons. It should be noted here that in a sustainable future for which such a water use system is meant , rainwater is assumed to be of drinking water quality (Meijer and van Leeuwen 1996). Furthermore, a household using a tumble dryer (with a current market penetration level of about 50%) could also produce another 5 litres of condensation water per week (based on a use frequency of 3 cotton cycles per week, a load of 3.3 kg and a fabric moisture content of 50%).

19.2.2 Water saving options

Water is used in the home for a variety of purposes. By increasing the efficiency of these processes water consumption can be decreased. Several water-saving options are being considered or have been adopted over recent decades (Zott 1984; Stamminger 1993), examples of which include the reuse of water in washing machines and dishwashers, decreased rinsing in washing machines and dishwashers, economy shower heads, toilets with flushing interception, etc.. Other saving options are at system level, such as the use of a second system of water mains with lower water quality and the utilisation of rain water (Kilian *et al.* 1996).

19.3 INTRINSIC WATER QUALITY AND WATER USE

Although the consumption rate of energy in our society is not sustainable, energy is essentially not consumed. After all, in a closed system the total amount of energy is constant (the first law of thermodynamics). What does occur is that when energy is used the exergy decreases. Exergy can be broadly interpreted as a measure of the number of functions for which the energy can be used; or in other words the quality of the energy. In this sense, electrical energy and fossil fuels are high quality energy sources and have a high exergy. Thermal energy with a temperature barely above that of the surrounding environment is a low quality energy source with little exergy. High quality energy can easily be converted into energy of a lower quality, but the reverse process cannot take place spontaneously (the second law of thermodynamics).

Similar rules, alas not thermodynamically proven, apply to water and the use of water. Just like energy, water is not permanently consumed and we can also make a distinction between high and low quality water. In our case the quality is determined by the quantity and the nature of the contamination. Pure water can thus be interpreted as that of the highest quality. As the purity decreases so do the quality and the number of potential applications. If through the use of water the loss in quality per individual domestic application is reduced to the bare minimum, it means that the intrinsic quality of the water is optimally exploited. This statement plays a central role in this chapter. The following examples illustrate how the loss of quality through use can be reduced. For this purpose four water quality classes have been defined in Table 19.2: clean water (class I), slightly polluted water (class II), polluted water (class III) and heavily polluted water (class IV).

Table 19.2. Water quality

Quality class	Type	Pollution
Class I	clean	no pollution
Class II	slightly polluted	low concentration of dissolved chemicals, small numbers of microorganisms
Class III	polluted	ditto class II + low concentration of finely suspended solids
Class IV	heavily polluted	high concentration of dissolved chemicals and bulk refuse

Note. Defining water qualities with more precision is not within the scope of this chapter. The specification is required to be geared to the purpose of the technical system and the purpose of water utilisation.

On reviewing the different domestic processes it would appear that not all processes need tap water (class I). For instance; much water is used up by a household for toilet flushing: approximately 30% of the household total. Currently class I water is used for this purpose. The quality of the drained water is class IV water. Essentially, class III water would be perfectly adequate for the flushing of toilets. In that way, less water quality would be lost than occurs in the present situation: the difference would be between class III and class IV instead of between class I and class IV.

An average of 30% of domestic water consumption is used for washing clothes and dishwashing. Particularly noteworthy is washing machines, which use 20% of the total water consumption (NIPO 1992).

The machine laundering process can quite simply be described as a separation of materials: water is the carrying agent that separates textile and dirt. The process goes through four to five phases. In each phase water is taken in and a mixture of water and dirt is discharged. The quality of the water that is discharged increases as the washing process progresses: the initial discharge is of a class III quality and the final discharge class II. If class II water were to be used for the first phases and class I for the later phases, the quality of the process would scarcely be affected. Technically speaking and from a hygienic point of view this does not detract from the washing efficiency (Terpstra 2000) If the measure described in the above example is applied, the total loss of intrinsic water quality would be less than it is currently, in which all the input class I water is converted into output class IV. Both situations are shown in Figure 19.1.

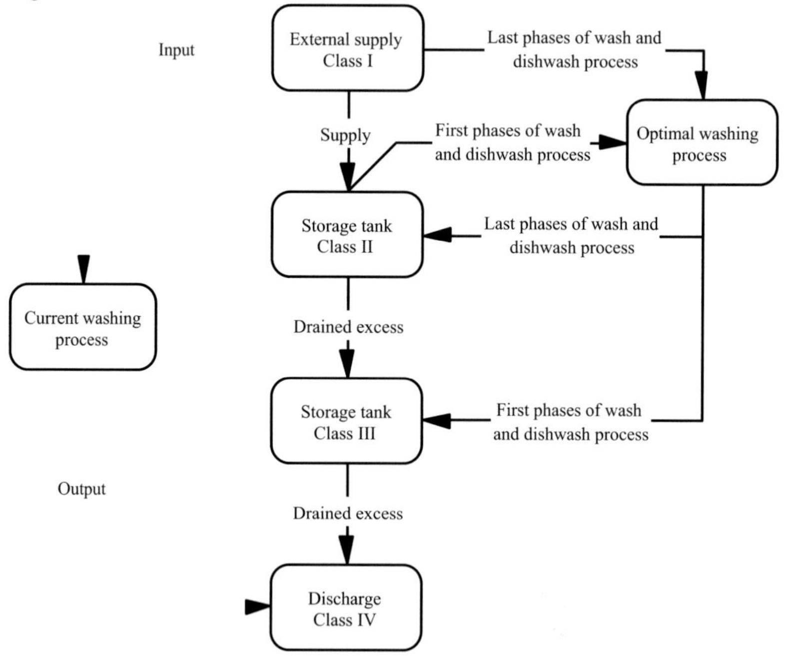

Figure 19.1. The washing process using water of different qualities.

The previous example shows that, for a particular process, the loss of intrinsic water quality can be minimised if the quality of the water used is taken into account. Here, the quality of input and output water differs. By integrating the various processes requiring water and by optimally balancing the quality of the input and output water, a model of a water use system can be formulated that would achieve a reduction in the consumption of class I water from the mains supply. To build this model it is essential to have an insight into the capacities and qualities involved in the different domestic processes.

19.3.1 Quality and quantity of the various domestic water flows

To develop a model of an optimal water use system it is vital to have an insight, both as to quality and as to quantity, into the water flows of the individual processes. Table 19.3 shows the chemical oxygen demand (COD), the Kjeldal nitrogen (N_{kj}) and the phosphate (P_{tot}) content of some of the important sources of the waste water discharged from a household. These data can be considered as indicators of water quality.

Table 19.3. Contamination of wastewater (g $p^{-1}d^{-1}$) and water flow classification (STORA 1985)

Facilities	COD	N_{kj}	P_{tot}	Quality class input/output
Toilet	51.1	10	1.4	III/IV
Shower/bath	1.2	± 0	± 0	I/II
Washbasin	7.2	0.06	± 0	I/II
Kitchen	20.5	0.3	± 0	I/IV
Machine washing	20.7	0.64	± 0	I–II/II–III
Total	100.7	11	1.8	

Based on this and on the efficiency and hygiene requirements of the processes, quality classes have been chosen for the water supplied and discharged. Note that this is a tentative categorisation that needs further scientific elaboration. The amount of water taken in and discharged from each quality class has been calculated as shown in Table 19.4.

Table 19.4. Water quality and use for domestic processes

Function/apparatus	Use * (l $p^{-1}d^{-1}$)	Input (l)	Class	Output (l)	Class
Personal hygiene					
Shower	40.0	63.5	I	63.5	II
Bath	20.4	120	I	120	II
Washbasin	3.9	4	I	4	III
Toilet flushing	42.8	7.2	III	7.2	IV
Food preparation	2.6	2.6	I	?	IV
Cleaning					
Washing, hand	2.4	10/30	I/II	20/20	II/III
Washing, machine	25.0	20/80	I/II	50/50	II/III
Washing up, hand	8.7	11.2	I	11.2	IV
Washing up, machine	5.5	10/15	I/II	15/10	II/III
Other	3.3	3.3	I	?	

19.4 MODELS FOR SUSTAINABLE WATER SUPPLY SYSTEMS

19.4.1 Model based on the current water supply system

An integrated model has been developed based on the current mains water supply system and on the given water flows and quality classes (Figure 19.2). In this model the water use system is entirely localised within the household: the boundaries of the system lie around the household. The effects of changes in the water use system in this subsystem can be deduced from the quality and quantity of the incoming and outgoing water. The model is based on the following assumptions:

- mains water is of drinking water quality
- buffers for class II and III water are not restrictive
- the household has the given facilities and equipment at its disposal
- consumption and frequency of consumption are in accordance with the NIPO data of 1992 (Table 19.1)

If all the processes mentioned are run parallel to each other, then the total amount of water consumed would currently be 154.6 l $p^{-1}d^{-1}$. Figure 19.2 shows the water flow that would be achieved with a water use system in accordance with the model. Here, the total water consumption is reduced to 86.7 l $p^{-1}d^{-1}$, a reduction of 44%.

When the average inflow of a reservoir is greater that the outflow, water is discharged unused into a reservoir containing lower quality water. This controlled discharge between the various reservoirs is a measure of the unused water quality.

If the household does not have a bath, then the water use system changes radically. As can be derived from Table 19.4 and Figure 19.2, in this case the mains water consumption is reduced by 20.4 l $p^{-1}d^{-1}$ to 134.2 l $p^{-1}d^{-1}$ for the current situation and 66.3 l $p^{-1}d^{-1}$ with the reuse system (a reduction of 51%). In the water use system without a bath the controlled water discharge is considerably less! In other words, the intrinsic quality of the supplied water is used more effectively.

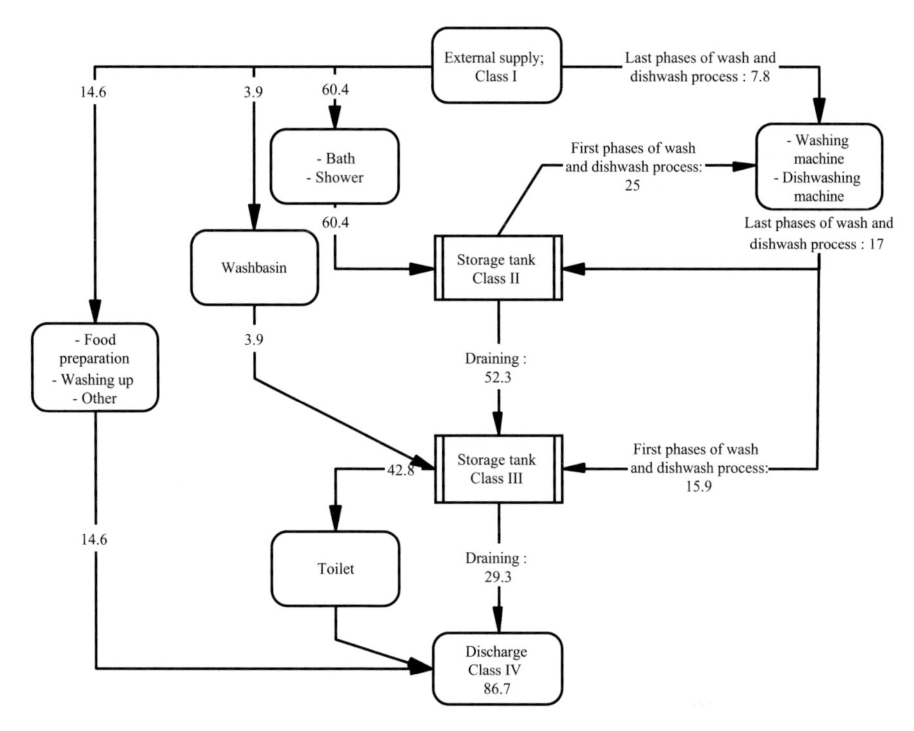

Figure 19.2. Water use system based on the current water supply system. Values represent the water consumption per person per day (l $p^{-1}d^{-1}$).

19.4.2 Model using rainwater

In a sustainable society serving as a basis for this study, according to Meijer and van Leeuwen rainwater may be clean and fit for drinking (Meijer and van Leeuwen 1996). If this rainwater is used in the household a new picture will emerge. The starting criteria for the water use system in this situation will be the same as those in the previous model with the exception of the following:

- the availability of 36 l p–1d–1 rainwater fit for drinking for the water use system
- the mains water will no longer be fit for drinking

The water quality and quantity for this set-up are shown in Figure 19.3. Here, the consumption of mains water has dropped to 50.7 l $p^{-1}d^{-1}$ (a 67% reduction).

In households without a bath the tap water consumption in this configuration is reduced even further to 30.3 l $p^{-1}d^{-1}$ (a 77 % reduction).

19.5 ESTIMATION OF WATER SAVING WITH DYNAMIC SIMULATION

In the previous examples the water-saving potential of the various models is estimated on the basis of an average water use per day for each process. Moreover, the capacity of the water storage tank is assumed to be such that it would never restrict water reuse.

In real life the household water use fluctuates substantially during the day. A number of processes, such as toilet flushing, use of the washbasin and showering, occur very often while other processes occur occasionally. Watering the garden and washing the car, which both require large quantities of tap water, are examples of such occasional events. And during a storm rainwater can come down in large quantities in a short period of time. Moreover the occurrence of household processes is not spread out evenly over time and the actual starting moment of each single process is a matter of coincidence (stochastic process).

In normal household situations it will not be possible to install storage tanks that can store enough water to be available for all circumstances. Because of cost and space restrictions, tanks have to be smaller and, because of this compromise, the water saving that was estimated earlier will not be achieved.

Because of the above issues the estimation of the water-saving potential of the different models is probably too optimistic. A more realistic estimation can be achieved using a dynamic computer simulation. In that case, variations in water use and in rainwater supply, as well as the effect of limited storage capacity, can be taken into account.

To find the saving potential under conditions that reflect the dynamic and stochastic nature of households, a series of dynamic computer simulations have been performed. These simulations give a better understanding of the relationship between water saving and the capacity of the storage tanks. The computer program for this task was based on the following starting points:

- the occurrence of a household process has a stochastic nature and is determined on the basis of a normalised probability function applicable for the particular process and household type and on the basis of an average occurrence frequency. The probability functions for the different processes are derived from previous research where household activity patterns were studied (Groot-Marcus *et al.* 2000)

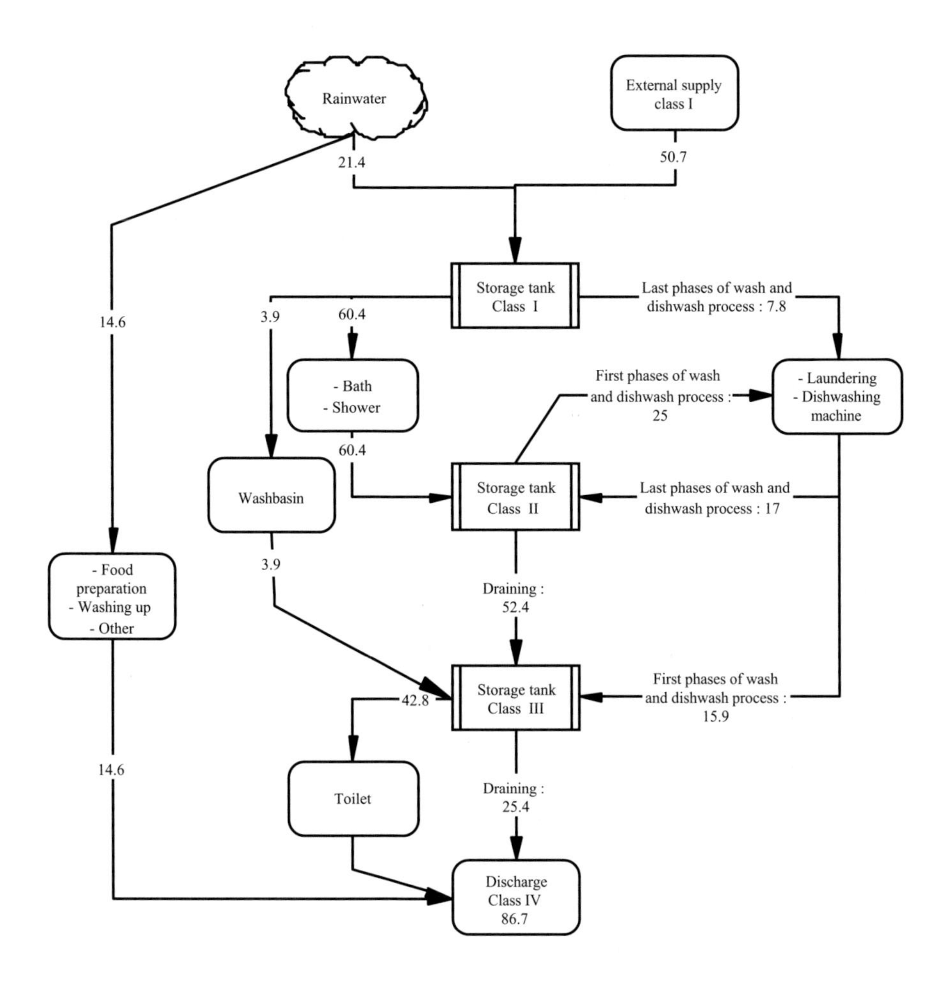

Figure 19.3. A water use system using rainwater.

- in all models rainwater is included in the reuse chain. The amount of rainwater recovered is based on the actual rainfall (over a whole year), a recovery factor of 0.75 and 25 m^2 for a two-person household and 55 m^2 for a four-person household. In the simulation the collected rainwater is treated as class II water. Rainwater is not used for food preparation, washing up, bathing and showering, and is fed into tank II

- washing and dishwashing are done with class II water, and drained as class III
- the time scale for the simulation is one year
- the minimum time period between processes is half an hour
- only water that fulfils the specific quality requirements for the process is reused
- the system boundaries are laid around one single household

Before a computer simulation can be executed a number of characteristics of the household have to be defined:

- the number of people in the household
- the processes included in the reuse chain
- the amount of water that is used and drained by the different processes
- the capacity of the storage tanks

In the present research the effects of family size, dimensions of the storage tanks and the effects of the application of water-efficient domestic technology were investigated, and some specific combinations of the previous parameters, the scenarios, were selected. The scenarios with which computer simulations were run are shown in Table 19.5.

19.5.1 Water saving

Water saving, as estimated with stochastic simulation, with the 11 different scenarios is shown in Table 19.6. All scenarios can achieve a water saving of ±40%. It is also shown that the capacity of the storage tanks can be substantially reduced without a severe reduction in the water saved.

In scenario 2, the scenario with the most up-to-date water-efficient technology, the relative water saving is lower than in the scenario with the appliances that are used in today's situation. The main reason for this is that the water demand for toilet flushing is steeply reduced to 4 litres per cycle instead of ±9 litres. Since there would be sufficient reuse water of the required quality available in the situation presented, the application of a water-efficient toilet system does not lower the water consumption in that case.

Table 19.5. Simulated scenarios (de Pauw and Terpstra 2000)

Scenario number	Processes included in the reuse chain	Family size	Storage capacity
1	all processes that are suitable[1], technology as market average[2]	2 persons	Unlimited
2	all processes that are suitable, modern (efficient) technology[3]	2 persons	Unlimited
3	all processes that are suitable, technology as market average	2 persons	Limited drainage of 'unused' water
4	processes with a substantial water use only[4]	2 persons	Unlimited
5	processes with a substantial water use only	2 persons	Limited drainage of 'unused' water
6	all processes that are suitable, technology as market average	4 persons	Infinite
7	all processes that are suitable, technology as market average	4 persons	Limited drainage of 'unused' water
8	processes with a substantial water use only	4 persons	Unlimited
9	processes with a substantial water use only	4 persons	Limited drainage of 'unused' water

[1] Processes that deliver water with a quality that can be used by one or more of the other processes and/or processes that require water with a quality that is supplied by other processes (including stormwater). The processes included are bathing, showering, floor cleaning, cooking, toilet flushing, dishwashing, fabric washing, hand washing, home cleaning and garden watering.

[2] The average of the processes presently used in Dutch households is taken for the water consumption of these processes.

[3] The water use of the best available technology is taken in this scenario.

[4] Only processes with a substantial water consumption are included: showering, bathing, toilet flushing and clothes washing

Table 19.6. Simulation scenarios (de Pauw and Terpstra 2000)

Scenario number	Water use[1] ($l\ p^{-1}d^{-1}$)	Water use with reuse system[1] ($l\ p^{-1}d^{-1}$)	Water saving (%)	Storage capacity Tank II/tank III (l)
1	154	71	54	9000/12000 (unlimited[2])
2	110	68	38	9000/12000 (unlimited)
3	153	83	46	2000/300
4	155	80	48	9000/12000 (unlimited)
5	155	85	45	2000/300
6	155	72	54	15000/1500 (unlimited)
7	155	86	44	2000/300
8	154	79	49	15000/1500 (unlimited)
9	157	89	43	2000/300

[1] Average of three simulation runs. A variation of about ±2 $l\ p^{-1}d^{-1}$ is observed.

[2] Unlimited implies that further increase of the container capacity does not lead to more water reuse.

The pairs 1–4, 3–5, 6–8 and 7–9 differ in the number of processes in the reuse chain. The second scenario of each pair includes processes with high water consumption only. For the pairs 1–4 and 6–8 where the storage capacity is large, a substantial reduction in water saving is found. For the situation where the storage capacity is limited, the effect is small. This can be explained by the fact that some of the processes that are excluded do not take place often but use large quantities of water. Examples are car washing and watering the garden.

Even the highest savings found in the simulations are smaller than savings in the static models. This is because, for reasons of hygiene, it was decided not to use stormwater for bathing, showering and dish-washing.

19.6 FEASIBILITY AND COMPATIBILITY

19.6.1 Feasibility

It is clear that the feasibility of the various systems has to be further investigated. The substantial water saving benefits are set against higher costs for purchasing and maintaining the technical system. Just how economically viable a water use system will be depends, among other things, on its design and the total costs involved.

The various systems call for different, more complex, technical facilities in the household. These systems have to be developed and tested in real-life situations for efficiency and reliability.

There will be a demand for more discipline and maintenance from the householders than with the current system. Although the water within the household will not be purified, the water use system will need more maintenance than a conventional system. From experience with domestic equipment (washing machines and dishwashers) with built-in recycling systems, it would be realistic to assume that several maintenance checks will be needed per year. Maintenance will consist partly of cleaning the system and the water reservoirs. Systems should be designed in such a way that the user could carry out periodical maintenance himself.

Initiatives for research projects, particularly to investigate overall feasibility, functional efficiency, user-friendliness and technical achievements are strongly recommended. The possibilities of interconnecting small-scale sanitary systems should also be studied, on an experimental scale as well as in practice.

19.6.2 Scaling up small-scale sanitation systems

In theory, the cascade system is suitable for scaling up to block or even district level. The system of reservoirs, valves, etc. would then have to be adjusted to serve a larger number of households and would need to be located outside. Scaling up would presumably mean an increase in the effectiveness of the system, the level of consumption of rainwater and the effectiveness of the buffer systems. The reason for this is that the technical system could be geared to an average impact instead of a fluctuating impact. Operational safety would increase because the cleaning and maintenance of the technical system could be placed in the hands of professional organisations.

A counter-argument would be that consumers could protest about the inflow of water originating from third parties into their households. There is also the risk that in this case the discipline required for carrying out maintenance checks would be less than that required for the water use system at household level. Scaling up also requires more emphasis on water hygiene and the intermediate purification of the various piped water inflows.

19.6.3 Waste separation and reuse

A number of different categories of substances are carried away in the water as domestic waste disposal. Quantitatively speaking, the most important are cleaning agents, organic kitchen refuse and urine/faeces. Substances such as bleaching agents and cosmetics are discharged occasionally and in small quantities. Table 19.3 shows that waste from toilets is quite substantial. It would therefore be logical to investigate whether separating toilet waste from other waste could lead to a reduction of the purification effort and a better reuse of waste substances.

A water use system can in theory be combined with small-scale, decentralised systems for wastewater treatment, especially since, with alternative systems, the quantity of waste water is much lower and the concentration of bulk waste higher than those in the current system. Smaller waste water treatment plants could be set up, which would be far easier to place in a residential setting. Also, there is also the possibility of interconnecting the urban wastewater treatment and water distribution systems with the water use system. This would then imply that the cascade principle would be wholly or partly abandoned.

Because the different qualities of effluent water would be collected in reservoirs, it could be possible to recover the thermal energy from the water. This applies especially to quality class II which has been used for personal hygiene, and class III which comes partly from cleaning processes. A water use system is therefore compatible with integral (domestic) energy management systems, those which transport energy from heat sources to processes requiring heat.

19.7 REFERENCES

Brundtland, G.H. (1987) *Our Common Future*, Oxford, Oxford University Press.

Groot-Marcus, J. P., Mey, S., Terpstra, P. M. J. (2000) Duurzame Waterhuishouding, Watergebruik in huishoudens. Wageningen, Wageningen University.

Hedberg, T. (1995) Urban water systems in a sustainable society; interface with everyday life., Report, Chalmers university of technology, Göteborg.

Kilian, R. M., R. M. M. Loos van der, *et al.* (1996) Water for the present and future. Utilisation of storm water and reuse of cleared waste water in households by application of small scale technologies, thesis, Wageningen Agricultural University, Wageningen.

Meijer, H. A. and J. J. W. Leeuwen van (1996) Sustainable urban water systems., Report, DTO, Delft.

NIPO (1992) Household water consumption, Report, NIPO, Amsterdam.

NIPO (1996) - Het waterverbruik thuis. VEWIN. Rijswijk.

Pauw, de I., Terpstra, P. M. J. (2000) Duurzame Waterhuishouding, Taak C1: Modelontwikkeling. Amsterdam, KIEM and Wageningen University.

Stamminger, R. (1993) 'Die neue Generation von Öko-waschmachinen.' Hauswirtschaft und Wissenschaft 6: 250 ev.

STORA (1985) Oxygen demand of household wastewater; assessment with production and use data, Report, Stichting Toegepast Onderzoek Reiniging Afvalwater, STORA, Rijswijk.

Terpstra, M. J. (2000) 'Assessment of cleaning efficiency of domestic washing machines with artificially soiled test cloth. Bonn, Shaker Verlag

Witteveen and Bos (1994) Duurzame stedelijke waterkringloop; verkennende studie DTO-water, Directoraat-Generaal Rijkswaterstaat RIZA.

Worp van de, J. J. and J. A. Don (1996) Definition of sustainability criteria for the urban water system., TNO Milieu, Energie en Procesinnovatie, Apeldoorn.

Zott, H. (1984) 'Washing processes of tomorrow.' Manufacturing Chemist: 42, -, Berlin.

20

Perspectives of nutrient recovery in DESAR concepts

Tove A. Larsen and Markus A. Boller

20.1 INTRODUCTION

Domestic sewage carries substantial amounts of valuable nutrients among which nitrogen (N) and phosphorus (P) are dominant in terms of detrimental effects in the receiving waters. In traditional waste management of industrial countries, these nutrients are wasted either to the effluent, to the sludge or into the air. Discharged into receiving waters, the nutrients start to boost nutrient cycles in the aquatic environment leading to adverse effects such as over-fertilization and eutrophication of lakes and reservoirs by phosphorus and nitrogen. This causes ecotoxicological problems in river waters caused by nitrogen compounds such as ammonia and nitrite. Aware of the fact that nutrients should not be discharged into receiving waters, wastewater managers in the 1990s designed treatment

Edited by P. Lens, G. Zeeman and G. Lettinga. ISBN: 1 900222 47 7

plants to enhance nutrient removal by introducing new treatment steps such as biological and/or chemical P and biological N removal. Thus, the N and P loads into receiving water bodies are minimized by converting ammonia (NH_4^+) into nitrate (NO_3^-) and subsequently to N_2 gas while P is retained in the sludge. However, nutrient removal in wastewater treatment plants caused a wave of tremendous and ongoing capital investment into the enlargement and enhancement of existing treatment facilities. The issue at stake here is to separate the nutrients from the wastewater by technical means in order to protect the aquatic environment. Possible alternatives for nutrient recycling are rarely considered in nutrient removal concepts. Apart from the enrichment of sewage sludge with phosphorus by chemical or biological P removal and consequent disposal in agriculture, additional treatment steps for concentrating and processing nitrogen and phosphorus for reuse were assessed as being non-economical.

A sound argument against nutrient processing and recycling from domestic sewage is the fact that the nutrients are present in an unfavorable mixture of other wastewater constituents. The presence of organic substances subject to decay, large volume chemicals such as detergents, organic micropollutants and heavy metals may indeed reduce technical possibilities and economical feasibility for nutrient reprocessing. These 'impurities' are also major reasons why the application of sewage sludge in agriculture has been continuously reduced over the last decade in most industrialized countries. In addition, the chemical properties of phosphorus and nitrogen species in wastewater differ from each other such that separation and concentration processes cannot be easily combined for both nutrients in a single treatment step.

In view of new perspectives based on 'waste design' expounded by Larsen and Gujer (2000), innovative concepts may arise in which the nutrients N and P are considered to represent (1) waste products of a value worthwhile to be reprocessed and recycled and, (2) major waste mass streams of our society for which new management may lead to a substantial gain in sustainability. Nutrient waste management starts by studying the nutrients' origin, followed by the existing mass flow pathways and exploration of alternatives to direct the nutrient streams in order to facilitate the use of the nutrients in agriculture.

20.2 ORIGIN AND SPECIATION OF WASTED NUTRIENTS

Both nitrogen and phosphorus are present in abundant amounts in domestic sewage. 'Abundant' means that both nutrients are not limiting factors for bacterial growth in treatment plants, which leads to some excess nutrient loads discharged to the receiving waters. Whereas the organic Carbon: Nitrogen: Phosphorus (C:N:P) weight ratios for aerobic bacterial growth are commonly 100:23:6, settled raw domestic sewage contains nutrients in the ratio of 100:36:7 in countries with P-prohibition in washing powders (for example, Switzerland and Canada) and 100:36:12 in other countries.

The daily per capita loads of nutrients in domestic sewage and its origin are summarized in Table 20.1.

Table 20.1. Annual per capita nutrient loads in domestic sewage (Siegrist and Boller 1999)

Raw wastewater	Nitrogen g (N_{tot}/person)	Phosphorus (no P-ban in detergents) (N_{tot}/person)	Phosphorus (with P-ban in detergents) (N_{tot}/person)
Urine	4450	450	450
Feces	550	200	200
Household waste	75	100	100
Textile washing	–	750	50
Other detergents	–	170	110
Surface runoff	700	50	50

As may be seen from Table 20.1, substantial nutrient input originates from urine. Compared to feces, urine is a nutrient-rich solution containing nitrogen (N), phosphorus (P), potassium (K), sulfur (S) and boron (B). Calcium (Ca), magnesium (Mg) and iron (Fe) are mainly present in feces (see Figure 20.1). N, P, and K are main constituents of conventional fertilizer, and S is increasingly used as fertilizer in Europe due to reduced anthropogenic emissions (Larsen and Gujer 1997).

Considering processing of the wastewater or nutrient streams in treatment plants, the chemical speciation of the nutrients is important. Concentrating on urine, the major fractions of nitrogen in raw sewage are in the form of urea (NH_2-CO-NH_2) and phosphorus in the form of ortho-phosphate (o-PO_4^{3-}). Because the major fractions of important nutrients are combined in a concentrated urine solution, this means that technical solutions for nutrient processing can be focused on a single waste stream with defined nutrient speciation. Therefore, to allow optimal nutrient recovery, all other nutrient streams should be avoided or treated separately. The use of P-free washing powders is considered to be a prerequisite for decentralized nutrient recovery systems. Also, the nitrogen and phosphorus flow contained in surface runoff is supposed to be diverted into rainwater infiltration facilities in future and is therefore not to be considered for nutrient recovery.

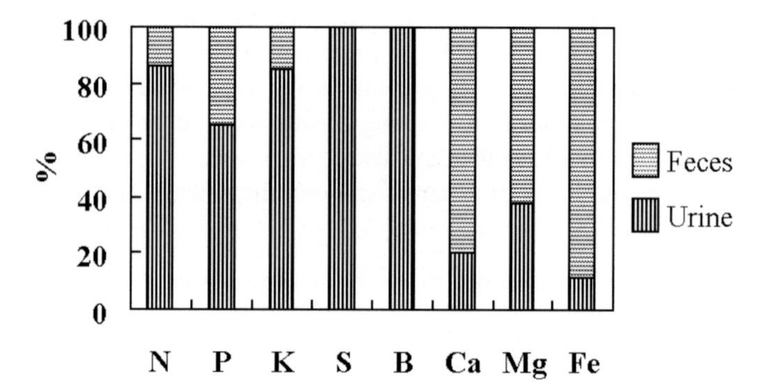

Figure 20.1. Distribution of nutrients from human metabolism (Larsen and Gujer 1996).

In domestic sewage urea and polyphosphates are readily hydrolyzed to ammonia (NH_4^+) and o-PO_4^{3-}, respectively (Hurwitz *et al.* 1965). Hydrolyzation starts in the sewer system and is nearly complete later in the treatment plant effluent. If nitrification takes place during treatment, the majority of NH_4^+ is transformed into nitrate (NO_3^-). In modern treatment plants NO_3^- is further transformed into molecular nitrogen gas by biological denitrification. Thus, nitrogen may leave the water cycle by stripping into the air. Residual fractions of organically bound N and P compounds are always present in small amounts in treatment plant effluents. Since the aim of nutrient recovery systems is not to protect receiving waters but to recycle as many nutrients as possible, denitrification is considered to be unsuitable for DESAR nutrient recovery concepts.

20.3 POSSIBLE WASTE STREAMS

Since most of the existing urban wastewater collection and treatment systems in industrialized countries were established in a typically centralized way, the transfer into possible DESAR systems can only be realized step by step. If decentralized activities are requested from a population connected to a centralized collection system, for example changing the type of toilet used or separate handling of biosolids, this is a partial DESAR concept. Full DESAR systems offer possibilities for the collection, treatment, recovery and reuse of water and nutrients on a household basis. The latter concept agrees with the ideas described by the International Water Supply and Sanitation Council (WSSCC) on household-centered environmental sanitation (1999) to tackle waste problems in urban and rural areas primarily on a household level. Only environmental sanitation problems that cannot be solved by the household are passed to the neighbourhood and community level. In the rapidly growing cities of developing countries, household-centered concepts seem to be the only way of dealing with the tremendous waste problems and the related hygienic hazards. Centralized collection for the larger part of the population is simply not feasible and innovative technologies on a household scale have to be developed which could probably also be transferred to DESAR concepts in industrialized countries.

The following classification of water and waste collection systems referring to different degrees of decentralization may be distinguished. The concepts are sketched in Figure 20.2 and show three different ways of handling nutrient streams from households.

20.3.1 Non-DESAR concepts

Combined wastewater is collected, including all waste in liquid form such as water from toilets, baths, washing activities, food preparation, etc. and rainwater runoff from roofs and other areas. The wastewater is collected in combined sewers and directed to a central treatment plant. Nutrients can be recovered from diluted or concentrated waste streams in the treatment plant.

20.3.2 Partial DESAR concepts (examples)

(1) Collection of household wastewater and discharge to a central treatment plant by separate sewers. All surface runoff is collected separately and infiltrated into the ground on site. This concept is presently a legal requirement in Switzerland.

(2) Collection of the urine stream by no-mix toilets and storage on a single household basis. Emptying and discharge of the urine is performed during the night through existing sewer systems and separate treatment and recovery at the wastewater treatment plant (WWTP). The other domestic wastewater flows are collected and treated in a conventional way. Surface runoff is discharged separately on-site into the ground.

20.3.3 Full DESAR concepts

The full DESAR concept implies small-scale closed cycles of water and material, with a minimum of inputs to and exports from some defined system. Such concepts make sense primarily for rural areas where nutrients can easily be absorbed by local agriculture, rainwater per capita is abundant, and long distance piping is very expensive. However, the main problems of water management and sanitation in DESAR systems occur in urban areas. Due to the necessary import of nutrients via foodstuff from agricultural areas, small, closed nutrient cycles are not possible in the urban environment.

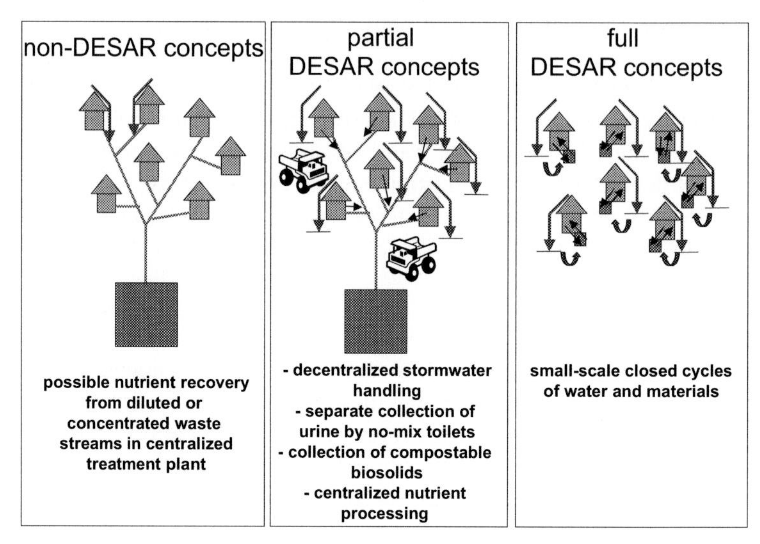

Figure 20.2. Centralized, partially decentralized and full decentralized nutrient recovery concepts.

With the concept of waste design (Henze 1997; Larsen and Gujer 2000), we can enlarge the focus of DESAR to encompass possible on-site technologies, remodeling the flow of water and material out of the system, allowing for optimized treatment and extended recycling. The example of urine separation, with the different possibilities of decentralized technology and larger scale recycling, illustrates the potential of such extended DESAR concepts (see Section 20.7 on Urine separation).

Treatment, water and nutrient recovery and recycling based on decentralized technologies still have to be developed and tested. Hiessl and Toussaint (1999) sketched very advanced scenarios for future decentralized urban water management approaching far-reaching decentralization by separating all sorts of water flows in a household by individual pipe systems. In addition, decentralized processing of waste is proposed without giving details on possible processes. Full-scale application in pilot and demonstration projects will have to show whether these complicated multi-pipe concepts can be considered as feasible ways to ensure hygiene and sustainable single house waste and water management.

20.4 OPTIONS FOR NUTRIENT RECOVERY

Before potential technical solutions for nutrient recovery are discussed, it is necessary to consider where and in what form the nutrients can be recycled and applied. N and P can be recycled to agri- or aquaculture in dry solid form or as concentrated nutrient solution. In most cases the nutrients have to be transported to the place of application. Recycling 'on the spot' for typical DESAR systems on a single house basis would require nearby agricultural activities in which the nutrients can be applied. For agricultural activities, a land area of at least 200–500 m^2 per person would be necessary for sustainable recycle of P and N (Otterpohl *et al.* 1999). Aquaculture for instance in the form of duckweed production and fish-farming only requires 6–7 m^2 per person but includes the construction of pond systems with a water depth of 0.5–1.0 m (Iqbal 1999).

The following possibilities exist for further N and P processing and reuse:

- P-enriched sludge in liquid or dried form from central WWTP used for land spreading in agriculture. Although land spreading is a cheap and easy way of nutrient recycling, there are several drawbacks which restrict land use more and more. The sludge is mixed with unfavorable substances, transport costs are high, large storage capacities are necessary, N recycling is low, hygienic aspects require additional investments (pasteurization).
- Nutrient reprocessing in the phosphate or fertilizer industry. The nutrients have to be transferred into a form which would allow easy transportation and could be further processed by the phosphate or fertilizer industry: This would offer one highly efficient method of nutrient recycle via marketed products. Unfortunately, the industry is not yet ready to cope with decentralized mass flows. Besides an attractive economy of nutrient recycling, structural, organizational and logistic changes would be necessary to deal with decentralized systems.
- Treatment of concentrated nutrient waste streams (digester supernatant, sludge hydrolysate, urine from no-mix toilets) in centralized or decentralized processing plants to produce reusable products for agriculture.

Nutrient recovery from concentrated streams leads to high-tech reprocessing technologies. In partial as well as in full DESAR concepts, these technologies have to be investigated and their suitability tested. Technologies on a small-scale household-based size do not yet exist for many processes. Nevertheless, this seems to be one way of moving from a water protection philosophy towards a waste management strategy.

20.5 TECHNOLOGIES FOR NUTRIENT RECOVERY

Most nutrient recovery technologies are currently in the research and development stage and few have been implemented in full-scale systems. Most systems are focused on the removal and further processing of nutrient-containing solutions in traditional wastewater treatment schemes. Consequently, many of the technical solutions are only suited for larger nutrient streams and do not fit into the perspective of DESAR concepts. Possible recovery technologies may be classified along two lines: (1) according to the type of waste stream, and (2) according to their suitability for DESAR. The first line includes systems which range from nutrient separation in traditional wastewater streams and nutrient-enriched solutions to nutrient recovery from new types of collection systems (for example, no-mix toilets). The second line may distinguish between systems which are only feasible for large nutrient flows or which are especially suited for DESAR concepts. There is no question that decentralized nutrient recycle concepts are easier to apply in rural areas – but what are the potentials of DESAR in the urban environment ?

Potential technologies for nutrient recovery from waste streams are:

- Calcium phosphate precipitation ($CaOHPO_4$), preferably in fluidized bed reactors (only for P recovery)
- Struvite (magnesium ammonium phosphate, $MgNH_4PO_4 \cdot 6H_2O$) precipitation, suited for concentrated solutions of NH_4^+ and PO_4^{3-}
- Ion exchange technologies, for either both or one of the nutrients, combined with concentrate treatment by struvite and other ammonium salt precipitation
- Membrane treatment with nanofiltration or reverse osmosis, eventually combined with pretreatment by micro- or ultrafiltration. Concentrate treatment by precipitation or drying
- Nutrient transfer into phyto-biomass such as plants and algae including harvesting and transfer into a usable form, for instance by composting
- Nutrient transfer into duckweed and use as feed supplement for fish and poultry.

20.5.1 Calcium phosphate precipitation

Phosphate precipitation has been practiced over more than four decades in central WWTP. In general, the precipitants used are Fe(II)–, Fe(III)–, or Al(III)–salts. For operational and economical reasons, hydroxyapatite ($CaOHPO_4$ or HAP) precipitation with lime is seldom applied. In most cases, chemical phosphate precipitation is performed as pre- or simultaneous precipitation leading to mineral solids-enriched bio-sludge with a P content of 3–4% d.w. Cheap reuse of the nutrients in the sludge is achieved by land spreading. In Switzerland, the large areas required for land spreading, the presence of micropollutants and hygienic requirements have decreased sludge application on agricultural land from 80% to under 40% between the 1970s and the 1990s.

One way of increasing the acceptance of nutrient reuse from wastewater is the production of a stable, dry solid which can easily be handled, stored and applied. In order not to mix the precipitates with unstable biosolids, post-precipitation or the treatment of concentrated streams should preferably be used for this purpose. HAP precipitation has been developed to full-scale application, mainly in Holland. HAP precipitation is performed at a pH of 9 or higher, including stripping of CO_2 and lime or base addition. Different seeding materials such as sand, calcite and apatite nuclei are used, mainly in fluidized bed reactors (for example, DHV Cristalactor®; Woods *et al.* 1999) to speed up the formation of readily settleable solids. In such processes HAP pellets of several millimetres diameter with a P content of 6–11% are produced.

The drawback of the HAP precipitation technology is that N components are not considered for recovery. This means that the combination of biological N removal and P precipitation can only be performed economically in central WWTP. Therefore, precipitation techniques for DESAR concepts have to be directed toward recovery systems including both P and N precipitation as purely physical-chemical treatments.

20.5.2 Struvite precipitation

Combined immobilization of soluble N and P components can be achieved by struvite precipitation (magnesium-ammonium-phosphate = MAP or $MgNH_4PO_4 \cdot 6H_2O$), a process which requires concentrated solutions of PO_4^{3-} and NH_4^+ and a low solids level. Supernatants of anaerobic reactors, hydrolyzed urine solutions or concentrates from membrane or ion exchange processes may be most suited. Optimal MAP production from digester supernatant aiming at high NH_4^+ incorporation into the precipitates involves complex processing such as the addition of phosphoric acid, magnesium oxide and NaOH as well as separation and drying of the precipitated salt (Siegrist *et al.* 1992). Figure 20.3 shows the flow scheme of a struvite precipitation plant tested for the treatment of digester supernatant (Siegriest *et al.* 1992).

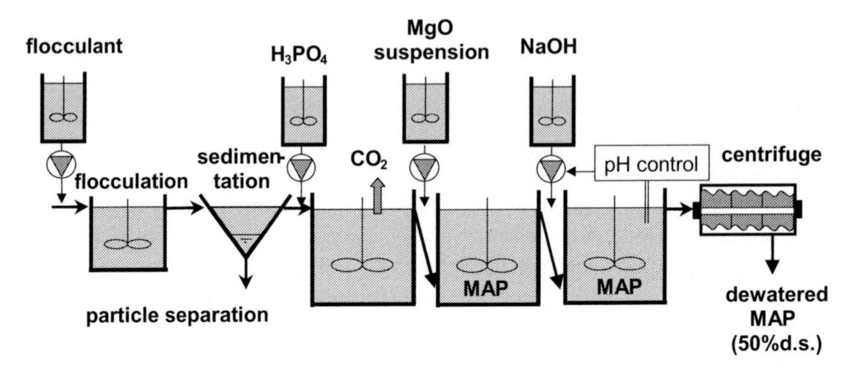

Figure 20.3. Flow scheme of a struvite precipitation plant treating digester supernatant. NH_4 was transferred up to 90% into MAP (according to Siegriest *et al.* 1992).

Battistoni *et al.* (1998) showed some simpler methods to reach the combined HAP and MAP precipitation in digester supernatant without chemical addition. The Mg and Ca content in the supernatant showed to be enough to guarantee the removal of P by a combination of air stripping for CO_2 removal and precipitates forming in a fluidized bed reactor. However, this process is not optimized for NH_4^+ removal, thus leaving most of the N in the effluent.

Adopting struvite precipitation in DESAR concepts restricts its potential use to the treatment of urine or concentrates from ion exchangers or membranes. All other applications such as the treatment of digester supernatant, concentrated streams in biological nutrient removal plants and sludge hydrolysates require centralized facilities. In view of the complex treatment technology for MAP precipitation, the processing of urine and concentrates would also have to be centralized, after the collection of waste streams from decentralized collection and storage facilities (partial DESAR).

20.5.3 Ion exchange technologies

Ion exchange is a treatment technology which is widely used in households for water softening. From an operation point of view, the ion exchange can be considered as a possible process for nutrient recovery suited for DESAR systems. It has been shown that ion exchange can be used for the ions PO_4^{3-}, NH_4^+ and NO_3^-. Now, there are different ion exchangers available which show especially high affinity to the ions in question. Because PO_4^{3-} and NH_4^+ are present in raw and anaerobically pretreated wastewater and the final product may lead to struvite (magnesium-ammonium-phosphate salt) and other ammonium salts, they represent the preferred species of P and N for ion exchange.

In the 1970s, Liberti *et al.* (1986) demonstrated that a combination of cationic and anionic exchangers is a possible way of recovering nutrients from secondary effluents. A natural zeolite (clinoptilolite) in Na^+ form served as an NH_4^+ exchanger while a strong base anionic resin in Cl^- form was used for PO_4^{3-} exchange. Thus, simple regeneration with NaCl (for example, from seawater) could be performed. The nutrient-enriched eluates from regeneration are subject to a complicated treatment in which the NH_4^+ solution is set to a pH of 11 by lime addition and NH_3 is stripped under vacuum. The NH_3 is mixed with the anionic eluate and an MgCl solution in stoichiometric amounts to form MAP. The excess NH_3 is absorbed in an acidic H_2SO_4 solution to form ammonium-sulfate. The precipitated salts can be reprocessed as fertilizers. The flow scheme of the nutrient recovery system if given schematically in Figure 20.4. It may be of interest to mention that the eluates contain 20–30 times fewer nutrients compared to the N and P concentrations in urine. In comparison to separate urine collection, the ion exchange is therefore a relatively costly and inefficient nutrient concentrating process.

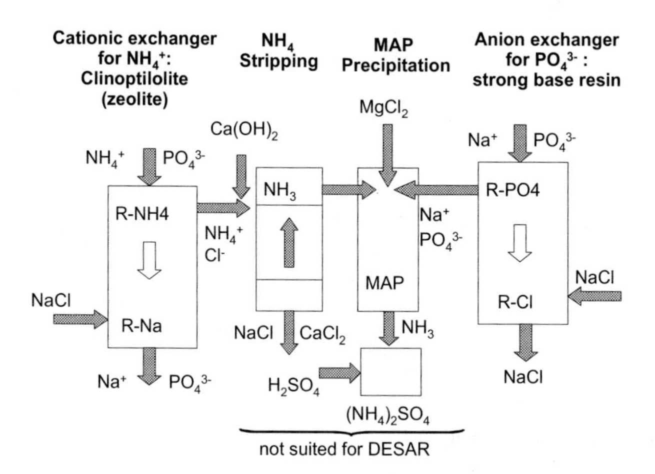

Figure 20.4. Scheme for nutrient recovery from wastewater by ion exchange and subsequent treatment of the eluates by NH_3 stripping and MAP precipitation (according to Liberti *et al.* 1986).

In the meantime, new and more specific ion exchangers for PO_4^{3-} on the base of polymeric ligand exchangers loaded with Cu^{2+} have been investigated and show some ways of increasing the efficiency and the economy of P fixation (Zhao and Sengupta 1998).

20.5.4 Membrane treatment

In the past, all forms of membranes have been tested to treat secondary effluents aiming at efficient solids separation (micro- and ultrafiltration) or at a recyclable water (nanofiltration and reverse osmosis). In most cases the recovery of nutrients was unfocused. Today, membranes in the form of micro- and ultrafilters are increasingly tested and used as sludge separation steps in biological treatment processes.

Membrane processes could be technically well suited for DESAR systems. They can be constructed as modular elements for very small flows and be fully automated. However, recent experience with membranes does not give any indication whether membrane processes would be feasible for the treatment of nutrient-enriched solutions in DESAR concepts. Very few studies on the treatment of human urine (Dalhammer 1997) and manure waste are available, so no conclusions for further development can be drawn. In order to come up with technically feasible solutions, much more research effort will have to be put into nutrient recovery by membrane filtration.

In principle, there are many different applications for membranes reaching from the preparation of solids containing nutrient solutions by micro- or ultrafiltration to the separation of larger molecules by nanofiltration and of ions by reverse osmosis. Also, the use of gas membranes could eventually be integrated into concentrating and separating NH_3. Apart from economic considerations, problems concerning membrane fouling and processing of the concentrates remain to be solved. Detailed investigations into the treatment of concentrated nutrient solutions such as urine will lead to more insights in the near future (see Section 20.7.4).

20.5.5 Nutrient transfer into plants

Reed beds and similar wetland systems represent a wastewater treatment technology typically suited to DESAR concepts. Wetland technologies are already widely applied for single houses and smaller communities in rural areas. The large area required restricts the application to rural areas and small wastewater flows.

For reed beds, a design area between 5–10 m^2/person is usually assumed to fulfill standard effluent requirements. The mass balance for N and P in such systems shows clearly that the level of nutrients taken up by the plants is relatively low. Taking the maximum nutrient uptake by optimized harvesting of reed measured by Hosoi *et al.* (1998) in Japan, only about 9% of N and 7% of P would be removed through plants on conventional reed beds. The majority of the nutrients would either be removed in the soil by adsorption of phosphate, denitrification of N or lost in the effluent.

Plant harvesting on reed beds may therefore be inefficient for nutrient removal and, as a consequence, unattractive for nutrient recovery.

20.5.6 Nutrient recycling in food chains by aquaculture

The use of domestic waste for the production of aquatic plants and fish farming is a form of reuse which will certainly increase in the future in many parts of the world. Aquaculture of this form is well suited to DESAR systems, and is easy to operate, but is labour-intensive and only applicable in rural areas. In addition, a strong dependency on climatic conditions may restrict its application to warmer climatic zones. Different aquaculture systems using domestic sewage are already practised in China, India, Bangladesh and the United States (Iqbal 1999).

A viable form of aquaculture is the production of protein-rich duckweed (protein content: 30–40% d.w.) which can be harvested and used for fish production or dried and used as animal feed. Figure 20.5 shows a frequently applied flow scheme of combined duckweed production and fish farming. Duckweed lagoons are applied for secondary and tertiary effluents aiming at the purification of the wastewater and the production of valuable biomass. In tropical and subtropical countries duckweed productivity reaches 2.7–8.2 g dry matter/m^2 and day. Compared to soybeans, the annual protein production in duckweed per m^2 is about ten times higher.

In correctly designed pond systems and with careful preparation of the fish, hygienic risks have showed to be small and the products are of high quality. Especially in two-pond systems where duckweed is grown in wastewater separate from fish production, the produced fish are ready for marketing and human consumption (Iqbal 1999).

Duckweed ponds are usually designed as serpentine plug-flow systems with multiple wastewater inlet points and recirculation facilities. The floating duckweed is regularly harvested mechanically or by hand. The duckweed is fed to animals in fresh or dried form. Fish production is the most frequent application of duckweed farming and includes one- or two-pond systems. It has to be pointed out that the combination of duckweed and fish cultivation requires a high degree of knowledge and skill.

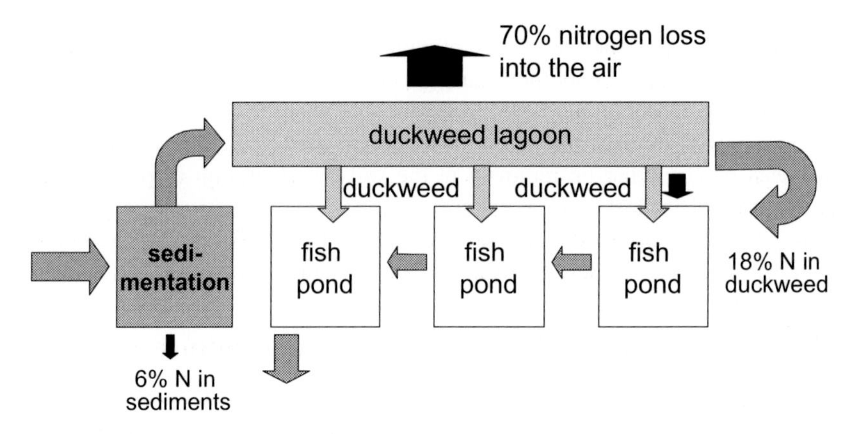

Figure 20.5. Typical flow scheme for duckweed production from wastewater in combination with fish farming and rough N balance.

The N balance of a duckweed–algae pond system, investigated by van der Steen *et al.* (1998), revealed that between 12–25° C, the major part of N (73%) was lost by volatilization of NH_3 into the air, some N (6%) settled to the sediments and 18% was incorporated into duckweed. Duckweed biomass production reached values between 4–16 g dry matter/m^2 and day and the N content of the duckweed amounted to 0.06 g N/g dry matter. With respect to wastewater purification, the nutrient removal may be satisfactory, but the nitrogen problem seems to be simply transferred from the water into the air. From a nutrient recovery point of view the N loss by stripping is severe and only a minor fraction of the nutrients is actually recycled into the food chain.

20.6 SUITABILITY FOR DESAR CONCEPTS

Summarizing the technical possibilities for recovering nutrients from domestic wastewater, DESAR concepts have the best perspectives in rural areas. In these regions, area requirements for low rate treatment technologies and nutrient application are usually not limiting factors, but technologies often applied in rural areas such as wetlands and soil filters are typical wastewater treatment steps which do not consider nutrient recovery. More efficient in this respect is the separate storage of urine and subsequent land spreading for agriculture (Hanäus *et al.* 1997).

The real challenge is to introduce DESAR in urban areas. At present, it is doubtful whether full DESAR systems can be realized in an urban environment. Multiple piping with its risks of wrong connections, not yet existing small-scale processing facilities with a high degree of automation, acceptance of system changes by the population and the lack of opportunities for local reuse of the recovered nutrients are just a few of the reasons why full DESAR reveals a low potential for future nutrient recovery.

However, partial DESAR systems in which concentrated streams of nutrient-rich solutions are produced, collected and treated in central processing plants on a household basis seem to offer feasible ways of turning wasted nutrients from existing centralized urban water systems into valuable fertilizer. Of various technical options, urine separation and processing is one of the most promising and meaningful alternatives. Therefore, the following sections will describe a project at EAWAG (Swiss Federal Institute for Environmental Science and Technology) focused on urine separation. The many aspects included in the planned studies should demonstrate the complex nature of a system transfer from centralized urban water management to partial DESAR concepts for nutrient recovery.

20.7 URINE SEPARATION (AN TECHNOLOGY): AN EXAMPLE

With the separation and reuse of urine and/or feces we can move towards closing the nutrient cycle. There exist a number of technical options for separating and reusing anthropogenic nutrients: compost, vacuum and urine-separating toilets. For various reasons, urine separation technology appears to be the most viable alternative, since urine separation can be achieved within the existing infrastructure, treatment is easier for the separated fractions, and there is a good chance that the technology will be generally accepted. By comparison, vacuum technology requires a new semi-decentralized infrastructure and composting of feces is only possible if it is not mixed with urine. Urine contains most of the nutrient emissions from the human metabolism (see Figure 20.1). If urine is kept out of the wastewater, nutrient removal at wastewater treatment plants will be obsolete, and nutrient emissions to the receiving waters would be dramatically reduced. The large amounts of nutrients contained in urine could satisfy a significant part of agricultural nutrient demand. Hence, using urine as fertilizer would be an important step towards closing the nutrient cycle. To emphasize the fact that urine is rich in nutrients, we use the term Anthropogenic Nutrients (AN), originally introduced in the form of ANS (Anthropogenic Nutrient Solution) by Larsen and Gujer (1996).

With urine separation – in this section referred to as AN technology – additional benefits for the receiving waters would be achieved. Recent research on receiving water quality has revealed that micropollutants, especially hormones and pharmaceuticals, may be a significant problem to fish populations due to their chronic toxicity. Some researchers estimate that most of the pharmaceuticals excreted from the human metabolism are contained in urine. Although beneficial for the receiving waters, experiments must be carried out to ensure that the proposed redirection of the urine flux from wastewater to agriculture is a safe alternative and not just the shifting of a problem between environmental recipients. In spite of the fact that nobody seems to be concerned with the same compounds in animal manure, there is a general uneasiness regarding accepting contaminants from humans.

Although AN technology is a simple concept, its translation into reality is a complex task that requires the cooperation and input of many different actors and areas of expertise. There are not only technical difficulties and objective risks for environment and human health at stake: consumer preference is a major aspect of every new technology, especially when it is as fundamental as sanitary technology.

When introducing AN technology, there are a number of aspects that deserve closer investigation. These aspects will be discussed in detail below, primarily based on Larsen *et al.* (2000). The numbering refers to the NOVAQUATIS project, an EAWAG project, the different aspects of which are covered in seven work packages (NOVA 1–NOVA 7).

20.7.1 Consumer attitude

The concept of AN technology begins and ends with consumers (Figure 20.6, NOVA 1). In which form will consumers accept necessary changes in the bathroom? Will consumers accept food from organic farmers who use AN as a fertilizer? Consumer attitude can be organized along three dimensions: attitude towards changes in bathroom technology, attitude towards quality of fertilizer, and attitude towards environmental and fiscal impacts. Consumer attitude may be investigated in different ways. Besides questionnaires and interviews, focus groups are an interesting alternative when complex environmental matters are concerned (Dürrenberger *et al.* 1997). In focus groups, information is presented and discussed in a structured way. Normally the moderator is not personally involved in the subject concerned. One way of quantifying the results of the focus groups is by means of questionnaires before and after the groups have been run; the questionnaires can be filled in personally or represent the concensus of the group. More involved methods may be used, for example, video-taping, allowing the reactions of the participants to be analyzed in detail.

Not all aspects of acceptance can be covered by theoretical studies. Some changes in everyday routines will be necessary for the successful introduction of AN technology. For example, the convenient discharge of toilet paper with at least 2 liters of water would endanger the excellent water-saving qualities of the No-mix (urine separation) toilet. Whether it is possible to dispose of the toilet paper in another way, for example, together with sanitary wastes, will have to be tested in pilot studies.

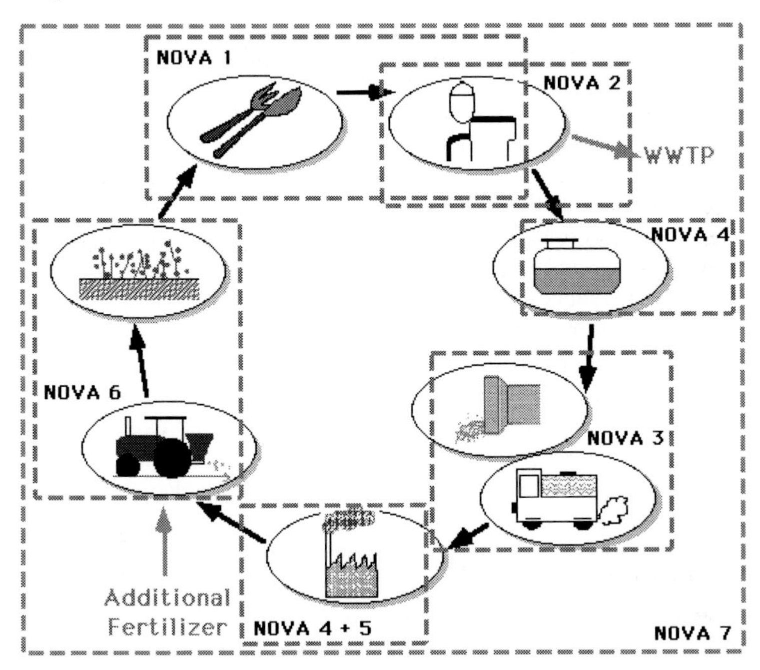

Figure 20.6. The anthropogenic nutrient cycle as it would arise if urine was separated at source and returned to agriculture as fertilizer (WWTP = Wastewater Treatment Plant). NOVA 1 to 7 refers to the work packages of the NOVAQUATIS project, an EAWAG project concerned with the different aspects of urine separation.

20.7.2 Development of new bathroom technology

Current sanitation technology is optimal in view of consumer interests: one flush and all problems of human excreta are solved. Clean, hygienic, and perfectly adapted to cultural habits in the Western hemisphere, the flushing toilet is one of the more pleasant achievements of civilization.

No-mix toilets as they exist today closely imitate the conventional flushing toilet. Due to the human anatomy, urine can be separately collected at the front of the No-mix toilet, which looks very similar to an ordinary flushing toilet. The main necessary change in toilet habits is the requirement that men should always sit. However, existing No-mix technology has flaws. The most important problem is the precipitation of solids when the urine is mixed with water. Developments in sanitary technology will be necessary in order to cope with the problem of clogging pipes (NOVA work package 2).

20.7.3 Transport of anthropogenic nutrients

The transport of AN is closely related to the treatment schemes adapted. Three methods of AN transport have been suggested:

(1) Storage in the household or semi-decentrally followed by truck transport (applied in Swedish pilot projects; Hanäus *et al.* 1997)
(2) Storage in the household followed by transport through sewers at night (suggested by Larsen and Gujer 1996)
(3) Transformation into a dry medium within the household followed by co-transport with the solid waste.

There are of course more ways of transporting AN, the most obvious being the construction of new piping. It is very doubtful whether this option will be economically viable – today it is even questionable whether the conventional piping system can be economically sustained.

The transport of AN through sewers at night has been assessed in more details at EAWAG (work package NOVA 3). Pilot studies with tracer material have shown that at least in the main sewers, dispersion is rather small (Huisman *et al.* 2000). The low dispersion coefficients observed pose a problem for the numeric simulation with existing software such as the sewer simulation program MOUSE, because numeric dispersion is always larger than the hydraulic dispersion observed in experiments. An important optimization problem is the size of the storage tanks. In order to prevent discharge of urine during rain events, it is necessary to design the storage tanks to be large enough such that several days of urine production can be stored in the household.

20.7.4 Processing of anthropogenic nutrients

The processing of AN always has to serve several purposes: ammonia stripping to the atmosphere during AN application must be prevented and we may have to remove or degrade organic matter with ecotoxicological or human toxicological potential. Additionally, the product must be attractive and acceptable to farmers. Transport and storage characteristics will play an important role in the evaluation of the product, as well as environmental impacts and economic costs. A recent survey among farmers showed that their requirements are highly diverse, ranging from ammonia nitrate to organically bound nitrogen (Haller 2000).

For the processing of AN, the following approaches are the most interesting (NOVA work package 4):

(1) Biological treatment (for stabilizing pH and degrading toxic components). Some basic experience on biological stabilization with partial nitrification is available, but is not yet published. This is a low-tech option which would have the potential of rapid implementation.
(2) Precipitation is a dominant problem for existing No-mix toilets due to clogging of pipes (Hanäus *et al.* 1997). However, controlled precipitation would be useful for recovering the phosphorous in a suitable form and possibly preparing the solution for membrane processing. These two aspects make precipitation one of the most interesting research topics for the development of AN technology.
(3) Membrane treatment with the purpose of avoiding urea hydrolysis, diminishing the amount of organic material in the fertilizer and possibly producing a very concentrated product which could easily be transformed into solid form by means of drying is a very interesting option. Although problems concerning fouling of the membranes and energy consumption must be anticipated, such a technology would be ideal from the point of view of infrastructure and further handling.

20.7.5 Hazard assessment of AN fertilizer

Referring to the quality of the fertilizer product, we are mainly concerned about the hazard potential of the organic micropollutants in AN. There is a risk that these micropollutants are taken up by plants and thereby enter the food chain, harm terrestrial ecosystems or, upon leaching, end up in groundwater or surface waters surrounding the fields. Currently, no quality requirements are defined, either for AN or for farm manure, which contains micropollutants similar to those contained in human urine. The development of methods for assessing the hazard potential of the raw product (urine) as well as of the processed product will be of great interest for practical application in agriculture (work package NOVA 5).

20.7.6 Assessment of agricultural nutrient demand and farmer attitude

The direct consumers of AN-based fertilizer are involved in agriculture and consequently farmers and their organizations play an important role in the development of the technology. Two aspects of farming are of primary importance: the requirement for and the acceptance of the fertilizer. There are many reasons that farmers should have an interest in AN fertilizer. In conventional agriculture, P fertilizer contaminated with heavy metals could be substituted with a product that does not contain these elements. In organic agriculture, the concept of closed nutrient cycles could be approached, eliminating the requirement for raw phosphate, which today is allowed as a fertilizer in most countries including Switzerland (FiBL 1999). Also, a lack of potassium will be a problem with time, at least in countries with low a potassium content in agricultural soil. Surprisingly, nitrogen seems to be the greatest problem in organic agriculture, although in principle nitrogen can be taken up biologically by certain crops.

The main problems with using human urine in agriculture are the potential hazards, discussed above. In order to further advance AN technology, we have to establish a basic knowledge of quality requirements for AN fertilizer, based on agricultural, natural and technical science as well as acceptance studies amongst the stakeholders, who are primarily farmers and retail organizations. For successful application in organic farming, a certification in the form of a 'Biocycle' label will be needed, certifying under which circumstances recycled human urine is safe and appropriate for use in organic agriculture. The work packages NOVA 5 and 6 will provide first indications of which parameters are required for such a label.

20.7.7 A comprehensive evaluation of AN technology

In discussing AN technology, there is a basic requirement to establish whether the technology will generally be more advantageous than current practice. Depending on the values of the people discussing the matter, environmental and/or socio-economic aspects are of importance. For this reason, there is a desire to collect, organize and aggregate available information on the advantages and disadvantages of urine separation technology along environmental and socio-economic dimensions. Such information could be organized around three themes: (a) nutrient economy, (b) AN technology itself, bathroom and infrastructure, and (c) impacts on receiving waters. An evaluation of the different aspects should draw on techniques from the fields of decision analysis, economic cost-benefit analysis, and life cycle assessment (NOVA 7).

20.8 HOW WELL DOES AN TECHNOLOGY CORRESPOND WITH THE DESAR CONCEPT?

All household-centered sanitation technologies will end up with some products which have to be removed, as a minimum the inorganic non-volatile components of human waste. For obvious reasons, it is not possible to use all the inorganic nutrients locally. Most people are concentrated in cities while the main production of food takes place in the countryside.

Which transport and central treatment facilities are compatible with the DESAR concept? Is some sort of static transport network acceptable; is centralized refinement of a fertilizer product possible?

If we look at the entire urban water discharge system, it is obvious that the sewers are at the center of this system. Despite the many positive aspects of sewers, they also have severe drawbacks. They cause an undesirable mixing of resources with pollutants, require large amounts of water to support their transport facility, it is more expensive to maintain the existing network than was the original construction, thus causing problems even for well-organized, rich Western countries. As a rough estimate, sewers are responsible for around 60% of the costs of urban water discharge (BUWAL 1994).

For DESAR concepts to make sense, they must counteract at least some of the disadvantages summarized above. The urine separation concept reduces mixing, but dependence on the centralized transport system remains. It is obvious that from the point of view of DESAR systems, urine separation is only a first step. However, it is a step with several advantages: urine separation immediately leads to water saving and rapidly brings advantages for treatment plants and receiving waters. The change can happen gradually, in the normal rhythm of changing sanitary devices in private households. Every single household can take the decision, allowing pioneer families to take the lead. Furthermore, the urine separation concept leads to a different type of research of significant relevance to DESAR processes: involvement of new actors such as households, producers of sanitary devices, and agriculture; development of small-scale technologies; development of the first 'biocycle' principles. For more details on the further development of household wastewater handling towards DESAR processes, see Larsen and Gujer (2000).

20.9 CONCLUSIONS

Various alternative pathways for household nutrient wastes can be identified and proposed which lead away from traditional centralized collection and treatment systems towards recovery and reuse of the valuable nutrients. Among possible solutions, fully decentralized systems including on-site waste processing and nutrient recycle reveal themselves to be impractical and unrealistic for urban areas in industrialized countries. Whereas in a rural environment simple recovery and recycle of nutrients may be possible, the larger nutrient loads produced in cities cannot be recycled in the same area. Therefore, some compromise of a partial DESAR concept must be applied which combines decentralized nutrient collection and centralized processing and distribution. Aiming at high nutrient recovery, nature-based processes such as nutrient transfer into bacterial or plant biomass (biosolids from activated sludge or biofilm processes, wetlands, reed beds, or ponds) have to be considered to be inappropriate because they incorporate only minor fractions of the wasted nutrients. Physical-chemical high-tech processes such as precipitation, ion exchange or membrane filtration may lead to considerably more efficient recovery systems which enable the production of concentrated nutrient streams for recycle. However, the technology is presently not developed to a stage that would guarantee technical feasibility and economy. Further development needs intense research in this field in the near future.

Analysis of household waste nutrient streams shows clearly that most nutrients are contained in human urine. The human body already concentrates the nutrients to a degree which could only be technically achieved by costly and complex processing. From a recovery perspective, all systems diluting urine by toilet flushing and other water consuming activities in a household and consecutive processing for the production of nutrient concentrates for recycle are inefficient. Why not collect the nutrients in an already highly concentrated form at source? Processing for nutrient recovery can then start on a concentrated level which does not require additional pretreatment steps for nutrient separation from wastewater. The idea is to separate urine by special toilets (no-mix toilets) and store it on a decentralized household basis. Collection of urine is performed at low wastewater flow through the existing sewer system and brought to a urine processing plant where a fertilizer product can be produced. The NOVAQUATIS project was established to investigate different aspects of urine separation and processing and to discover which problems could arise along the new nutrient pathways. The concept of urine separation may look relatively simple: however, its translation into practice is a complex matter which requires the cooperation and input of many different actors and areas of expertise. Many technical problems remain to be solved and objective risks for environment and human health have to investigated and, above all, the consumers' preference for new technologies should be carefully examined. The success of proposed system changes relies heavily on the consumers' attitude to accept or reject new household technologies. Therefore, it is important that system changes in urban water management, like changes from centralized to partial or full DESAR systems, are tackled in an integrated way including all actors involved.

20.10 REFERENCES

Battistoni, P., Pavan, P., Cecchi, F. and Mata-Alvaraez, J. (1998) Phosphate removal in real anaerobic supernatants: modeling and performance of a fluidized bed reactor. *Wat. Sci. Tech.* **38**(1), 275–283.

BUWAL (1994) Daten zum Gewässerschutz in der Schweiz. *Umwelt-Materialien* 22.

Dalhammer, G. (1997) Behandling och koncentrering av humanurin, Bilaga 4, Lägesrapport, September 1997, Källsorterad humanurin I kretslopp, Kungl Tekniska Högskolan, Sweden. (In Swedish.)

Dürrenberger, G., Behringer, J., Dahinden, U., Gerger, A., Kasemir, B., Querol, C., Schüle, R., Tabara, D., Toth, F., van Asselt, M., Vassilarou, D. and Willi, N. (1997) Focus Groups in Integrated Assessment: A Manual for Participatory Research. Darmastadt: Center for Interdisciplinary Studies in technology, Darmstadt University of Technology, Germany.

FiBL (1999) Zugelassene Hilfsstoffe für den biologischen Landbau. Forschungsinstitut für biologischen Landbau (FiBL), Ackerstrasse, Postfach, CH-5070 Frick, Switzerland. (In German.)

Haller, M. (2000) Düngeverhalten von Bio- ind IP-Landwirten. Umfrage zur Akzeptanz des Projektes NOVAQUATIS. Semester work at the Swiss Federal Institute of Technology, UNS, D-UMNW ETHZ. (In German.)

Hanäus, J., Hellström, D. and Johansson E. (1997) A study of a urine separation system in an ecological village in northern Sweden. *Wat. Sci. Tech.* **35**(9), 153–160.

Henze, M. (1997) Waste design for households with respect to water, organics and nutrients. *Wat. Sci. Tech.* **35**(9), 113–120

Hiessl, H. and Toussaint D. (1999) Szenarios für Stadtentwässerungs-Systeme. *GAIA* **8**(3), 176–185.

Hosoi, Y., Kido, Y., Miki, M. and Sumida, M. (1998) Field examination on reed growth, harvest and regeneration for nutrient removal. *Wat. Sci. Tech.***38**(1), 351–359.

Huisman, J.L., Burckhardt, S., Larsen, T.A., Krebs, P. and Gujer, W. (2000) Propagation of waves and dissolved compounds in sewers. *Journal of Environmental Engineering* **126**(1), 12–20.

Hurwitz, E., Beaudoin, R. and Walters, W. (1965) Phosphates – their fate in a sewage treatment plant–waterway system. *Water and Sewage Works* **112**, 84.

Iqbal, S. (1999) Duckweed Aquaculture. SANDEC Report No. 6/99.

Larsen, T.A. and Gujer, W. (1996) Separate management of anthropogenic nutrient solutions (human urine). *Wat. Sci. Tech.* **34**(3–4), 87–94.

Larsen, T.A. and Gujer, W. (1997) The concept of sustainable urban water management. *Wat. Sci. Tech.* **35**(9), 3–10.

Larsen, T.A. and Gujer, W. (2000) Waste design and source control lead to flexibility in wastewater management. Accepted for the first IWA conference, Paris 2000, submitted to *Wat. Sci. Tech.*

Larsen, T.A., Peters, I., Alder, A., Eggen, R.I., Maurer, M. and Muncke, J. (2001) Urine source separation: a step towards sustainable wastewater management. Submitted to *Environmental Science and Technology.*

Liberti, L., Limoni, N., Lopez, A., Passino, R., Kang, S.J. and Horvatin, P.J. (1986) The RIM-NUT process at West Bari for removal of nutrients from wastewater: first demonstration. *Resources and Conservation* **12**, 125–136.

Otterpohl, R., Albold, A. and Oldenburg, M. (1999) Source control in urban sanitation and waste management: ten systems with reuse of resources. *Wat. Sci. Tech.***39**(5), 153–160.

Siegerist, H.R. and Boller, M. (1999) Auswirkungen des Phosphatverbots in den Waschmitteln auf die Abwasserreinigung in der Schweiz. *Korrespondenz Abwasser* **46**(1), 57–65. (In German.)

Siegrist, H., Gajcy, D., Sulzer, S., Roeleveld, P., Oschwald, R., Frischknecht, H., Pfund, D., Mörgeli, B. and Hungerbühler, E. (1992) Nitrogen elimination from digester supernatant with magnesium-ammoinum-phosphate precipitation. In *Chemical Water and Wastewater Treatment II*, Gothenburg Symposium, 28–30 September, Nice, Springer-Verlag, Berlin.

Van der Steen, P., Brenner, A. and Oron, G. (1998) An integrated duckweed and algae pond system for removal and renovation. *Wat. Sci. Tech.* **38**(1), 335–343.

Woods, N.C., Sock, S.M. and Daigger, G.T. (1999) Phosphorus recovery technology modeling and feasibility evaluation for municipal wastewater treatment plants. *Env. Tech.* **20**, 663–679.

WSSCC (1999) Household-centred environmental sanitation. Report of the Hilterfingen Workshop.

Zhao, D. and Sengupta, A.K. (1998) Ultimate removal of phosphate from wastewater using a new class of polymeric ion exchangers. *Wat. Res.* **32**(5), 1613–1625.

21

Potentials of irrigation and fertilization

Gideon Oron, Lailach Ben-David, Leonid Gillerman, Rony Wallach, Yossi Manor, Tova Halmuth and Ludmilla Kats

21.1 INTRODUCTION

The growing demand for water and increasing environmental awareness reinforces today's intensive efforts towards improving the treatment and reuse of domestic effluent (Shelef 1991; Sarikaya and Eroglu 1993; Angelakis *et al.* 1999). These efforts include additional studies to examine pathogens (bacteria, viruses and parasites) in the effluent and their related impact on the soil and plants that will be used for human and animal consumption (Rose and Gerba 1990; Powelson *et al.* 1990; Tanaka *et al.* 1998; Young *et al.* 1992). The

Edited by P. Lens, G. Zeeman and G. Lettinga. ISBN: 1 900222 47 7

nutrient content (primarily ammonia, phosphate and potassium) and additional constituents in the effluent might have adverse effects on agricultural productivity, both in the long and short term. The additional constituents that have to be seriously considered include sodium, calcium and manganese, allowing the assessment of the Sodium Absorption Ratio (SAR) of the effluent and related impact on the soil. Heavy metals (e.g. selenium and boron) that are occasionally contained in the effluent might destroy the soil structure and ultimately cause a reduction in productivity. The long-term effects are primarily related to the accumulation of dissolved solids in the soil, plants and groundwater (Banin *et al.* 2000).

Consequently, research has to focus simultaneously on several related areas to increase the availability of water resource. These include improved treatment of domestic, industrial and agricultural wastes, obtaining effluent with minimal health and environmental risks. The advanced treatment methods should be based on combined biological, chemical and mechanical processes, including methods of membrane technology and disinfection processes with minimal by-products. Methods, primarily drip irrigation for agriculture, for the improved disposal and reuse of the effluent should also be examined.

Several things can be done to solve water shortage problems, and to close the gap between supply and demand:

(1) Use water more efficiently by adapting advanced utilization technologies and water charge policies for use. These include drip irrigation (as oppposed to open-surface and/or sprinkler irrigation methods) and gradually increasing the charge for domestic water consumption.
(2) Import water from external sources. Water from the Sierra Mountains is transported to the southern part of California via national carriers or by other means such as pipes or floating plastic containers. Similarly, wastewater from Tel Aviv, which is treated in the Dan Region plant, is reused in the south of Israel after a soil-aquifer-treatment (SAT) stage (Banin et al. 2000; Ho et al. 1991; Nasser et al. 1993).
(3) Use extra (marginal) water sources such as saline water, run-off water (Pasternak and DeMalach 1987).
(4) Store effluent and run-off water in reservoirs.
(5) Implement advanced treatment methods such as membrane technology.

21.1.1 Characteristics of 'extra' water resources

Several issues must be taken into account when considering extra (marginal) water for utilization:

(1) The potential of the water sources which is characterized by available amounts, current and future anticipated qualities and location, primarily over the long term.
(2) Consumer characteristics as given by seasonal demand distribution, quality level requirements and location of consumption sites.
(3) Watershed management: potential amounts of waters, soil erosion, the migration of nutrients and contaminants.
(4) Environmental control of the use of treated wastewater and/or saline water which might damage the soil, aquifers and, ultimately, the environment.
(5) Extra treatment phases required to meet environmental and health safety standards.
(6) Storage of water required to meet the peak demand periods.

21.1.2 Treated wastewater renovation

Treated domestic wastewater is a relatively stable water source and can be utilized for agriculture, industry, recreation, gardening, industrial plant cooling, and recharge of groundwater (Cromer *et al.* 1984; Oron *et al.* 1986; Burau *et al.* 1987; Asano and Mills 1990). Practically, it is associated with implementing advanced irrigation technologies and water harvesting, for reducing demand and satisfying water requirements. Furthermore, drought events that occur in dry regions amplify water management problems and require long-term measures to reduce the water system's vulnerability, and short-term measures to mitigate drought impacts. Water resources management in arid and semi-arid regions is a complex, multi-faceted task because of the many hydrological, environmental, economic, social and managerial factors that need to be integrated. The holistic approach is appropriate in order to satisfy the demands of users for adequate water quality while ensuring the necessary level of environmental protection.

The use of treated wastewater for agricultural irrigation is attractive for several reasons (Gamble, 1986; Chang *et al.* 1990): (a) water shortage problems can be resolved; (b) large amounts of water can be disposed of during the year with or without storage (under some circumstances water storage can be thought of as an extra treatment phase) with minimal environmental risks (Shelef 1991; Oron *et al.* 1992; Juanico and Shelef 1994); (c) there are economic benefits due to the nutrients that increase the fertilizing properties of the effluent (Oron *et al.*

1986; Nielsen *et al.* 1989). Utilization of sludge as an alternative disposal solution is an additional economic benefit.

Disposal of wastewater and sludge for agricultural utilization diminishes (due to absorption by plants) the potential of pollution and exemplifies the constructive approach of controlling the risk of environmental pollution. Combining effluent and sludge disposal and reuse for agricultural purposes is flexible and can be adjusted according to local conditions (Sarikaya and Eroglu 1993). Despite the worldwide practice of effluent reuse, mainly for agricultural irrigation, there are still conflicting opinions regarding the possible adverse environmental and health impacts (Smith 1982; Nellor *et al.* 1985; Ward *et al.* 1989; Rose and Gerba 1990; Farid *et al.* 1993). These dichotomous views and related considerations frequently stem from the wastewater treatment level and quality control measures. Decisions made regarding effluent reuse usually consider global aspects of the problem only to a very limited extent.

Adequate wastewater treatment, proper reuse and real time flow and quality control of the complementary water and wastewater systems allow environmental and health risks to be minimized. Potential directions for treated wastewater reuse include:

(1) Irrigation, which includes agriculture, parks, cemeteries, green belts, and others.
(2) Industrial use for cooling, boiler feed, concrete mixing, etc.
(3) Under unique conditions treated wastewater can be an alternative source of potable water supply.
(4) Non-potable urban uses such as fire protection, pavement cleaning, air conditioning, toilet flushing, etc.
(5) Use in recreation, nature reserve and environmental sites (lakes and ponds, nature reserve areas, snow making etc.).
(6) Groundwater recharge.

Depending on its quality, effluent can be used, according to regulations and control, for a broad range of applications.

21.1.3 Reuse criteria – the barrier approach

Reuse of effluent, primarily for agricultural irrigation, is commonly subject to reuse criteria. Many of the countries and states in which effluent is reused issue reuse criteria which refer primarily to health and environmental issues. The main control parameters in the reuse criteria include BOD_5 and the suspended matter content, concentration of coliforms and the dissolved oxygen content. However, several biological agents are still not included in the various reuse criteria, partially due to technical and financial monitoring constraints. Also, parameters which are not included are total dissolved solids (TDS), boron, sodium and heavy metals.

The barriers for effluent reuse, which are essentially a series of safety factors, include the wastewater treatment level, the additional treatment due to open-surface storage, the disinfection stage, type of crops, technology of application (e.g. drip versus open surface irrigation) and the timing of harvesting. Each of these barriers is associated with expenses that affect the productivity of the integrated system.

Reuse of effluent for irrigation should be based on the barrier approach, namely a consecutive series of treatment phases that will guarantee minimal risks and a higher productivity.

21.1.4 Subsurface Drip Irrigation (SDI) – the safe effluent reuse technology

One possible option for minimizing health and environmental risks during effluent reuse is to use subsurface drip irrigation (SDI) or onsurface drip irrigation (ODI) systems. The soil under SDI operates as a complementary biological filter, functioning as an additional barrier which minimizes pollution (Guesab 1993; Oron *et al.* 1999). The SDI has the following advantages, primarily because the soil surface layer remains dry:

(1) reduced evaporation
(2) reduced generation of run-off and higher water saving
(3) better control of weeds due to their low germination rate
(4) saving of herbicides, which are commonly required for weed control, thus also a reduction in pollution hazard
(5) improved traffic and maneuver conditions for agricultural machinery
(6) minimal environment pollution due to the restricted flow of effluent to groundwater
(7) improved nutrients uptake (Chase 1985)
(8) minimal environment and health risks (because there is no contact with the applied effluent).

The main limitations of SDI are the need to use extra mobile irrigation systems for the germination of seasonal crops and damage, which might be due to the presence of rodents in the soil, who often puncture the subsurface polyethylene pipes.

21.2 MATERIALS AND METHODS

Field experiments are currently in progress confirming the potential of SDI systems for effluent reuse. The improved yields obtained under SDI can be attributed to the high availability of nutrients and efficient water uptake, as well as to reduced competition with weeds and minimal upper soil and fruit contamination.

21.2.1 The experimental site

Field experiments are in progress in the commercial fields of the Chafets-Chaim kibbutz. Mean annual precipitation is around 600 mm between October and March. The mean minimal ambient temperature is around 8°C during January and around 17°C during August. The mean maximal temperature reaches 20°C during January and 31°C during August. Maximal class 'A' evaporation is around 8 mm/day. The soil in the experimental site consists of about 36% clay, 17% silt and 47% sand.

21.2.1.1 The crops

Cultivation in commercial fields is based on conventional crop rotation that includes cotton, corn, wheat and various vegetables (such as cabbage and paprika). Paprika has been raised almost every year in some of the fields since 1996. Cabbage (*Fictor sp*) was planted on 1st May 1998 at a load of 15,000 plants per hectare. The planting was under wet soil conditions in a field in which wheat was previously cultivated. The planting was at 0.5 m spacing, three rows on one bed, 1.92 m wide. The cabbage was harvested at the beginning of September. Paprika was cultivated during the summer of 1996, 1997 and 1999 while corn was raised during 1996 and cabbage during 1998. Wheat was the winter crop prior to the cultivation of paprika.

Corn is commonly raised at a row spacing of 0.96 m and at a plant spacing of 10–15 cm. One drip lateral serves two adjacent cornrows. The corn is usually planted from April to June and grows for about 100 days. The amount of effluent applied for germination is about 800 m^3/ha (sprinkler irrigation) and the total seasonal amount applied (including drip irrigation) is about 5500 m^3/ha.

21.2.1.2 Irrigation and effluent quality

The vegetable fields are usually irrigated several times for germination (as well as to preserve an upper wet layer and to prevent soil crust formation) by sprinkler irrigation, applying 600 m^3/ha to 800 m^3/ha. All crops are drip irrigated three to six times per week. The total amount applied by drip irrigation to the cabbage (both SDI and conventional onsurface drip irrigation) was around

5500 m^3/ha. The effluent applied is obtained from the adjacent effluent reservoir. effluent reservoir. The nutrient content of the effluent applied for irrigation (mainly ammonia and phosphate) means that the mininum of artificial fertilization to be used (Table 21.1).

In all experiments 2.3 l/hr compensating emitters are used. The emitters are one metre apart on the drip laterals. The subsurface drip systems are at a depth of 40 cm. The entire system is controlled by a series of filters, automatic (computer controlled) valves and injection units for the intermittent addition of chlorine (5 to 15 mg/l every second irrigation) for diminishing emitter-clogging risks.

Table 21.1. Characteristics of the effluent applied for cabbage irrigation at the Chafets-Chaim kibbutz (mg/l) during 1998 (Oron *et al.* 1999)

Parameter*	Range of concentration (7 samples)
pH (–)	7.59–9.15.
EC (dS/m)	1.8–1.9
TSS	29–54
BODt	6.6–23
BODf	6.6–16
NH_4	16.6–49.5
PO_4	11.9–78.7
Na	195–224
K	35.0–39.0
Ca	34.7–51.0
Mg	24.0–32.0
SAR (–)	6.01–7.24

* TSS – total suspended solids; BOD – biological oxygen demand; SAR – sodium absorption ratio.

21.2.1.3 Soil and plant sampling

Soil samples for analyzing the nutrient and pathogen content were taken several times during the growing season. Soil samples were taken close to the emitters and at two additional distances (25 cm and 50 cm). These samples were taken at the soil surface and at depths of 30 cm, 60 cm and 90 cm and occasionally up to a depth of 150 cm. The soil samples were subject to conventional nutrient and pathogen analysis (APHA 1995). On harvesting, the plants and fruits were subject to pathogen analysis.

21.3 RESULTS

21.3.1 Irrigation

The cabbage field was irrigated three times per week, based on data obtained from class 'A' pan evaporation (Figure 21.1). Although the relatively high content of suspended matter (29–54 mg/l, see Table 21.1) in the effluent applied was due to chlorination, no emitter clogging was encountered.

21.3.2 Soil moisture and salinity

The soil moisture profile indicates that under conventional DI the water content close to the soil surface is relatively high (Figure 21.1). Maximal water content under SDI was around the emitter (Coelho and Or 1996). The lowest soil salinity was detected near the emitter location, both for the DI and SDI systems. The highest salinity was observed for SDI near the soil surface. Therefore, the SDI for plants with a deep root system is advantageous (for example, cotton and alfalfa) due to the high water content and low salinity near the emitter. However, for plants with a shallow root system (for example, cabbage), the increased salinity at the soil surface might be a limiting factor, primarily for salt-intolerant plants (Figure 21.2).

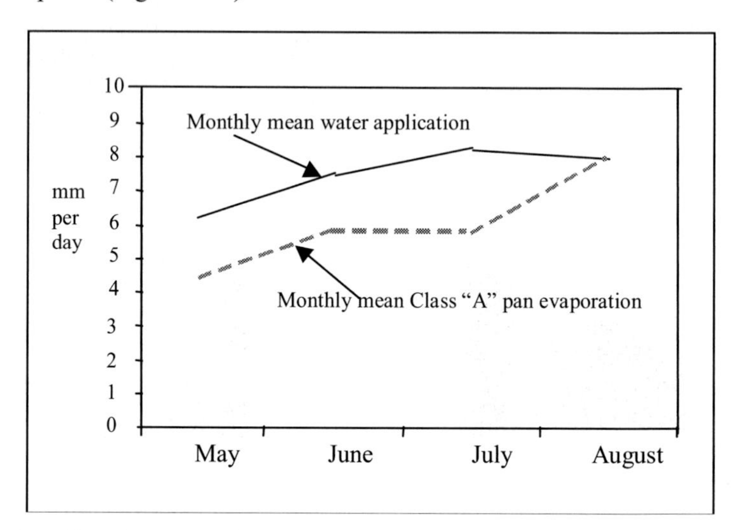

Figure 21.1. Class 'A' evaporation and effluent application rate for cabbage irrigation, 1998.

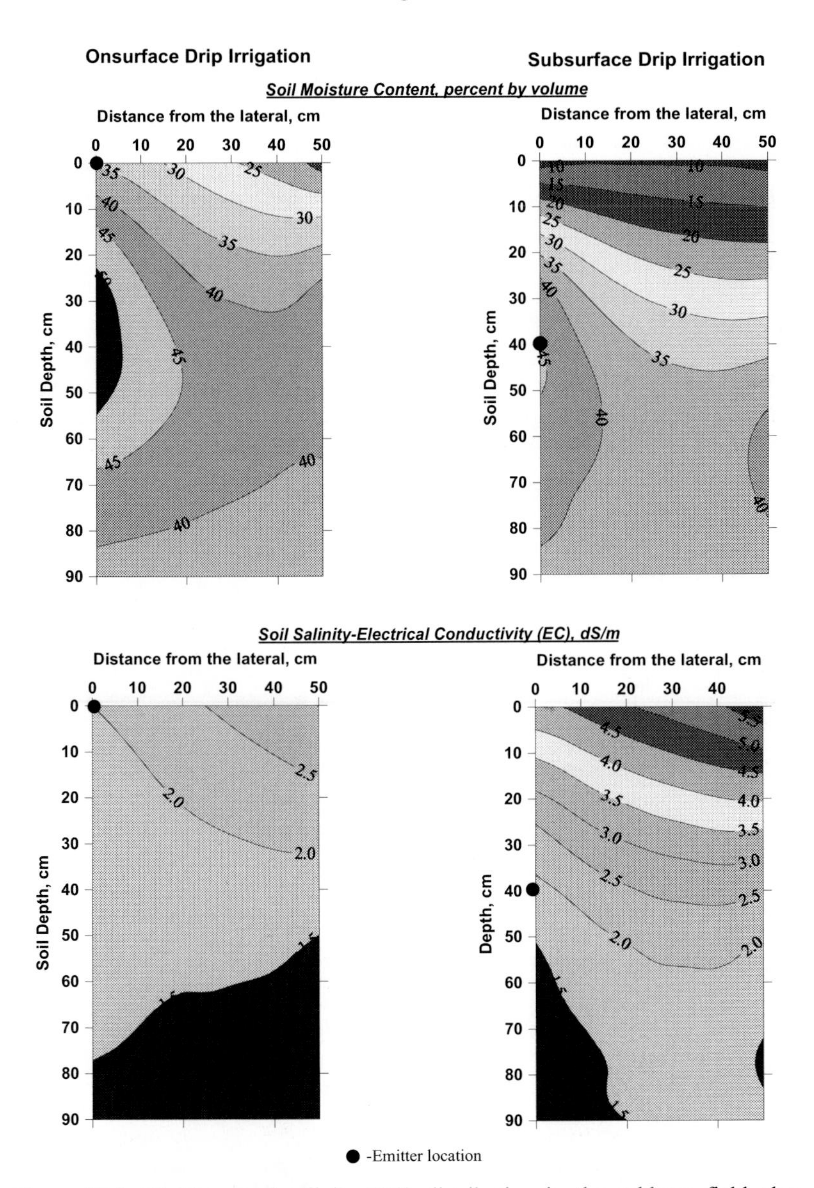

Figure 21.2. Moisture and salinity (EC) distribution in the cabbage field clay soil, Chafets-Chaim kibbutz, 16 July 1998.

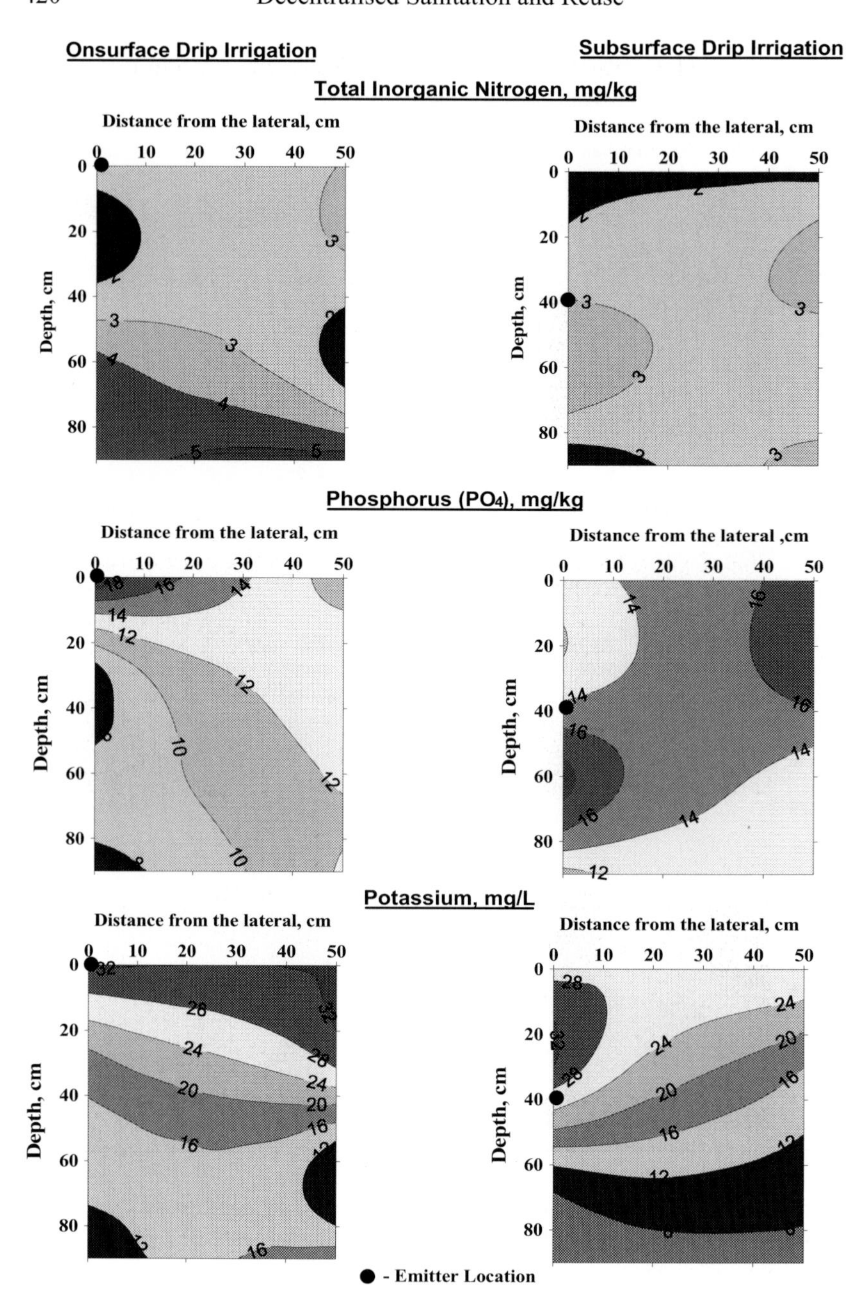

Figure 21.3. Nutrients distribution in the corn field clay soil, Chafets-Chaim kibbutz, 16 June 1996.

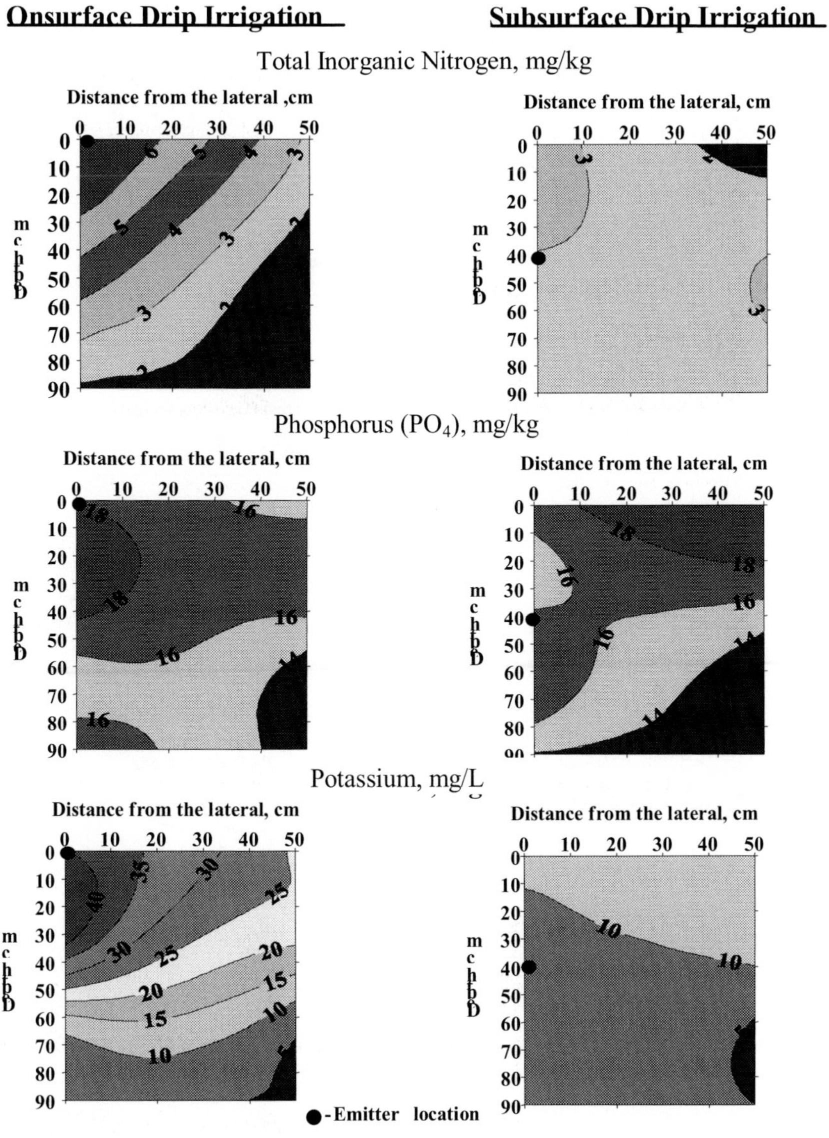

Onsurface Drip Irrigation | **Subsurface Drip Irrigation**

Total Inorganic Nitrogen, mg/kg

Distance from the lateral ,cm | Distance from the lateral, cm

Phosphorus (PO_4), mg/kg

Distance from the lateral, cm | Distance from the lateral, cm

Potassium, mg/L

Distance from the lateral, cm | Distance from the lateral, cm

● - Emitter location

Figure 21.4. Nutrient distribution in the in the paprika field clay soil, Chafets-Chaim kibbutz, 22 July 1996.

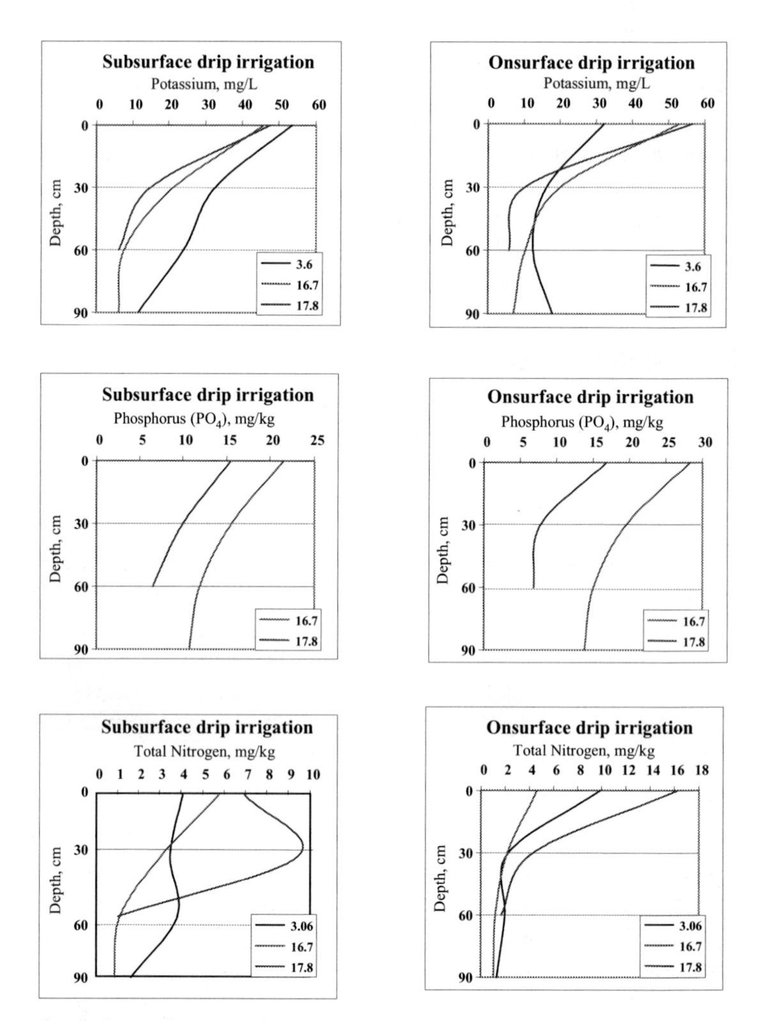

Figure 21.5. Comparing onsurface and subsurface drip irrigation nutrient profiles in the clay soil under effluent application, cabbage field, Chafets-Chaim kibbutz, 1998.

21.3.3 Nutrients in the soil

The distribution of the main nutrients in the soil for DI and SDI is shown in Figures 21.3 and 21.4. Only negligible differences in the nitrogen profiles were detected for the two application systems. Similarly, negligible differences were encountered for potassium for the two drip systems. The nutrient profiles in the soil are in agreement with other studies (Mendham *et al.* 1997). The phosphorus profile differs for the two application systems. Frequently reduced phosphorous (total inorganic phosphorus) can be found near the subsurface emitter and a relatively low content close to the soil surface. A high phosphorus content was detected for the DI system layout, close to the soil surface (Figures 21.3 and 21.4). The findings referring to the distribution of nutrients in the soil are verified by describing the changes with depth (Figure 21.5). Main variations refer to the phosphorous, although for SDI a real shift was also detected for nitrogen once during the 1998 cabbage experiment.

21.3.4 The yields

Improved yields were obtained in the 1996 corn experiment (Table 21.2). The results indicate the potential of attaining increased productivity under SDI. The cabbage yield was measured on 31 August 1998 by harvesting the cabbage heads in two plots of 2m × 2m. The mean yield for DI was 55,500 kg/ha and only 45,800 kg/ha for SDI. The reason for the lower yield under SDI is probably due to the shallow root system, which could not use the water efficiently as could the plants under the conventional DI system. Although the yield was important, the main objective was to examine the effect of the irrigation technology on the soil and plant contamination. Further work is in progress to obtain higher yields by adaptation of an improved application schedule.

The paprika yield during the 1999 experiment (the dry fruits are further processed for spices) was around 8200 kg/ha for SDI and around 5600 kg/ha for DI. The yield result is based on the manual harvest of three sections of 1.0 m × 1.0 m.

Table 21.2. The corn yield (mg/ha) of the 1996 experiment, Chafets-Chaim kibbutz

Irrigation method	Green matter	Cobs
Onsurface drip	5.4 ± 0.3	7.9 ±1.6
Subsurface drip	11.4 ± 1.4	16.2 ± 1.2

21.3.5 Soil and plant contamination

Conventional and modified methods were used to analyze the microorganism content in the wastewater, soil and plants (APHA 1995; Doane and Anderson 1987). Soil and plant contamination depends to a large extent on the applied effluent quality, soil conditions (moisture content; organic matter, salinity), and the technology of application. The microorganism content in the applied effluent is relatively low in comparison with other sites, probably due to the extended detention in the open-surface reservoirs (Table 21.3). The main factors affecting survival of microorganisms are the soil characteristics and moisture (Figures 21.6 and 21.7). This finding holds for the fecal coliforms, and two types of coliphages (F+ and CN-13) were examined (Oron *et al.* 1999). The coliphages can be used as indicators of the virus content in the soil.

Consequently, the fruits under SDI were not contaminated compared with those under DI. This finding strengthens the belief that spray and sprinkler irrigation with effluent may be associated with high health and environmental risks.

Table 21.3. Pathogen content in the effluent applied for cabbage irrigation (count/100 ml)

Date	Fecal coliforms	Coliphage F+	Coliphage CN-13
3 June	2.6×10^4	6.0×10^2	6.2×10^3
11 June	4.6×10^4	2.5×10^2	9.5×10^2
3 July	4.4×10^4	4.2×10^3	7.4×10^3
16 July	3.5×10^4	6.2×10^3	2.4×10^3

21.4 DISCUSSION

Field experiments were conducted in commercial fields with different crops. The fields were irrigated with secondary wastewater from a local stabilization pond system and temporary storage in an open surface reservoir. In all experiments the fields were irrigated by onsurface and subsurface drip systems. Emitters were installed at a depth of approximately 40 cm. Evaluation criteria included the yields, constituents in the soil and contamination indicators such as fecal coliforms and coliphages in the soil and the plants.

It can be assumed that under SDI the soil medium establishes an excellent environment for a variety of complex, complementary, biodegradation processes. The unique soil environment induces the various microorganisms to go through biodegradation processes or to be adsorbed by the soil particles (Taylor 1978; Gerba *et al.* 1981). The effluent and the transportation of the microorganisms through the porous media is restrained by several complex and frequently interactive, physical, chemical and microbial processes (Hickman *et al.* 1989). The efficiency of the microorganism adsorption, inactivation and binding to the soil surface particles depends on the characteristics of the porous media. These properties define the biofiltration efficiency and the potential for microorganisms removal.

No unexpected appearance or plant reaction was discovered in any of the experiments. However, plants with a shallow root system, such as cabbage, might develop less under SDI than under DI. This can be attributed to the depth of the subsurface drip laterals (approximately 40 cm), which prevent adequate water transport in the soil.

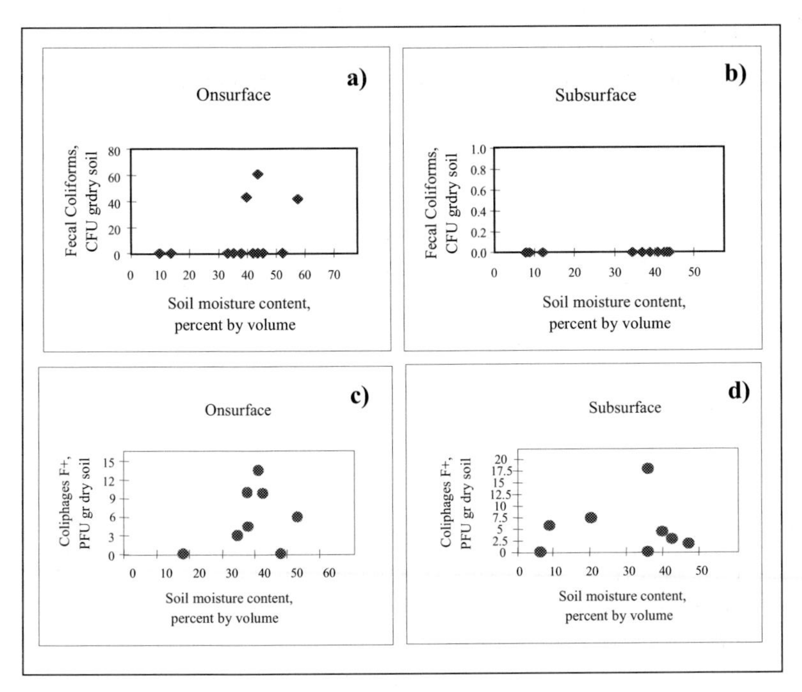

Figure 21.6. Fecal coliforms (a) and (b) and coliphage F+ (c) and (d) profiles in the clay soil after applying effluent for cabbage irrigation, 16 July 1998.

Outdoor experiments are still in progress in various fields with different effluent qualities and crops. Results indicate that improved yields are obtained under SDI, probably due to several agronomic advantages such as high nutrient availability and reduced salinity effects near the point source. In addition, health and environmental risks are diminished due to minimal contact of the disposed effluent with onsurface agro-technology activities. This also includes no direct contact of the above-surface foliage parts of the plants and fruits and the applied effluent. No specific problems of emitters clogging were encountered due to adequate filtering of the effluent at the head control.

Acknowledgments

Partial financial support for this study received from BARD research fund, (project IS-2552-95) and EC-Copernicus research fund (project number IC15-CT98-0105) is gratefully acknowledged. The support of NETAFIM, a drip and irrigation systems manufacturing company, is also very much appreciated.

21.5 REFERENCES

Angelakis, A.N., Marecos do Monte, M.H.F., Bontoux, L. and Asano, T. (1999) The status of wastewater reuse practice in the Mediterranean Basin: need for guidelines. *Water Research* **33**(10), 2201–2217.

APHA (1995) Standard methods for the examination of water and wastewater, 19th edn. American Public Health Association, Washington D C.

Asano, T. and Mills, R.A. (1990) Planning and analysis for water reuse projects. *Journal American Water Works Association* January, 38–47.

Banin, A., Greenwald, D., Negev, I. and Yablekovic, J. (2000) The phenomenon of seasonal decrease in the soil infiltration rate in the recharge basin in the Shorek site (Israel): Investigating the factors and reasons and development of treatment methods. Final report submitted to MEKOROT, the Hebrew University, the Faculty of Agriculture, Rechovot. (In Hebrew.)

Burau, R.G., Sheik, B., Cort, R.P., Cooper, R.C. and Ririe, D. (1987) Reclaimed water for irrigation of vegetables eaten raw. *California Agriculture* July/August, 4–7.

Chang, L.J., Yang, P.Y. and Whalen, S.A. (1990) Management of sugar cane mill wastewater in Hawaii. *Wat. Sci. Tech.* **22**(9), 131–140.

Chase, R.G. (1985) Phosphorus application through a sub-surface trickle system. *ASCE Proceedings of the III International Drip/Trickle Irrigation Congress*, Fresno, California, 18–21 November, Vol. II. I, 393–400.

Coelho, F. E., and Or, D. (1996) A parameter model for two-dimensional water uptake intensity by corn roots under drip irrigation. *Soil Science Society of America Journal* **60**(4), 1039–1049

Cromer, R.N., Tompkins, D., Barr, N.J. and Hopmans, P. (1984) Irrigation of Monterey Pine with wastewater: effect on soil chemistry and groundwater composition. *Journal Environmental Quality* **13**(4), 539–542.

Doane, F.W. and Anderson, N. (1987) *Microscopy in Diagnostic Virology,* Cambridge University Press, Cambridge.

Farid, M.S., Atta, S., Rashid, M., Munnick, J.O. and Platenburg, R. (1993) Impact of the reuse of domestic wastewater for irrigation on groundwater quality. *Wat. Sci. Tech.* **27**(9), 147–157.

Gamble, J. (1986) A trickle irrigation system for recycling residential wastewater on fruit trees. *HortScience* **21**(1), 28–32.

Gerba, C.P., Goyal, S.M., Cech, I. and Bogdan, G.F. (1981) Quantitative assessment of the adsorptive behavior of viruses to soils. *Environmental Science & Technology* **15**(8), 940–944.

Guessab, M., Bize, J., Schwartzbrod, J., Maul, A., Morlot, M., Nivault, N. and Schwartzbrod, L. (1993) Wastewater treatment by infiltration-percolation on sand: results in Ben-Sergao, Morocco. *Wat. Sci. Tech.* **27**(9), 91–95.

Hickman, G.T., Novak, J.T., Morris, M.S. and Rebhun, M. (1989) Effects of site variations on subsurface biodegradation potential. *Journal Water Pollution Control Federation* **61**(9), 1564–1575.

Ho, G., Gibbs, R.A., and Mathew, K. (1991) Bacteria and virus removal from secondary effluent in sand and red mud columns. *Wat. Sci. Tech.* **23**(1–3), 261–270.

Juanico, M. and Shelef, G. (1994) Design, operation and performance of stabilization reservoirs for wastewater irrigation in Israel. *Water Research* **28**(1), 175–186.

Mendham, D.S., Smethurst, P.J., Moody, P.W. and Aitken, R.L. (1997) Modeling nutrient uptake: a possible indicator of phosphorus deficiency. *Australian Journal Soil Research* **35**, 313–325.

Nasser, A.M., Adin, A. and Fattal, B. (1993) Adsorption of polio virus 1 and F^+ bacteriophages onto sand. *Wat. Sci. Tech.* **27**(7–8), 331–338.

Neilsen, G.H., Stevensen, D.S., Pitzpatrick, J.J. and Brownlee, C. (1989) Yield and plant nutrient content of vegetables trickle-irrigated with municipal wastewater. *HortScience* **24**(2), 249–252.

Nellor, M.H., Baird, R.B. and Smyth, J.R. (1985) Health effects of indirect potable water reuse. *Journal AWWA* January 88–96.

Oron, G., DeMalach, Y. and Bearman, J.E. (1986) Trickle irrigation of wheat applying renovated wastewater. *American Water Resources Association (AWRA) Water Resources Bulletin* **22**(3), 439–446.

Oron, G., DeMalach, Y., Hoffman, Z. and Manor, Y. (1992) Effect of effluent quality and application method on agricultural productivity and environmental control. *Wat. Sci. Tech.* **26**(7–8), 1593–1601.

Oron, G., Gerba, C.P., Armon, R., Manor, Y., Mandelbaum, R., Enriquez, C.E., Alum, A. and Gillerman, L.(1999) Optimization of secondary wastewater reuse to minimize environmental risks. Final report, submitted to BARD (Volcani center, Bet-Dagan, Israel).

Pasternak, D. and DeMalach, Y. (1987) Saline water irrigation in the Negev Desert. Paper Presented at the Regional Conference on Agriculture and Food Production in the Middle East. Athens, Greece, January 21–26.

Powelson, D.K., Simpson, J.R. and Gerba, C.P. (1990) Virus transport and survival in saturated and unsaturated flow through soil columns. *Journal of Environmental Quality* **19**(3), 396–401

Rose, J.B., and Gerba, C.P. (1990) Assessing potential health risks from viruses and parasites in reclaimed water in Arizona and Florida, USA. Paper presented at the fifteenth biennial conference of the IAWPRC, Kyoto, Japan, 29 July–3 August, 2091–2098.

Sarikaya, H.Z. and Eroglu, V. (1993) Wastewater reuse potential in Turkey: legal and technical aspects. *Wat. Sci. Tech.* **27**(9), 131–137.

Shelef, G. (1991) Wastewater reclamation and water resources management. *Wat. Sci. Tech.* **24**(9), 251–265.

Smith, M.A. (1982) Retention of bacteria, viruses and heavy metals on crops irrigated with reclaimed water. Australian Water Resources Council, Canberra, p. 308.

Tanaka, H., Asano, T., Schroeder, E.D. and Tchobanoglous, G. (1998) Estimating the safety of wastewater reclamation and reuse using enteric virus monitoring data. *Water Environment Research* **70**(1), 39–51.

Taylor, D.H. (1978) Interaction of bacteriophage R17 and reovirus type III with the clay mineral allophane. *Water Research* **14**(2), 339–346.

Ward, R.L., Knowlton, D.R., Stober, J., Jakubowski, W., Mills, T., Graham, P. and Camann, D.E. (1989) Effect of wastewater spray irrigation on rotavirus infectionrates in an exposed population. *Water Research* **23**(12), 1503–1509.

Young, R.N., Mohammed, A.M.O. and Warkentin, B.P. (1992) Principles of contaminant transport in soils. *Developments in Geotechnical Engineering* **73**, 327.

22

Potentials of urban and peri-urban agriculture in Africa by the valorization of domestic waste in DESAR

F. Streiffeler

22.1 THE GROWTH OF URBAN AGGLOMERATION IN AFRICA

The countries of sub-Saharan Africa which have long been characterized by the lowest degree of urbanization in all world regions, now show the highest rate of growth in urbanization. According to the statistics of the United Nations

Edited by P. Lens, G. Zeeman and G. Lettinga. ISBN: 1 900222 47 7

Development Programme (UNDP 1996, 74 ff.), urban growth rates in Africa between 1995 and 2000 are, on average, 4.3% whereas the average for Asia is 3.2%, for Latin America 2.3%, for Europe 0.5% and for North America 1.2%. Most of this growth is due to migration.

This urban growth is not homogenous in Africa. The urbanization rate is highest in the southern part of Africa (48%) and lowest in the eastern part of Africa (21%); between these regions, northern Africa has an urbanization rate of 45%, western Africa one of 36% and central Africa one of 21%. By the year 2025, it is predicted that 800 million people will live in cities (Deutsche Gesellschaft für die Vereinten Nationen 1996, p. 25). Also, the population is becoming concentrated in the larger cities. In 1950, 80% of the urban population lived in cities with fewer than 500.000 inhabitants. This percentage had dropped by 1994 to 60%, and it is projected that it will drop further by 2015 to 54%. On the other hand, in 1994, 8.1% of all Africans lived in cities with over 5 million inhabitants; this percentage is predicted to reach 19% by 2015.

22.2 ENVIRONMENTAL PROBLEMS IN AFRICAN CITIES

With the growing urban population, urban environmental problems are also growing. But these problems are not the same as those of cities in industrialized countries. Industrial pollution is not as important in sub-Saharan Africa; air pollution by motor vehicles is only important in inner cities but not in the towns where the bulk of the population is living. The main environmental problem is domestic waste. With current levels of population growth, domestic waste is growing proportionately – in sub-Saharan Africa, between 0.6–0.8 kg (UNDP 1996) domestic solid waste is generated per person per day. However, a large part of this domestic waste is not evacuated, and central and richer parts of cities are often the only parts which are served by the municipal refuse collection service.

The reasons for these low collection rates are different: municipalities cannot afford to pay for the waste evacuation of the whole city; so, evacuation is concentrated on certain areas, generally more affluent areas and the suburbs. Often, squatter settlements do not have the financial means to pay for waste evacuation; in these quarters, the streets are also often unpaved, and can be so narrow that a dust cart cannot drive through.

The result is that many cities in sub-Saharan Africa, especially the poorer areas, are very dirty. There are different ways that people use to evacuate their waste from their homes: waste may be thrown on to vacant places, or thrown into rivers or the sea, or they may be buried in the earth.

Figure 22.1. An unofficial dump in a poor quarter of Conakry/Guinea. (Photo: author)

Figure 22.2. Unemptied waste bins at Conakry, Guinea. (Photo: author)

Even if some areas are served by the municipal evacuation service, this service is often irregular, and people leave extra waste beside full dustbins.

However, as well as the official waste evacuation service, there are many small refuse collectors, often as an informal economic activity. So, in a project in Kinshasa, Democratic Republic of Congo which no longer exists, a certain number of households were serviced, for cash, by 'pousse-pousseurs' (handcar-drivers) who transport refuse material to official rubbish dumps or unofficial tips. Later, refuse collection in Kinshasa was transferred to the 'Programme National d'Assainissement' (PNA), but the system collected only 7% of estimated waste material because it received only 5% of the financial means provided for its functioning and only 9% of the means provided for further investments (Lubuimi 1995, p. 133). This also meant that the task of refuse collection was largely taken over by handcar drivers who, in 1992, removed 8% of household waste.

As well as individual refuse collectors, groups may also collect waste. These groups, often in the form of non-governmental organizations (NGOs) have become more common in recent years. They can be inhabitants of an area, often women, who collect waste in order to ensure that the area becomes cleaner. This is to avoid poor hygiene which is a main cause of the rapid spread of infectious diseases.

22.3 URBAN AGRICULTURE IN AFRICA

One astonishing fact is that migrants coming to towns from the countryside do not cease their agricultural activities when they arrive in the city. Many continue to practise their agricultural activities to maintain their livelihood. It is untrue that migrants going from villages to cities are attracted by the 'bright lights' of the town and they all have jobs lined up for when they reach the city. In reality, migrants hope that by coming to the towns, they can increase the number of activities which, in various combinations, can guarantee their survival. Basically, the logic behind this is not so very different from that of the subsistence farmer who combats the natural risks of agriculture by spreading production over different crops, different sites and different growth periods. In towns, however, the strategy of risk spreading is maintained, but it goes beyond agriculture, and the inclusion of other activities from the secondary and tertiary sectors is possible.

But urban agriculture is not only practised by migrants, but also by people who have already lived in the city for a long time; many have no engagement in the formal sector but only in the informal one. Since informal activities are often very competitive, households cannot survive by these alone. They must be complemented by urban agriculture which is generally practised partly for subsistence and partly to sell crops. Even many people engaged in the formal sector practise urban agriculture because their salaries are insufficient to keep their household. Finally, people living in the suburbs of cities also practise peri-urban agriculture as their main activity and consider themselves as 'peasants'.

Land accession for inhabitants in certain urban areas depends on the history of settlement and traditional and modern land rights. However, land use without formal land rights is also widespread.

Figure 22.3. Urban agriculture in Nairobi, Kenya.

A survey of the frequency of urban agriculture in some African cities is presented in the following list quoted from UNDP (1996, p. 55). The presence of urban farmers in selected cities is as follows:

- Burkina Faso: 36% of families in Ouagadougou are engaged in horticultural cultivation or keeping livestock.
- Cameroon: In Yaounde 35% of urban residents are farmers.
- Gabun: 80% of families in Libreville engage in horticulture.
- Kenya: 67% of urban families farm (80% of which are low-income units) on urban and peri-urban sites; (29% of these families farm in the urban areas where they live. 20% of urban dwellers in Nairobi grow food in urban areas).
- Mozambique: 37% of urban households surveyed in Maputo produce food, while 29% raised livestock.
- Tanzania: 68% of families in six Tanzanian cities engage in farming; 39% raise animals.
- Uganda: 33% of all households within a 5 km radius around the centre of Kampala engaged in some form of agricultural activity in 1989.

- Zambia: A survey of 250 low-income households in Lusaka showed that 45% grow horticultural crops or raise livestock in their backyards or gardens on the periphery of the city.

22.4 DIFFERENT LOCATIONS OF URBAN AGRICULTURE IN AND AROUND THE CITY

Before presenting a typology of urban agriculture based on location of plots, something must be said of the special structure of the fast-growing African cities.

Some towns exist with one centre, generally the historically oldest part of the town with the highest population and housing density; but many more cities are less centralized and more 'pluralistic', in the sense that they consist of several more or less centralized quarters or districts. In such cases, we have a declining population and building density with increasing distance from these sub-centres with vacant land, green belts and zones of lower building density between them. One implication of this multicentred structure is that the spatial distribution of urban agriculture does not correspond to the von Thünen model which predicts declining intensity in land use in proportion to the distance from one presumed central point. There is also a tendency toward the development of spontaneous 'satellite-settlements' of the poor migrants around very big cities (such as Kulinda, west of Nairobi); we thus have a metropolis as well as spaces with low density between the new settlements.

It is possible to distinguish three types of urban agriculture where location is the distinguishing criterion: intra-urban agriculture, household gardens and peri-urban agriculture.

The category of intra-urban agriculture covers spaces between and at the peripheries of the different areas or agglomerations constituting the whole city, 'no man's land' (though not from a legal point of view) along roads, rivers, railways, and so on, as well as seasonally flooded land or land on slopes unsuitable for buildings and parks. Whereas these spaces can form distinctive areas, there is still another category of squatter land which may be public property (e.g. land around public buildings) or privately owned vacant land.

Generally, people who practice urban agriculture on these spaces are the poorest. The fact that they often have no land rights also has implications for the type of crops that are cultivated. When there is no land security, no perennial or even annual crops are cultivated, but only fast-growing short-cycle crops such as leaf vegetables. However, the designation of the legal situation as a total absence of land rights would only be partly true; often, a local legitimacy develops with respect to users' rights, when a person in this area grows plants with a certain continuity and visibly invests labour in order to maintain his plot in a good state. Land can also be rented.

The importance of household gardens in many African countries cannot be overestimated. Unlike the situation in rich countries where gardens are often only cultivated for pleasure, in developing countries cultivating urban household gardens may constitute an essential component of the spectrum of activities ensuring a plentiful source of food and thus survival. Production is generally for consumption. Vegetables and fruit are typically grown in household gardens, as dietary supplements. The advantages of household gardens are the following:

- Usually, land is more secure in that there are fewer problems with property rights. At least, the household garden is no less secure than the house or the hut. Because in many traditional legal African systems the right to plant trees is closely connected to land rights, it is logical that many plots are not only used for vegetables but also for trees and bushes.
- Work in household gardens is less time-consuming because the distance from house to field is short. In most cases the cultivation of household gardens is the work of women, who use kitchen waste as compost.
- Household gardens can be irrigated, unlike, for instance, roadside fields.
- There are fewer problems with theft which is one of the major problems of intra-urban agriculture in general, particularly with roadside fields.

Naturally, the number and size of household gardens depends on the town and location. In Lusaka, even in high density areas, half of the residents maintained a vegetable garden with an average size of 30 m^2 (Jaeger 1985).

The category of peri-urban agriculture is the most heterogeneous. First, there is the sub-type of village agriculture which is generally transformed by the inclusion of villages into the urban periphery. If there is no demand for land in peri-urban areas, land can be leased, often by the traditional custodians of the land, to urban-based people for one or two years.

Since colonial times, the state has organized agricultural projects in peri-urban areas in order to feed the cities and to create jobs. These projects were generally organized as cooperatives and used land which, due to the land legislation of the colonial and post-colonial periods, was declared to be state-owned: one example is the great agricultural project in the Ndjili valley near Kinshasa, Democratic Republic of Congo. On the other hand, such projects in the urban periphery, which are often technically highly developed and often supported by external assistance, also exist in a form of private ownership.

Often, the people working on these capitalist farms are given an allowance to cultivate the land of the capitalist farm. However, because of the disappearance of clear traditional land rights in peri-urban areas, such as in the supplanting of 'traditional' ownership by 'modern' ownership, land is also used by poor urban residents, especially recent migrants, who have not found intra-urban spaces for agriculture because they have not yet acquired informal land rights. This is the most underprivileged group because they have no land security as well as long distances to travel between their homes and fields.

22.5 PROBLEMS OF URBAN AGRICULTURE

Urban agriculture encounters many problems, land access being the most important. This chapter will focus on two problems, plant nutrition and plant disease.

The first problem arises from the fact that because of the restricted space for agriculture in urban and peri-urban areas, agriculture is more continuous than in rural areas where shifting cultivation with regular fallow periods after some years is usual. Moreover, in urban agriculture the integration of plant production and animal husbandry does not exist. Finally, the use of chemical fertilizers is more the exception than the rule, and their use is generally found only in 'professional' cultivators. The only type of urban agriculture which encounters no problems with plant nutrition are house gardens where household waste is used to irrigate and feed plants.

The second problem is the frequency of plant disease, which also plays an important role for urban producers. I carried out a survey in the town of Kisangani (now in the Democratic Republic of Congo), where I asked 426 cultivators about the main problems they experienced with urban agriculture. Plant diseases were mentioned most often (30.5%) (Streiffeler 1994). There are many reasons which can explain the high frequency of plant disease in urban areas:

- the often poor air quality
- imported pests have a greater role in urban than in remote areas
- pesticides are often used in peri-urban agriculture, but not correctly, so that plant diseases may transfer to other types of plants
- traditional plant protection techniques such as mixed culture are not as widespread as in rural areas
- a lack of plant rotation (which is common in rural areas).

22.6 AN INTEGRATIVE SOLUTION

Given the importance of urban agriculture for the survival of urban households and the high quantities of waste generated in many urban quarters, it is a tempting idea to try to resolve the plant nutrition problem in poor parts of African cities by integrating it with the hygiene problem and using suitable waste as compost for plots in urban agriculture. This integrative solution to two problems would also seem to be applicable to waste water. This is important because water is becoming increasingly scarce in many cities, especially in the poor areas. Certainly, this integration of waste recycling and plant nutrition has always existed on a small scale and without external promotion in the form of plant nutrition in kitchen gardens by using kitchen refuse and by means of 'unorganized' refuse collectors who collect the organic material to bring it to places where it can be composted.

The idea of promoting this integration in a organized manner and also in external development cooperation projects was pioneered in the 1980s by the Food Energy Nexus Programme of the United Nations University at Tokyo, Japan (Sachs and Silk 1990). Later, this solution was also seen in the perspective of 'sustainable cities' which developed in the 1990s, and was also recommended in the Habitat conference in 1996 and in its precursors (such as Abidjan in 1995). This perspective was also included in the 'brown agenda' of the World Bank (Leitmann 1994).

22.7 DECENTRALIZED OR CENTRALIZED SOLUTIONS?

In the past and in large European cities, only one alternative existed: centralized composting stations (or urban gardeners could themselves compost the organic waste individually, or for a shared garden of a house). These composting stations have sophisticated technology and are expensive. These technologies were transferred to large African cities to discover if the technology would work there, but these trials were, on the whole, unsuccessful: the technologies were generally too expensive for the municipalities which receive a low share of the national budget, and there were often problems in obtaining spare parts for these imported technologies because they were not locally available. Thus, decentralized composting using simple technologies is still more relevant and useful to developing countries.

22.8 PROBLEMS WITH AN INTEGRATED SOLUTION

Despite the successes that this solution has achieved over time there exist some problems, which are mentioned in the following sections.

22.8.1 The safety of the composting material

This topic is in relation to the composition of the compost material and to the process of composting. The first problem stems from the fact that the domestic waste is often not separated and can include toxic and infected material, and the producers of this waste are often not informed about possible dangers in utilizing it. This is also true for peasants using waste material, as has been found, for instance, in the peri-urban area of Kano, Nigeria. Also, organic waste is often collected by informal collectors without any knowledge of how to produce safe compost, and this can be problematic.

22.8.2 Administrative problems

As to the waste collection, there is the problem that the financial resources of the municipalities are generally only sufficient to carry out waste evacuation of a minority of urban inhabitants.

Urban agriculture is rarely promoted by local government. Whereas in the past, urban agriculture, especially on areas without land title (this form of urban agriculture is very common) was often stopped or prevented by the local government, it is now more tolerated; but this toleration does not mean that agriculture on this land receives the necessary services, such as high-growth seeds, agricultural advice, and so on.

22.8.3 Too few refuse collection trucks

Even in the capitals of sub-Saharan Africa the number of dust carts is totally insufficient, and a high proportion of these dust carts are broken. In Dar es Salaam, it is estimated that, due to the lack of sufficient and functioning trucks, only 22% of urban waste is collected.

22.8.4 Economic problems

In various African countries, the economic situation vis-a-vis the commercialization of composted organic waste is such that the price received from the sale of compost is not sufficient to cover the costs of collection and of composting the organic material. For farmers the transport costs for compost are higher than for mineral fertilizers, since compost is voluminous. The production of chemical fertilizers is also sometimes subsidized.

So, households that are served by the waste collection must make their own contribution. However, this can be a problem for very poor households; thus, the participation rate of urban households in waste collection is often only 50% or less.

Another problem is the fact that peasants are not interested in buying compost all year round. In the dry season, for instance, or after the harvest, there is no demand for compost. Theoretically, compost could be stored or transformed into peat, but fresh compost is more popular.

22.8.5 Cultural problems and advantages

The acceptance of the use of compost for plant nutrition depends on cultural values and traditions. In tribal languages of Senegal like those of the Fulbe or the Wolof welcome the introduction of compost as fertilizer because manure is traditionally used to fertilize fields. On the other hand, there is confusion when it comes to differentiating between compost and manure.

Cultural barriers also exist to the use of composted waste for plant nutrition. In Yaounde, Cameroon some farmers were found to be against the use of urban waste, especially waste that contained human excreta. These cultural attitudes must be taken into consideration in the use of organic waste for agricultural purposes

22.9 A CASE STUDY OF WESTERN AFRICA

This section presents information from a research project on the use of composted organic wastes from urban households for phyto-sanitary purposes in the peri-urban agriculture of Western Africa, financed by the European Union. This project was realized in collaboration with a phyto-pathologist from the Faculty of Agriculture and Horticulture at the Humboldt-University at Berlin, Germany, another phyto-pathologist from the Scottish Agricultural College at Ayr, United Kingdom and National Agricultural Research Institutions in Senegal, Togo and Guinea. The objective of this research project was to enhance the resistance of plants to plant diseases by specific composts.

In Western Africa so far, most centralized composting waste projects have broken down for economic reasons once external funds have stopped coming in (for example, Dakar in 1965, Cotonou in 1999). Compost is more expensive to produce than manure. The quality of compost from centralized compost stations is low. In Senegal, the transport costs of compost are too expensive for agriculture more than 7 km away, therefore decentralized compost stations seem to be a solution.

In Senegal, the project continues the former activities of the RUP department (Relay for participatory urban development) of the ENDA Third World which no longer exists. This ENDA project began in 1994 and had the main objective of cleaning two urban areas. It was carried out in the town of Rufisque, 27 km from Dakar in Senegal.

In the beginning, a forum made up of members of the municipality, local authorities, NGOs, local people and peri-urban cultivators was organized. Each household that participated in the waste evacuation had to pay 500 FCFA (approximatively 0.90 US$), whereas communal taxes for waste evacuation were usually unpaid.

In Conakry, all small enterprises helped to realize the waste collection. Households had to pay waste fees directly to the enterprises involved, who collected from house to house and stocked the waste in containers to be emptied by the government-run waste transport service.

In Senegal, in the beginning of the project, the waste was not separated on the level of the household but in the compost station. Three collectors engaged by the project took the waste from households on barrows pulled by donkeys. At the compost station, six sorters separated the usable organic material from other material.

The composition of waste varies in African cities depending on climatic conditions, the paved surface and the soil type in backyards, feeding habits and economic activities. So, in Lomé, Togo sand made up 49% of waste, and this could be explained by sweeping sandy soils in non-cemented backyards. On the contrary, in Conakry, the percentage of biomass in waste is 50–66% due to high biomass production and rocky soils. Richer households produce more packing waste and less sweepings as their backyards are often paved. Metals and plastic, so long as they are usable, are recycled at household and at collector level, and are even sold at market.

The quality of the raw material naturally influences the quality of the compost. Therefore the project began to advise households to separate the waste into organic and non-organic waste. This was reinforced by the distribution of dustbins. The separation of waste at household level still has to be checked at the compost station.

22.9.1 Processes in the compost station

After separation of the different components, the usable organic material is laid on a cemented platform which contains small channels conducting liquid substances into a collecting basin; then, the heap is irrigated; afterwards, the liquid collected is poured onto the organic material at regular intervals.

Figure 22.4. A selection of compostable and non-compostable material.

Then the fermentation process begins. Changes in temperature are indicated in Figure 22.6.

During the process the heap heats up to 70°C, and this temperature is sufficient to kill the pathogens. If the temperature exceeds 70°C, the separators turn the heap to avoid leaks of transformation.

After 14 days, the compost heap is turned over. All the 15 days, tests of phyto-toxity are made. So, e.g., it is tested if the tabaco-mosaic-virus is not more living in the organic material; normally, this virus doesn't survive the peak of 70. After 2–3 months, the compost is ready and sold to peri-urban cultivators or to a tree-nursery.

Figure 22.5. The ripe compost on the compost station.

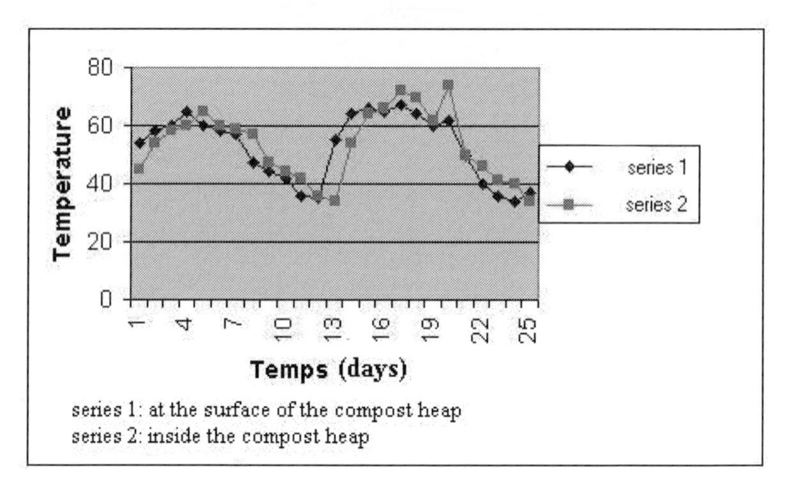

Figure 22.6. Temperature changes during composting (series 1: on the surface of the compost heap; series 2: inside the compost heap).

In the research project, the plants that were most susceptible to disease were first identified. These plants were the tomato in Senegal, the cassava in Togo and the potato and the sweet potato in Guinea. Then, compost extracts at different doses were applied, and laboratory and field experiments were carried out to study the application of this compost in reducing the rates that plants succumbed to disease. As the project is still continuing, final results are not yet available, but first results are promising.

The aim of our research project is to look for an alternative valorization of compost. If compost can be used to treat plant diseases it will be just like using pesticides. Since pesticides are more expensive than manure, chemical fertilizers and garden mould, the income from the sale of the compost will rise.

22.10 THE USE OF LIQUID WASTES

In dry areas and in areas characterized by seasonal dryness which are far from rivers and streams, the reuse of urban sewage has potential for urban agriculture. On the one hand, there is a huge quantity of urban sewage that could be utilized in such a way, and on the other hand, drinking water is generally too expensive to use for agriculture. This interest in sewage is also reflected in the fact that urban cultivators often bore illegally into wastewater channels in order to steal the water to irrigate their fields.

However, the use of non-treated wastewater to irrigate crops can be very dangerous: it is absolutely necessary to treat the wastewater before it is used in such a way. Generally, since experiences with highly sophisticated systems for wastewater treatment have not worked well in many African states and in other parts of the developing world, there is now a tendency towards decentralized systems. These systems work with local available technologies and do not depend on foreign countries for their repair, and they are cheaper than more sophisticated systems. As well as anaerobic techniques, many other techniques exist for treating waste and wastewater, such as lagoons, quagmires for treating sewage, sprinkling systems using wood plantations, meadows using *pistia stratiotes* or gravel (see Niang 1999). In experimentation at Cambaréné in the peri-urban region of Dakar, these techniques were compared, to see which system produced the best quality treated water. The result of these experiments was that the sprinkling system using gravel, the traditional lagoon and the meadow using *pistia stratiotes* produced the highest quantities of water (95%, 82% and 70% respectively). As to the quality, the result was that no technique corresponded to the requirements of the World Health Organization (WHO). It was also found that no one technique out of all those tested was superior in all parameters (such as the removal of faecal coliforms, streptococci and other parasites).

22.11 CONCLUSIONS

Despite its popularity, practising urban agriculture in order to clean African cities does not constitute a miracle solution. The development of an integrated urban solution with the collection of organic urban wastes and their use in urban and peri-urban agriculture is still only in its infancy. At present, only a small percentage of urban waste is collected, and this is mainly in better-off areas. Poorer areas are very dirty, and they do not have other infrastructure such as water, wastewater channels, electricity and so on. There is now a certain tendency in international organizations to argue that the public sector should retire from these services and let private enterprises provide them, ranging from large private organizations to small individual waste collectors with a handcar. However, it seems that such a complete privatization would be no better than a complete public service system: Without any public intervention, this private system could not function throughout all urban areas; as well as financial support, coordination and sensitization campaigns are also needed. So, public–private cooperation is needed. This cooperation structure should also include external agencies which should make their contribution in a material way – by using their own products in environmental technology, by giving financial support for the development of local and locally adapted environmental technology – but also in a non-material way by communicating their own experiences and scientific results. In this concept of an union, the researcher would also have an important role in the development of a better urban environment.

However, the best technical solutions can fail if they are planned and executed without popular participation. As Mabogunje (1990), the well-known Nigerian specialist in urban development, described, predatory and individualistic class attitudes prevail in many urban residential areas of Nigeria which have created a sense of lack of rights, and means that public institutions are misused by those in power for their own personal purposes. In order to change this tendency and reinstall a sense of community, the author suggests transferring the British model of neighbourhood councils to Nigerian cities. The functions of these neighbourhood councils are the followings:

- to organize or stimulate self-help within the local community to improve the quality of life for the residents as a whole (for instance by clearing dumped materials from derelict sites)
- to help those in the community in need of special facilities
- to represent the needs and wishes of the local community to central and local government and local firms.

It seems likely that such institutions would be highly effective in ensuring a more equal distribution of the benefits of a developed urban agriculture. Moreover, such institutions could also help to overcome the overall lack of cooperation and coordination which is one of most fundamental problems of urban agriculture.

22.12 REFERENCES

Deutsche Gesellschaft für die Vereinten Nationen (1996) *Weltbevölkerungsbericht,* UNO-Verlag, Bonn. (In German.)

Jaeger, D. (1985) Subsistence food production among town dwellers – the example of Lusaka, Zambia. Research paper, Royal Tropical Institute, Amsterdam.

Leitmann, J. (1994) The World Bank and the Brown Agenda. *Third World Planning Review* **16**(2), 117–127.

Lubuimi M.L. (1995) Exemple de la stratégie de lutte contre la pauvreté et le développement efficace: le rôle du secteur informel dans la gestion des déchtes à Kinshasa. In Marysse, St. (ed.) *Le secteur informel au Zaire. Partie I: Concept, ampleur et méthode,* Universitaire Faculteiten Sint Ignatius, Antwerp. (In French.)

Mabogunje, A.L. (1990) The organization of urban communities in Nigeria. *International Social Science Journal* **125**, 355–366.

Niang, S. (1999) Utilisation des eaux usées brutes dans l'agriculture urbaine au Sénégal: bilan et perspectives. In Smith, O.B. (ed.) *Agriculture urbaine en Afrique de l'Quest*, International Development Research Center, Ottawa. (In French.)

Sachs, I. and Silk, D. (1990) *Food and Energy. Strategies for Sustainable Development*, United Nations University Press, Tokyo, S. 34–84.

Streiffeler, F. (1994) L'agriculture urbaine en Afrique: la situation actuelle dans ses aspects principaux. In *International Foundation for Science: Systèmes Agraires et Agriculture Durable en Afrique sub-Saharienne.* Fondation Internationale pour la Science, Stockholm, pp. 437–454.

United Nations Development Programme (UNDP) (1996) *Food, Jobs and Sustainable Cities. Urban Agriculture*, UNDP, New York.

23

Guidelines and regulations on wastewater reuse

M. Salgot and A.N. Angelakis

23.1 INTRODUCTION

Advances in the effectiveness and reliability of wastewater treatment technologies have improved the capacity to produce reclaimed wastewater that can serve as a supplemental water source, in addition to achieving water quality protection and pollution abatement requirements. In developing countries, particularly those in arid parts of the world, reliable low-cost technologies (both for treatment and reuse) are needed for acquiring new water supplies and protecting existing water sources from pollution (Angelakis *et al.* 1999). The implementation of wastewater reclamation, recycling, and reuse promotes the preservation of limited water resources in conjunction with water conservation and watershed protection programs. In the planning and implementation of

Edited by P. Lens, G. Zeeman and G. Lettinga. ISBN: 1 900222 47 7

water reclamation and reuse, the intended water reuse applications dictate the extent of wastewater treatment required, the quality of the water produced, and the method of water distribution and application (Asano 1998).

Only in few countries and states has water reclamation and reuse become well-established and the value of reclaimed water been fully recognized. In these countries and states laws and regulations exist that mandate water reuse under certain conditions. In some US states (such as Texas) regulations require that a study should be conducted to investigate the possibility of using reclaimed water for applications that currently use potable water or freshwater (Crook and Surampalli 1996). In the United States, as of March 1992, 18 states have adopted regulations regarding the reuse of reclaimed water, 18 states had guidelines or design standards, and 14 states had no regulations or guidelines (US EPA 1992). In states with no specific regulations or guidelines on water reclamation and reuse, programs may be permitted on a case-by-case basis. In addition, various countries (such as Israel, South Africa, and Tunisia) have also established regulations or guidelines. Finally, in some other countries (such as Cyprus, Spain, Italy and Greece, regulations for using reclaimed water for irrigation are being prepared. Regulations refer to actual rules that have been passed and are enforceable by governmental agencies. Guidelines, on the other hand, are not enforceable but can be used in the development of a reuse program (Angelakis and Asano 2000; Angelakis and Bontoux 2000).

Standards, criteria, rules, guidelines, good practices and other measures which try to regulate wastewater reclamation and reuse can be prepared, as for any activity related to the environment, and made public before they are adopted. This may generate comments from the public and subsequent modifications that can influence, often decisively, the type of criteria which will be finally published and enforced.

When requirements or norms related to water reuse are fixed, wastewater reclamation is necessary, thus guaranteeing the desired quality of water. Reclamation and reuse go together in any type of operation of this kind which is made legally or planned.

The objective of this chapter is to present the basic concepts in developing guidelines or regulations on wastewater reclamation and reuse. In addition, it includes the development of such criteria in various countries with brief reviews of existing criteria, guidelines and regulations.

23.2 USES FOR RECLAIMED WASTEWATER

In the planning and implementation of wastewater reclamation and reuse, the reuse applications (see Table 23.1) usually dictate the wastewater treatment needed and the degree of reliability required for the treatment processes and operations.

In modern society, water is vital for agricultural irrigation. For this reason, it seems logical that in water reuse, irrigation predominates. As a consequence, regulations issued for wastewater reuse are specially developed for agricultural irrigation. In this case, reclamation and reuse criteria focus principally on sanitary and environmental protection, and usually refer to (Crook 1998): (a) wastewater treatment, (b) reclaimed wastewater quality, (c) treatment reliability, (d) distribution systems, and (e) control of areas where reclaimed wastewater is reused.

In spite of this, existing criteria can change according to the main objective of the wastewater application to soil: (a) additional wastewater treatment by soil (with or without water reclamation), (b) water elimination, and (c) irrigation water for agricultural areas, golf courses and public access areas.

For uses other than irrigation, guidelines or regulations are not so well developed, mainly because there are less reuse cases and opportunities. There are certain records in relation to aquaculture and cooling or other industrial uses (Asano 1998). Regulations can also be found for recreational uses and for various non-potable urban uses.

In very few cases, regulations have been issued for reuse as tap water, as is the case for Windhoek in Namibia (Odendaal *et al.* 1998). Wastewater reuse regulations have been proposed in California targeted at artificial groundwater recharge applications (Asano 1998). It is possible that in future, when reuse fields are expanded, new regulations to cover different reuse possibilities will be further developed.

23.3 REUSE CONDITIONS

The acceptability of reclaimed wastewater for any specific use depends on its physical, chemical and microbiological quality; and mainly on the sanitary risk related to this quality. In any case, an adequate infrastructure for reuse must exist. This infrastructure includes the water treatment and wastewater reclamation facilities, the distribution network, and storage facilities if necessary or compulsory .

One obvious control measure, forgotten from time to time, is the assessment of treatment reliability, and of the entire reuse infrastructure. The design and performance of distribution systems is important to guarantee that reclaimed wastewater does not degrade before its use and is not used improperly. Open-air storage can result in water quality degradation due to microorganisms, algae or suspended solids; and it can cause bad odours or give colour to reclaimed wastewater. Nevertheless, if they are properly managed, open storage systems can improve the quality of the resource.

Table 23.1. Categories of municipal wastewater reuse and potential issues/constraints (Source: Tchobanoglous and Angelakis 1996)

Wastewater reuse categories	Issues/constraints
Agriculture irrigation Crop irrigation Commercial nurseries Landscape irrigation Parks School yards Freeway medians Golf courses Cemeteries Greenbelts Residential	(1) Surface and groundwater pollution if not managed properly, (2) marketability of crops and public acceptance, (3) effect of water quality, particularly salts, on soils and crops, (4) public health concerns related to pathogens (bacteria, viruses, and parasites), (5) use for control of area including buffer zone, (6) may result in high user costs
Industrial recycling and reuse Cooling water Boiler feed Process water Heavy construction	(1) Constituents in reclaimed wastewater related to scaling, corrosion, biological growth, and fouling, (2) public health concerns, particularly aerosol transmission of pathogens in cooling water
Groundwater recharge Groundwater replenishment Salt water intrusion control Subsidence control	(1) Organic chemicals in reclaimed wastewater and their toxicological effects (2) total dissolved solids, nitrates, and pathogens in reclaimed wastewater
Recreational/environmental uses Habitat wetlands Lakes and ponds Marsh enhancement Streamflow augmentation Fisheries Snowmaking	(1) Health concerns of bacteria and viruses, (2) eutrophication due to nitrogen (N) and phosphorus (P) in receiving water, (3) toxicity to aquatic life
Miscellaneous uses Fire protection Air conditioning Toilet flushing	(1) Public health concerns on pathogens transmitted by aerosols, (2) effects of water quality on scaling, corrosion, biological growth, and fouling, (3) cross-connection
Aquaculture Potable reuse Blending in water supply Pipe to pipe water supply	(1) Constituents in reclaimed wastewater, especially trace reservoir organic chemicals and their toxicological effects, (2) aesthetics and public acceptance, (3) health concerns about pathogen transmission, particularly viruses

Control of the areas where wastewater is reused is paramount to reduce sanitary and environment risks. It must be reiterated that risk reduction to acceptable levels is the final objective of all guidelines and regulations related to the reuse of water.

When considering wastewater reclamation and reuse, every prospective user must be aware of the legal and economic limitations existing in his country.

Regulations can be based on the establishment of the end-product (reclaimed wastewater) quality criteria or in the definition of the reclamation equipment of wastewater (compulsory or as a reference). In both cases, equipment and regulations could be complemented with the definition of Good Reuse Practices or similar indications.

In some non-agricultural wastewater reclamation practices, different legal problems can crop up, usually related to the water or resources legislation of every country. For example, when groundwater is recharged with reclaimed wastewater, it must be clear who is the owner of the water in order to avoid problems. In the United States, these problems can lead to diverse lawsuits (National Research Council 1994). In other cases, such as in Spain, when an aquifer is recharged, groundwater belongs to the State.

In non-coastal areas and arid climates; it may be that urban treated or untreated wastewater is the only water that flows in streams. The users downstream rely on that flow and have rights to it. In those cases, it is not feasible to reuse wastewater upstream for other purposes.

23.4 FACTORS AFFECTING DEVELOPMENT OF EFFLUENT QUALITY CRITERIA

Legal dispositions could have a different rank, so certain differences must be established among them. It is to consider that in some cases, regulations have law status (California) and are more enforceable than recommendations, theoretically without legal coverage (WHO 1989; US EPA 1992; for reports on Andalusia and Catalonia (1994) see Salgot and Pascual 1996). In any case, a country by country, or lower administrative level, study of the legal specificity must be performed.

Usually, reclaimed wastewater quality is traditionally fixed, independently from other considerations, using standards. Standard figures depend on several concepts such as:

(1) economic and social circumstances
(2) legal capacity from different entities and implicated administrations
(3) human health/hygienic degree (endemic illnesses, parasitism)

(4) technological capacity
(5) previously existing rules and/or criteria
(6) crop type
(7) analytical capacity
(8) risk groups possibly affected
(9) technical and scientific opinions
(10) other miscellaneous reasons.

Three types of factors can be distinguished: technical or technological (analytical, treatment methods and capacity, knowledge, etc.); legislative and economical (criteria, socio-economical, legal competence, etc.) and health-related (sanitary state, disease, risk groups, etc.).

All of these standards and quality regulations have been a matter of discussion among scientists, health and legislation officers and engineers, because of the numerical expression of such standards and secondarily because of the parameters to be controlled. Much discussion has taken place among research teams and regulating bodies, even from the same country, on the quality that reclaimed wastewater must meet for reuse without risk, or with an acceptable risk degree. Researchers should be divided into three categories according to the origin of the standards they apply, especially when dealing with reuse for irrigation purposes. The other possibilities for reuse (groundwater recharge, industrial uses, and others) are less common and follow mainly the pattern established for irrigation use.

WHO-related researchers rely on the document 'Health Guidelines for the Use of Wastewater in Agriculture and Aquaculture' issued in 1989 for defining their standards. California-related teams have literature regarding the Standards of the United States, but the 1978 (California) and 1992 (US EPA) publications of that State are the most important. Other countries (France, Israel, and the former USSR) have developed their own standards.

(a) Socio-economic factors. Apart from the standards, additional factors must be considered when trying to implement a wastewater reuse program, among them the socio-economic situation of the relevant country. This leads to a main consideration, which is the economic one. Obviously, when 'first-hand' water is available at a reasonable price or for free, it is not worthwhile to reuse wastewater unless a special protection of the environment is needed or there are other valid reasons.

Nevertheless, the main budgetary limitation for reclamation and reuse is the cost associated with the advanced treatment methods needed to reach the qualities established in the regulations and also associated with the costs for water storage and distribution. In this sense, legislation can establish priority uses for wastewater. For example, in the Balearic Islands (Spain), it was declared by law that golf courses must use reclaimed wastewater for irrigation. Another economic constraint or concern is the cost related to the compulsory analytical tasks in order to guarantee that reclaimed wastewater fulfils the desired and required quality. For instance, microbiological tests to determine the presence of pathogens are expensive, resulting in poor monitoring schemes of reused water.

Finally, the existence of a consensus – meaning a real acceptance of reclamation and reuse by public, scientific and technicians – is paramount in the selection and establishment of criteria. It is expected that farmers or users will pay part of the reuse costs and that the use of wastewater implies a reduction in the number of crops that can be grown.

(b) Administrative factors. Any reuse process implies administrative procedures. It usually includes the water concessions or permits, disposal authorizations and the definition of the necessary control tasks. It is also usual to define the project form and the authorities to whom technical projects and control plans must be presented.

(c) Health and hygienic factors. The sanitary condition of the population where reuse will be carried out is very important, since the biological quality of the wastewater depends on it. The incidence and prevalence of parasitic, viral or bacteriologic illnesses is reflected in the wastewater quality, even in reclaimed wastewater (Touyab 1997).

This consideration must be extended to wastewater discharged from health centres. It is important to establish a pre-treatment or an adequate treatment of wastewater from hospitals and similar installations, to obtain a reduction in pathogenic organisms at the origin, and not affect further operations from the hygienic point of view. The alternative is not to reuse such wastewater.

It is necessary to distinguish clearly the different groups of risk that people can be directly or indirectly exposed to by reclaimed wastewater. The WHO guidelines (1989) carried out a first approach to that risk concept, pointing out the difference between workers and public. It is necessary to further develop this aspect, especially if it is considered that the degree of infection, even for a single infection, can depend on the group of population age (Moukrim 1999). Regarding the risk related to health it is necessary to differentiate between the possibility of direct contact with wastewater (such as workers) and indirect contact (such as a worker's family, or crop consumers).

Health education, especially for the population that will use the reclaimed water, can contribute to the reduction of the risk inherent to the practice. Catalan recommendations include this type of training (Generalitat de Catalunya 1994).

(d) Technical or technological factors. It is important to identify, in a case-by-case basis, realistic outcomes of wastewater treatment processes; both conventional (secondary) and advanced (including disinfection) must be considered. They should be studied from different points of view, especially technological and technical capacity.

With regard to the type of treatment, several standards and recommendations indicate that lagooning is the method of reference for the treatment of wastewater, capable of reaching a microbiological quality good enough for reuse without additional treatment. In other cases, physico-chemical treatments plus disinfection are advocated; but other types of treatment are being studied, especially disinfection ones, in order to obtain efficiencies equivalent or exceeding to the lagooning systems.

(e) Other considerations. It is important to define the specific points where reclaimed water quality criteria must be met.

When studying the existing regulations (see Table 23.2) it becomes clear that the control parameters that have been considered until now are the biological ones. Now, attention is paid only to fecal coliforms (total coliforms in California and other US states) and nematode eggs. In contrast, viruses are not included. Coliform analysis is cheap and easy to perform, while the determination of nematode eggs, unless performed in really contaminated wastewater, is a frustrating and expensive control (Asano 1998). Perhaps testing for viruses has not been considered until now because their monitoring and control is even more difficult and expensive. The use of bacteriophages as virus indicator is promising in the future.

In relation to biological parameters, it seems necessary to consider if other non-biological parameters exist that can influence the health risk derived from reuse practices. Several studies have been performed in this sense (for example, WHO 1989) and a description can be found in Crook (1998). Crook states that industrial wastes discharged in municipal sewerage systems can introduce chemicals that can adversely affect wastewater biological treatment processes and also the final effluent quality.

For this reason it seems necessary to consider the implementation of control parameters related to the presence of chemicals and their toxicity. Occasionally, physical contamination (i.e. temperature increase or radioactivity presence) can be described, but these are not usual cases. Chemicals seem to cause more concern, and, for example, in the Catalan recommendations (Generalitat de Catalunya 1994), several limits are fixed for the heavy metal content.

The concept of use regulation is being included in several regulations. This means that this should also be considered when controlling the whole reclamation systems. Then, authorized use (irrigation of vegetables, fodder, fruit trees, industrial crops and others) must be checked, and good reuse practices (such as night time irrigation, sprinkler irrigation forbidden in heavy winds, etc.) should be controlled. Also, the education of people involved with wastewater reclamation and reuse facilities is increasingly needed.

Table 23.2. Some regulations and recommendations for wastewater reuse for irrigation

Country/state	Main features	Comments
California/USA	22 to 23 TC/100 mL depending on the type of irrigated crop	States the treatment method depending on the use. Advanced treatments (tertiary) required.
France	200 or 1000 FC/100 mL depending on the irrigation type and crop. Nematode egg limitation specified	Follows WHO recommendations. Being revised.
Israel	12 TC/100 mL to 250 FC/100 mL. Regulations for BOD5, SS, DO and Residual Chorine. Includes contact time	Quality stated according to the crop to be irrigated. Underwent several revisions, finally not published until now.
WHO	200 or 1000 FC/100 mL depending on the irrigation type and crop. Nematode egg limitation appears	Exposed group indication. Lagooning as reference treatment.

FC: fecal coliforms; SS: suspended solids; DO: dissolved oxygen

Traditionally, guidelines and regulations have been issued by water-related administrations, both from an engineering and a health point of view. The legislative capacity of every administration will mark the boundary of validity of every rule, the evaluation and quality criteria, and the typology of infractions derived from non-compliance.

Nevertheless, the development and application of rules from their initial context must be considered. Californian regulations have been applied in many countries, adopted by the corresponding authorities. This was done without considering that the circumstances of sanitary culture, economy, knowledge of the practice, etc. are specific to a given territory and in no case must be applied to other territories, without further considerations.

In the United States, since the origin of reuse practices, the legislative authority corresponds to the States (California, Arizona, Florida, and others) while in other countries it is the Central Government (France, Italy, and others) who sets legislation or recommendations. In countries associated with supranational communities, i.e. the European Union (EU), the possibility of compulsory legislation should not be ignored. This is not yet the case in the EU, but there are several movements that propose the setting down of reuse directives at a Europe-wide level (Bontoux 1998).

In some cases, and due to diverse circumstances, regional administrations have recognized the legal vacuum that threatens good reuse management and prepared their own guidelines or regulations (for example, this has taken place in diverse regions of Italy; Balearic Islands, Andalusia, and Catalonia in Spain, and others).

23.5 HISTORICAL DEVELOPMENTS

The evolution of reuse rules cannot be completely understood without reviewing these standards. In 1918 the legislative fever on reuse started. A summary of this evolution can be found in Table 23.3.

For many years, the State of California regulations were the only legal valid reference for reclamation and reuse. During the 1970s and 1980s, other states, countries and international agencies were proactive in this department. After the appearance of the US EPA recommendations in 1992, few changes have been made. In Europe, as mentioned earlier, moves have been made for reclamation and reuse legislation in the EU.

As explained before, California's regulations were the only regulations for a long time, and are thus considered the best by many technicians and scientists. Nevertheless, these regulations were enormously restrictive and, as mentioned before, were implanted in temporal, legal and socio-economic circumstances very different from the present ones. This made different supranational entities discuss the possibility of implementing new ones or suggesting modifications. The WHO (1989) and the World Bank (Bartone 1991) sponsored several studies on this subject. Later, the US EPA also carried out several studies and compared the existing state laws, issuing recommendations in 1992 (US EPA/US AID 1992).

At present, both California and Israeli regulations are under revision. In addition, a committee of experts have been established for an initial revision of the WHO guidelines. Finally, various studies are in progress to establish guidelines or regulations in various countries and in the EU.

In addition, and in several regions and countries in southern Europe, work on regulations has been carried out; thus creating on a smaller scale several reuse schools which refer to guidelines and/or regulations (Asano 1998).

23.6 EXISTING REGULATIONS

23.6.1 Basic concepts

Existing wastewater reuse regulations for irrigation purposes are based mainly on biological quality considerations, crops to be irrigated, and risk groups. Details of the main features of such legal pieces can be examined in Table 23.2. It should be stated again that, for the time being, biological indicators are the only ones taken into account.

Table 23.3. Historical data of water quality for unrestricted irrigation

Year	Data and quality criteria
1918	California State Board of Public Health set up the Regulation governing use of sewage for irrigation purposes 2.2 TC/100 mL
1952	First legislation in Israel
1973	WHO 100 FC/100 mL, 80% of samples
1978	State of California Wastewater Reclamation Criteria: 2.2 TC/100 mL
1978	Israel: 12 FC/100 mL in 80% of samples; 2.2 FC/100 mL in 50% of samples
1983	World Bank Report (Shuval *et al.* 1986)
1983	Florida State: No *E.coli* detection in 100 mL
1984	Arizona State: Standards for virus (1 virus/40 L) and *Giardia* (1 cyst / 40 L)
1985	Feachem *et al.* 1983 report
1985	Engelberg report (IRCWD)
1989	WHO Recommendations for wastewater reuse: 1000 FC/100 mL; <1 nematode egg/L
1990	Texas State: 75 FC/100 mL
1991	Sanitary French Recommendations based on WHO
1992	US EPA Guidelines for water reuse: No fecal coliform detection in 100 mL (7 days median. No more then 14 FC/100 mL in any sample)

WHO (1989) discusses two approaches for developing guidelines on wastewater reuse for irrigation purposes: the setting of numerical standards based on technical decisions and the epidemiological approach. As several authors comment (e.g. Shuval *et al.* 1986), the epidemiological approach is not really useful for setting regulations. On the other hand, the 'standard' approach is not based on 'real' circumstances, which makes it highly objectionable. Nevertheless, in the present circumstances it seems to be the only feasible approach. In the 190s, starting with a study by Haas (1983), several studies performed by Haas and others (Regli *et al.* 1991; Asano *et al.* 1992; Haas 1996; Haas *et al.* 1996a and b; Gerba *et al.* 1996) were carried out, using risk assessment approaches to calculate health risks in the field of wastewater reuse. This alternative should be considered in future in order to establish numerical standards more in line with reality. However, at present, it is necessary to rely on other, more 'classical', methods.

23.6.2 Recommendations, guidelines and regulations worldwide

Quality criteria used for unrestricted irrigation in various parts of the world are summarized in Table 23.4. A description of the existing situation in various regions of the world follows:

Table 23.4. Quality criteria for the reuse of reclaimed wastewater for unrestricted irrigation (Angelakis 1997)

Agent or state	Type	Quality required in terms of public health
US EPA (1992)	Guidelines	Fecal coliforms should not exceed 14 MPN/100 mL in any sample, which in practice means not detectable. Secondary treatment should be used followed by filtration (with prior coagulant and/or polymer addition) and disinfection.
Arizona	Regulations	Fecal coliforms should not exceed 2.2/100 mL (median) and 25/100 mL (single sample).
California CA /T-22 (1978)	Regulations	Total coliforms should not exceed 2.2 MPN/100 mL (the number of coliform organisms should not exceed 23/100 mL in more than one sample per month). Secondary treatment is required followed by filtration and disinfection.
Colorado	Guidelines	Total coliforms should not exceed 2.2/100 mL (median). Effluent used should be oxidized, coagulated, clarified, filtered, and disinfected.
Florida	Regulations	Fecal coliforms should not exceed 25/100 mL over 30 day period in 75% of the samples. Secondary treatment with filtration and high level disinfection is required. Also, concentrations of 20 mg/L COD (annual average) and 5 mg/L TSS (single sample) in the effluent are required .
Georgia	Guidelines	Fecal coliforms levels not to exceed 30/100 mL. Biological treatment (30 mg BOD/L and 30 mg TSS/L) is required.
Idaho	Regulations	Total coliforms should not exceed 2.2/100 mL (median). Effluent used should be oxidized, coagulated, clarified, filtered, and disinfected.
Illinois	Regulations	Minimum treatment required is: two cell lagoon system with sand filtration and disinfection or mechanical secondary treatment with disinfection.
Indiana	Regulations	Fecal coliforms should not exceed 1000/100 mL (median) and 2000/100 mL (single sample). Disinfection is required if these limits are exceeded.
Michigan	Regulations	Treatment requirements are governed by Michigan Water Resources Commission issued NPDES (National Pollutant Discharge Elimination System) permits
N. Carolina	Regulations	Fecal coliforms not to exceed 1/100 mL. Tertiary treatment (5 mg TSS/L, monthly avg. and 10 mg TSS/L, daily max.) is required.

Agent or state	Type	Quality required in terms of public health
Nebraska	Guidelines	Biological treatment and disinfection prior to application are required.
New Mexico	Guidelines	Fecal coliforms should not exceed 1000/100 mL. Adequately treatment with disinfection is required.
Oregon	Regulations	Total coliforms should not exceed 2.2/100 mL (median) and 23/100 mL (single sample). Biological treatment including coagulation, filtration, and disinfection is required.
Texas	Regulations	Fecal coliforms should not exceed 75/100 mL. Minimum treatment is required for obtaining30 mg/L and 10 mg/L BOD with pond system and other than pond system, respectively.
Utah	Regulations	Total and fecal coliforms should not exceed 2000 and 200/100 mL (30d mean), respectively. Minimum required treatment is secondary with concentrations of 25 mg/L BOD and TSS (30d mean).
Washington	Guidelines	Total coliforms should not exceed 2.2/100 mL (mean) and 24/100 mL (single sample) Minimum treatment required is secondary including filtration.
West Virginia	Regulations	Minimum treatment required is secondary with disinfection and concentrations of BOD and TSS 30 mg/L.
Wyoming	Regulations	Fecal coliforms should not exceed 200/100 mL. BOD concentration in the effluent should not exceed 10 mg/L (daytime).
Canada (Alberta)	Regulations	Total coliforms not exceed 1000 geometric mean/100 mL and fecal coliforms not to exceed 200 gm/100 mL, in more than 20% of the samples. Also, total coliforms not to exceed 2400gm/100 mL on any given day, for vegetables to be irrigated
Cyprus (1997)	Provis. standards	Fecal coliforms not to exceed 50/100 mL and 100/100 mL in 80% of the samples per month and as a maximum value allowed, respectively. Also, intestinal nematodes not to be > 1 egg/L. Tertiary treatment should be used followed by disinfection is required.
France (1991)	Guidelines	As those of WHO with additional rules.
Israel (1978)	Regulations	Total coliforms should not exceed 2.2 and 12 MPN/100 mL in 50% and 80% of the samples, respectively. Secondary treatment or equivalent (such as long storage process) followed by disinfection

Agent or state	Type	Quality required in terms of public health
Japan	Criteria	Total coliforms and BOD should not exceed 50 count/mL and 20 mg/L, respectively.
Jordan	Regulations	Fecal coliforms < 200/100 mL and nematodes <1 egg/L for reuse in public areas. Recommended: Fecal coliforms for unrestricted irrigation of 1000 MPN/100 mL and BOD_5 for public parks and artificial groundwater recharge 50 mg/L. Only fruit trees, forests and fodder crops can be irrigated. Necessary presence of residual chlorine in wastewater.
Kuwait	Criteria	Total coliforms should not exceed 100 count/100 mL. Advanced level treatment is required with effluent BOD and TSS not to exceed 10 mg/L
NSW, Australia	Guidelines	Thermotolerant coliforms should not exceed 10/100 mL (median). Minimum treatment required is secondary plus filtration with ≤ 2 NTU in the effluent.
Saudi Arabia	Regulations	Total coliforms should not exceed 2.2 count/100 mL. BOD and TSS concentrations in the effluent should not exceed 10 mg/L.
South Africa	Guidelines	Maximum fecal coliforms should be 0.0 count/100 mL. Also, minimum level treatment required is standard primary, secondary and tertiary.
Tunisia (1975)	Regulations and/or Law	Intestinal nematodes should be ≤ 1 egg/L. Minimum treatment processes required are stabilization ponds or equivalent.
Victoria, Australia	Guidelines	For unrestricted irrigation (no public access): pH= 6.5–8.0, BOD_5 <10 mg/L, TC <1 org./100 mL, virus <1 org/50 mL, parasites <1 org./10 L, Cl residual >1 mg/L (after 30 min contact or equivalent disinfection). Advanced treatment processes are used where high quality reclaimed water is required for uses such as for irrigation.
WHO (1989)	Guidelines	Limited health risk from unrestricted irrigation water having less than fecal coliforms 200/100 mL and intestinal nematodes ≤1 egg/L. Primary and secondary treatment should be used, preferably followed by filtration or polishing and disinfecting.

(a) Europe. A very limited number of European countries have guidelines or regulations on wastewater reclamation and reuse. Most of the northern states do not have any specific legislation on wastewater reuse because they do not need to reuse water, and their rivers have a sufficient dilution factor.

The only reference to wastewater reuse at European level is article 12 of the European Wastewater Directive (91/271/ECC): 'Treated wastewater shall be reused whenever appropriate'. In order to make this statement reality, common definitions of what is 'appropriate' are needed. Bontoux (1997) analyzes the current situation in Europe about the need of a common legislation on wastewater reuse. This matter is approached in a sanitary and commercial way and points out the importance of this problem in the European water policy.

In 1977 Italy adopted guidelines for wastewater reuse in the framework of its National water law CITAI (Commitato Interministeriale per la Tutela delle Acque dal Inquinamento 1977) These guidelines follow the California regulations, and in the Italian context proved to be inappropriate and were not followed. Nevertheless, local regulations were published, and were also published in Sicily in 1989. These regulations have very different criteria than national regulations, and are very close to the WHO guidelines.

The different recommendations from other Italian regions are a combination of the WHO and the California guidelines (Bontoux 1997).

France enacted a national code of practice, under the form of recommendations from the Conseil Supérieur d'Hygiène Publique de France (CSHPF 1991). These recommendations use the WHO guidelines as a basis, but complement them with strict rules of application. CSHPF (1991) calls for a strict observation of these restrictions to ensure the best possible protection of the public health (Table 23.5).

In Greece a preliminary study for establishing quality criteria is in progress (Angelakis *et al.* 2000).

Portuguese legislation has allowed reuse of correctly reclaimed wastewater in a large number of crops by Law 74/90 (art. 32) since 1990 (Anonymous 1990). However, microbiological criteria and criteria related to the irrigation systems are still needed. Important projects on wastewater reuse for irrigation are also being carried out, but guidelines have not yet been adopted.

In Cyprus, provisional standards related to the use of treated wastewater effluent for irrigation purposes are being established. Stricter standards than the WHO ones were adopted with the aim of covering the specific conditions of Cyprus (Table 23.5). Additionally, there is a practical code for irrigation with reclaimed wastewater.

Table 23.5. Wastewater reuse recommendations in France (CSHPF 1991)

Treatment	Criteria	Irrigation type	Type of vegetation
None	None	Localized	Industrial cereal Fodder Fruit trees Forest and green areas
Stabilization ponds 8-10 days retention time	≤ 1 nematode egg/L	Sprinkling (aerosol propagation limited)	Fruit trees Cereal Fodder Nurseries Green areas with restricted access
Stabilization ponds 20-30 days retention time	≤ 1 nematode egg/L ≤ 103 FC/100 mL	Low pressure sprinkling Furrows	Fruit trees Pasture Vegetables Leguminous
Tertiary and disinfection Stabilization ponds 20-30 days retention time	≤ 1 nematode egg/L ≤ 200 FC/100 mL	Low pressure Sprinkling	Green areas with public access

In Spain, existing laws state only that (Salgot and Pascual 1996):

(1) Every potential water user should ask for administrative concession of a given amount of water, even reclaimed wastewater.
(2) For reclaimed wastewater, before giving permission, a 'compulsory' report should be issued by the health authorities. The conclusions given in this report are to be forcefully implemented in order to obtain the concession.
(3) The Government must develop a Reclamation and Reuse Regulation (not yet in force, but being prepared).

Due to the lack of any government regulations, the health authorities of some autonomous regions have been forced to develop their own guidelines in order to face the demands for approval. Three such guidelines have so far been published (Salgot and Pascual 1996). The Canary Islands have their own water law and regulations. In these Islands, the Government does not own water resources, as is the case in the rest of Spain. Reclaimed wastewater is in the water market and is sold at a good price: 0.4 US$/m^3.

(b) Southern Mediterranean. Water demand in Southern Mediterranean countries is usually more stringent than in Northern Mediterranean countries. These requirements are currently increasing because of economic development, a rise in tourism and population growth. Moreover, in these countries water resources are limited because of their arid climate. As a result it is necessary to exploit all available water resources, including wastewater.

Morocco does not yet have any specific regulation with regard to wastewater reuse. Reference is usually made in the projects to WHO recommendations.

Agricultural wastewater reuse in Tunisia is regulated by the 1975 Water Law, and by a 1989 Decree. This Water Law provides a legal framework for reclaimed wastewater use. This code prohibits the use of raw wastewater in agriculture, and irrigation with reclaimed wastewater of any vegetable to be eaten raw. This law stipulates that wastewater for irrigation must have a quality, which does not allow the transmission of diseases. The 1989 decree specifically regulates reuse of wastewater in agriculture.

(c) Near East. In Israel about 72% of the wastewater collected by municipal sewers is used for irrigation or groundwater recharge. In 1952 Israel published a regulation based on the California standards. Local, regional and national authorities must approve the use of reclaimed wastewater. Effluent used for irrigation must meet water quality criteria set by the Ministry of Health (Oron 1998).

In Jordan the standards for wastewater treatment and reuse were introduced in 1982 by martial law. In 1989 a more liberal version of the martial law was passed.

In Lebanon wastewater treatment and reuse are regulated by a 1930 legislation.

(d) South Africa. South Africa adopted a water quality management strategy much more complex than other countries with plenty of water resources. Wastewater reuse has been very significant in the national budget for water (Water Act 1956). This act has turned into the most important rule in wastewater reuse policy.

(e) South America. No specific legislation is known in South American countries, but WHO standards are adopted when possible. Several studies of the World Bank are available to consult (Bartone 1991).

(f) North America. In the United States no federal standards exist for wastewater reuse. On the contrary, many states have developed their own wastewater reuse recommendations and regulations (Table 23.4). Usually, regulations are based on the expected contact degree with this water.

As mentioned earlier, California was the pioneer state in the development of standards. In 1918, the state normative established a maximum total coliform content of 2.2 TC/100 mL. These regulations are as strict as potable water regulations (Asano *et al.* 1992). The Clean Water Act (1977) promoted wastewater treatment and the elimination systems that use soil for wastewater reclamation and recycling. The US EPA study (1992) describes all the state guidelines and their main characteristics.

23.7 REFERENCES

Angelakis, A.N. and Asano, T. (2000) Wastewater reclamation and reuse in Eureau countries. Necessity for establishing EU guidelines. Eureau, Brussels, p. 54.

Angelakis, A.N. (1997) Development of wastewater reclamation and reuse practices for urban areas of Yemen. Food and Agriculture Organization of UN, Rome, Italy, p. 144.

Angelakis, A.N. and Bontoux, L. (2000) Wastewater reclamation and reuse in Eureau countries. *Water Policy Journal* (accepted).

Angelakis, A.N., Marecos do Monte, M.H., Bontoux, L. and Asano, T. (1999) The status of wastewater reuse practice in the Mediterranean basin. *Wat. Res.* **33**(10), 2201–2217.

Angelakis, A.N., Tsagarakis, K.P., Kotselidou, O.N. and Vardakou, E. (2000) The necessity for the establishment of Greek regulations on wastewater reclamation and reuse. Report for the Ministry of Public Works and Environment and Hellenic Union of Municipal Enterprises for Water Supply and Sewage. Larissa, Greece, p. 110. (In Greek.)

Anonymous (1990) Portuguese legislation on irrigation water quality. Decree-Law 74/90. Journal of the Republic, I series no 55, 1990.03.07. Lisbon, Portugal.

Asano, T. (1998) (ed.) Wastewater reclamation and reuse. Water quality management library, Vol. 10. Technomic Publishing, Lancaster, PA, USA.

Asano, T., Leong, L.Y.C., Rigby, M.G. and Sakaji, R.H. (1992) Evaluation of the California wastewater reclamation criteria using enteric virus monitoring data. *Wat. Sci. Tech.* **26**(7–8), 1513–1524.

Bartone, C.R. (1991) International perspective on water resources management and wastewater reuse-appropriate technologies. *Wat. Sci. Tech.* **23**, Kyoto, 2039–2047.

Bontoux, L. (1997) Aguas residuales urbanas. Salud Pública y medio ambiente. *The IPTS report*, 19, 6–13. (In Spanish.)

Bontoux, L. (1998) The regulatory status of wastewater reuse in the European Union. In *Wastewater Reclamation and Reuse* (ed. T. Asano), Technomic Publishing Company Inc., Lancaster, PA, USA, pp. 1463–1475.

CITAI (1977) Smaltimento deli liquami sul suolo e nel sottosuolo. Altegato 5, Delibera 4.2.1977. GURI, no. 48.S.O. 21 Febbraio. Roma, Italy. (In Italian.)

Crook, J. (1998) Water reclamation and reuse criteria. In *Wastewater Reclamation and Reuse* (ed. T. Asano), Technomic Publishing Company Inc., Lancaster, PA, USA, pp. 627–705.

Crook, J. and Surampalli, R.Y. (1996) Water reclamation and reuse criteria in the USA. *Wat. Sci. Tech.* **33**(10–11), 475–486.

CSHPF (1991) Recommandations Sanitaires Concernant l'Utilisation, après Épuration, des Eaux Résiduaires Urbaines pour l'Irrigation des Cultures et des Espaces Verts. Circulaire DGS/SD1.D./91/N° 51, Paris, Conseil Supérieur d'Hygiène Publique de France (In French.)

Feachem, R.G., Bradley, D.J., Garelick, H. and Mara, D.D. (1983) Sanitation and disease–health aspects of excreta and wastewater management. World Bank Studies in Water Supply and Sanitation 3. Published for the World Bank by John Wiley & Sons, Chichester, UK.

Generalitat de Catalunya. Departament de Sanitat i Seguretat Social. Direcció General de Salut Pública (1994) Prevenció del risc sanitari derivat de la reutilització d'aigües residuals com a aigües de reg/Guia per al disseny i el control sanitari dels sistemes de reutilització d'aigües residuals. Barcelona, Spain. (In Catalan.)

Gerba, C.P., Rose, J.B., Haas, C.N. and Crabtree, K.D. (1996) Waterborne rotavirus, a risk assessment. *Wat. Res.* **30**(12), 2929–2940.

Haas, C.N. (1983) Estimation of risk due to low doses of microorganisms: a comparison of alternative methods. *American Journal of Epidemiology* **118**(4), 573–582.

Haas, C.N., Crockett, C.S., Rose, J.B., Gerba, C.P. and Fazil, A.M. (1996) Assessing the risk posed by oocysts in drinking water. *Journal AWWA* **88**(9), 131–136.

Haas, C.N. (1996) How to average microbial densities to characterize risk. *Wat. Res.* **30**(4), 1036–1038.

IRCWD (International Reference Center for Waste Disposal) (1985) Health aspects of wastewater and excreta use in agriculture and aquaculture: The Engelberg report. IRCWD News, No. 23, Dubendorf, Switzerland.

Moukrim, A. (1999) Wastewater reuse in Morocco: the Agadir case. In *Water Resources* (coord.. M. Salgot), edited by Fundación AGBAR, Barcelona. (In Catalan.)

National Research Council (1994) Groundwater recharge using waters of impaired quality. National Academy Press, Washington, DC.

Odendaal, P.E., van der Westhuizen, J.L.J. and Grobler, G.J. (1998) Wastewater reuse in South Africa. In *Wastewater Reclamation and Reuse* (ed. T. Asano), Technomic Publishing Company Inc., Lancaster, PA, USA, pp. 757–779.

Oron, G. (1998) Water resources management and wastewater reuse for agriculture in Israel. In *Wastewater Reclamation and Reuse* (ed. T. Asano), Technomic Publishing Company Inc., Lancaster, PA, USA, pp. 757–779.

Regli, S., Rose, J.B., Haas, C.N. and Gerba C.P. (1991) Modeling the risk from *Giardia* and viruses in drinking water. *Journal AWWA* **83**(11), 76–84.

Salgot, M. and Pascual, M.A. (1996) Existing guidelines and regulations in Spain on wastewater reclamation and reuse. *Wat. Sci. Tech.* **34**(11), 261–267.

Shuval, H., Adin, A., Fattal, B., Rawitz, E. and Tekutiel, P. (1986) Wastewater irrigation in developing countries: health effects and technical solutions. World Bank Technical paper 51, The World Bank, Washington, DC, USA.

State of California (1978) Wastewater Reclamation Criteria, An Excerpt from the California Code of Regulations, Title 22, Division 4, Environmental Health, Dept. of Health Services, Sacramento, California.

Tchobanoglous, G. and Angelakis, A.N. (1996) Technologies for wastewater treatment appropriate for reuse: Potential for applications in Greece. *Wat. Sci. Tech.* **33**(10–11), 17–27.

Touyab, O. (1997) Thèse Doct. Faculté des Sciences, Université Ibn Zohr, Agadir, Morocco.

US EPA (1992) Guidelines for Water Reuse: Manual. US EPA and US Agency for Internal Development. EPA/625/R-92/004, Cincinnati, Ohio, USA.

US EPA/US AID (1992) Manual: guidelines for water reuse. EPA/625/R-92/004, Washington, DC, USA.

WHO (1989) Health guidelines for the use of wastewater in agriculture and aquaculture. Report of a WHO Scientific Group, Geneva, Switzerland.

Part IV

Environmental and public health aspects of DESAR

24

Hygienic aspects of DESAR: water circuits

M. Salgot

24.1 INTRODUCTION

Although the main books and courses on wastewater treatment systems are principally directed towards large scale plants, the reality is that in every developed country almost all of the large treatment plants have been built, and a high number of small systems are still not constructed or even unplanned. This requires the development of specific training material and courses.

In Europe, Directive 91/271 fixed the need to build all treatment plants for towns with over 2000 p.e. before the end of 2005. It means that a huge job (and consequently an active market) is still open for smaller systems. As well as this, appropriate technologies have to be developed for the replacement of old systems and new plants to be built for new urban developments.

Edited by P. Lens, G. Zeeman and G. Lettinga. ISBN: 1 900222 47 7

Nevertheless, it is time to consider how the planning and construction of new treatment plants will be addressed in the future. There are two tendencies, as stated by Wilderer and Schreff (2000). The first is the traditional wastewater management concept (urban wastewater collection systems plus treatment of wastewater in a central treatment plant) that has been successfully applied over many decades in densely populated areas of industrialised countries. The second is treatment of the wastewater close to where it is generated as an alternative to the traditional centralised system. Decentralised reclamation after a centralised secondary treatment, in order to obtain different water qualities for different uses, is also to be considered.

The objectives of small and decentralised wastewater management systems are (Tchobanoglous and Angelakis 1999):

- protecting public health
- protecting the environment from degradation or contamination
- reducing costs of treatment by retaining water and solids near their point of origin through reuse.

King (2000), when reporting the findings of a meeting of the Environmental Sanitation Working Group of the Water Supply and Sanitation Collaborative Council held in Milan in February 2000, stated the new strategies for environmental sanitation, clearly indicating that it is essential to ensure that:

- people lead healthy and productive lives, and
- the natural environment is protected and enhanced.

Additionally, it was said that environmental sanitation is much more than the safe disposal of human wastes. The agreed definition of this sanitation is "Interventions to reduce people's exposure to disease by providing a clean environment in which to live, with measures to break the cycle of disease. This usually includes disposal or hygienic management of human and animal excreta, refuse, wastewater, stormwater, the control of disease vectors and the provision of washing facilities for personal and domestic hygiene. Environmental sanitation involves both behaviours and facilities which work together to form an hygienic environment."

Usually, all these features are established in a context where societal water use is supposed to be linear. Nevertheless, and due to the increasing water demand and the point scarcity of the resource, there are usually several loops in water use (Salgot and Vergés 1999). This means that water is used (reused) several times along a catchment area, especially in arid and semi-arid areas of the world. We can find "indirect reuse" worldwide. In some cases, reuse is practised on purpose, and then treated wastewater is reused directly, without being disposed of previously in the environment (see Figure 24.1).

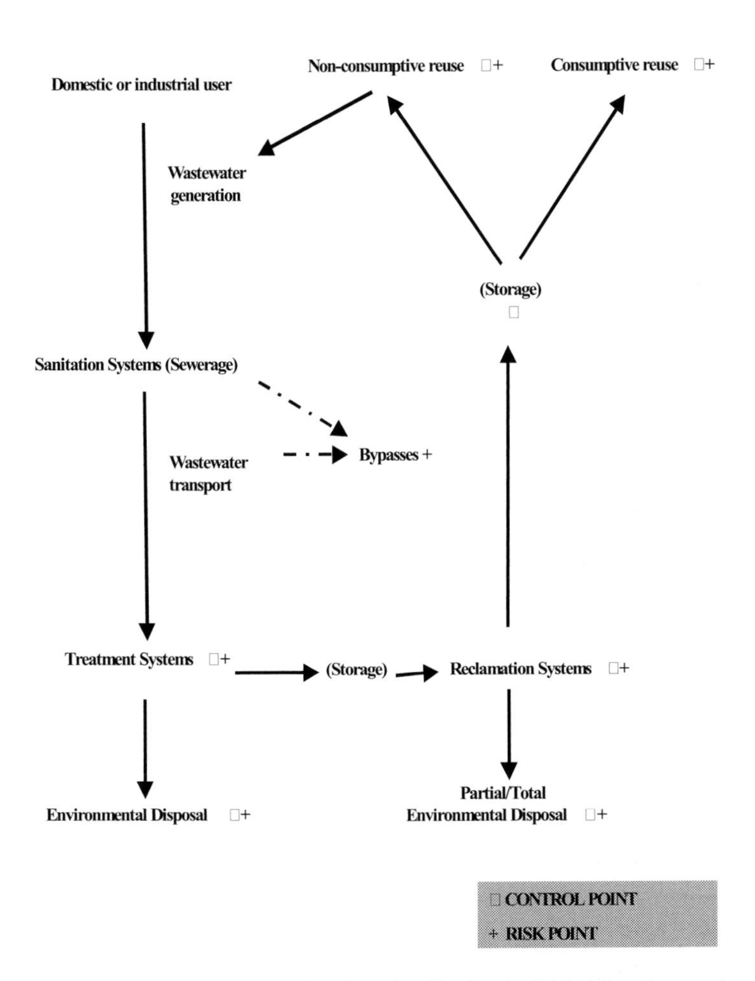

Figure 24.1. Sanitation treatment and reclamation line/cycle. Risk (+) and control points (□) are also indicated.

Reuse practices require, by law, safety measures, usually established with standards. There are other possibilities, such as the analysis of risk and the establishment of control points in the reuse schemes.

Nevertheless, it seems clear that there is a hygienic concern related with wastewater treatment and reuse. In general, there is a need to demonstrate the safety of sanitation and reclamation facilities and activities. It should be noted that hygienic problems of sanitation (centralised and decentralised) and reuse are different, but have common features.

24.1.1 Common sanitation and reuse hygienic concerns

Wastewater generation and disposal is the most common way to spread and expand water-related infections and illnesses (more specifically waterborne diseases, but since we are dealing only with wastewater we will use the first term). Accordingly, there is much concern about how treated and untreated wastewater reaches the environment, especially natural waters. The factors influencing the degree of impact of the pathogens also receive a lot of attention.

First of all, it should be noted that infection is not an equivalent term for illness or disease. An individual can be infected without developing an illness; although an individual affected by a pathogen-related illness needs to have been infected previously.

We can consider two types of wastewater-related diseases: those associated with pathogens (appearing in the short term) and those associated with chemicals (usually appearing in the long-term). It is usual to consider only pathogen-related illnesses when dealing with sanitation and hygiene, but although these appear mainly over the longer term, chemical-related illnesses (long-term toxicity) could also be important (see Figure 24.2).

Diseases originating from pathogens are communicable (from one person to another) while diseases originating from chemicals are not communicable. The latter are not traditionally considered as water-related diseases.

24.1.1.1 Sanitation systems

The two critical points when considering sanitation globally are sewerage system and wastewater treatment plant management. Sewerage systems are the initial points for wastewater management. It is necessary to control wastewater generated at the source, in order to achieve a good final effluent quality and to obtain reusable treated wastewater. It will be only emphasised here that in small wastewater treatment systems, the effect of dilution does not appear, which makes such systems more susceptible to improper disposals (e.g. industry or hospital wastewater).

Usually, municipalities are responsible for wastewater treatment plant management. Small plants may experience problems due to the lack of qualified staff. In this case, it is usual to transfer plant management to specialised companies.

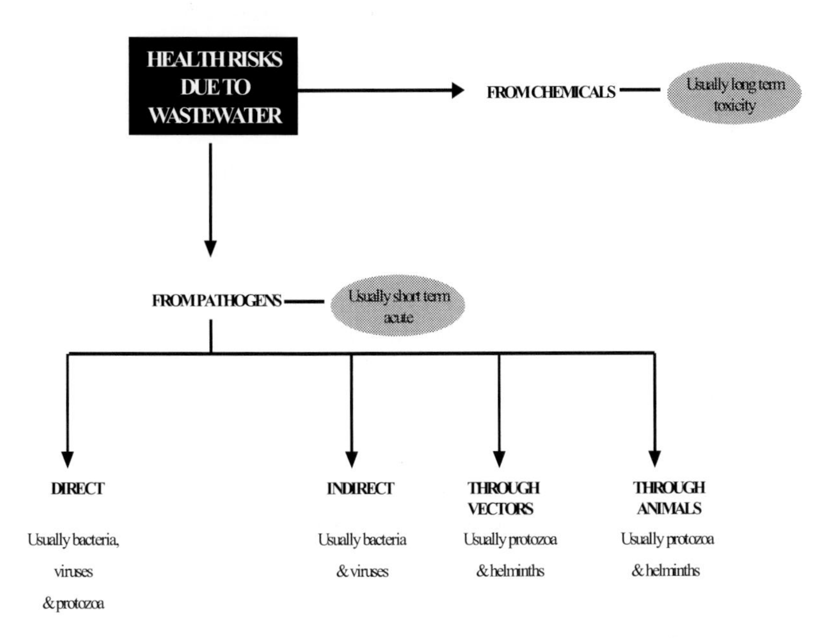

Figure 24.2. Types of health risk, and methods of transmission of pathogens.

24.1.1.2 Small wastewater treatment plants

There is an increasing tendency to build small wastewater treatment plants using extensive (natural) technologies, because of the reduced management expenses and theoretical management simplicity. In some way, this tendency is correlated with the "sustainability" promoted by non-governmental organisations and municipalities.

Apart from the absence of the dilution effect, there are two main features of small and decentralised systems that are considered to be disadvantages:

- It is more difficult to pay professional attention to on-site and small scale treatment facilities.
- Building and operating a great number of small plants is considered to be more expensive than building one single large facility.

Nevertheless, it is not clear if these drawbacks still hold, given the recent advances in management tools and technological improvements in small wastewater treatment facilities.

24.1.2 Reclamation and reuse features

Crook (2000) explains several of the problems encountered when trying to reuse reclaimed wastewater:

- Regulators are confident that water reuse standards provide enough safeguards to ensure that reclaimed water reuse will not impose undue health risks. Usually they are unwilling to relax such standards without supportable data.
- Utility personnel (and others) often consider the standards to be too stringent and, therefore, overprotective of public health.
- The existing definitive data on reclaimed water are insufficient to convince certain segments of the public that water reuse is safe.
- More definitive data need to be collected – through research or other means – to allow regulators to develop scientifically sound water reuse criteria.

All of this means a strict control of reclamation and reuse facilities. Obviously there is a need to know what happens with pathogens and chemicals supposed to generate a health risk all along the sanitation line.

24.2 THE CONCEPT OF RISK

Risk (Rowe and Abdel-Magid 1995) is defined as the probability of injury, disease, or death under specific circumstances. In quantitative terms, risk is expressed in values ranging from zero (that is, harm will not occur) to one (that is, the certainty that harm will occur).

It is clear that from wastewater disposal or reuse, some degree of risk can be generated, and the obvious desire is to keep it as close to zero as possible. This risk originates from both pathogens and chemicals (see Table 24.1).

It seems evident that the degree of risk suffered by an individual is variable. It can be related to the degree of exposure, personal characteristics and circumstances (see Table 24.2). In the case specified here, the risk can also be classified as direct (direct contact with wastewater) or indirect.

Usually, risk can be assessed and/or analysed. There are several methods for performing such an assessment or analysis, but usually they comprise four steps:

(1) hazard identification

(2) dose and response assessment/analysis
(3) exposure pathway and scenario analysis
(4) risk characterisation – prediction of occurrence.

A good description of these steps can be found in Rowe and Abdel-Magid (1995), Chang *et al.* (1998), and Sakaji and Funamizu (1998).

Table 24.1. Sources of risk associated with the presence of pathogens and chemicals in wastewater

	Pathogens	Chemicals
Risk derived from		
	Single or repeated "consumption" or contact	Repeated "consumption"
Due to		
	Bacteria Viruses Helminths Protozoa	Heavy metals Nitrates and nitrites Organic microcontaminants
Through		
	Drinking, Aerosols Eating vegetables, shellfish, etc. Direct or indirect contact, vectors	Drinking Eating several types of food
If the individual is affected:		
	Usually immediate effects	Usually medium- and long-term effects

Table 24.2. Examples of risk groups in DESAR systems depending on the relation with wastewater treatment or reclamation facilities

Direct risk	Users of reclaimed wastewater Neighbours Workers People visiting the facility or walking nearby
Indirect risk	Worker's family Groundwater users Citizens

24.2.1 Biologically-related risk

When dealing with pathogens and the impact they have on mankind, wastewater or reclaimed wastewater related diseases (or infections) may be grouped into five main categories, as summarised in Table 24.3.

The final sanitation target should be to reach a zero or near zero risk level of pathogens such that they are incapable of originating illnesses.

Considering the same disinfection capability for both centralised and decentralised sanitation systems, we can establish theoretical differences between the two systems from a hygienic point of view. If wastewater treatment is focused on a single point (centralised), the disposal of treated wastewater is usually also centralised and wastewater (and pathogens) are eliminated at that single point.

Table 24.3. Categories of diseases associated with water and/or reclaimed wastewater (modified from Rowe and Abdel-Magid 1995)

Categories	Definitions/observations/examples
Waterborne diseases	Infections that may be spread through a water supply system Water acts exclusively as a passive vehicle for the pathogen Typhoid fever, cholera, giardiasis, dysentery, infective hepatitis
Water-washed diseases	Caused by a shortage of water for personal hygiene Affect the body exterior surfaces Conjunctivitis, trachoma, leprosy, tinea, ascariasis, yaws, giardiasis, cryptoporidiasis
Water-based diseases	Infections transmitted through an aquatic invertebrate host, usually an animal An essential part of the life cycle of the infecting organism takes place in this aquatic animal Schistosomiasis, guinea worm, filariosis
Water-related insect vectors	Infections spread by insects which rely on or live near a surface water system Trypanosomiasis, yellow fever, dengue, onchocerciasis, malaria
Infections due to poor sanitation	Spread within a community usually due to the absence of suitable sanitation facilities Hookworm, roundworm, ascariasis

If equal amounts of wastewater are treated in a decentralised way, we must consider that treated wastewater is spread throughout a larger area, and the same number of pathogens is distributed through several points. Thus, a dilution effect is to be supposed. The same effect happens, for example, in reuse for agricultural irrigation. Consequently, and theoretically, pathogen decay in the environment should be improved using decentralised systems. Nevertheless, this does not seem clear in practice, and there are several additional factors to take into consideration.

1. From the treatment point of view:

(1) When disposing of wastewater in a sensible area, the law asks for supplementary wastewater treatment (i.e. nutrients elimination)
(2) If wastewater is going to be treated further then we must consider that an additional reduction of pathogens is obtained, even if disinfection is not the main purpose of the new treatment. Theoretically, the risk is reduced.
(3) If wastewater is to be reused, again a supplementary treatment is usually provided. In this case, a disinfection procedure is usually required. Nevertheless, this is an important point, trying to define a degree of disinfection that ensures a minimal or zero risk.
(4) The reliability of wastewater treatment plants should be further studied and ensured.

2. From the pathogen point of view:

(1) Not all pathogens are equally sensitive to environmental conditions and disinfectants.
(2) Several pathogens have life forms capable of resisting environmental conditions.
(3) There is a lack of knowledge on pathogen indicators adapted to water reuse conditions.
(4) It is not clear what pathogens are capable of generating an infection or a disease.

In such circumstances, several considerations arise:

(1) Economical considerations, depending on the country's characteristics (climate, wealth, society, and so on).
(2) Reuse can help to spread pathogens directly (for example, by the consumption of irrigated vegetables) or indirectly (for example, by indirect contact of workers' families).
(3) To some extent, reuse improves the quality of the receiving waters, because of the reduction in the incoming amount of wastewater. If the resource is used for water supply, there is a better raw water quality.

Obviously, there is a need to balance the above-mentioned considerations. As indicated, biological risk is related to the pathogens that are present in a community. That presence depends mainly on local hygienic practices and on environmental factors. Nevertheless, it is necessary to state that infections or diseases in a given population depend on the:

- concentration of the infectious agent
- amount of pathogens entering the body
- duration of exposure to the agent
- properties of the exposed microbial cells
- host characteristics.

Table 24.4. The most common wastewater associated disease-causing agents. From Rowe and Abdel-Magid (1995), Metcalf and Eddy (1991) and Yates and Gerba (1998)

	Agent	Disease
Bacteria	*Salmonella typhimurium*	Salmonellosis
	Salmonella typhosa	Typhoid fever
	Salmonella paratyphi	Paratyphoid fever
	Shigella spp	Bacillary dysentery
	Vibrio cholera	Cholera
	Mycobacterium tuberculosis	Tuberculosis
	Campilobacter jejuni	Diarrhoea
	Pathogenic *Escherichia coli*	Diarrhoea
	Yersinia enterocolitica	Diarrhoea and septicemia
	Legionella pneumophila	Legionellosis
	Leptospira icterohaemorrhagiae	Leptospirosis
Viruses	Poliovirus	Poliomyelitis
	Hepatitis A virus	Infectious hepatitis
	Hepatitis E virus	Hepatitis
	Rotavirus	Diarrhoea/gastroenteritis
	Adenovirus	Respiratory disease
	Norwalk agent	Gastroenteritis
	Reovirus	Gastroenteritis
	Astrovirus	Diarrhoea, vomiting
	Calicivirus	Diarrhoea, vomiting
	Coronavirus	Diarrhoea, vomiting
	Coxsackie A	Meningitis, fever, respiratory illness, herpangina
	Coxsackie B	Myocarditis, rash, meningitis, fever, respiratory illness
	Echovirus	Meningitis, encephalitis, respiratory illness, rash, diarrhoea, fever
Protozoa	*Entamoeba histolytica*	Amoebiasis (Amoebic dysentery)
	Giardia lamblia	Diarrhoea
	Cryptosporidium parvum	Diarrhoea
	Balantidium coli	Diarrhoea, dysentery
	Cyclospora cayetanensis	Intestinal diseases
	Toxoplasma gondii	Toxoplasmosis
	Phyllum *microspora*	Microsporidiosis (intestinal and nervous diseases)

Table 24.4. (*cont'd*)

	Agent	Disease
Helminths	*Schistosoma haematobium* (T)	Schistosomiasis (Bilharziasis)
	Schistosoma mansoni (T)	
	Schistosoma haematobium (T)	
	Ascaris lumbricoides (N)	Ascariasis
	Ancylostoma duodenale (N)	Anemia, intestinal diseases
	Necator americanus (N)	Anemia, intestinal diseases
	Clonorchis spp. (T)	Clonorchiasis
	Taenia spp.(C)	Taeniasis
	Enterobius vermicularis (N)	Enterobiasis
	Hymenolepis nana (C)	Hymenolepiasis
	Trichuris trichura (N)	Trichurasis
	Strongyloides stercoralis (N)	Diarrhoea, abdominal pain, nausea
	Toxocara canis (N)	Fever, abdominal pain
	Toxocara cati (N)	Fever, abdominal pain

N = Nematodes; T = Trematodes; C = Cestodes

Wastewater secondary treatment is usually not intended to eliminate pathogens. Not all the pathogenic agents act in the same way when they reach the environment, as stated in Table 24.3. The dependence of a non-human host modifies the degree of risk and the method of transmission. For this reason, and others, it is very difficult to establish the presence of a single pathogenic agent in a wastewater or reclaimed wastewater affected environment, especially taking into account the high number of pathogenic agents that can be present in wastewater (see Table 24.4).

Indicator organisms have been used in an attempt to simplify the analytical work, but the most common indicators cannot describe or represent all pathogens present (Martín *et al.* 1999). Until the present day, faecal and total coliforms, used since the beginning of the twentieth century, obtained an almost unanimous consensus as faecal contamination indicators. Nevertheless, they have several drawbacks, the main one being their scarce relationship with the fate of viruses and other organisms in the environment (Campos 1998).

During recent years, there have been indications that coliform determination should be substituted or complemented by other organisms better indicating the water microbiological contents and the efficiency of wastewater treatment facilities. Such new organisms must comply with the classical requirements for indicators.

Now, it seems that bacteriophages could be a good indicator not only for viruses but also for bacteria. *E. coli* and *Bacteroides fragilis* phage seem to be the possible future substitute or complement to *E. coli* or coliforms (Campos 1998).

Parasitological quality of wastewater is also a matter of concern. The determination of nematode eggs as proposed by the WHO recommendations presents several problems, especially the concentration phase and the evaluation of the eggs' viability. Additionally, the presence of protozoa (*Giardia* and *Cryptosporidium*) cysts/oocysts needs to be determined.

24.2.2 Non-microbiological risk/chemically-related diseases

Diseases may also be caused by the presence of chemical substances in wastewater. Chemical substances may be found alone or in combination. Usually, industrial discharges are the major source of chemicals in raw wastewater. If DESAR systems receive such products, it is necessary to exert control in the source, dilute until a convenient level or reduce the chemicals to an acceptable level using another method. The risk generated by chemicals is the risk that they will enter the human food chain and negatively affect human health (see Table 24.5).

Table 24.5. Chemicals present in wastewater/reclaimed wastewater that can cause toxicity

Group	Chemical	Effect
Inorganic	Heavy metals	Depends on the metal and the possibility of bioaccumulation: cancer, nervous system effects
	Boron	Plant toxicity
	Residual chlorine	Aquatic biota toxicity
	Nitrates	Methemoglobinemia, cancer
Organic	Organic halogens	Cancer
	Pesticides	Cancer, teratogenic effects, nervous system effects
	Polynuclear aromatic hydrocarbons	Cancer

Usually, chemicals that pose a health hazard or a significant health risk exist in wastewater at very low concentrations (Rowe and Abdel-Magid 1995). Consequently, the main problem is the ingestion of the chemical during long periods of time. There is an indirect method of ingestion, modified by the environmental conditions. For example, under acidic soil conditions, the bioavailability of heavy metals for plants increases. Then if wastewater is reclaimed and reused, toxic metals may build up in plants. The availability, uptake and accumulation of toxic metals in crops and consequently in animals and humans is a matter for concern.

The addition of chloride for disinfection has two main negative effects. One is the increase in formation of halogenated compounds, and the second is the subsequent toxicity for aquatic biota. It means that usually a dechlorination treatment needs to be performed before environmental disposal. The use of chlorine and derivatives for wastewater disinfection is currently being discussed and, in some countries, alternative methods are used.

The presence of nitrates in potable water causes concern, especially if this anion reaches groundwater. The European Environment Agency (1998) has tried to quantify this problem, extended in all European countries. Cancer and children's health problems are the main concerns in this case.

The presence of boron in wastewater is mainly derived from detergents and the food processing industry. Concern about this element is mainly due to its toxicity for plants.

Many of the organic chemicals found in wastewater are stable and persistent, and could be carcinogenic and mutagenic. Olivieri and Eisenberg (1998) published a comprehensive study on the presence of organic chemicals in reclaimed wastewater for potable purposes, and concluded that the cancer risk is negligible when using advanced wastewater treatments. Nevertheless, the degree of treatment is too high if wastewater is to be reclaimed for other purposes. Another approach is described by Chang *et al.* (1998), indicating that conventional wastewater treatment systems are not designed to remove potentially toxic chemicals (we did previously the same observation for pathogens). Another interesting comment from these authors is that chemical agents are only one factor in a complex multiple causal relationship. They also state that chemicals are present in environmental media in low concentrations and have long latency periods between exposures to the chemicals and expression of the symptom. As a result, it is difficult to establish a baseline exposure level and to distinguish exposures received from multiple pathways.

The human health issues involving toxic chemicals during the land application of wastewater are cases in point. Toxic chemicals are omnipresent in wastewater. There is, however, no unequivocal epidemiological evidence to demonstrate any harm caused by even one potentially toxic chemical present in irrigation-bound wastewater. It is a challenge to develop criteria that are not overly restrictive to the beneficial use of wastewater and yet that will protect humans from potential harm that could be caused by the hundreds of different toxic chemicals that may be present in wastewater (Chang *et al.* 1998).

24.3 DISCUSSION

After several years of controlled wastewater reclamation and reuse in several countries, hygienic problems related with wastewater treatment, reclamation and reuse have been clearly established, although some problems need further study. From the hygiene point of view, the main problems could be identified as:

- Lack of useful microbiological indicator tools, especially for viruses, helminths and protozoa.
- Need for quicker determination of microbiological quality.
- High cost of existing analytical methods, especially for protozoa.
- Lack of enough data for a good understanding of the emergent indicators for viruses.

The number of analytic tests legally necessary for the knowledge of microbiological risk levels and reliability of treatments is high, and can cause economic and practical difficulties if the treatment and reuse is performed at a decentralised level. The analytical work necessary to determine safety from a chemical point of view is even more expensive than from a microbiological point of view.

The influence that natural wastewater reclamation treatments (that is, lagooning, SAT or soil-plant systems) can exert on hazardous chemicals is determined. There are now promising research tasks in this sense (Salgot *et al.* 1999; Downs *et al.* 2000).

It is necessary to exert a higher control at the source, especially for the contents of nutrients, boron and hazardous chemicals. Several practices related to wastewater treatment for example, disinfection) need to be improved.

Suitable control and reuse practices should be established. Figure 24.3 summarises such practices.

24.4 CONCLUSIONS

Hygienic practices must be considered when planning, building, carrying out and managing small and decentralised wastewater treatment, reclamation and reuse systems. The success of such practices is closely related to a lack of sanitary problems.

A further knowledge of hygienic, microbiological and parasitological characteristics of DESAR systems is especially important in order to gain a better knowledge of health hazards and risks.

The hygienic aspects of DESAR must be studied and solved from an interdisciplinary point of view.

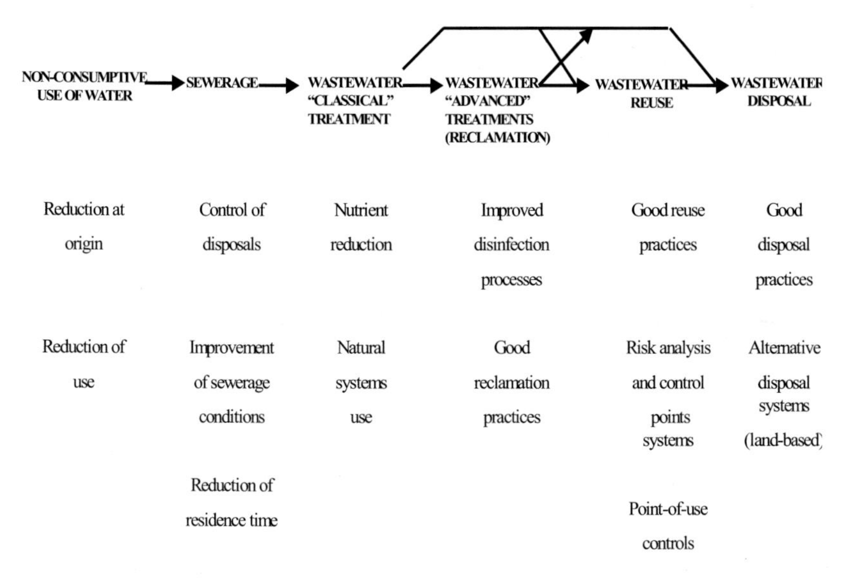

Figure 24.3. Hygienic practices in wastewater management aimed at reducing risk.

24.5 REFERENCES

Campos, C. (1998) Indicadores de contaminación fecal en la reutilización de agua residual regenerada en suelos. PhD thesis, University of Barcelona. (In Spanish.)

Chang, A.C., Page, A.L., Asano, T. and Hespanhol, I. (1998) Evaluating methods of establishing human health-related chemical guidelines for cropland application of municipal wastewater. Chapter 13 in *Wastewater Reclamation and Reuse* (ed. T. Asano), Technomic, Lancaster, PA.

Crook, J. (2000) Research needed to demonstrate safety of reclaimed water. *Water Environment & Technology* **12**(1), 8.

Downs, T.J., Cifuentes, E., Ruth, E. and Suffet, I. (2000) Effectiveness of natural treatment in a wastewater irrigation district of the Mexico city region: a synoptic field survey. *Water Environment & Research* **72**(1), 4–21.

European Commission (1991) Council Directive concerning urban wastewater treatment, 91/271/EEC, May, OJ N° L135/40, May.

European Environment Agency (1998) *Europe's Environment: The Second Assessment.* Elsevier, Oxford.

King, N. (2000) New strategies for environmental sanitation. *Water21,* April, 11–12.

Martín, J., Matia, L.l., Ventura, F. and Campos, C. (1999) La qualitat dels recursos no convencionals. Chapter 7 in *Recursos d'aigua* (coord. M. Salgot), Fundació AGBAR/Universitat de Barcelona/Generalitat de Catalunya, Barcelona, Spain. (In Catalan.)

Metcalf and Eddy (1991). *Wastewater Engineering: Treatment, Disposal, Reuse*, 3rd edn, McGraw Hill, Singapore.

Olivieri, A.W. and Eisenberg, D.M. (1998) City of San Diego health effects study on potable water reuse. Chapter 12 in *Wastewater Reclamation and Reuse* (ed. T. Asano), Technomic, Lancaster, PA.

Rowe, D.R. and Abdel-Magid, I.M. (1995) *Handbook of Wastewater Reclamation and Reuse*. CRC-Lewis, Boca Raton, Florida.

Sakaji, R.H. and Funamizu, N. (1998) Microbial risk assessment and its role in the development of wastewater reclamation police. Chapter 10 in *Wastewater Reclamation and Reuse* (ed. T. Asano), Technomic, Lancaster, PA.

Salgot, M. and Vergés, C. (1999) Recursos hídrics no convencionals. Chapter 3 in *Recursos d'aigua* (coord. M. Salgot), Fundació AGBAR/Universitat de Barcelona /Generalitat de Catalunya, Barcelona, Spain. (In Catalan.)

Salgot, M., Anderbouhr, T., Pascual, L. and Folch, M. (1999) DRAC reclamation project, Palamós (Girona province, Spain) Unpublished.

Tchobanoglous, G. and Angelakis, A.N. (1999) Small and decentralized wastewater management systems. An overview. In *Management of Wastewater and Solid Wastes, with Emphasis on the Wastewater Collection, Treatment and Disposal, and the Management of Produced Biosolids* (eds A.N. Angelakis and E. Diamadopoulos), Hellenic Union of Municipal Enterprises for Water Supply and Sewerage, Larissa (Greece), July, pp. 33–48.

Wilderer, P.A. and Schreff, D. (2000) Decentralized and centralized wastewater management: a challenge for technology developers. *Wat. Sci. Tech.* **41**(1), 1–8.

Yates, M.V. and Gerba, C.P. (1998) Microbial considerations in wastewater reclamation and reuse. Chapter 10 in *Wastewater Reclamation and Reuse* (ed T. Asano), Technomic, Lancaster, PA.

25

Hygienic aspects of solid fractions of waste water

A. E. Stubsgaard

25.1 INTRODUCTION

In waste water treatment, it can be an advantage to separate solid waste fractions from liquid waste, as water and solid waste involve different treatment methods and have different reuse potentials.

The solid fractions contain relatively large amounts of easily degradable organic compounds. As they decompose, heat is produced. The heat raises the temperature of the material, which increases the rate of reduction of initial populations of organisms in the solid waste, and, to a certain degree, increases the rate of decomposition. Thereby, the temperature rises further and the cycle continues.

 Decentralised Sanitation and Reuse: Concepts, Systems and Implementation. Edited by P. Lens, G. Zeeman and G. Lettinga. ISBN: 1 900222 47 7

Those fractions of waste water containing important concentrations of easily degradable organic compounds thus, among other advantages, permit a self-created hygienization. Parameters such as the amount of essential nutrients and moisture, the pH, and the presence/absence of oxygen etc. crucial for the process rates of waste treatment and hygienization.

In this chapter, the hygienic aspects of Decentralized Sanitation and Reuse (DESAR) solutions with separation of solid fractions will be discussed. Solutions without reuse will only be discussed if needed for comparison.

25.2 TYPES OF SOLID FRACTIONS OF WASTE WATER

Table 25.1 briefly describes separation methods for solid and liquid waste, and the corresponding treatment methods.

Table 25.1. Separation methods with the corresponding treatment methods for solid fractions of waste water in DESAR

	Separation method	Treatment method
Separation at source	Dry sanitation: no water, or very little, is used	Composting, drying, or high pH
	Bin for organic kitchen waste	Composting
	Black waste water and (maybe ground) kitchen waste pumped into container	Anaerobic or aerobic treatment
Separation after source	'Coffee filters' after vacuum toilets	Filtration of black waste water, followed by drying and/or composting
	Röttebehalters	Filtration of all waste water, followed by drying and/or composting
	Septic tanks	Sedimentation, followed by drying beds or transport to waste water treatment plant or landfill

These methods will be described from a hygienic point of view, after giving some examples of the influence of each treatment method on the degree of hygienization.

25.3 REDUCTION RATES OF DIFFERENT TREATMENT METHODS

Time and temperature are the main factors affecting the reduction of the initial populations of potentially harmful organisms. As illustrated by Figure 25.1, there is a synergistic correlation between time and temperature. The higher the temperature, the less time is needed for elimination; and the longer time the pathogens are left, the lower the temperature needed.

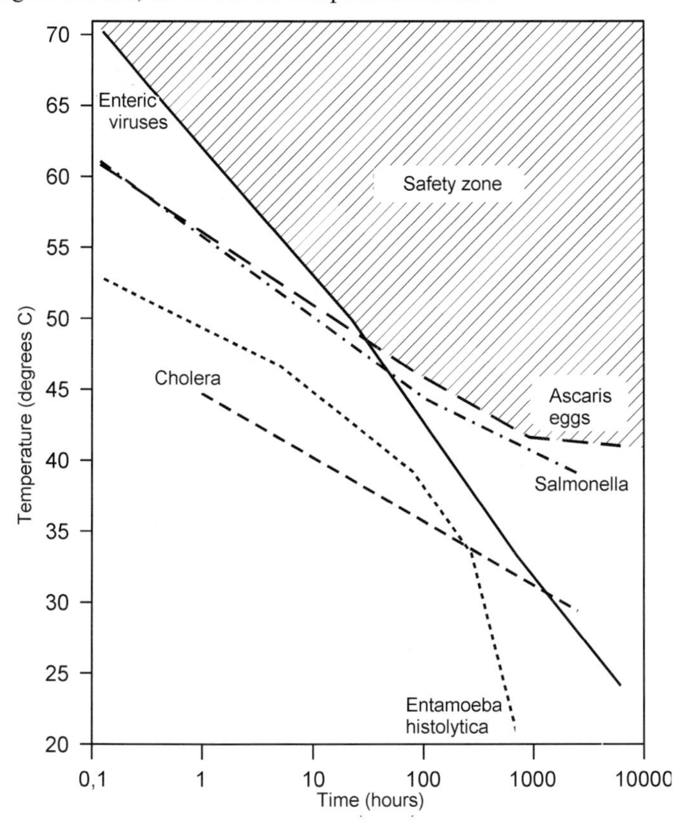

Figure 25.1. The influence of temperature and time on elimination of some common pathogens. The lines represent the necessary combination of time and temperature for total loss of each pathogen's capacity of infection. Thus, the hatched zone represents the combinations of time and temperature that are estimated to be lethal for all pathogens (Feachem *et al.* 1980).

In DESAR solutions the pathogen species and levels in the waste water can be expected to vary considerably from one location to another and from one period to another. Because analysis of pathogens is relatively expensive, and the chance of meeting a pathogen small, indicator organisms are often used instead. In water, different groups of coliform bacteria are the most commonly used indicators (Council Directive 1975; WHO 1998), whereas in solid waste, no specific indicator has been chosen.

Figure 25.2 compares the commonly observed reductions of the group of indicator organisms, faecal streptococci (FS). The arrows indicate the intervals of commonly observed reductions. Data for all DESAR solutions does not exist.

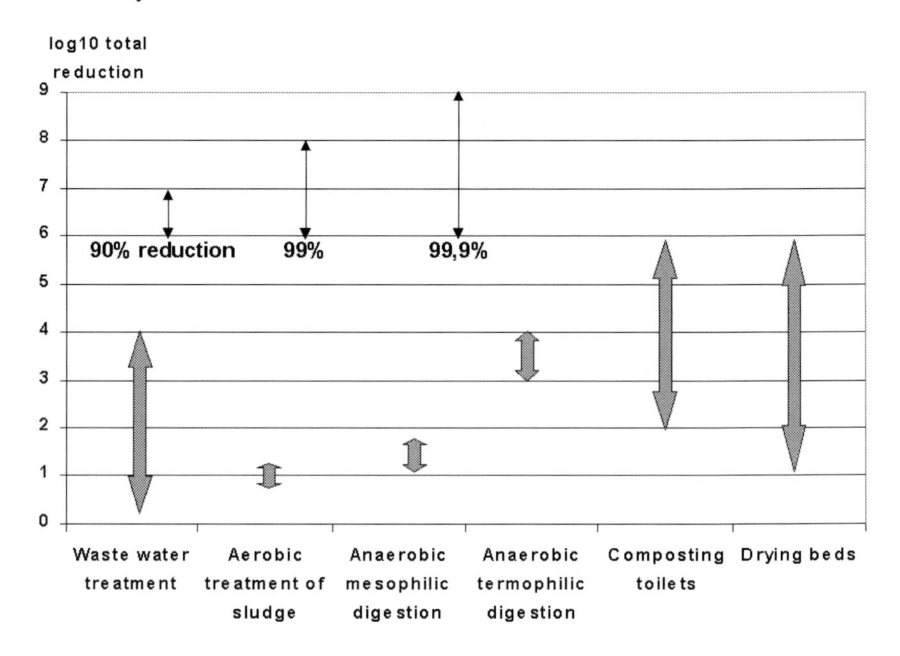

Figure 25.2. Common intervals of reduction of faecal streptococci for different treatment methods (Danish EPA 1997; Ilsoe 1993; Stubsgaard 1996, 2000).

As shown in Figure 25.2, waste water treatment reduces the concentration of FS in the waste water by 0.15–4 $\log_{10}$ units, depending on the initial FS content and the degree of treatment. Aerobic and mesophilic anaerobic stabilization of sludge generally reduces FS by, respectively, 1 and 1.5 $\log_{10}$ units (Strauch 1988; Danish EPA 1997). The interval of possible reduction is quite small, as the processes are well defined and controlled.

Composting of organic wastes and sludge reduces FS by very varying $\log_{10}$ units, depending on the starting material and the degree of control of the composting process, that is, certain parameters such as temperature, pH and access of oxygen through appropriate bulking agents.

In existing European all-year composting toilets, a 3–4 $\log_{10}$ unit reduction in FS is normally observed after sufficient composting time.

A reduction of 3–4 $\log_{10}$ units in biosolids is generally considered sufficient for achieving a reusable product with an acceptably low risk of infection. In material with high initial levels of FS, further reduction might be desired for sufficient hygienization.

25.4 DECENTRALIZED SLUDGE TREATMENT

For nearly all waste water treatment systems, a prior settling of suspended matter is required, either in a septic tank or in a Röttebehalter.

Many microorganisms and parasites will adhere to the sludge particles or they will sediment by themselves (Danish EPA 1997). Thereby, the potential of pathogens in sludge is higher than in waste water, which means that one should be very careful to assure sufficient reduction before reuse.

25.4.1 Septic tank

In a septic tank, temperatures are often low because the septic tank is buried in the ground. Furthermore, the strongly reducing conditions in the tank resemble conditions in our intestines. Therefore, pathogens are in an environment where they can persist for a relatively long time (Danish EPA 1997; Stubsgaard 1996).

Thus, the reduction rate in a septic tank is negligible. For safe reuse of the sludge, some type of further treatment is needed.

25.4.2 'Röttebehalters' and 'coffee filters'

The Röttebehalter and the coffee filter are both macro filters (pore size approximately 1.5 mm), more or less shaped like a bag, which filter black waste water or all waste water from the user. These systems replace the septic tank. In the Röttebehalter, a bulking agent is added before filling, which increases the porosity and thereby the transport of gases into and out of the material. Still, the retained solid fraction of these systems needs some sort of aerobic or anaerobic treatment before becoming hygienically safe.

25.4.3 Drying beds

In drying beds, hygienization can be compared to a slow composting process, as there is a lack of porosity in the sludge. Therefore, the reduction rate is relatively low (Bruce and Fisher 1990). Drying beds planted with reed, willow or other plants will accelerate exsiccation, and increase porosity, thereby the reduction rates.

There are examples of high reduction rates in drying beds in areas with high solar radiation (Esrey *et al.* 1998).

When drying beds are considered as part of a DESAR solution there is, however, the risk of bad odours to take into account, since DESAR solutions will often be located close to where the waste water is produced, which means close to humans. Thus, the placement of the drying beds must be chosen carefully. There is much less risk of odour problems with sludge from systems such as the Röttebehalter and the coffee filter, as the sludge has been exposed to a higher degree of gas exchange in the filtrating systems than sludge from the bottom of a septic tank.

25.4.4 Anaerobic or aerobic treatment of nearly solid fractions

Instead of drying beds, sludge may be anaerobically digested or aerobically composted, either by liquid composting or by composting with a bulking agent. Reductions according to the values in Figure 25.3 can be obtained.

25.5 DRY SANITATION

In dry sanitation little or no water is used at the source. Thus, the solid fractions never mix with the water phase, and active separation is rendered superfluous.

In developing countries infectious diseases caused by direct or indirect exposure to faeces have a severe impact on the economy, as illness and death cause millions of lost working years worldwide. Rapidly growing urban areas with dense populations without proper sanitation exacerbate this problem.

Therefore, one of the final overall conclusions at the 9th Stockholm Water Symposium in 1999 was as follows: 'To close the enormous sanitation gap in the developing world today, where several billions lack safe sanitation, water-borne sanitation has to give way to dry sanitation as a tested and valid alternative. Dry sanitation consisting of urine-separating toilets, with no or little water added, is a very elegant way to minimize water use and to create usable fertilizers both from urine and solid faecal matter. Human faeces can be seen as a resource that can be safely reused after reduction of pathogens and heavy metals. At the same time, poorly planned latrines – where insufficient attention is paid to hydrologic conditions – may generate groundwater pollution and make the city an unsafe and insecure place of residence and human activity' (SIWI 1999).

In the industrialized world, too, dry sanitation is arousing increased interest, since it lowers water consumption and waste water production, as well as facilitating the reuse of nutrients. Furthermore, by proper reuse, the risk of contaminating groundwater or surface water is almost eliminated. These are considered valuable qualities in certain regions.

In today's dry sanitation, the solid fraction of the black waste water is separated at source. The solid fraction – the faecal matter – falls into a container under the toilet, to be composted or dried.

25.5.1 Drying

In many sub-tropical and tropical countries, faeces have traditionally been air-dried. Some of the separation toilets on the European market also dry faeces. Drying faeces involves a relatively high consumption of energy, unless it is sun- and/or air-dried in dry parts of the world. Dried faeces contains no or very few pathogens (Esrey *et al.* 1998; Stenstroem 1999). However, handling the dusty material implies a risk of pathogenic aerosols. The dry matter should therefore, at least, be handled outdoors, because wind helps to diminish the concentration of aerosols.

25.5.2 Burning

In several countries, dried faeces are burnt (Svensson 1993). Burning, of course, eliminates latent human health hazards. It is a form of reuse, as the energy potential is exploited and the ashes can be spread on arable land.

25.5.3 Alkaline treatment

In several countries, faecal matter is mixed with ash or lime during filling, resulting in an important rise in pH (Esrey *et al.* 1998). Under certain mixing and storage conditions, the high pH eliminates pathogens within four to six months (Stenstroem 1999). Thereafter, the dry matter can be handled and reused as fertilizer or fuel without hygienic risks.

25.5.4 Composting

Compost toilets aim to produce hygienically safe organic fertilizer from faeces. The reduction of the initial populations of microorganisms in the compost depends, as shown in Figure 25.1, on time and temperature.

Figure 25.3 gives some examples of FS reduction in the Snurredassen composting system. A Snurredassen composting system contains four compartments for compost. When one is filled, it is left to be composted until the three others are filled and the first has to be emptied. In Figure 25.3 the results from all compartments belonging to one household have the same symbol.

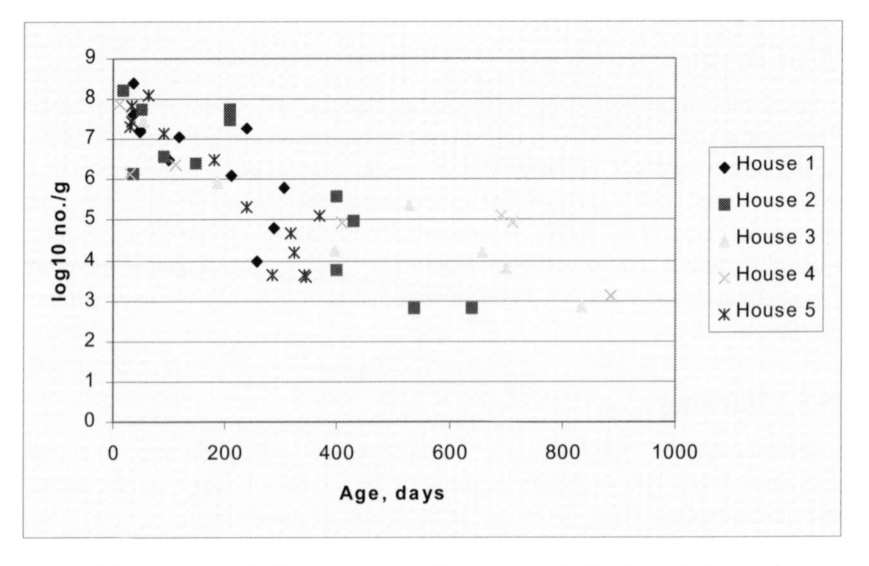

Figure 25.3. Reduction of FS in composting faecal matter in five households, each with a Snurredassen with four compartments for compost (Stubsgaard 1996).

As is shown in Figure 25.3, initial populations of FS are decimated over the course of time. The reduction rate is significantly dependant on time and temperature (95% confidence level, based on ash content).

If the results were shown for each compartment it would reveal that the first compartments filled had a poorer reduction rate than the following compartments. This is because with experience, using habits changed. For example, the importance of the right amount of bulking agents became evident, as did minimizing water use for cleaning procedures.

If the results from the first compartment of each house are ignored, the reduction of the indicator organisms in the composted faecal matter is approximately 4–5 orders of magnitude, which is more than the 3–4 orders of magnitude needed for reaching a reusable product with an acceptably low risk of infection.

The compost was recomposted in outdoor containers. An initial regrowth of FS was observed in all the cases analysed, followed by a relatively high reduction rate. The literature gives several examples of bacterial regrowth, some of them pathogens (Burge *et al.* 1978; Löfgren *et al.* 1978; Pereira *et al.* 1987; Strauch 1987; Ilsøe 1993; Engen *et al.* 1994). Thus, regrowth of FS can indicate regrowth of pathogens. Regrowth is possible at certain conditions, for example, in new, anoxic zones, created by the high oxygen consumption caused by remixing nutrients and bacteria during replacement from one compost container to another.

Regrowth reveals that the composting matter is not stabilized. As long as the compost contains easily degradable organic matter, there is a risk of favourable conditions for regrowth.

Instead of using precious space inside the house to compost faecal matter, some systems are designed for outdoor composting. Composting matter from one of these systems was analysed in Denmark over six months. At the same time, a buried latrine was analysed, to compare the hygienic effect of composting with this common way of handling latrines. The results are illustrated in Figure 25.4.

Figure 25.4 shows that the FS in the two portions of buried latrine were reduced by approximately two orders of magnitude during the experimental period, whereas FS in the four composts were reduced by approximately four orders of magnitude. These results are in accordance with the results of several other investigations (Ilsøe 1993). Thus, composting latrines lead to a better reduction in faecal microorganisms than burying latrines. In the experiment no oxygen was measured by Unisense microelectrodes in the buried latrine, whereas the oxygen profile gradually moved downwards through the four composts. The latter suggests a relatively quick stabilization process in the composts. Furthermore, lysimeter experiments showed that burying increases percolation, which implies a risk of ground water pollution.

Until now, only the reduction of FS in compost bins has been discussed. Still, concerning hygiene, there are risks related to indoor dry sanitation systems. For example, under certain inappropriate conditions, flies can propagate and find their way from the compost bin to the inhabitant's toothbrush and to food in the kitchen, thus creating a direct human health hazard. These findings show that some of the existing dry sanitation systems for indoor use still need technical development and/or the users need to be educated. Composting is complicated.

Good habits or professional maintenance is required to obtain odour-free compost with an acceptably low risk level.

25.6 EXISTING POTENTIAL VECTORS

A risk assessment of the described DESAR solutions could be useful. Still, local or national habits, religious barriers, hygienic awareness and public health are so decisive to the outcome of a risk analysis of DESAR solutions that it is considered impossible to elaborate a risk assessment of general validity.

However, potential vectors in existing waste water treatment and sludge handling systems can be defined for a first comparison. Here, some examples are given for kitchen waste, sludge from centralized waste water treatment plants and waste water effluents.

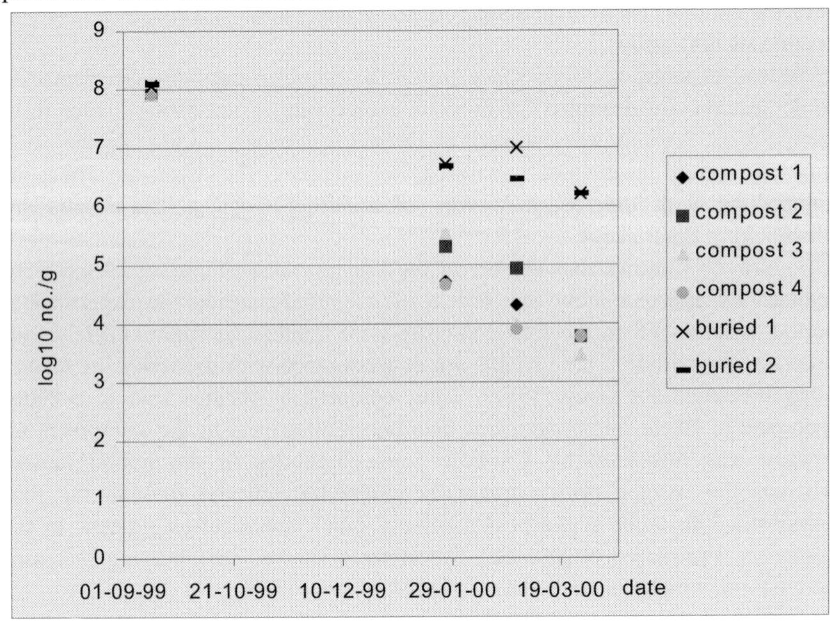

Figure 25.4. Reduction of FS in compost and buried latrine over one winter in Denmark (Stubsgaard 2000).

25.6.1 Kitchen waste

The organic matter in kitchen waste is a considerable part of the solid fraction of waste from households.

The kitchen waste accumulates in the waste bin or in the compost bin. Nappies, food scraps, leftover food, earth and vegetable and fruit peels are all mixed up with other wastes, creating good conditions for microorganism propagation. Table 25.2 compares measured concentrations of different microorganisms and parasite eggs in organic kitchen waste with concentrations in other organic-rich materials.

Table 25.2. Initial numbers of certain organisms in waste. Number pr. g wet weight or pr. ml. 'n.d.' = not detected (Feachem *et al.* 1980; Strauch 1986; de Bertoldi *et al.* 1985; Ilsøe 1993; Ernøe 1995; Stubsgaard 1996, 2000)

	E. coli	Faecal Strep.	*Salmonella*	Viruses	Parasite eggs
Waste water	10^3–10^6	10^4–10^6	10^{-3}–10^2	10^1–10^3	10^{-3}–10^{-1}
Sludge	10^3–10^9	10^4–10^9	10^1–10^3	$<10^4$	10–10^2
Latrine	10^7–10^9	10^7–10^9	n.d.–10^4	n.d.–10^3	n.d.–10^3
Organic kitchen waste	10^7–10^9	10^6–10^8	1/3 pos.	n.d	n.d

The concentrations listed in Table 25.2 show that there are relatively high numbers of indicator organisms in kitchen waste, close to where we prepare our food and eat. Therefore, kitchen waste is a potential vector in existing waste management and future DESAR systems.

25.6.2 Sludge

There are several examples of sludge-mediated contamination of animals and humans with bacteria, viruses and parasites (Danish EPA 1997). This has led to improved sludge treatment and restrictions on spreading periods, amounts and methods. Now, in the European Union (EU), the risk of human infection caused by sludge from waste water treatment plants is very low, especially considering the amounts of sludge disposed of. Ameliorated sludge handling is an example of how time and human efforts can reduce human health hazards, whilst maintaining the same degree of reuse.

25.6.3 Waste water effluents

As illustrated in Figure 25.2, reduction of FS in waste water can be limited in centralized waste water treatment plants. Furthermore, microorganisms stay suspended and survive relatively well in surface water and seawater (Nickelsen *et al.* 1995). This indicates that humans bathing close to effluents from waste water treatment plants are exposed to a human health hazard, when the skin, eyes, hands, especially the mouth, comes into contact with water. Therefore, EU directives on these concerns have been and are continually being issued.

Still, the risk is impossible to assess, as the incubation period and other potential sources of contamination often hinder the sick person in determining the cause of the illness. Even when the cause is known, it might not be registered (Nickelsen *et al.* 1995).

Generally, the most important source of waste water pathogen content is human faecal matter. By separation and local treatment of the solid fractions, the level of pathogens in the effluent will diminish, thereby reducing the risk of contaminating people bathing in the vicinity of the effluent. Dry sanitation solutions are especially prophylactic to the risk of pathogens in waste water effluents.

25.6.4 Human beings as vectors

The routes of infection with faecal microorganisms are relatively well known. The routes from faeces to face are illustrated in the F-diagram (Figure 25.5). The arrows indicate potential routes of infection, while the crossing bars illustrate possible prevention mechanisms. Waste water or water containing waste water are marked as ‘Fluids’.

The routes and barriers of Figure 25.5 seem evident. Nevertheless, the barriers are not respected. This implies, for example, that in 1990 more than 152 million persons suffered from high-intensity infection with the human hookworm (Murray and Lopez 1996). Thus, even in societies with relatively high standards of hygiene, hygienic habits in everyday life are not sufficiently implemented to prevent infection. Therefore, physical barriers are crucial in the decentralized handling of the solid fraction of waste water.

25.7 LEVEL OF PATHOGENS NECESSARY FOR DISEASE TO OCCUR

There is an individual immune system threshold as regards to pathogens. Beyond this threshold, disease symptoms set in. The limit varies with age, situation, general health, and other factors.

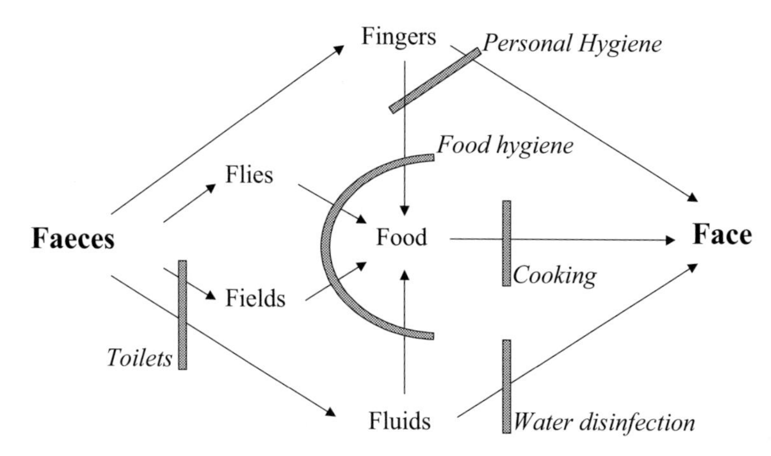

Figure 25.5. The F-diagram showing routes of infection with faecal pathogens, together with physical and habitual barriers (Esrey *et al.* 1998).

Table 25.3 shows some variations in individual limits to pathogen exposure without disease occurring. Table 25.3 also shows the lower limit for any of the participating persons becoming ill.

Table 25.3. Numbers of pathogenic bacteria required per healthy person for illness to occur (n.d. = not determined) (Kowal 1982).

	Percentage ill			
Bacteria	1–25	26–50	51–75	76–100
E. coli	10^6	10^8	10^8–10^{10}	10^{10}
Salmonella typhi	10^5	10^5–10^8	n.d.	10^8–10^9
S. meleagridis	10^6	10^7	10^7–10^8	n.d.
S. derby	n.d.	10^7	n.d.	n.d.
S. pullorum	n.d.	10^9	n.d.	10^9–10^{10}
Eneteroccus faecalis	10^9	10^{10}	n.d.	n.d.

The number of pathogens needed for people to become ill is relatively high, even for 1–25% of people, compared to the probable number of these organisms present in waste water, sludge or composting faecal matter. Thus, a person has to ingest at least 1 cm^3 of solid fractions of waste water to exceed his limit value of the examined bacteria.

However, viruses and parasites are not included in Table 25.3. These organisms can have other limit values. In theory, one infective virus or parasite spore or egg is enough to infect a person.

25.8 DESAR RISK MANAGEMENT

Concerning DESAR, the hygienic risk is different from the risk associated with centralized waste water treatment. In DESAR systems, there is no dilution effect, since the locally treated waste water and solid fractions are not mixed with waste water from other sources. Therefore, there is an elevated risk of exposure to high levels of potentially pathogenic organisms during use and operation; that is, before reduction has been completed.

Another risk-increasing factor is that considerably more people will be in physical contact with potentially pathogenic solid fractions of waste water because:

- DESAR solutions are usually placed physically closer to the users than centralized waste water treatment plants
- DESAR implies a much higher number of waste water treatment systems per inhabitant, often relatively easy to access
- at least some DESAR systems will be operated by their owners.

Therefore, it is important to establish hygienic barriers through design, use and operation. Physical barriers should be:

- robust constructions
- closed constructions preventing aerosols and animal vectors
- long storage = sufficient volume
- (re-)mixing systems without physical contact between solid waste and humans
- no manual digging out from containers
- no possibility of children or animals falling into containers.

The following are all points to be aware of when considering the implementation of a DESAR solution:

- As explained in Section 25.6.4, everyday life implies a considerable risk of a person being infected by his environment. If users operated the very small DESAR systems, for example a system for one household, themselves, the risk of being infected with a pathogen that hasn't already infected the users is relatively small. Therefore, it is appropriate that a family operates its own system.
- It is necessary that users become involved at the decision-making stage. If a decentralized solution requiring user involvement for proper

operation is not welcome, appropriate habits will not be established. Risks to hygiene can be the result.

- Training and a help desk will often be necessary. Even if the operational phase is professionally attended, some DESAR systems require certain maintenance in the utilization phase.

As described, the hygienic barriers illustrated in Figure 25.5 do not always exist. This could result in widespread disease. Therefore, the hygienic barriers must never stand alone. As outlined above, several physical barriers have to be implemented in every DESAR solution to assure prophylaxis.

25.9 REFERENCES

Bruce, A.M. and Fisher, W.I. (1990) A Review of Treatment Process Options to Meet the EC Sludge Directive. *J.IWEM.*4.

Burge, W.D., Cramer, W.N. and Epstein, E. (1978) ASAE, Soil and Water Division. Paper no. 76–2559, pp. 510–514.

Conlan, K. and Van Maele, B. (2000) Trial protocol for the revised bathing water directive. Bathing season 2000. During a workshop in Brussels on 10/4/2000. Developed on behalf of the European Commission DG Environment – Water protection unit.

Council Directive 76/160/EEC of 8 December 1975 concerning the quality of bathing water

Danish EPA (1997) Hygiejniske aspekter ved behandling og genanvendelse af organiske affald. *Environmental project no. 351. (In Danish.)*

De Bertoldi, M., Frassinetti, S., Bianchin, L. and Pera, A. (1985) Sludge hygienization with different compost systems. In *Inactivation of Microorganisms in Sewage Sludge by stabilization Processes* (eds Strauch *et al.*) Proc. of a CEC seminar, Hohenheim, October, Elsevier, 64–76

De Bertoldi, M., Cicilini, M. and Manzano, M. (1991) Sewage sludge and agricultural waste hygienization through aerobic stabilization and composting. In *Treatment and use of Sewage Sludge and Liquid Agricultural Wastes*. Proc of a CEC Symposium, Athens, 1–4 October, Elsevier, 212–226.

Engen, Ø, Hansen, J.F., Linjordet, R. and Ånestad, G. (1994) Hygiejniske Aspekter ved Hjemmekompostering av Hage- og latrinavfall. *Jordforsk*, Norway. (In Norwegian.)

Esrey, S.A., Gough, J., Rapaport, D., Sawyer, R., Simpson-Hébert, M, Vargas, J. and Winblad, U. (1998) *Ecological Sanitation*, Department for Natural Resources and the Environment, SIDA, Stockholm.

Feachem, D.J., Bradley, H., Garelick, H. and Mara, D.D. (1980) Appropriate Technology for Water Supply and Sanitation. *Health aspects of Excreta and Sullage Management: A State-of the Art Review.* The World Bank, Washington DC.

Ilsøe, B. (1993) Smitstofreduktion ved affaldsbehandling. Work report from the Danish Environmental Protection Agency, No. 43. (In Danish.)

Murray, C.J.L. and Lopez, A.D. (1996) *Global Health Statistics. Global Burden of Disease and Injury Series*, vol. 2. Harvard University Press, Cambridge, MA.

Nickelsen, C., Ernø, H., Møller-Larsen, A., Kerzel A. and Kerzel, H.M. (1995) Bathing Water – Microbiological Control – Literature Study. Environmental Project No. 314, Danish Environmental Protection Agency.

Pereira, N., Stentiford and Mara, D.D. (1987) in *Compost: Production, Quality and Use* (eds d.B.e.al), Elsevier Applied Science, pp. 276–295.

Stockholm International Water Institute (SIWI) (1999) Urban Stability Through Integrated Water-related Management. The 9th Stockholm Water Symposium, 9–12 August, SIWI, Sweden, p. 14.

Stenstroem, T.A. (1999) Sustainable sanitation. In *Urban Stability Through Integrated Water-related Management*, Proc of Stockholm Water Symposium, 9–12 August, Stockholm International Water Institute, Sweden, 353–356.

Strauch, D. (1987) in *Compost: Production, Quality and Use (eds* d.B.e.al.), Elsevier Applied Science.

Strauch, D. (1989) Improvement of the quality of sewage sludge: Microbiological aspects. In *Sewage Sludge Treatment and Use: New developments, Technological Aspects and Environmental Effects*, Proc of a conference held in Amsterdam, 19–23 September, Elsevier, 160–179.

Stubsgaard, A.E. (1996) Hygiejniske og miljømæssige aspekter af komposttoiletter. *Special report*. Microbial Ecology. Biological Institute. University of Aarhus. (in Danish).

Stubsgaard, A.E. (2000) Composting and burying faecal matter. The environmental impact. DHI, Institute for Water and Environment. (In Danish).

Svensson, P. (1993) Nordiska erfarenheter av källsorterande avloppssystem. Institutionen för samhällsbyggnadsteknik.

World Health Organisation (WHO) (1998) Guidelines for Safe Recreational Water Environments: Coastal and Freshwaters. Draft for consultation. Geneva, October, EOS/DRAFT/98.14.

26

The environmental impact of decentralised compared to centralised treatment concepts

L. Reijnders

26.1 INTRODUCTION

In cities in industrialised countries, centralised urban sanitation systems (CUS) dominate. The starting point for their emergence were hygienic considerations. Decentralised sewage discharges and handling human excrement were found to cause major outbreaks of infectious diseases such as cholera. The water closet that discharges into a sewer system was argued to be the solution to these problems (de Jong *et al.* 1998). As they developed, sewer systems not only took

Edited by P. Lens, G. Zeeman and G. Lettinga. ISBN: 1 900222 47 7

in discharges from water closets but several other inputs, such as excess rainwater and wastewater from industry.

Wastewater treatment systems originated as end-of-pipe measures to reduce the negative impacts of organic matter on surface waters into which these sewers emitted. Here physical pre-treatment and aerobic biological treatment of wastewater dominate. In some cases additional post-treatment steps are included, such as disinfection or phosphorus removal.

In industrialised countries CUS are often said to have solved problems associated with agents of infectious disease in human excrement and to have reduced surface water pollution. Indeed, problems such as massive outbreaks of waterborne diseases and 'dead' and foaming waters are largely a thing of the past. Thus the idea that connecting wastewater outlets to CUS essentially solves wastewater problems has become widespread in industrialised countries. This idea is, however, not necessarily based on a realistic evaluation of what is actually happening. If we look at centralised urban sanitation systems, it is not difficult to find a number of weaknesses. In part these weaknesses are related to high investment costs precluding their application in situations where the actual availability of capital is low, as in many developing countries.

Weak points in CUS partly have an environmental character. Reductions of treatment efficiency associated with poorly maintained sewer systems, industrial discharges and heavy rainfall are common. Centralised urban sanitation systems can also lead to major losses of nutrients such as phosphate (Meganck and Faup 1989; Niemczynovic 1993, Björklund *et al.* 2000). Large amounts of relatively clean wastewater furthermore can also become polluted. CUS are substantial users of energy (Infomil 1997). Moreover, aerobic biological treatment is not well suited to dealing with a number of persistent micropollutants such as heavy metals, poorly degradable synthetics and volatile petrochemicals. This leads to significant amounts of toxicants in discharges of wastewater treatment facilities (Tonkes and Balthus 1997) and to emissions into the air of volatile organics (Clapp *et al.* 1994). Toxicants in turn may give rise to insufficient quality of surface waters and sediments (Schoot-Uiterkamp 1994; de Wit 1994). Much of the persistent toxicants partition to sludges (Wong and Henry 1988; Ure and Davidson 1995). In urban settings these may become so polluted that they cannot be sustainably used as fertiliser (Björklund *et al.* 2000).

Several of the environmental weaknesses noted above stem from mixing a variety of dissimilar wastewater in centralised urban sanitation systems. Such mixing may be complete or partial (de Jong *et al.* 1998). Mixing discharges from industry and excess rainwater from pavements and gutters can lead to a significant loading with heavy metals and problematic organocarbon compounds (de Wit 1994; Schoot-Uiterkamp 1994). This in turn negatively affects sludge, surface water and sediments. Because heavy metals are long-lasting in the environment this contributes heavily to the time-integrated environmental impacts of CUS (Huijbregts *et al.* 2000). Mixing in discharges of organic solvents or bactericides ('disinfectants') may occasionally lead to significant deterioration of the wastewater treatment process. Mixing is furthermore problematic in view of the limited capacity of CUS; when there is heavy rainfall, CUS capacities may be exceeded, in turn leading to the discharge of untreated water via weirs.

DESAR systems can evade or mitigate the problems associated with mixing different wastewater. However, they may lack the physical and financial economies of scale that can be exploited in centralised treatment plants (Lundin *et al.* 2000). This in turn may be argued to give rise to relatively poor removal rates or unfavourable differences as to the hardware requirements per financial unit of 'treatment service'. Also, different treatment processes may be involved in CUS and DESAR systems that in turn may lead to different environmental impacts.

This then gives rise to questions as to the relative environmental impact (including risk) of centralised and DESAR systems. For instance: is a compost toilet or a small anaerobic sludge blanket reactor (ASBR), environmentally speaking, better or worse than CUS providing a similar service?

There is also another question that merits attention in this context. The basic concepts used in the treatment of wastewater are very old indeed. The septic tank, the most common type of DESAR currently in use, dates from the nineteenth century. The system of water closets plus sewers dates from the same century. The basics of aerobic treatment of wastewater and anaerobic treatment of sludge that currently dominate CUS go back to around 1900 (Higgins *et al.* 1985).

Since 1900, major changes have occurred. Environmental knowledge has increased. Heavy metals and a number of toxic organics have emerged as problematic contaminants in sewage sludges and sediments. Nitrogen compounds such as ammonia and dinitrogenoxide (Seitzinger and Kroese 1998) are currently linked with environmental problems such as soil acidification and the 'greenhouse effect'. Eutrophication of waters is now considered to be more of a problem than it was in 1900. Energy use is increasingly viewed as a problem.

New methodologies for wastewater treatment, including new anaerobic reactors (Seghezzo *et al.* 1998), ion exchange (Mels and van Nieuwenhuijzen 2000) and membrane filtration (STOWA 1998) have emerged. For membrane filtration and ion exchange, economies of scale are less important than for the aerobic treatment of wastewater as used in CUS (STOWA 1998). Furthermore, process control has changed dramatically due to the increased availability of relatively cheap real time monitoring and automated control devices. This is favourable to relatively small treatment facilities.

Looking back, while using current insights and technology, one may ask the question whether – if starting now from scratch – we would make the same water treatment choices.

26.2 ENVIRONMENTAL IMPACTS TO BE CONSIDERED

This question and the matter of the relative environmental impacts of DESAR and CUS are more easily raised than answered. In dealing with these questions, methodological problems also emerge.

First, which definition of environmental impact should be used in the evaluation of different approaches to wastewater? I feel that any impacts following the release of substances and biological agents (pollution) and the use of natural resources should be considered. Spatial impacts and the effect on nature and living things should also be included. Negative impacts, moreover, should be expressed in terms of present effects and in terms of sustainability (Reijnders 2000).

To give an example of the latter aspect, the use of fossil fuels should be considered as unsustainable because of the limited nature of fossil fuel reserves.

Table 26.1 lists the factors that in my view should be considered in evaluating environmental impact, including risk. Such impacts may be positive, neutral or negative. They should be evaluated in view of the actual performance of wastewater treatment systems, thus reflecting system vulnerability.

Table 26.1. Aspects to be considered in the evaluation of environmental impact or risk of wastewater treatment

- Depletion of scarce, virtually non-renewable abiotic resources
- Depletion of renewable abiotic resources
- Depletion of biotic resources
- Impact on water levels in soils
- Contribution to climate forcing (enhanced ‘greenhouse effect’)
- Human toxicity
- Human infection risk
- Pest/disease risk for animals and plants
- Ecotoxicity
- Contribution to photochemical smog by volatile organics and nitrogen oxides
- Acidification (aquatic and terrestrial)
- Eutrophication
- Bad smells
- Emission of heat
- Biological nuisances (such as flies)
- Noise
- Damage to ecosystems and landscape
- Spatial requirements

Unfortunately the availability of reliable data on the aspects listed in Table 26.1 is such that evaluating the environmental impact of wastewater treatment options can often only be done on the basis of highly incomplete data (Emmerson *et al.* 1995; STOWA 1996; Zhang and Wilson 2000).

On the other hand, incomplete data may be less of a problem when the different aspects mentioned in Table 26.1 have highly unequal weights. In such cases a limited number of aspects may be sufficient to pass judgement. If data reliability for these major aspects is sufficient, lack of data concerning minor aspects is relatively unimportant.

Answers to the question of relative environmental merit are possible at two levels: the strategic level that concentrates on the major *potential* benefits, and the tactical level that focuses on the *actual* relative merits of existing installations. These will be dealt with in order in the rest of this chapter.

26.3 A STRATEGIC VIEW OF WASTEWATER TREATMENT

Experience of dealing with wastes in general is that, from an environmental point of view, preventing waste is to be preferred over treating waste (Niemczynowicz 1993; Reijnders 1996). Also, mixing chemically dissimilar wastes before treatment is usually disadvantageous, environmentally speaking. In view of the problems associated with mixing in CUS pointed out above, there seems to be no reason why mixing dissimilar wastewater would be an exception to this rule (Niemczynowicz 1993; Lundin *et al.* 2000).

Thus from the strategic point of view it would seem wise to favour wastewater prevention and to unbundle the treatment of domestic wastewater, rainwater and wastewater from industries that are not fully dominated by degradable organics. It should be noted though, that there are limits to the environmental wisdom of unbundling. This follows from the environmental impacts associated with additional hardware and transport. So there is an optimum unbundling.

The precise determination of this optimum requires detailed knowledge of the economic activities, treatment systems and reuse potential involved and their environmental implications. This is beyond the scope of this chapter. However, substantial unbundling is an important strategic ingredient of environmental compatibility of wastewater handling.

A further strategic consideration is the choice of important environmental aspects. This can only be based on subjective choice.

My subjective choice as to the strategic evaluation is to limit discharges into water, conserve nutrients for reuse, and concentrate on energy and materials efficiency associated with the life cycle of the wastewater treatment infrastructure. This life cycle begins with primary production such as mining and ends with final disposal (ISO 1997). The strategic importance of discharges into water is clearly linked to the purpose of wastewater treatment. Life cycle energy efficiency, life cycle materials efficiency and the conservation of nutrients are highly important for resource conservation and the prevention of persistent pollution (Brownlow 1996; Reijnders 2000).

On the basis of the previous considerations a first approximation of a preferable strategy seems possible.

Rainwater should be handled in such a way that it does not become substantially polluted. So it should preferably not be collected in (corroding) zinc gutters, and should for instance not be mixed with detergent solutions derived from washing cars in the street. Similarly, strict requirements as to industrial and combustion processes should prevent dust that contains substantial loads of heavy metals being mixed with rainwater (Ure and Davidson 1995) or organocarbons such as polycyclic aromatic hydrocarbons and chlorinated dioxins. Rainwater should preferably be used for feeding surface water or should be infiltrated in the soil.

To the extent that rainwater becomes polluted, treatment should be fitted to the pollutants present. If we look at the original service that CUS intended to perform, dealing with excrement, the water toilet should be re-evaluated. There are several possible alternatives. First, the compost toilet, which can function at the level of an individual household. Requirements as to hardware are moderate, environmental costs associated with compost collection are limited, and conservation of nutrients is relatively high. A second possibility is a vacuum toilet, combined with the primarily anaerobic digestion of excrement. This in turn can be combined with the conversion of kitchen waste. This treatment system generates compost and biogas. It is in principle well fitted for higher numbers of people, for example in a hospital, a large office or a neighbourhood. Hardware requirements are larger than for a compost toilet, but biogas production leads to a (small) energy advantage. A third possibility is the separate collection of urine followed by its use as a fertiliser (Tillman *et al.* 1998; Lundin *et al.* 2000; Björklund *et al.* 2000).

Mixed domestic wastewater should be treated by the anaerobic degradation of organics combined with aeration and membrane filtration (Vigneswaran and Ben Aim 1989; STOWA 1998) while optimising the conservation of nutrients. The argument in favour of this option is mainly related to considerations of energy efficiency and conservation of nutrients for reuse. In the latter context conservation of phosphates should be prioritised in view of the relatively limited natural resources and their importance in eutrophication in fresh water systems (Brownlow 1996). High efficiency conservation of phosphate is possible by anaerobic treatment (Meganck and Faup 1989). Recovering N-compounds by means of, for example, ion exchange may also be environmentally beneficial (Mels and van Nieuwenhuijzen 2000). Apart from very small installations, energy required for membrane filtration may be partially derived from methane generated by anaerobic conversion. Household products that may end up in wastewater should be designed to be compatible with this treatment (Reijnders 2000). The sludge is to be used as a source for fertiliser and/or soil conditioner, while taking adequate steps to eliminate biological agents that may give rise to infectious disease. Industries should concentrate on pollution prevention and be fitted with tailor-made wastewater treatment depending on their remaining wastewater pollutants. If watery industrial wastes are essentially organic, combining them with other wastewater dominated by organics may be beneficial.

Another strategic matter is the extent to which unbundled wastewater treatment systems should be centralised or decentralised. Here there are two important considerations. On the one hand, the physical infrastructure (such as piping) necessary for centralised treatment is usually a relatively heavy environmental burden. Also, reuse may favour decentralised treatment (Fujita 1989; Lundin *et al.* 2000). On the other hand, centralised treatment may have physical economies of scale, and has traditionally provided better process control. Looking ahead in the context of treating wastewater, the advances in automated monitoring and control lessen the advantages of centralised treatment. Also, membrane filtration and ion exchange are characterised by limited economies of scale. The relevant factors favour the decentralisation of treatment where the population density is relatively low; reuse is an important consideration. Centralised treatment is perferred in areas of high population densities (Tillman *et al.* 1998). The optimum extent of centralisation at high population densities may well be less than in the past due to the availability of new technologies for both treatment and process control.

26.4 COMPARISON OF CURRENT WASTEWATER TREATMENT SYSTEMS

We will now discuss the relative environmental impacts and risks of current wastewater treatment systems. In dealing with this topic a number of methodological matters arise. The definition of environmental impact (including risk) has already been dealt with (see Section 26.2). There is, however, also the question of which systems should be compared. Centralised sanitation systems take in a variety of waste streams whereas DESAR systems may be specialised. This means that they provide different services and that a direct comparison of the two may be considered unfair. One may argue that centralised systems should be compared with an array of DESAR and/or more dedicated centralised systems that provide a similar service. If we agree with this argument, there is unfortunately a lack of well-documented DESAR treatment facilities which are currently handling highly mixed industrial, domestic and rain-derived wastewater. However, for the purpose of comparing CUS and DESAR treating domestic wastewater the situation is different. Here a fair comparison is possible (STOWA 1996; Tillman *et al.* 1998; Lundin *et al.* 2000, Björklund *et al.* 2000).

We now have to decide whether to take into account only the direct effects of centralised and DESAR systems or the life cycle effects. Here we will take a life cycle view of CUS and DESAR, including the hardware involved in these systems. This precludes missing important environmental impacts that remain hidden when concentrating on direct effects. Such 'hidden effects' may outweigh indirect effects concerning specific aspects of environmental impact (Emmerson et al. 1995). The best way to study life cycle effects is by life cycle assessment (LCA). The standard way to perform LCA includes most of the aspects mentioned in Table 26.1 (ISO 1997). Table 26.2 outlines the steps in LCA.

Table 26.2. Steps in life cycle assessment (LCA) (European Environment Agency 1997; ISO 1997)

- Goal and scope definition: choice of functional unity as object of LCA (e.g. a certain wastewater treatment service: for instance, reducing the amount of phosphates discharged by 200 Irish urban households by 80%). Defining the intended scope and boundaries of the system to be considered.
- Inventory: establishment of process tree, including all processes of the life cycle, gathering of empirical data pertinent to environmental pressures generated by the processes involved, including resource use and environmental releases.
- Impact assessment: classification of aspects of environmental pressure to be considered (see Table 26.1), calculating pressures associated with the life cycle based on the inventory.
- Evaluation and interpretation analysis of validity and uncertainty, interpretation of results in view of the goal defined.

Performing only the first two steps of full LCA and categorising the data on the basis of the environmental pressures that they may cause leads to a LCI (life cycle inventory). This inventory does not contain the calculated environmental pressures, but shows environmentally relevant interventions such as environmental emissions and uses of ores.

In performing LCA, a methodological problem arises. Outcomes of LCA often do not allow for realistic estimates of environmental impacts. This concerns such effects as eutrophication, acidification, ecotoxicity, human toxicity and the risk of infection. The lack of time- and site-specificity of LCA is a main cause for this deficiency, while infection risk is not covered at all by LCA. An alternative approach is to use LCA for a first approximation and afterwards to correct the outcomes with available data on real life impacts.

A second methodological problem that arises concerns the system boundary of the object of LCA. I feel that such boundaries should not be too narrow if the study is to have relevance in comparisons between CUS and DESAR. Physical infrastructure such as piping should, for instance, be included, as should fertiliser production.

Limited use has been made of LCA and related methodologies in evaluating wastewater treatment. Zhang and Wilson (2000) looked at the life cycle energy consumption of a large sewage treatment plant in south-east Asia. Emmerson *et al.* (1995) carried out an LCI of small-scale aerobic wastewater treatment plants. Nichols (1997) and Dennison *et al.* (1998) made LCI-type inventories for different ways of handling sludge. Sonesson et al. (1997) used an LCA-type evaluation for different options for handling organic wastes, including organic wastes involved in wastewater treatment. Tillman *et al.* (1998) made LCIs for different municipal wastewater systems, contrasting pumping to a CUS and to DESAR options. Lundin *et al.* (2000) studied the influences of scale and technology on a number of calculated environmental loads of wastewater systems for housing developments in Lulea and Horn in Sweden. Björklund *et al.* (2000) looked into the management of biodegradable wastes in Stockholm using LCA methodology and considering some of the impacts mentioned in Table 26.1. STOWA (1996) analysed a number of wastewater treatment options using LCA methodology, again looking some of the environmental impacts given in Table 26.1.

Since the studies of Tillman *et al.* (1998), Lundin *et al.* (2000) and Björklund *et al.* (2000) are the most relevant to the evaluation of DESAR options, brief summaries of these studies will be given in Box 1 below.

Box 1. Summary of LCAs and LCIs relevant to the evaluation of DESAR and CUS

Tillman *et al.* (1998) Life cycle assessment of municipal wastewater systems, *International Journal of Life Cycle Assessment* **3,** 145–157

Wastewater treatment was evaluated using LCI methodology for the village of Hamburgsund and a suburb of Gothenburg, in Sweden. Existing CUS were compared with, among others, anaerobic digestion followed by treatment of the liquid fraction in sand filter beds and separate collection of urine to be used as fertiliser. In Hamburgsund, the authors state that alternatives to CUS showed a lower environmental impact than existing CUS. In Gothenburg, some impacts were lower and others were higher. Assumptions as to energy technologies were found to be very important in determining environmental impacts.

Björklund *et al.* (2000) Planning biodegradable waste management in Stockholm, *Journal of Industrial Ecology* **3,** 43–58

Several waste management options were evaluated for Stockholm, Sweden, including anaerobic digestion and the separate collection and utilisation of urine using LCA methodology. It was concluded by the authors that increased nutrient recycling can reduce the net environmental impact but that separating urine and spreading it as a fertiliser leads to increased acidification, due to the emission of ammonia. Transportation was found to have low importance for overall environmental impact at high rates of nutrient recycling. Ancillary systems such as the generation of electricity, however, were found to be very important for net environmental impact.

Lundin *et al.* (2000) Life cycle assessment of wastewater systems: influence of system boundaries and scale on calculated environmental loads, *Environmental Science and Technology* **34**, 180–186

Wastewater treatment options were studied for a projected housing development in Lulea (2700 inhabitants) and Horn (200 inhabitants), in Sweden, using LCI methodology. A CUS was compared with small- and large-scale systems separately handling urine or treating blackwater by liquid composting. According to the authors, the separation systems outperformed CUS by having lower emissions and higher nutrient conservation. However, this conclusion was (for liquid composting) critically dependent on the way in which electricity was generated. Using data on the generation of electricity from other industrialised Western countries, liquid composting became much less preferable.

The LCAs and LCIs that have been performed show major shortcomings. Uncertainties in data have generally been neglected, although it is known that several environmentally relevant aspects of wastewater treatment show large variability. These include the presence of toxicants in emissions of wastewater treatment facilities (Tonkes and Balthus 1997; Clapp *et al.* 1994), partitioning of heavy metals and arsenic during wastewater treatment (Gommers and Rienks 1999), partitioning and degradation of polycyclic aromatic and chlorinated compounds (Gommers and Rienks 1999), emissions of untreated wastewater associated with storm surges and leakage of sewers, as well as emission of compounds that give rise to bad smells (Peek 1991).

None of the studies deals with the problem of lack of site- and time-specificity in LCA, with nuisances, with infection risk for humans or pest/ disease risk for animals and plants.

Emmerson *et al.* (1995), Nichols (1997), Dennison *et al.* (1998) and Zhang and Wilson (2000) use a limited number of the environmental aspects outlined in Table 26.1.

STOWA (1996), Sonesson *et al.* (1997), Tillman *et al.* (1998), Björklund *et al.* (2000) and Lundin *et al.* (2000) include more of the environmental aspects outlined in this table. Even in these studies, however, there are major shortcomings in dealing with environmental aspects. For instance, while discharges of heavy metals into water are evaluated by STOWA (1996) on the basis of their toxicity in water, their impact as a constituent of sludge is only evaluated on a weight basis (in terms of contributing to an amount of waste), thereby obscuring toxicity associated with the application of sludge on soils. The impacts on soils and sediments of toxicants that survive wastewater treatment or leak from sewer systems are also neglected in the STOWA (1996) study. This study furthermore neglects energy because it argues that the share of energy consumption in total societal energy consumption is small, whereas it can also be argued that, due to the scarce and non-renewable nature of fossil resources used for energy generation, energy consumption should be a major issue (Reijnders 2000). Sonesson *et al.* (1997), Tillman *et al.* (1998) and Lundin *et al.* (2000) all neglect problems associated with the presence of heavy metals or problematic organics in wastewater, including eco- and human toxicity. This is a shortcoming because such compounds are heavily weighted in LCA-type evaluations of toxicity (Huijbregts *et al.* 2000). Björklund *et al.* (2000) consider the impact of heavy metals, but do this separately from their LCA. Moreover they do not go into the exotoxicity of organic toxicants or resource depletion; neither do Sonesson *et al* (1997) or Lundin *et al.* (2000).

Finally one may note that in considering urine as a fertiliser, Tillman *et al.* (1998), Björklund *et al.* (2000) and Lundin *et al.* (2000) did not consider injecting instead of spreading urine, though the former may substantially reduce ammonia emissions. Thus I find that available LCAs and LCIs leave much to be desired. However, an interesting point has emerged from the studies by Tillman *et al.* (1998), Lundin *et al* (2000) and Björklund *et al.* (2000). They show that the outcomes of life cycle studies critically depend on ancillary systems such as energy production.

So far the findings of published LCAs and LCIs tend to agree with the strategic approach outlined in Section 26.3, although future studies should bear in mind that it is important to consider ancillary systems involved in the generation of energy. However, there is clearly much work to be done before existing CUS and DESAR can be properly compared using LCA methodology.

26.5 REFERENCES

Björklund, A., Bjuggren, C., Dalemo, M. and Sonesson, U. (2000) Planning biodegradable waste management in Stockholm. *Journal of Industrial Ecology* **3**, 43–58.

Brownlow, A.H. (1996) *Geochemistry*. Prentice-Hall, NJ

Clapham, Jr., W.B. (1981) *Human Ecosystems*. Macmillan, New York.

Clapp, L.W., Talarczyk, M.R., Park, J.K. and Boyle, W.C. (1994) Performance comparison between activated sludge and fixed film processes for priority pollutant removals. *Water Environment Research* **66**, 153–160.

De Jong, S.P., Geldof, G.D. and Dirkzwager, A.H. (1998) Sustainable solutions for urban water management. *European Water Management* **1**(5), 47–54.

Dennison, F.J., Azepagic, A., Clift, R. and Colbourne, J.S. (1998) Assessing management options for wastewater treatment works in the context of life cycle assessment. *Wat. Sci.Tech.* **38**(11), 13–20.

De Wit, J.A.W.(1994) *Watersysteemverkenningen: Emissiebeleid zonder grenzen*. RIZA, Lelystad. (In Dutch.)

Emmerson, R.H.C., Morse K.K., Lester, J.N. and Edge, D.R. (1995) The life cycle analysis of small-scale sewage treatment processes. *J .CIWEM* **9** (June), 317–325.

European Environment Agency (1997*) Life Cycle Assessment. A guide to approaches, experiences and information sources*. Environment Issues Series No 6. Copenhagen.

Fujita, K.(1989) Application of deep bed filtration in wastewater treatment. In *Water, Wastewater, and Sludge Filtration* (eds S. Vigneswaran and R. Ben Aim), CRC Press, Boca Raton, FL.

Gommers, P. and Rienks, J. (1999) *'Gezuiverde' cijfers over zuiveren*. RIZA, Lelystad. (In Dutch.)

Higgins, L.J., Best, D.J. and Jones, J. (1985) *Biotechnology*. Blackwell Scientific Publications, Oxford.

Huijbregts, M., Thissen, U., Guinée, J.B, Jager, T. and van de Meent, D., Ragas, A.M.J., Wegener Seeswijk, A. and Reijnders,L. (2000) Priority assessment of toxic substances in life cycle assessment. *Chemosphere* **41**, 119–151.

Infomil (1997) *Energie, Rioolwaterzuiveringsinrichtingen*. Infomil, Den Haag. (In Dutch.)

ISO (1997) *Environmental Management. Life Cycle Assessment. Principles and Framework*. ISO, Geneva.

Lundin, M., Bengtsson, M. and Molander, S. (2000) Life cycle assessment of wastewater systems: influence of system boundaries and scale on calculated environmental loads. *Environ. Sci. Technol.* **34**, 180–186.

Meganck, M.I.C. and Faup, G.M. (1989) *Enhanced Biological Phosphorous Removal from Waste Water,* CRC Press, Boca Raton, FL.

Mels, A., and van Nieuwenhuijzen, A. (2000) *Physical-chemical pretreatment of wastewater*. www.ct.tudelft.nl/wmg/sanitary/stowares.htm

Nichols, P. (1997) Applying life cycle methodology to the treatment and disposal of sewage sludge. *Proc. Life Cycle Assessment.* SCI Environment and Water Group, London.

Niemczynowicz, J. (1993) New aspects of sewerage and water technology. *Ambio* **22,** 449–455.

Peek, C.J. (1991) *Rioolwaterzuiveringsinrichtingen*. RIVM, Bilthoven. (In Dutch.)

Reijnders, L. (1996) *Environmentally Improved Production Processes and Products.* Kluwer, Dordrecht, the Netherlands.

Reijnders, L. (2000) A normative strategy for sustainable resource choice and recycling. *Resources, Conservation and Recycling* **28**, 121–133.

Schoot-Uiterkamp, J. (1994) *Watersysteem verkenningen: Vele kleintjes*. RIZA, Lelystad. (In Dutch.)

Seghezzo, L., Zeeman, G., van Lier, J.B., Hamelers, H.V.M. and Lettinga, G. (1998) A review. The anaerobic treatment of sewage in UASB and EGSB reactors. *Bioresource Technology* **65**, 175–190.

Seitzinger, S.P., and Kroese, K. (1998) Global distribution of nitrous oxide production and nitrogen inputs in freshwater and coastal marine ecosystems. *Global Biogeochemical Cycles* **12**, 93–111.

Sonesson, U., Dalemo, M., Mingarini, K. and Jönsson, H. (1997) ORWARE – A simulation model for organic waste handling systems. Part 2: Case study and simulation results. *Resources Conservation Recycling* **21** , 39–54.

STOWA (1996) *Het zuiveren van stedelijk afvalwater in het licht van duurzame milieuhygiënische ontwikkeling*. Hageman, Zoetermeer

STOWA (1998) *Mogelijkheden voor toepassing van membraanfiltratie op rwzi's.* Hageman, Zoetermeer. (In Dutch.)

Tillman, A.M., Svingby, S. and Lundström, H. (1998) Life cycle assessment of municipal wastewater systems. *Int. J. LCA* **3**, 145–157.

Tonkes, M. and Balthus, C.A.M. (1997) *Praktijkonderzoek aan complexe effluenten met de totaal effluent milieubezwaarlijkheid methodiek*. RIZA, Lelystad. (In Dutch.)

Ure, A.M. and Davidson, C.M. (eds) (1995) *Chemical Speciation in the Environment.* Blackie Academic & Professional, London.

Vigneswaran, S. and Ben Aim, R. (eds) (1989) *Waste, Wastewater, and Sludge Filtration*. CRC Press, Boca Raton, FL.

Wong, L.T.K. and Henry, J.G. (1988) Bacterial leaching of heavy metals from anaerobically digested sludge. In *Biotreatment Systems* (ed. D.L. Wise), CRC Press, Boca Raton, FL.

Zhang, Z. and Wilson, F. (2000) Life cycle assessment of a sewage treatment plant in south-east Asia. *J. CIWEM* **14**, 51–56.

Part V

Sociological and economic aspects of DESAR

27

The role of public acceptance in the application of DESAR technology

H. Mattila

27.1 INTRODUCTION

Wastewater has a worldwide reputation for being dirty, no matter how well it is treated. On the other hand, greywater is often considered "clean" and safe to be recycled for different purposes, even if it might contain considerable quantities of impurities including, for example, faecal organisms. These are factors that affect successful implementation of decentralised sanitation and reuse (DESAR). The opinions of professionals in various fields and the public about DESAR concepts have to be taken into consideration. Alternatives that are technically efficient and economically optimal might not be accepted because of other non-technical or non-economical criteria (Spulber and Sabbaghi 1998).

Edited by P. Lens, G. Zeeman and G. Lettinga. ISBN: 1 900222 47 7

The reuse of wastewater, even for potable purposes, is accepted by both professionals and the public if it is for so-called indirect use, as is the case, for example, in some cities in the United States. Indirect use means the purposeful augmentation of a potable water supply source with highly treated reclaimed water derived from treated municipal wastewater. In all cases, reuse projects should be implemented only after a thorough, project-specific assessment of health concerns and measures to mitigate them. (Crook *et al.* 1999).

The success of DESAR concepts depends heavily on local conditions. The need for reusing wastewater varies depending, for example, on the weather conditions in an area. Vision 21, which emerged from the Second World Water Forum, held in The Hague in March 2000, states that each country should establish minimum standards of service in the water sector by which it will measure progress in achieving the Vision of having water and sanitation services for all, by 2025. The targets and the standards should be adopted by each community, city or country to meet its own situation (WSSCC 2000).

27.2 THE NEED FOR THE ONSITE TREATMENT OF WASTEWATER

Finland will be used in this chapter as an example because the DESAR question has become very prominent in the country during the last years. The situation in the Lake Pyhäjärvi area is described more thoroughly to shed light on the implementation of a project. It is essential to note that the conditions and needs related to wastewater treatment vary from one country to another. Finland, with a population density of 17.3 persons/km^2, has good experience from small wastewater treatment plants which could be regarded as DESAR systems in another country. In this chapter, the DESAR concept involves house-specific wastewater treatment – in some cases small treatment units for a few houses. In some other countries (for example, the United States and Japan) treatment plants serving hundreds of houses are considered to be DESAR systems.

27.2.1 The evolution of wastewater treatment in Finland

The Finns have loaded their watercourses relatively heavily with different pollutants during the past decades. From about 1950 onwards, industrial development began, and the pulp and paper industry badly polluted certain rivers and lakes. At the same time, cities grew and the expansion of water distribution networks resulted in an extra load of wastewater on the same receiving waters. The use of artificial fertilisers in agriculture and forestry and subsequent leaching and runoffs into receiving waters has also increased. All these causes of pollution are now controlled to some extent.

Successive governments have supported rural areas in constructing drinking water supplies up to now, but wastewater treatment has been largely neglected. Now developers are working to keep the countryside alive, by trying to slow down migration from rural areas to cities. That is why farmers are supported in their efforts to process their products locally: to turn their berries into wine or jam, to can their vegetables, and so on. The downside of this activity is that it increases the wastewater load on the watercourses.

Wastewater treatment by industry, cities and even villages developed rapidly from the late 1960s to the early 1980s (see Figure 27.1). Today, diffuse pollution is the major concern in water protection. Watercourses downstream of big cities and industries are improving, but other waters are in danger of becoming contaminated, because they are receiving more nutrients, solids and even bacteria, than they can tolerate.

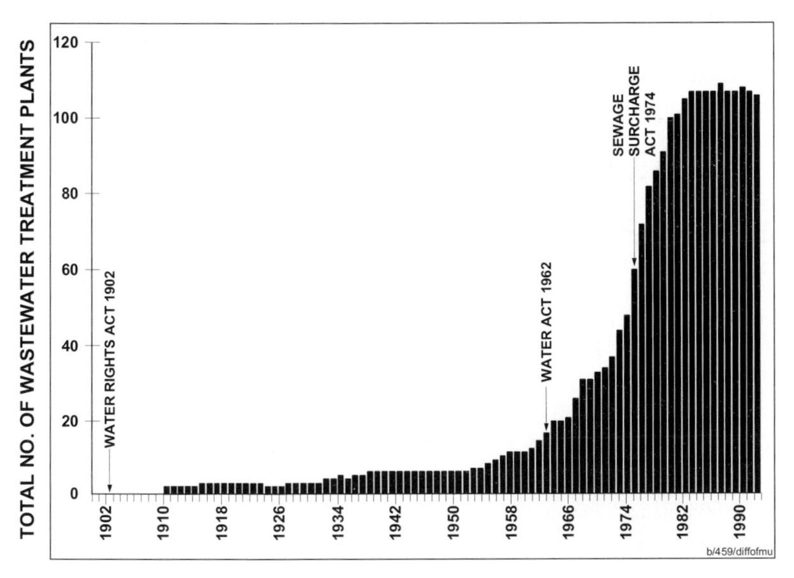

Figure 27.1. Number of wastewater treatment plants in cities in Finland (1900–1993) (Katko and Lehtonen 1999).

Because of the major sources of pollution mentioned above, the wastewater load from individual households outside sewer networks has so far been neglected. Even the Finnish Water Act has become outdated in this regard. Until February 2000, it was only acceptable to treat wastewater in septic tanks. Yet, it is known that even properly working septic tanks can only remove a maximum of 70% of the solid matter in wastewater (Mäkinen 1983; Santala 1990).

Dissolved impurities flow freely into the environment, often directly into a ditch or a river.

National targets for water protection in Finland by 2005 were set in 1998. The targets for scattered settlements are ambitious. The BOD load should be reduced by 60% and the phosphorus load by 30% (Ministry of the Environment 1999). The above legislation was amended on 1 March 2000 to meet the targets for 2005. The new Environmental Protection Act states that wastewater in rural areas must be treated to the extent that it cannot have a negative impact on the environment. The treatment technology, or even the methodology, is not specified in law, but municipalities are given the right to set local requirements on these matters depending on local circumstances. On the other hand, municipalities are also responsible for controlling the quality of wastewater treatment in their areas.

The application of this concept to practice is difficult in some Finnish rural areas where municipalities have few inhabitants (Figure 27.2). Thus, it may not always be easy for local politicians to set strict enough requirements: on-site wastewater treatment always means extra costs to households. Therefore, it may have been better to give Regional Environmental Centres the right to set local requirements. These organisations have the best knowledge of the regional environment, the area's soil quality, the groundwater deposits and other local conditions. This expertise could prevent local sociological problems from hindering the implementation of DESAR concepts.

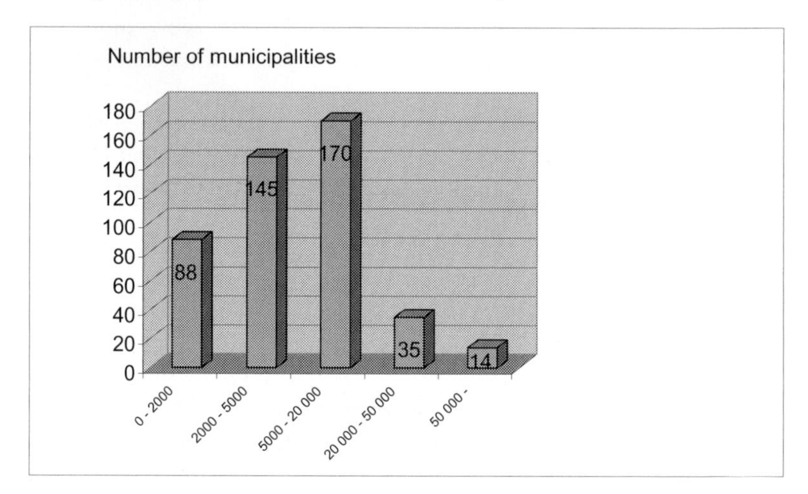

Figure 27.2. Population distribution in Finnish municipalities in 1999 (Association of Finnish Local and Regional Authorities 2000).

The need for decentralised wastewater treatment in Finland has been discussed so far mainly from the water protection point of view. This is mainly due to the fact that the country has thousands of relatively shallow lakes which are in danger of over-eutrophication. There is also a real need for DESAR in Finland for technical and economical reasons. With a population density of only about 17.3 people/km^2 it is clear that in many places the centralised systems would be either too complicated, too expensive, or both.

The idea behind describing the need for and the benefits of DESAR in Finland, or any other part of the world, is that people are generally curious about the effects of their activities. People are usually reasonable: if they are told of the positive effects of their behaviour, and are thus prepared to act accordingly.

27.2.2 The need for on-site treatment of wastewater in the Lake Pyhäjärvi catchment area

The Lake Pyhäjärvi Protection Project is presented here to describe an example of DESAR implementation in a typical rural area in Finland. As mentioned before, Finland has thousands of lakes. The value of these lakes was previously taken for granted, but now people realise that the quality of the water in the lakes cannot be maintained without efforts to minimise the incoming load.

Lake Pyhäjärvi is the most important lake in Southwest Finland (Figure 27.3). It has great recreational value, and some communities and industries depend on the lake as a source of raw water. About twenty professional fishermen work on the lake, and the value of their annual fish catch exceeds one million Euros.

The eutrophication of the lake has proceeded quickly during the past twenty years. This is caused by highly intensive farming, the large number of houses without a connection to a sewer network within the catchment area, and summer cottages on the lake. The incoming load of Lake Pyhäjärvi is completely from diffuse sources. It originates from cultivated fields, forests, swamps and ditches as well as cowsheds, poultry farms and pig houses, septic tanks and saunas.

Water protection projects within the catchment area have been implemented over the past few years. The Southwest Finland Regional Environment Centre (SFREC) began water protection work in 1992 in both of the major river catchment areas from which water flows into the lake. To expedite the restoration of the lake, and to secure the funding required, five municipalities and some of the industries and organisations situated around the lake established the Lake Pyhäjärvi Protection Fund (LPPF) in 1995. The main focus of the Lake Pyhäjärvi Restoration Project (LPRP) so far has been to reduce its annual incoming phosphorus load, which was over 20,000 kg/a in the 1990s. It has been estimated that if the incoming phosphorus load can be reduced by 40% and if the internal load does not increase at the same time, eutrophication of the lake can be stopped (Malve *et al.* 1994).

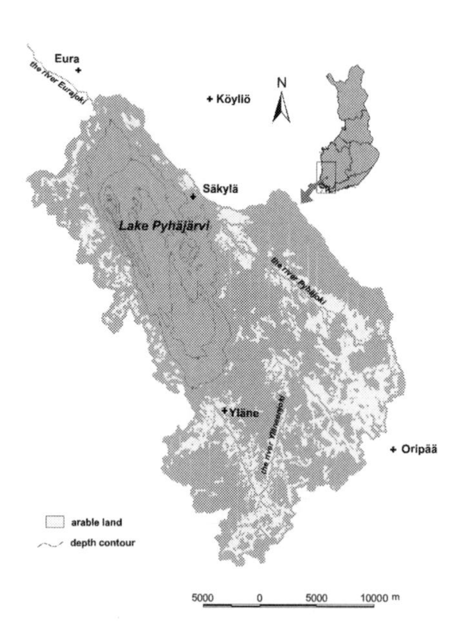

Figure 27.3. Lake Pyhäjärvi and its catchment area. (Southwest Finland Regional Environment Centre, © National Land Survey of Finland 7/MYY/01). The brownish areas are cultivated fields. Farms and other settlements are situated mainly in the river valleys. There are over 1000 summer cottages along the 80 km lake shoreline.

Agricultural water protection in Finland is based on an agreement with the European Union. Since 1995, farmers have worked according to the Finnish Agri-Environmental Programme (FAEP), which was signed by the Government of Finland and the European Union. The farmers within FAEP are bound to particular methods of agro-production and must use methods that promote environmental protection. These include, for example, limiting the use of artificial fertilisers, buffer strips along ditches and rivers, keeping thorough records of environmental actions taken, and training in the use of pesticides.

The treatment methods for runoff waters from agriculture are being further developed in the Lake Pyhäjärvi Restoration Project – tool development section (July 1996–October 2000). It is run by SFREC and LPPF and financed (50%) by EU-Life environment funds. The main focus of the project is the treatment of runoff waters from agriculture, but some activities are also related to domestic wastewater. Activities include the preparation of guidelines for wastewater treatment in rural areas for the entire project area and education on the subject.

Another undertaking, the Hajasampo project, concentrates on problems related to domestic wastewater in rural areas. The project was launched at the start of 1998. It is financed by the Ministry of Agriculture and Forestry, Ministry of the Environment, National Technology Agency, municipalities, LPPF, households and companies manufacturing equipment for wastewater treatment. The Finnish Environment Institute also finances the Hajasampo project and provides project management. It is worth mentioning that households are also participating by paying cash for the treatment units.

Since the water entering Lake Pyhäjärvi is from diffuse sources, education and information activities play an important role in lake restoration (Figure 27.4). The project has organised study tours for local people to show them what kind of measures they could take to save the lake. The project staff write regularly for local newspapers on water protection and environmental matters. Some public lectures have been given, a couple of videotapes have been prepared, local fairs have been attended, and so on.

Figure 27.4. Visiting one of the wastewater treatment units with the press.

Lake Pyhäjärvi is one of the most actively researched lakes in Finland (Malve *et al.* 1994; Ventelä 1999; Ekholm 1998; Helminen 1994). The Lake Pyhäjärvi Restoration Project also supports many studies and research activities concerning the catchment area and/or the lake. The aim is to determine the effectiveness of the water protection measures taken, to monitor the state of the rivers and the lake, and to better understand the reasons behind the eutrophication process.

LPRP carried out a wastewater survey in the catchment area of the lake in 1995. The municipalities conducted the actual survey while the Finnish Environment Institute published the results (Pyy 1996). There were two main objectives for the survey: first, the magnitude of the wastewater load into the lake was to be estimated; and second, municipalities were interested in discovering the most problematic households with regards to wastewater. The homeowners' interest in improved systems was also studied, and information about improved systems was distributed. According to the survey there are 2500 houses and summer cottages in the catchment area of the lake. Of these, about 2000 were visited and the owners interviewed.

The reason why the summer cottages are treated separately is the difference between the technology used in them and the houses used for all year round. Summer cottages do not normally have flush toilets, but a traditional earth toilet or a modern composting toilet. On the other hand, 90% of the houses occupied all year around are equipped with flush toilets.

It is estimated that roughly 15% of the phosphorus load of Lake Pyhäjärvi is from wastewater (Pyy 1996). The rest comes from agriculture and forestry, natural leaching, and as deposits from the air. The figure is probably slightly higher than the Finnish average due to the large number of summer cottages in the Lake Pyhäjärvi area. In any case, it is high enough to warrant special attention. The need for DESAR in the area is indisputable, and there is a common will to implement it.

27.3 DEVELOPING DESAR IN THE LAKE PYHÄJÄRVI AREA

27.3.1 The introduction of on-site wastewater treatment as demonstration units

The SFREC constructed five different kinds of soil filters in the Pyhäjärvi area in 1992. Three of them are traditional sand filters improved with a layer of phosphorus adsorbent. As P-adsorbents, two filters were packed with Fosfilt material while one had gypsum for adsorption purpose. The other two filters were constructed above ground as test filters. One had peat as the filter medium. The peat filter became clogged within a couple of months. Another filter type consisted of a chamber filled with sand and lime. The medium of that filter was changed in 1994.

The SFREC constructed eight more soil filters in the area in 1994. One of them became clogged after a couple of months and was later rebuilt. All of the above-mentioned filters were designed with a flow of 50 l/person/day and constructed for a single household. The required surface area is 15–20 m^2 (Santala 1990). The following conclusions can be drawn from the water analyses (water samples have been taken in all seasons, and the results given are averages):

- septic tanks remove only 15–20% of organic and nutrient loads
- the more thoroughly the solid material is removed from the wastewater, the better the soil filter functions. The best results have been attained when septic tanks remove more than 70% of the solids
- BOD reduction has been very good (88–97%)
- the ability to remove phosphorus has declined continuously, but was as high as 70% after 5 years of operation
- nitrate removal capacity has not been as good, only about 25–35%
- the quality of the wastewater has improved greatly. Coliforms have been removed with an efficiency of above 99% (Elomaa 1998).

The soil filters constructed for research purposes have worked properly. The basic criterion for a proper soil filter is careful design and construction. So long as the septic tanks are big enough (at least 600 l/person, minimum 2.5 m^3) and the design is followed in detail during construction, a soil filter can operate for years without any need for reconstruction.

All changes in incoming wastewater quality disturb the functioning of the soil filter. The risk of disturbance is obvious if the septic tanks are not working due to poor maintenance or some other reason. Households should also remember to use non-toxic detergents when treating wastewater biologically. Otherwise, the efficiency of the treatment unit will be disturbed every time a house is cleaned. This fact was clearly shown by one of the soil filters researched in the Lake Pyhäjärvi area. Its performance varied considerably from one sampling to another at the start of the research, but after biodegrable detergents were used in the household, the soil filter started to have a better removal efficiency.

It must be noted that these experimental soil filters were constructed by SFREC, and the only contribution by the homeowners was their permission for constructing such units in their yards. In some cases the farmers participated in the construction with their tractors and labour. All the homeowners with soil filters received reports on the efficiency of their treatment units. Groups of visitors come to see the filters every now and then, and the feedback of the functioning of water treatment is exchanged.

For the success of the project, it was important to first introduce some properly functioning wastewater treatment units to people living in the area. Without evidence of improved water protection provided by more advanced treatment techniques than mere septic tanks, it would have been more difficult to persuade homeowners to invest in the project. It is also evident that the DESAR concept will be accepted only if it is proved to be safe and technologically reliable.

27.3.2 Technical and financial support to households to implement DESAR

After the above-mentioned wastewater survey had taken place, the Management Group of LPPF decided that the efficiency of wastewater treatment in the Pyhäjärvi area should be improved.

As mentioned, the Water Act was outdated in this regard. Thus, LPPF made a proposal to the Ministry of the Environment in June 1996 to amend the law. According to the proposal, more effective methods to treat wastewater than septic tanks would be required in areas such as the catchment area of Lake Pyhäjärvi. Because it was obvious that these types of changes are impossible in the short term, a plan was prepared to get people to improve wastewater management of their own volition.

The first obstacle for not investing in a soil filter, or any other wastewater system, is lack of knowledge. People generally believe that their wastewater load is negligible compared to other sources. When they see the results of a wastewater survey, people see that they can have an impact on their own lake by taking better care of their waste. It should be remembered that the wastewater impurities have the most pronounced effect near the point of discharge. Nutrients are dissolved and easy to use by algae. Organic material settles on the bottom of a watercourse causing oxygen demand. Moreover, there are plenty of faecal bacteria that die as time passes by and sunlight disinfects the water. They can nevertheless cause health risks close to the point of wastewater discharge.

The second obstacle is posed by the need to select a suitable treatment method for households. Due to the high number of households without a connection to a sewer network in Finland (about 200,000), and because of changes in legislation, rapid development is taking place in this field. Companies are looking for emerging markets and new types of treatment units are invented every year. In 1996–97 the project employed an engineer capable of making designs and cost calculations for the homeowners. As a result, homeowners got free tailor-made plans to improve their wastewater management. An example of a simplified plan is presented in Figure 27.5.

The third obstacle is, of course, financing. The municipalities reserved some money (ranging from 6000 Euros from the municipality of Yläne to 18,000 by the municipality of Säkylä) from their annual budgets for this purpose. Homeowners could apply for financial support from their municipality when investing in improved wastewater treatment. The support was, and is even today, 50% for houses in all year round use and 30% for summer cottages.

This resulted in the installation of about 30 different types of wastewater treatment units in the Lake Pyhäjärvi area in 1996–97. Some were traditional soil filters or infiltration fields, but others were new (Figure 27.6). The performance of the new units will be reported in June 2001, when there will be sufficient follow-up data to draw conclusions. The reporting will also include a follow-up of the "old" soil filters installed by SWFREC and described earlier.

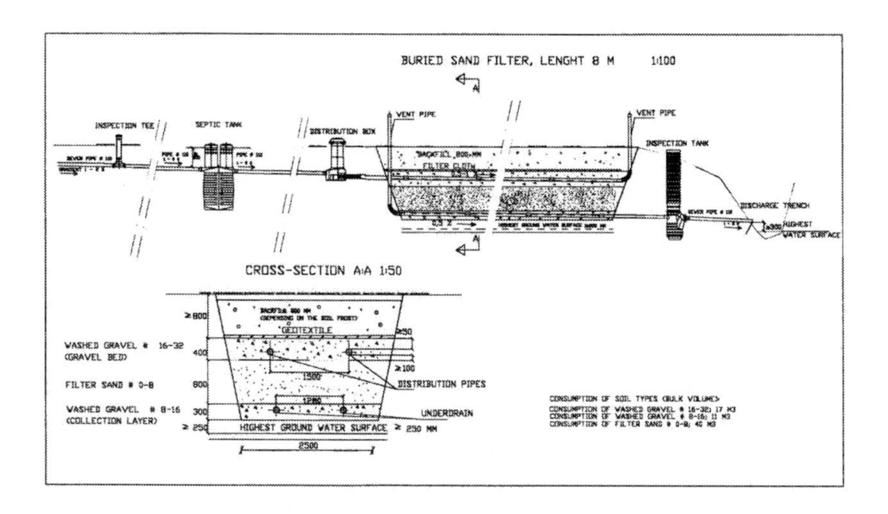

Figure 27.5. An example of a wastewater treatment plan for a single household (Saralahti 2000).

Figure 27.6. New types of wastewater treatment units for single households. Green Pack (above), septic tank with 1400 Filt (top right) and Sakofilter (bottom right).

Municipal engineers have not been keen to participate in the process. They are of the opinion that they have enough to do with centralised sewer systems and treatment plants. On the other hand, officers in charge of environmental protection activities have taken part enthusiastically.

27.3.3 Enforcement of DESAR by a special project

Work to improve the quality of wastewater treatment has become even more effective since 1998. The Hajasampo project was launched jointly by several parties (described earlier in Section 27.2.2).

At the beginning of the Hajasampo-project, homeowners received the same support as before. The only difference was that this project also involved the manufacturers of treatment units. Their interest was aroused as a result of the research work going on within the Hajasampo project: they wanted reliable, credible and impartial results concerning the performance of their treatment units. The positive results gained could be used in marketing and the negative results in product development.

During the first year of the Hajasampo project, 46 treatment units were installed in the municipalities around Lake Pyhäjärvi, followed by 30 additional units in the second year. Some of them are soil treatment or infiltration units, but there are also new types of treatment units (see Figure 27.6).

Little can be said at present about the performance of these new types of units due to the small number of water samples so far analysed. In general, all units can treat wastewater to some extent, if they are installed carefully and used correctly.

It is only natural that there have been some problems in installing the new equipment. Most of the problems have been related to unclear instructions for installation, mistakes made by inexperienced contractors, or unexpected circumstances at construction sites. However, there have also been technical problems in operating the new units, such as difficulties in getting wastewater distributed equally into filter media. Because the Hajasampo project is also a product development project, the manufacturers are given time to improve all the unsuccessful units and resolve all problems.

One lesson has clearly been learnt so far: there is no standard solution for small-scale wastewater treatment, but each individual case needs a tailor-made design. Each location has its own topography, soil quality, constructions, wells, and so on, which have to be taken into account in designs.

When the Hajasampo project was planned and scheduled, it was agreed that about 100 treatment units were needed to get reliable operating results, and to test also the operation and maintenance organisations that will take care of the units. The 46 + 30 units already mentioned, plus those constructed before the Hajasampo project, add up to about 100 units. Therefore, project engineer designs have not been available free of charge to homeowners since September 1999.

After a couple of months of discussions and negotiations, it was agreed that the services earlier provided by the project engineer of the research project must be secured in some way. The existing consulting firms were not interested in designing these kinds of small wastewater treatment units. In other words, people were not willing to pay enough to get the designs done by the consulting firms. As a consequence, in December 1999 eleven homeowners had a meeting and established the Water Supply Co-operative of Southwest Finland.

The voluntary co-operative employed a water adviser to serve the members in operating and maintaining the existing wastewater treatment units and in designing and installing new ones. The water adviser is supposed to assist also with problems, such as getting clean water from wells or springs. Sufficient experience from the co-operative has not yet been gained to be able to analyse the success of the organisation. One thing that can be said is that people's interest in the subject has not waned even though the design work is no longer free of charge. The water adviser got 17 orders from new customers during the first months of the co-operative.

The main finding in the Lake Pyhäjärvi area in Finland is that people are willing and able to improve the standard of wastewater treatment within their own compounds if they are given assistance with design, selection of a suitable treatment method and construction/installation works. Another observation is that a population's interest in making the necessary investments can be raised through financial support by the municipality.

The experiences described above concern the use of ordinary flush toilets. The Hajasampo project is also conducting a follow-up study on about 100 composting toilets. Most of these are associated with summer cottages, but there are also some in a military camp in the project area, and a couple also serve houses occupied all year round. In all cases, urine is separated from the solid waste and grey waters are infiltrated into the soil.

People who are ecologically minded enough may be interested in waterless toilets (Skjelhaugen and Saether 1999). However, if they are not aware of how they are expected to manage their systems based on environment-saving technology, they might be disappointed. This became clear in connection with another project in the Finnish municipality of Merimasku, where people who moved from the city were so enthusiastic about the beautiful scenery of their new home by the coast that they, without thinking about the consequences, accepted waterless toilets for their houses. After a few months, they started to request sewers after having experienced smells and other problems related to the operation of the toilets. Most of the problems could have been avoided by careful use of the toilets. In one case people had poured water in the toilet causing flooding in the house, and in another case they had not used enough supporting material for the composting process, causing smells and a huge number of flies in the building (Olenius 1999).

27.4 OPERATION AND MAINTENANCE

The problems related to wastewater in rural areas cannot be solved simply by constructing treatment units. The units must also operate. So far, about 100 treatment units have been built in the project area, and one of the tasks is to test suitable ways to organise the operation and maintenance of these units.

It is obvious that the homeowners' own efforts to maintain the wastewater unit cannot be relied on. It has been noted that sometimes even the septic tanks are not regularly maintained. The reason for this is not necessarily the cost of the work but is more likely to be ignorance. By forgetting to maintain the septic tanks, one can ruin the treatment unit in a relatively short period of time. Although wastewater treatment can be decentralised, management of the systems may have to be centralised to some extent (Figure 27.7)

Figure 27.7. The maintenance work of small wastewater units is sometimes quite heavy and dirty and cannot be left to the homeowner, if the treatment units are to function well. Here, a maintenance crew is replacing a Green Pack filter.

There are several options for solving the operation and maintenance problems of the units. Maintenance activities can be taken care of by municipalities, manufacturers, other private companies or water co-operatives. The two last options will be tested in the Lake Pyhäjärvi area.

Whatever the method chosen to organise operation and maintenance activities for small-scale wastewater units, homeowners will have to pay some extra costs. These costs should remain reasonable. One criterion for the upper limit could be the charges that people connected to a sewer network are paying to the same municipality. Charges for the service could be collected, for example, on a monthly basis as in Georgetown, US (Dix and Nelson 1998) or based on a separate agreement with the customer as in the case of the co-operative in Southwest Finland, where the water adviser invoices the customer every time separately depending on the actual work done.

These examples from many parts of the world show that DESAR systems can function only if the operation and maintenance work has been implemented by a trained professional in the field (Ohmori 1996). Not many of the general population are interested or committed enough to take care of their wastewater treatment units to keep them running year after year.

27. 5 REFERENCES

Association of Finnish Local and Regional Authorities (2000) Population in the Finnish Municipalities in 31.12.1999 (cited 10 September 2000), available from http://www.kuntaliitto.fi

Crook, J., MacDonald, J.A. and Trussell, R.R. (1999) Potable use of reclaimed water. *Journal AWWA 91*(8), pp. 40–49.

Dix, S.P. and Nelson, V.I. (1998) The Onsite Revolution: New Technology, Better Solutions. *WATER/Engineering & Management*, October, 20–26.

Ekholm, P. (1998) Algal-available phosphorus originating from agriculture and municipalities. PhD thesis, University of Helsinki.

Elomaa, H. (1998) Efficiency of sand filters in the catchment of Lake Pyhäjärvi. *Finnish Journal of Water Economy, Water Technology, Hydraulic and Agricultural Engineering and Environmental Protection* **3**, 5–7. (In Finnish.)

Helminen, H. (1994) Year-class fluctuations of vendace (Coregonus albula) and their consequences in a freshwater ecosystem. Reports from the Department of Biology, University of Turku, No. 37.

Katko, T. and Lehtonen, J. (1999) The evolution of wastewater treatment in Finland. *Vatten 55(*3), 181–188.

Mäkinen, K. (1983) Structure and operation of septic tanks – A literature review and experiments. Publication no. 227, National Board of Waters, Helsinki. (In Finnish.)

Malve, O., Ekholm, P., Kirkkala, T., Huttula, T. and Krogerus, K. (1994) Nutrient load and trophic level of lake Pyhäjärvi (Säkylä). A study based on the water quality data for 1980–92 using flow and water quality models. Publications of the National Board of Waters and the Environment – Series A181, Helsinki. (In Finnish, abstract in English.)

Ministry of the Environment (1999) Water Protection Targets for the Year 2005, Publications of Ministry of the Environment – The Finnish Environment, no. 340, Edita, Helsinki.

Ohmori, H. (1996) Maintenance and management of johkasou systems. P*roceedings of Japan–China symposium on environmental science*, 211–214.

Olenius, J. (1999) Village of the archipelago. Final report, 12 October, unpublished. Original in Finnish.

Pyy, V. (1996) Wastewater treatment inventory in the rural areas of the catchment of Lake Pyhäjärvi. Publications of Finnish Environment Institute – Duplication series no. 15, Helsinki. (In Finnish.)

Santala, E. (ed.) (1990) Wastewater treatment in soil – small applications. Publications of National Board of Waters and the Environment – Series B1, Helsinki. (In Finnish.)

Saralahti, K. (2000) Simplified plan of wastewater treatment plant for a single household – An example drawing. Unpublished.

Skjelhaugen, O.J. and Saether, T. (1999) A case study of a single house installation for source sorting the wastewater. *A conference paper at the 4th International Conference on Managing Wastewater Resources*, 7–11 June, Ås, Norway.

Spulber, N. and Sabbaghi, A. (1998) *Economics of Water Resources: From Regulation to Privatization*, Kluwer Academic Publishers, Boston.

Ventelä, A-M. (1999) Lake Restoration and Trophic Interactions: Is the classical Food Chain Theory Sufficient? Publications of University of Turku, Annales Universitatis Turkuensis Series AII, no. 121. Biologica-Geographica-Geologica, Turku.

WSSCC (2000), Water Supply and Sanitation Collaborative Council, Vision 21: A Shared Vision for Hygiene, Sanitation and Water Supply. The Second World Water Forum, March 2000, The Hague.

28

Public awareness and mobilization for sanitation

M. Wegelin-Schuringa

28.1 INTRODUCTION

Sanitation programmes depend critically for their success on effective public awareness and mobilization through information, education and communication. Experiences over the past decades demonstrate that even the technically best-designed programmes fail or produce meagre results, because decision makers and intended beneficiaries are not adequately consulted, informed, educated or mobilized.

One of the problems with sanitation is that it is rarely a strongly felt need, especially in rural areas. Few people realize that many diseases are caused by poor hygiene and sanitation, and neither do they understand the way these diseases are transmitted. Although health considerations are rarely a motivating

© IWA Publishing. Decentralised Sanitation and Reuse: Concepts, Systems and Implementation. Edited by P. Lens, G. Zeeman and G. Lettinga. ISBN: 1 900222 47 7

factor for a community to construct sanitation facilities, it is for health reasons that good hygiene behaviour and sanitation are promoted. For the community, various other factors such as privacy, convenience and status are more important. The key to getting people motivated to improve sanitation is to understand these factors and to use them as a basis for the development of an intervention and communication strategy (Wegelin-Schuringa 1991).

Another of the big challenges in mobilization for sanitation is that human waste disposal is on the one hand an extremely individual issue as the use of toilets and hygiene behaviour is a private subject in most cultures. On the other hand, the lack of sanitation management is a public issue, with repercussions far beyond the level of an individual user. Finding the right carrot (and stick) for the right audience is the key to success.

There is a distinct difference between communication and mobilization for sanitation in rural areas and in low-income areas in cities. Rural areas tend to be characterized by relative social cohesion and homogeneity, where it is relatively easy to reach audiences through traditional and participatory means of communication. The environmental conditions, moreover, are generally supportive to on-site sanitation solutions that can be managed at individual household level. Except for cement, construction materials are likely to be available in the surroundings and at specific periods (such as after harvesting or sowing), people have time to spend on construction.

The reverse is often true in low-income urban areas. These tend to be characterized by high population densities, where it is difficult to find room for individual toilets or sewerage systems; social cohesion can be quite low and it may be very difficult to get people to organize themselves for a communal activity. In addition, the proportion of the population that rents their dwelling may be high and hence willingness to get involved in sanitation improvements may be low, since this is considered the task of the landlords. On the other hand, motivation for sanitation may well be high, especially for women, because lack of latrines are a severe problem with respect to convenience, privacy and safety. A final difference is that people in urban areas are generally fully occupied with earning money or finding work and this is likely to influence their willingness to improve sanitation conditions.

The first section of this chapter will give some background on communication and behaviour change theories. This is followed by a description of a systematic approach to develop a strategy for awareness raising and communication to mobilize different segments of society for sanitation improvements. This approach consists of a number of components in a process:

(1) Assessment of main risk factors and problems in environmental sanitation
(2) Assessment of current knowledge, attitudes and practices
(3) Finding the right incentives
(4) Audience segmentation
(5) Monitoring and setting verifiable goals
(6) Financing, cost recovery and willingness to pay

In the second section of this chapter, some tools for communication and mobilization are discussed.

28.1.1 Understanding public awareness, communication and mobilization

Public awareness is only one element in a wider continuum of a communication process that includes advocacy, social mobilization and programme communication (McKee 1992). Creating awareness and getting the commitment of decision-makers for a social cause is the first component in the continuum and is called advocacy:

Advocacy consists of the organization of information into an argument to be communicated through various interpersonal and media channels with a view to gaining political and social leadership acceptance and preparing a society for a particular development programme. The goal of advocacy is to make the issue a political or national priority. Advocacy, in the first instance, may be carried out by key people in international agencies, as well as special ambassadors, but is gradually taken over by people in national and local leadership positions and the print and electronic media. Advocacy leads directly to social mobilization.

Social mobilization is the process of bringing together all feasible and practical inter-sectoral social allies to raise people's awareness of and demand for a particular development programme, to assist in the delivery of resources and services and to strengthen community participation for sustainability and self-reliance. In McKee's opinion, the concept of social mobilization is the glue that binds advocacy activities to more planned and researched programme communication activities.

Programme communication is the process of identifying, segmenting and targeting specific groups/audiences with particular strategies, messages or training programmes through various mass media and interpersonal channels, both traditional and non-traditional. Communication is an instrument based on a two-way dialogue, where senders and receivers of information interact on an equal footing, leading to interchange and mutual discovery. Planners, experts and field workers must learn to listen to people about their concerns, needs and possibilities. Policy makers need to be personally contacted to benefit from dialogue and influence decisions.

Communication and mobilization for behavioural change is a complicated process of human actions, reaction and interaction. It involves looking at situations from other people's viewpoints, and understanding what they are looking for. It means understanding obstacles to change. It means presenting relevant and practical options, and it means telling people what the effect will be of the choices they make. Communication can work towards a situation where policy makers, the private sector and the population/communities become committed to programmes and this helps to prevent expensive mistakes.

People tend to change when they understand the nature of change, and view it as beneficial, so that they make an informed and conscious choice to include it in their list of priorities. Unless their circumstances are taken into account, and their needs are met, no effort for change will be successful. People need to be informed and convinced, or they do not feel part of the effort.

The activities of advocacy, social mobilization and programme communication do not necessarily happen consecutively. In general, advocacy begins the process and leads to social mobilization and programme communication. But advocacy is needed at various times in a programme, not only at the beginning. Social mobilization benefits from a thorough analysis of who the best partners are for a particular programme, and what they potentially may contribute. There is a need for a greater emphasis on promotion and social marketing to provide incentives and benefits that are important to the success of sanitation improvement programmes. In social marketing the 'Four Ps' of marketing are often cited – Product, Price, Place and Promotion. The concept of 'product' in social marketing does not necessarily mean a physical product. It can also mean a change in behaviour such as using a latrine, or stopping dumping waste in a river. Which incentives work best depends very much on the individual area. Different incentives are required for different stakeholders. Users need incentives to use sanitation facilities and to follow good hygiene practices. Providers require incentives to deliver better services.

28.1.2 Understanding attitudes and behaviour change

What messages influence people's knowledge and attitudes, and how does that contribute to changes in behaviour? Research from communication and behavioural change sciences makes clear that this is a complex issue and evidence shows that the clearer the message on a concrete topic, the more the audience can relate to it and the higher the chance that knowledge increases. Research in social sciences has shown that knowledge on a topic may increase, people may even change attitudes, but that the step to improved behaviour and practices depends on a complex set of social and psychological factors. Hubley introduced the BASNEF model (Beliefs, Attitudes, Subjective Norms and Enabling Factors; see Table 28.1) for understanding behaviour in health communication (Hubley 1993).

Individual beliefs about the consequences of certain behaviour and the value placed on each consequence lead to personal attitude or judgement. Attitudes combined with subjective norms contribute to behavioural intention. Subjective norms are beliefs about what behaviour other influential people would wish the person to perform. Enabling factors such as income, housing, water supply, agriculture and sanitation have to be available so that the intention leads to a change in behaviour. Table 28.1 explains the influences on behaviour and communication and the actions needed in the BASNEF model.

Table 28.1. The BASNEF model (Hubley 1993)

	Influences	Actions needed
Beliefs, **A**ttitudes (individual)	Culture, values, traditions, mass media, education, experiences	Communication programmes to modify beliefs and values
Subjective **N**orms (community)	Family, community, social network, culture, social change, power structure, peer pressure	Communication directed at persons in family and community who have influence
Enabling **F**actors (inter sectoral)	Income/poverty, sanitation services, women's status, inequalities, employment, agriculture	Programmes to improve income, sanitation provision, situation of women, housing, skill training

The starting point is the individual person's behaviour. However, an understanding of the influences on behaviour can lead to interventions that go beyond the individual to include programmes at family, community and national levels and involve educational, social, economic and political change.

28.2 STRATEGIES, APPROACHES AND STEPS

A systematic approach to plan and implement a strategy for public awareness, communication and mobilization is needed to mobilize different segments of society to support the development of sustainable sanitation management. The different components in this process are discussed below.

28.2.1 Assessment of main risk factors and problems in environmental sanitation

Before it is possible to develop a strategy for an intervention in environmental sanitation, it is necessary to get an overview of the present conditions with regard to environmental sanitation. An assessment will focus on the different technologies that are in use for sanitation, the reasons for selection of these technologies, experiences with the technology (construction, cost, use, problems, repairs, maintenance) as well as the technical know-how and financing mechanisms that have been available. At the same time an inventory of the main risk factors and problems associated with the sanitation practices and technologies must be made. For instance, in Zambia, roaming pigs would eat human faeces left in the bush and, with their dirty snouts, would come back and contaminate the compound, utensils or plates lying around and children that played with them. Because hygiene behaviour is a major determinant for health risks connected to sanitation and latrine use, availability of water for hand washing, fly control and animals with access to the compound (such as chicken, pigs or goats that may transfer faeces into the compound) must also be included.

The assessment in itself can be a powerful tool for public awareness raising as it can focus the attention of authorities and communities on existing environmental sanitation conditions. Depending on the situation, this can be carried out by local government staff with the support of some members of the community, or by the community itself with the use of participatory methodologies (see Section 28.3.2). One way of assessing local conditions and problems is by walking around the community. During the walk, the environmental cleanliness of the area can be seen and potential risks of contamination can be identified, such as faeces lying around, water bodies being polluted directly or indirectly by faeces, the existence of communal defecation areas, sanitation around water points and solid waste disposal mechanisms.

The information gathered during this assessment is likely to indicate differences within the community, not only in facilities and practices used, but also in the attitudes of the people. On the basis of this, a rough classification of risks, problems and options for technical intervention in environmental sanitation can be made for the purpose of follow-up planning with regard to communication and mobilization, but also with regard to technology choice and implementation.

For instance, it may be that part of the population is interested in sanitation improvements; existing systems are bad from a hygiene point of view, but it is clear how they can be improved. Under such conditions, it will be relatively simple to work with the people to improve or construct sanitation systems to be more hygienic. It may also be that the majority of the population is not interested in sanitation improvements, while at the same time health risks are high and cholera is endemic. Under such conditions, hygiene education and mobilization has to precede any intervention in terms of construction until such time that people are motivated.

28.2.2 Assessment of current knowledge, attitudes and practices

Sanitation is to a large extent a social phenomenon, rather than a technical one, and therefore it is essential that background information on cultural, social, economic and environmental factors influencing sanitation is acquired before actual planning can start. This is especially true when a new technology is to be introduced, but it is also necessary to develop a communication and mobilization strategy and at a later stage to develop a strategy for hygiene education.

Sanitation behaviour is based on ideas and taboos associated with defecation and on traditional habits that originated in local cultural, social and environmental conditions. The table below gives an illustration of the extent of cultural variations in defecation practices, which can be seen as a continuum. These variations will eventually determine which technology options will be acceptable to the people. For instance, in a culture where the handling of faeces is acceptable (as is common in Vietnam or China), composting technologies are much more likely to be accepted than in cultures where handling of faeces is regarded to be impure (as for instance in Pakistan and Zimbabwe). Similarly, religion can be extremely influential in sanitation practices; for instance, in Islamic communities, a latrine can never face Mecca and communal facilities may be less acceptable because it would entail women leaving the house or compound to defecate.

Sanitation practices are not only based on cultural and environmental conditions, but also on access to sanitation technology in terms of knowledge, materials and funds. In addition, awareness of health aspects of sanitation behaviour is important because it determines the degree of sustainability of an intervention in sanitation. When new latrines are constructed in a programme and sanitation behaviour is not addressed at the same time, people are unlikely to support the improvements with the sustained behaviour change needed for improved health. The reverse, however, is also true: conventional health messages may be widely known and largely understood, but these messages by themselves may not persuade people to implement desired changes because of other constraints, such as inappropriate technology in areas with a high water table or unstable soils.

Table 28.2. Cultural variations in defecation practices (Wegelin-Schuringa 1991)

Sanitation aspect	Extent of cultural variation	
Choice of preferred site		
1. Location	Open field	Sheltered
	Near or in water	No water contact
2. Visibility of (intention of) use	Within the house	Away from the house
	Socially prescribed	Individually selected
3. Direction of latrine	Allowed	Not allowed
	Prescribed	Not prescribed
Preferred posture	Squatting	Sitting
	Ritually prescribed	Individually preferred
Preferred times of defecation	Sunrise or sunset	Whenever the need arises
Frequency of defecation per day	Once or less	More than four times
Anal cleansing material	Only water used	Paper, leaves, sticks, corncobs, stones etc. used
Cleaning after defecation	No cleaning	(Ritual) bathing
Social organization around defecation and use of latrines	Strict male/female separation	No specific rules on separation
	Communal defecation accepted	Communal defecation not tolerated
	Avoidance rules for latrine use within family	All members of the family can use latrine
Attitude to human faeces	Cannot be handled	Seen as useful resource for composting or feeding animals
	Children's faeces considered harmless	Children's faeces considered harmful

28.2.3 Finding the right incentives

Because health considerations are rarely a reason to be interested in sanitation, it is necessary to find the reasons that do motivate people for it. At user level, these may be convenience, safety, privacy, status or economic incentives. It is more *convenient* to go to a latrine near or in the house than to have to walk to the bush, especially during the rainy season. The *safety* aspect is especially important in urban slum areas where social control is low. For women, going to a latrine at night may become almost an invitation for rape. Similarly, at night evil spirits abound and snakes or wild animals are not seen.

During the sanitation and communication situation analysis that was carried out in Zambia as part of the development of a sanitation strategy, many different attitudes and practices towards latrines were found.

The most common reasons for not using or having latrines are:

- there is no room
- do not want to share a latrine with in-laws
- do not want to share a latrine with the opposite sex
- do not want to share a latrine with non-related people, but cannot deny access to neighbours
- bad smell
- fears of safety for elderly and young children
- faeces is food for the pigs
- fear of snakes in a dark latrine
- do not want to be seen using a latrine

Reasons why people are interested in having a latrine are:

- there is insufficient cover in the neighbourhood
- the public latrine is too far away
- population and settlement densities are too high to provide privacy for outside defecation
- health reasons, especially cholera
- being modern
- convenience
- able to get one with a subsidy
- ability to take a bath in the (improved) latrine (Wegelin-Schuringa and Ikumi 1997)

Box 28.1. Attitudes and practices towards sanitation in Zambia.

The most common need with respect to defecation is probably the desire for *privacy*, although the level of privacy needed may vary according to sex, age or social status. Generally, women have more need for privacy than men and often it is this aspect of a latrine that they like most, especially if the latrine can also be used for bathing. Another important factor influencing interest in latrines, especially with men, is connected with *status* and *prestige*. Usually, people who already have a latrine constitute the upper layer of the community, they are likely to be more 'modern', have an education and have seen the outside world: all attractive aspects in the eyes of the rest of the community. In densely populated areas, the aspect of a *clean environment* is often cited as a positive aspect of sanitation, not only by men and women, but also by young people for the purpose of sports activities. Finally, reuse of excreta may be an *economic* incentive either for people for their own use or for sale to farmers.

It should be noted, however, that if status or prestige are the motivating factors, this does not imply that people also use the latrine. There are many examples of latrines being used as storage rooms, or reserved only for visitors or certain members of the family. This implies that for effective and sustained use, hygiene education is a crucial aspect of sanitation improvements (Wegelin-Schuringa and Ikumi 1997)

Just as with the communities themselves, it is unrealistic to expect other stakeholders such as government staff at different levels or the private sector to become interested in the improvement of sanitation conditions if they do not get anything out of it that they see as a profit. Obviously, such incentives are different for stakeholders at different levels. But it is necessary to find the right incentive for the right target group. At national level, these may be exposure as a good example at international fora; being quoted in the international media and literature or being at a good 'level' in international health or environment statistics. At municipal or district level, these may be elections for the 'sanitation' town of the year; access to (regional) training for the municipal/district engineers that win the election or matching funds for cost recovery.

28.2.4 Audience segmentation

Segmentation of audiences and their communication needs is essential for effective communication and mobilization. Without understanding the differences among various segments, or sub-segments, it is difficult to design effective messages that call for change. The process of audience segmentation has to be based on the outcome of the assessments of main risk factors and problems, the current knowledge, attitudes and practices as well as on the incentives that have already been identified. Target audiences for sanitation improvements range from community level to national level. In the process of audience segmentation, research has to be carried out to find the most efficient and effective way to reach each target group with respect to place, time and channel of communication. It is, for instance, not very effective to conduct a public awareness campaign on television if the target group does not watch this medium regularly.

At *community level*, different target groups that can be identified are men, women, youth, children, the rich, the poor, ethnic minorities, and so on. This is known as audience segmentation by gender. Gender is a specific parameter for socio-economic analysis. All groups have different roles and responsibilities in society and may attach different values to services and the benefits to be derived from them (Dayal *et al.* 2000). Consequently, their demand for and access to services and their economic behaviour differ and hence so should messages for their mobilization. In addition to these different segments of the community, community level organizations, traditional chiefs/community elders, churches, schools and health centres are target groups at community level.

At *district or municipal level*, the target group for advocacy and awareness raising are district/municipal planners, staff of different departments involved in sanitation management (such as public works, water, sewerage, health), the private sector (formal and informal), the political representation (councillors, local chiefs), professional associations and non-governmental organizations (NGOs)s. They must be informed about current environmental conditions, about health statistics at local level, about developments in the sanitation sector and about the integrated nature of water and environmental sanitation. The main aim of the messages is to motivate the target group to take the initiative or support efforts at a local level, with respect to planning, construction, operation and maintenance as well as with financial and human resources. Other aims may be to show the importance of hygiene behaviour in combating sanitation-related diseases; to give examples of how without community involvement programmes fail; the need to put economic value on latrine use (what it costs to be ill) and sustainability elements at community level.

At *national, regional and provincial level*, people make policy decisions and/or influence development. This is an important target group, since one reason why sanitation is receiving little attention is because it has not been given any priority at this level. Included in this target group are politicians (ministers, members of parliament, councillors), professional associations, educational institutions, donors, NGOs, churches and the media. To mobilize them, it is important to have data and information that they need to discharge to their respective audiences, such as telling them what it costs the nation if people get sick with dysentery, cholera or other water- or sanitation-related diseases due to a lack of sanitation.

28.2.5 Monitoring and setting verifiable goals

In order to direct the communication strategy and mobilization efforts, it is necessary to have an agreement on the specific operational goals of the intervention. These goals have to be set together with the main stakeholders involved and will be different for the different target groups. In the communication strategy, these will concern the effectiveness of the messages that are being communicated as well as the effectiveness of the channel that is being used. Thus, for each segment that is targeted, a goal has to be set, with a time span and an indicator that is to be measured and that is verifiable. The same applies to the mobilization effort. Indicators need to be set with government staff and programme staff, to assess if the mobilization efforts that have been designed at the start of the programme indeed have the desired effect. Traditionally, these efforts would be monitored by counting the number of activities having taken place at community level, according to the plan. This, however, in no way assesses the impact of the activities, although these are at this stage most important because they determine the interest that the community will eventually have in becoming involved in sanitation improvements. Therefore, the indicators have to be set in such a way that they monitor the effectiveness of the mobilization (Shordt 2000). The actual collection of monitoring data, in addition, should not be done by those who carry out the mobilization activities – most likely district/municipal government staff – but by those people or organizations that have an interest in sanitation improvements being carried out in a sustainable manner. Table 28.3 gives an illustration of indicators that monitor effectiveness of a sanitation intervention, and hence its sustainability, after construction and education activities have been carried out.

28.2.6 Financing, cost recovery and willingness to pay

Financing and cost recovery for sustainable sanitation schemes on the one hand and ensuring equity on the other are key issues which any sanitation programme needs to address. This concerns local community-based sanitation initiatives as well as large-scale programmes funded by international donor organizations.

The cost of on-site sanitation programmes can be divided into three categories. These are institutional and project delivery costs, material and labour costs and operation and maintenance costs. The first category includes the cost of community mobilization and development, communication, information and training, as well as technology delivery costs such as engineering supervision and logistic support. These costs are usually paid by the government or external support agencies.

Material and labour costs have to be paid by the community, at least to a large extent. This may be paid partly in cash and partly in kind, depending on the provision of appropriate financing and credit facilities and the total cost of the proposed sanitation intervention. At the mobilization stage, the community needs to be aware of the various components that make up the total costs and the parts that are covered by grants or subsidies. Generally, most government supported programmes do not include substantial grants or subsidies, hence targeted subsidies may be necessary from the rich to the poor, who cannot afford the costs of a latrine. Often the provision of credit schemes poses problems.

Table 28.3. Indicators for community managed sanitation programmes and services (Dayal *et al.* 2000)

Variables	Indicators and sub-indicators
1. Effectively sustained	Functioning programme ▪ Coverage levels for safe excreta disposal, drainage and solid waste disposal ▪ Upkeep of coverage levels ▪ Level of quality of installation and upkeep Effective financing ▪ Degree of autonomous financing of household facilities and community services ▪ Coverage of costs ▪ Degree and timeliness of payment Effective management ▪ Level and timeliness of repairs of community systems ▪ Budgeting and accounting for service to M/W/R/P
2. Effective use	Safe and environmentally sound use ▪ Degree and nature of access (R/P) ▪ Change in disposal practices by and within households (M/W/C/R/P) ▪ Environment free from human waste risk
3. Demand responsive service	User demands ▪ User contributions during implementation Project responsiveness to demand ▪ User voice and choice in planning and design ▪ Satisfaction of user demand ▪ Ratio of user-perceived costs/benefits for M/W, R/P
4. Division of burdens and benefits	Gender and poverty focus during establishment and operations ▪ Nature of payments ▪ Cost sharing in community and households ▪ Division of labour between M/W and R/P households ▪ Division of functions and decision making between M/W, R/P

M = men; W = women; C = children; R = rich; P = poor

The final component is the cost of operation and maintenance, which has to be fully borne by the users. Since the choice of technology will to a large extent determine the cost of operation and maintenance, this has to be clearly communicated with the community at an early stage.

Willingness to pay for sanitation improvements, if people can opt for the sanitation system that they want and are willing to pay for, is found to be much higher than expected. This has been proven in many well-known case studies such as Prosanear in Brazil, Baldia Pilot Project and Orangi Pilot Project in Pakistan and the Kumasi Sanitation Project in Ghana (Wright 1997). The key features to success in this willingness to pay are again highly dependent on an effective communication strategy:

(1) Community members make informed choices on:
- whether to participate in the project
- technology and service level options based on willingness to pay – based on the principle that more expensive systems cost more
- when and how their services are delivered
- how funds are managed and accounted for
- how their services are operated and maintained

(2) An adequate flow of information is provided to the community and procedures are adopted to facilitate collective action decisions within the community and between the community and other actors.

(3) Governments play a facilitative role, set clear national policies and strategies, encourage broad stakeholder consultation and facilitate capacity building and learning.

(4) An enabling environment is created for the participation of a wide range of providers of goods, services and technical assistance to communities, including the private sector and NGOs (Sara *et al.* 1998)

28.3 METHODS AND TOOLS FOR COMMUNICATION AND MOBILIZATION

28.3.1 Mass media

Media and other channels of communication have to be selected on the basis of what is appropriate, considering the preferences and characteristics of whoever is going to use the information. This means that television exposure is only effective in places where watching is regular. Radio is in many developing countries a much more effective medium because it is much more common. Awareness-raising films may also be shown with success in the 'open air' cinema, as a 'pre-programme' for the main film, as was done with great success in the Bhaktapur Development Project in Nepal in the early 1980s (Lohani and Guhr 1985). Also, theatre can be used very effectively for communication and mobilization since it can easily be adapted to the target audience, for instance to children.

The effectiveness of the use of written media depends not only on the literacy rate, but also on the circulation figures of local newspapers, although this may not mean much. In Kenya, for instance, newspapers are read widely at street corners where the papers are sold, but where the reading of unsold papers is also permitted. Similarly, newspapers are likely to be shared among the literate people within a community. What has to be kept in mind in using mass media is that this method of communication informs people, but is unlikely to effect a change in behaviour. For this to happen, participatory methods are more effective.

28.3.2 Participatory approaches

In participatory approaches, people are helped to analyse their own situation and to come up with solutions that are appropriate for their circumstances. Many such approaches are used in water and sanitation programmes and, by involving users/communities/customers/beneficiaries from the start of a programme, the ownership is vested with them, which enhances sustainability. These participatory approaches can be applied at all phases in the project cycle and for different purposes. In the context of this chapter, they are used as a tool for public awareness raising for mobilization and for the development of a communication strategy. But they can also be used for implementation and construction, for operation and maintenance and for monitoring and evaluation. There are many manuals describing participatory approaches such as Srinivasan (1990), WHO (1996), Dayal *et al.* (1999) and Shordt (2000), some of which are described briefly below. For the analysis of risk factors and problems in the sanitation environment, the most appropriate methods are community mapping and transect walks. The assessment of knowledge, attitudes and practices in sanitation is best carried out through focus group discussions, three pile sorting cards and the sanitation ladder.

Community mapping: groups of men and women draw a map of the local settlement including roads, houses, health facilities, all water sources and all latrines (public and houses with private latrines). The map usually includes other information needed for the project, such as water sources. Through the mapping, information can be obtained on access to water and sanitation, settlement patterns and divisions between different groups that make up the community. Information can also be obtained on radios or televisions present and on the division of different segments within the community.

Transect walks: these are systematic walks with key informants through the area of interest while observing, asking, listening and seeking out problems and solutions. Walking through the community leads to an understanding of the power divisions, environmental sanitation, risk practices and problems, sanitation technologies in use, construction quality and environmental conditions of importance to technology selection.

Focus group discussions: these are discussions with a small group of people in the community, either mixed or separate with the different segments of the community, on a specified topic. The aim of these discussions is to get a deeper understanding of the issues that are being confronted.

Three pile sorting cards: cards that contain pictures, words or sentences, depicting negative, positive and neutral aspects of a certain topic (sanitation) are given to the group for sorting in three piles (positive, negative and neutral). The discussions during the sorting will give insight into knowledge, attitudes and practices of hygiene behaviour.

Sanitation ladder: different sanitation technologies are depicted on cards. Groups are asked to sort the cards according to their level of technology (from outside defecation to a VIP latrine or small-bore sewerage system) and to indicate where people are at present in the ladder and where they want to go (see Figure 28.1). This is a good tool for discussing the upgrading of sanitation technologies and to assess what people like about which technologies.

Figure 28.1. Pictures used in a sanitation ladder in Zambia (CMMU 1996).

28.3.3 Community events

Community events can be organized to raise awareness about possibilities for action at a neighbourhood level. Examples of such events are a sanitation day in which the neighbourhood gets involved in a clean-up campaign. Activities may be coupled to other community events such as religious feasts, school inaugurations, elections or other community awareness-raising events.

Figure 28.2. Community cleaning in Kibera, Nairobi, Kenya (photograph by author).

28.3.4 Environmental profiles

The community mapping mentioned in Section 28.3.2 is one way of making an environmental profile at a community or neighbourhood level. But there are many different tools for making such profiles and this can be done at different levels: municipal, neighbourhood or village. The environmental profile can focus on environmental sanitation in a particular area, but can also focus on city-wide sewerage management or on city-wide sources of pollution. In rural areas, the environmental profile can be done to get an overview of the water resources management situation. The profile is an extremely useful tool to raise awareness on sanitation or environmental issues at a district or municipal level.

28.5 REFERENCES

Community Management and Monitoring Unit (1996) Introducing WASHE at district level. Programme Co-ordination Unit, Water and Sanitation Development Group, Government of Zambia.

Dayal, R., Van Wijk, C. and Mukherjee, N. (2000) Methodology for Participatory Assessments. Linking Sustainability with Demand, Gender and Poverty. The World Bank Water and Sanitation Program, Washington, DC.

Hubley, J. (1993) Communicating health : an action guide to health education and health promotion, Macmillan, London.

Lohani, K. and Guhr, I. (1985) Alternative sanitation in Bhaktapur, Nepal: an exercise in community participation, Deutsche Gesellschaft für Technische Zusammenarbeit, Eschborn, Germany.

McKee, N. (1992) Social mobilization and social marketing in developing communities: lessons for communicators, Southbound, Penang, Malaysia.

Sara, J., Garn, M. and Katz, T. (1998) Some key messages about the Demand Responsive Approach. The World Bank, Washington, DC.

Shordt, K. (2000) Action monitoring for effectiveness: improving water, hygiene &environmental sanitation programmes. Technical paper, no. 35E, IRC International Water and Sanitation Centre, Delft, The Netherlands.

Srinivasan, L. (1990) Tools for Community Participation. PROWESS/UNDP Technical Series Involving Women in Water and Sanitation, US.

Wegelin-Schuringa, M. (1991) On-site sanitation: building on local practice. Occasional Paper, no. 16, IRC International Water and Sanitation Centre, The Hague, The Netherlands.

Wegelin-Schuringa, M. and Ikumi, P. (1997) Sanitation and communication analysis for peri-urban and rural areas in Zambia (final report). IRC International Water and Sanitation Centre, The Hague, The Netherlands.

WHO (1996) The PHAST Initiative: Participatory Hygiene and Sanitation Transformation. A new approach to working with communities. WHO, Geneva.

Wright, A.M. (1997) Toward a Strategic Sanitation Approach: Improving the sustainability of urban sanitation in developing countries. The World Bank, UNDP-World Bank Water and Sanitation Program, Washington, DC.

29

Perspectives and hindrances for the application of anaerobic digestion in DESAR concepts

Hartlieb Euler, Look Hulshoff Pol and Susanne Schroth

29.1 INTRODUCTION

The protection of natural resources is a major task on the way to sustainable development. Reuse of wastewater and waste play a key role in securing human life and production processes.

Increasing contamination by the uncontrolled disposal of wastewater and organic waste puts the supply of potable water to the population at risk and

Edited by P. Lens, G. Zeeman and G. Lettinga. ISBN: 1 900222 47 7

endangers the environment in numerous developing countries. This inhibits productive utilisation of aquatic ecosystems and consequently economic development. Insufficient hygiene can risk quality of life and increase illness in the population with consecutive social costs.

Although reduction and reuse possibilities of wastewater and solid waste are increasingly discussed, the fate of the remaining quantities that it cannot be avoided wasting still needs to be addressed. For treatment, a wide variety of concepts and technical solutions is available. However, they still require optimisation and adaptation measures with respect to implementation in developing countries. Here, cheap and adapted systems with both small and large capacities are in demand. In line with the dominating warm climatic conditions, anaerobic processes are of specific interest.

Anaerobic technology is one of the oldest technologies applied to the treatment of organic residues. Almost any type of wastewater, sludge or solid waste containing biologically degradable substances can be treated anaerobically. Due to their manifold environmental benefits, anaerobic processes are well suited to sectoral, environmental and regional programmes for sustainable development, for water protection and energy projects, sanitation measures or activities aimed at the reduction of greenhouse gases.

For years, however, anaerobic technologies were neglected in many areas compared to artificially aerated activated sludge processes, oxidation ponds, incineration, composting plants and landfills. With growing environmental and climatic concerns and increasing awareness of the limited availability of fossil fuels and the need to implement improved resource-saving technologies, anaerobic processes are now gaining importance in many countries for a range of substrates. There is, however, still a strong need for further technical development of reliable, sustainable, simple and more cost-efficient anaerobic municipal and industrial systems, in particular for small- and medium-scale application. The implementation and dissemination of appropriate treatment systems for the removal and recovery of nutrients, the reuse of water and production of compost as well as an increased and improved utilisation of the biogas produced are required. A combination of anaerobic processes with other treatment and disposal options is attractive to many decision makers to improve reuse and recycling activities and to save money.

Although often discussed as the central issue of anaerobic or combined anaerobic/aerobic systems in the sector of domestic wastewater treatment, the energy and climate aspects of the technology play only a subordinate role in investment decisions. Focal points are often the simplicity of the technology and more attractive treatment costs, especially if a specific technology in developing countries has reached maturity. Besides the comparably low land demand, a low sludge production rate is another important factor. According to a World Bank estimate, sludge treatment and disposal accounts for 50% of the cost of existing wastewater treatment plants and for 90% of the problems occurring during their routine operation.

The information given below is based on years of construction and consultancy experience in the anaerobic technology sector by TBW GmbH Frankfurt. Data were collected within numerous projects. One important source is a four years supraregional sectoral project for the promotion of anaerobic technology in European countries, Africa, Asia and Latin America, carried out on behalf of GTZ – Deutsche Gesellschaft für Technische Zusammenarbeit (GTZ/TBW: Supraregional Sectoral Project 'Promotion of Anaerobic Technology for the treatment of Municipal and Industrial Sewage and Wastes'). In the project, state-of-the-art anaerobic technology for municipal and industrial wastewater and waste treatment was assessed and discussed, including their economics, environmental effects and implementation conditions. Data banks on anaerobic plants and relevant institutions with activities in the field of project financing, research and technology implementation were created. A further base was a number of climate-related projects integrating anaerobic technology into greenhouse gas mitigation efforts (Measures for the Implementation of the Framework Convention on Climate Change, Regional Study Quang Ninh Province, Vietnam, on behalf of GTZ, Eschborn; Greenhouse Gas Emission Control: Measurements of methane emission from pond system for effluent of sugar factory, Bolivia, on behalf of the Ministry for the Environment (BMU), Bonn/Berlin; Preparation of a project for Joint Implementation of Climate Protection Measures on behalf of BMU, Bonn/Berlin).

29.2 THE ANAEROBIC TREATMENT PROCESS

29.2.1 Features of anaerobic treatment

With anaerobic wastewater treatment in general, the first treatment step is run anaerobically (in the absence of oxygen), treating both wastewater and sludge, that is, the raw sewage, jointly. Given suitable temperatures and the necessary knowledge, these processes are considered to be cheaper, as they generally have fewer process steps, demand a lower degree of mechanisation, produce less sludge and demand less energy. Whereas the sludge is already well stabilised in the process, the treatment efficiency that can be achieved in the anaerobic steps usually requires an aerobic (in the presence of oxygen) post-treatment of the wastewater to meet common effluent discharge standards.

Anaerobic and aerobic processes complement each other in wastewater treatment. In fact, in most cases, anaerobic and aerobic processes are combined.

Figure 29.1 shows the simplified configuration of aerobic and anaerobic municipal wastewater treatment.

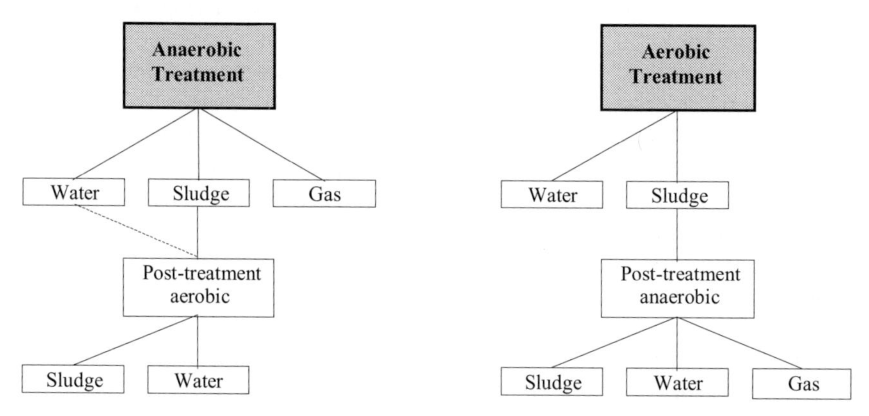

Figure 29.1. The fundamental differences between anaerobic and aerobic processes for wastewater treatment.

This schematic configuration is still unusual in developing countries since post-treatment of sludge (anaerobic) and wastewater (aerobic) are rarely applied. Anaerobic processes are often implemented without post-treatment as a cheap and decentrally applicable alternative, with larger treatment capacities, but at the expense of treatment efficiency. Comparably, aerobic processes are mainly implemented without sludge digestion in developing countries.

In the solid waste sector, the anaerobic digestion step is likewise usually followed by an aerobic composting process – processes which are also applied in parallel.

29.2.2 Reactor technology

29.2.2.1 Anaerobic treatment of municipal wastewater

The development of anaerobic sewage treatment is still relatively new. The first pilot scale attempts were only been made 10–15 years ago. At present, a breakthrough in technology can be observed. Its rapid development in developing countries can contribute considerably to a reduction in investment and operation costs, space required, sludge volume and emissions. Off-site and decentralised treatment systems even within 'megacities' are increasingly becoming an attractive alternative to traditional solutions.

Domestic sewage treatment is at present restricted to sewage with temperatures of at least 15°C (from the results of the GTZ/TBW sectoral project

'Promotion of anaerobic technology'), although treatment at lower temperatures is also possible, but this option has not yet been fully assessed. Treatment applications can therefore only be found in countries with tropical and sub-tropical climates.

With decentralised treatment concepts for domestic wastewater, low maintenance processes will preferably be applied, especially as their shortcomings still include the low level of knowledge and a lack of professional training. Gas collection and use are also still deficient for many of the systems implemented in recent years.

29.2.2.2 Anaerobic treatment of industrial wastewater

In the industrial wastewater sector, anaerobic treatment has developed into a well accepted technology with many applications and a multitude of high-rate processes as well as low-rate reactors.

Extensive applications can be found in the agriculture industry, but increasingly also in other industrial sectors such as breweries, food industries, paper plants, distilleries and even chemical industries. Since industrial wastewater treatment facilities are often on-site systems with a high technological level, their technology is mostly unsuitable for integration in decentralised sanitation concepts. Thus, there is a need for the development and dissemination of cheap, reliable and simple systems for small- and medium-sized industries.

Figure 29.2 shows an example of a high-rate anaerobic treatment system reducing mechanical equipment and construction costs to a minimum, with a modular construction system which is easily adaptable to the needs of the respective industry and location.

29.2.2.3 Anaerobic treatment of organic solid waste

Anaerobic technology for the treatment of solid waste is still comparably new, but is presently increasingly used in Germany (see Figure 29.4) and Denmark as well as a few other European countries. To our knowledge, there is no operating example of anaerobic digestion of solid waste in developing countries.

At present, anaerobic processes in the solid waste sector are mainly implemented:

- as part of mechanical-biological treatment of unsorted and separately collected municipal waste
- for the treatment of separately collected municipal or industrial biowaste

- in agricultural or municipal co-fermentation plants for municipal or industrial solid waste
- for crop residues from agriculture or specifically grown energy crops (grass, corn, and so on).

Figure 29.2. Anaerobic treatment of sugar cane wastewater in Jamaica: ponds transformed into modular anaerobic reactors with upflow-regime (UASB) and gas collection. The treatment system is suitable for both industrial and domestic wastewater. The modular construction also enables decentralised applications (Design and Construction: TBW GmbH, Frankfurt).

Anaerobic processes for biowaste in most cases, if not built close to agricultural land, integrate an aerobic post-treatment step (composting) after anaerobic digestion. Due to the high organic matter content in most municipal solid wastes (50–85%) and other factors, the technology is particularly suitable for implementation in developing countries. Figure 29.3 illustrates the relationship between the moisture content of organic wastes and their suitability for anaerobic or aerobic treatment.

The anaerobic digestion of organic solid waste offers interesting potentials as it combines the generation of biogas from waste with the production of a compost that can be used in agri- and horticulture or agroforestry. However, the first pilot plants for anaerobic waste treatment and co-fermentation are only just beginning to be adapted to conditions in developing countries.

Another potential field for this application is the treatment of organic waste from agro-industry.

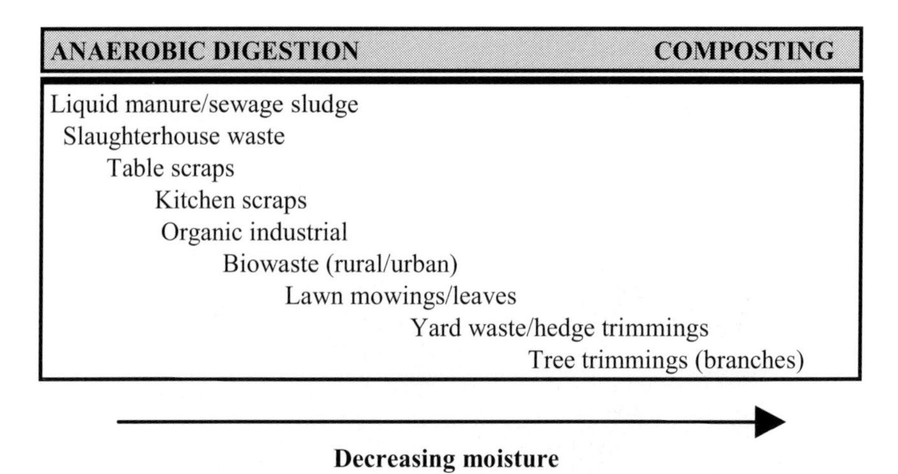

Figure 29.3. The suitability of organic wastes for anaerobic digestion or composting.

29.2.3 The suitability of anaerobic technology

Due to the wide variety of suitable substrates, anaerobic technology is increasingly utilised for the treatment of liquid and solid wastes of various origins in developing and industrialised countries.

Proven implementation areas in industrialised countries are sludge treatment, which is standard in larger sewage treatment plants, and the treatment of industrial wastewater and of manure and dung in agriculture. Meanwhile, anaerobic processes are also applied in the municipal wastewater sector in centralised and decentralised treatment plants, mainly in Mediterranean and developing countries, and for solid waste treatment, mainly in mid-European industrialised countries.

The most varied areas of implementation are to be found in industry (Figure 29.5). In recent years, combined systems for the treatment of substrates of different consistencies and origin (co-fermentation) have become popular in Europe, especially in Germany and Denmark. Here, manure and sewage sludge are predominantly used as the basic (mono-)substrate for the co-fermentation of, for example, municipal biowaste, industrial sludge and wastewater.

Figure 29.4. Combined anaerobic–aerobic treatment of solid organic waste, in Teugn, Germany (Design and Construction: TBW GmbH; Frankfurt).

29.2.4 State of implementation of anaerobic technology

Figures 29.6–29.8 indicate in which countries anaerobic technologies have been most frequently applied. Japan has the most anaerobic reactors for the treatment of industrial wastewater (Figure 29.6). Interest in anaerobic technology has spread rapidly in Japan and many companies are active in this field. Perhaps more indicative for the market potential is the 'anaerobic reactor density', defined here as the number of plants per million inhabitants. The ten countries with the highest density are shown in Figure 29.9. The Netherlands, with a value of 5.83, is clearly leading. As the market for the application of anaerobic treatment in the Netherlands is almost saturated, the reactor density value gives a rough indication of the market potential of anaerobic treatment of industrial effluents in other countries. This strongly depends, of course, on the degree of industrialisation in the particular country. The values for Mexico and Brazil, leading countries in the application of anaerobic technology in Latin America, are 0.46 and 0.40 respectively, whereas India, the leading country in Asia, has only 0.06 anaerobic plants per 1 million inhabitants. These values clearly show that there is an enormous untapped market potential for anaerobic technology. Moreover, anaerobic treatment of domestic sewage, which is (still) not an option in temperate climates, is not included in these considerations. India's leading position in the total installed COD-removal capacity is due to its large number of plants for the treatment of highly concentrated distillery wastewater (Figure 29.10).

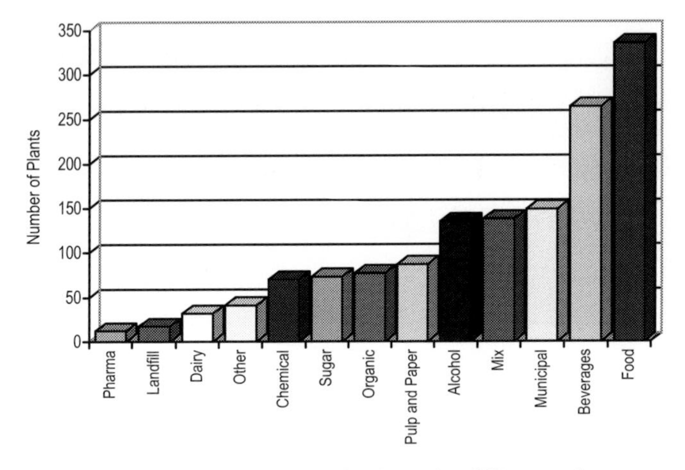

Figure 29.5. The implementation of anaerobic plants for different substrates worldwide up to 1998 (agriculture not included; mix: co-fermentation plants).

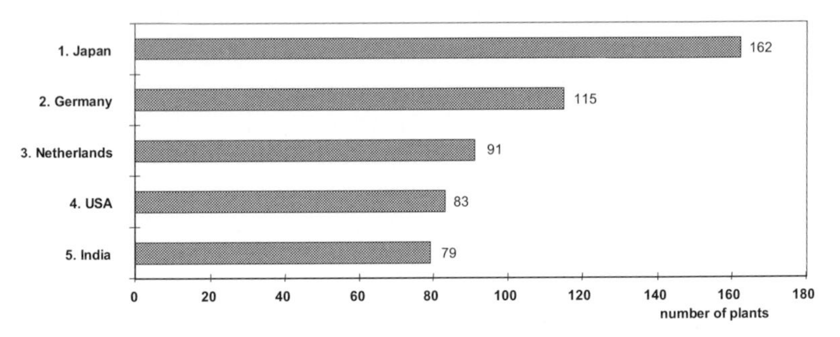

Figure 29.6. The five leading countries in the implementation of anaerobic treatment for industrial wastewater in 1998.

In developing countries, the application of anaerobic digestion for wastewater treatment is more widespread in Latin America than in south-east Asia, although there are considerable differences per region. Brazil, Mexico and

Colombia are clearly ahead in development, while India, Thailand and China are the leading countries in south-east Asia. Anaerobic treatment in Thailand is, however, mainly restricted to (agro-)industrial effluents. The share of sewage receiving anaerobic treatment is however less than 1% in all countries.

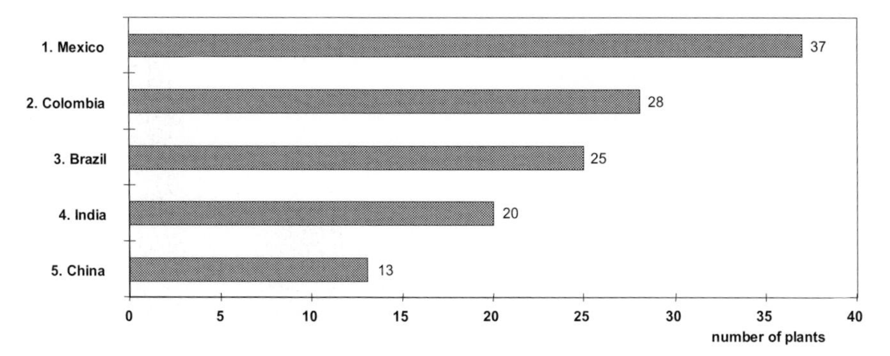

Figure 29.7. The five leading countries in the implementation of anaerobic treatment for domestic sewage in 1998.

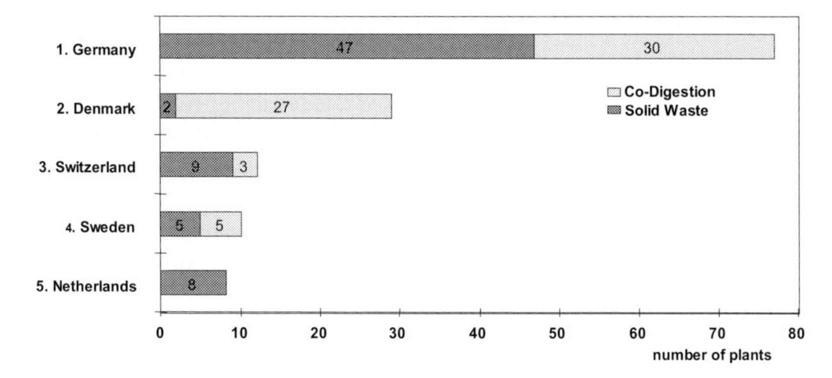

Figure 29.8. The five leading countries in the implementation of anaerobic treatment for organic solid waste in 1998.

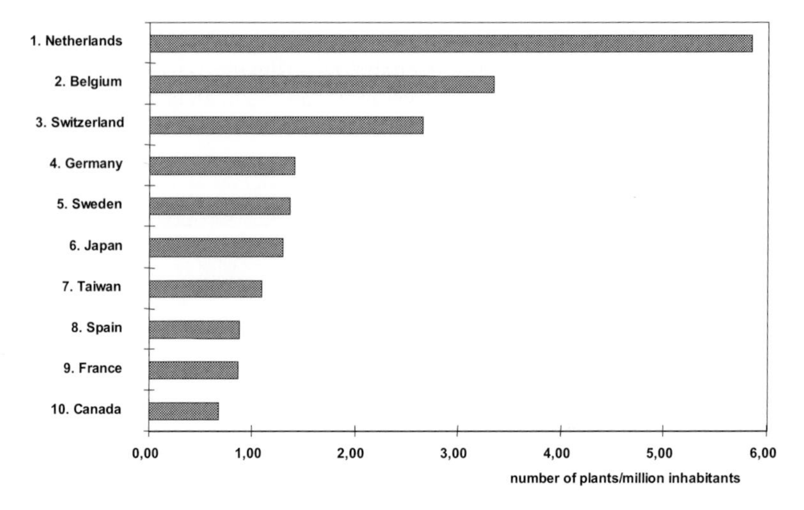

Figure 29.9. The 10 countries with the highest relative 'anaerobic reactor density' in number of plants per million inhabitants in 1998.

29.3 ECONOMIC ASPECTS

Obviously, financial means in developing countries are very limited. Existing legal regulations are frequently not applied due to lack of funds for the payment of necessary treatment facilities. Local industry is often outdated and financially weak and, as a result, cannot afford expensive treatment systems. Adequate incentives for implementation of environmental technologies are lacking, since subsidies keep the prices for water and energy low. Penalties for not complying with regulations, if applied at all, are usually not high enough to have a deterring effect.

Although the trend in the field of industrial wastewater treatment is set by international firms offering anaerobic technologies, it can be observed that there is an expanding number of local companies specialising in anaerobic treatment technologies. Due to urbanisation there is an increased demand for treatment systems with low space requirements. It is recognised that in this respect anaerobic technologies are more economical than aerobic systems. Moreover, operational costs of anaerobic systems are clearly lower than those of aerobic systems, due to the negligible energy demand of anaerobic reactors and the possibility of utilising biogas as a renewable energy source.

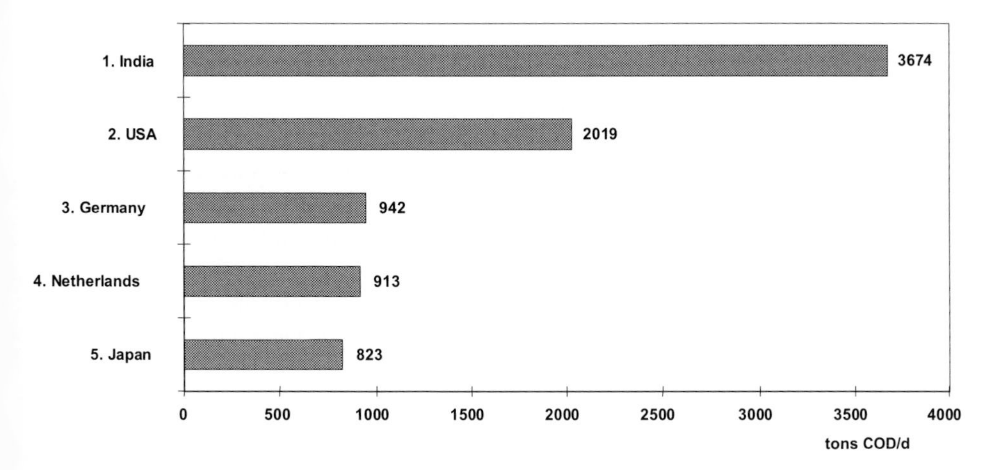

Figure 29.10. The five countries with the highest COD-removal capacity (in tons COD/d) installed in 1998.

Motivation for private investment in anaerobic treatment usually results from the higher costs for the current consumption of natural resources, fees for disposal/discharge or other taxes, or from the higher costs of alternative processes. Anaerobic plants have profits such as returns from disposal fees, sales of energy and compost, own utilisation of the products and resulting cost savings. The overall costs of the different treatment systems are also a result of the frame conditions, investment incentives and other specific factors related to waste and wastewater treatment.

Beyond financing of investments, payment for plant operation has to be clarified in the planning phase of a project. Anaerobic treatment is rarely economically viable due to the returns from energy and compost sales alone: these may pay for the operational costs of a plant, but cannot fully recover the investment costs. These returns may however be of utmost importance since, particularly in developing countries, a high number of treatment plants fail, independent of the applied treatment process, mostly due to insufficient allotment of operational resources (in particular energy costs and spare parts), forcing them to close down.

There are options for financing investment costs through programmes of international donors (such as IDB or the World Bank) and bilateral co-operation, so that the difference in operational costs can be decisive. As well as international and national development banks, the Dutch, German (KfW) and Japanese (JICA) financing institutions especially are becoming aware of this.

29.3.1 Wastewater

The relative and absolute (economic) advantages of anaerobic and combined processes are determined by costs of land, energy, water supply and health services, fees for waste and wastewater and costs of sludge disposal.

Compared to *pond systems*, anaerobic processes are only of economic advantage if land availability is scarce and land prices are above US$10–12/m² (this value depends on the desired effluent quality), or if climate and health-related factors (for example, mosquitoes) are taken into consideration.

In comparison to *activated sludge systems*, anaerobic or combined processes usually perform better with respect to investment and operational costs.

A dynamic cost comparison between the activated sludge process, pond systems and UASB reactors with and without post-treatment was made with the following assumptions:

- design capacity for all processes: 50,000 p.e. (population equivalent)
- final effluent concentration: 20 mg/l BOD_5 (Z1) or 50 mg/l BOD_5 (Z2)
- assumed land price: US$25/m²
- constant electricity cost: US$0.1/kWh
- utilisation period: 20 years
- interest rate: 8%

It showed that the total costs for activated sludge processes are more than twice as high as those for UASB with post-treatment pond for a final effluent quality of 50 mg/l BOD_5. For a final effluent quality of 20 mg/l BOD_5, the total costs were almost twice as high. If returns from energy utilisation and feeding to public grids are valorised, this ratio further changes to the advantage of UASB reactors (see Table 29.1 and Table 29.2). As these tables show conservative estimates, UASB reactors appear to be even more favourable if recent practical experiences are taken into account.

The high cost of technical equipment for the activated sludge process (approximately ten times that for UASB systems) mainly derives from costs for aerators and equipment for primary and secondary settlement tanks as well as for the sludge digester and thickener (more than 80% of the total for technical equipment). These plant components are not necessary for UASB systems, for which mainly screens, pumps and other electrical appliances are calculated.

Table 29.1. Dynamic cost comparison (costs in 1000 US$) of three wastewater treatment methods without power generation (for assumptions, see above)*

Basic scenario	Pond system		UASB + facultative pond	UASB	Activated sludge with sludge digestion	
	Z 1	Z 2	Z 1	Z 2	Z 1	Z 2
Investment costs						
Construction costs	369	276	950	766	1.026	951
Technical equipment costs	25	21	48	45	585	506
Land costs	2.125	1.300	625	175	525	500
Total	2.519	1.597	1.623	986	2.136	1.957
Investment costs (US$/p.e.)	50	32	32	20	43	39
Annuity of capital costs	211.8	135.6	154.1	98.8	233.8	212.4
Annuity of operational costs	*74.8*	*67.6*	*82.3*	*74.1*	*220.0*	*203.8*
Total annual costs	*286.6*	*203.1*	*236.3*	*172.8*	*453.8*	*416.2*
Costs in US$/m³	*0.098*	*0.070*	*0.081*	*0.059*	*0.155*	*0.143*

* Data base: DHV Consultants (1993); cost assessment from 12 case studies conducted within GTZ/TBW sectoral project 'Promotion of anaerobic technology'; own calculations (1998)

The land prices proved to be a decisive factor in the comparison between pond and UASB systems. Figure 29.11 presents the land demand for the above processes and assumes preconditions used in the calculations. If the final effluent concentration is not taken into consideration, pond systems can generally be assumed to be the most economic alternatives if land prices are below US$10/m². If energy recovery from UASB processes are considered, the economic threshold values expressed as land prices will be reduced. For effluent concentrations of 20 and 50 mg BOD_5/l, the threshold value of land prices is about 10–12 and 14 US$/m² respectively.

For industrial wastewater and liquid manure, the concentration of organic substances in the substrate to be treated and energy prices often play a decisive role in the resulting financial profitability. Sometimes energy or compost revenues alone can compensate for investment and operation costs and thus allow for a profitable operation.

Table 29.2. Dynamic cost comparison (costs in 1000 US$) of three wastewater treatment methods with power generation, considering returns from electricity feeding (for assumptions, see above)*

Basic scenario	Pond system		UASB + facultative pond	UASB	Activated sludge with sludge digestion	
	Z 1	Z 2	Z 1	Z 2	Z 1	Z 2
Annuity of capital costs	211.8	135.6	154.1	98.8	233.8	212.4
Annuity of operational costs	*74.8*	*67.6*	*82.3*	*74.1*	*220.0*	*203.8*
Proceeds (negative costs) for power generation	0	0	–11	–11	–27	–23
Total annual costs	*286.6*	*203.1*	*224.9*	*161.5*	*42.,0*	*393.2*
Costs in US$/m³	*0.098*	*0.070*	*0.077*	*0.055*	*0.146*	*0.135*

* Data base: DHV Consultants (1993); cost assessment from 12 case studies conducted within GTZ/TBW sectoral project 'Promotion of anaerobic technology'; own calculations (1998)

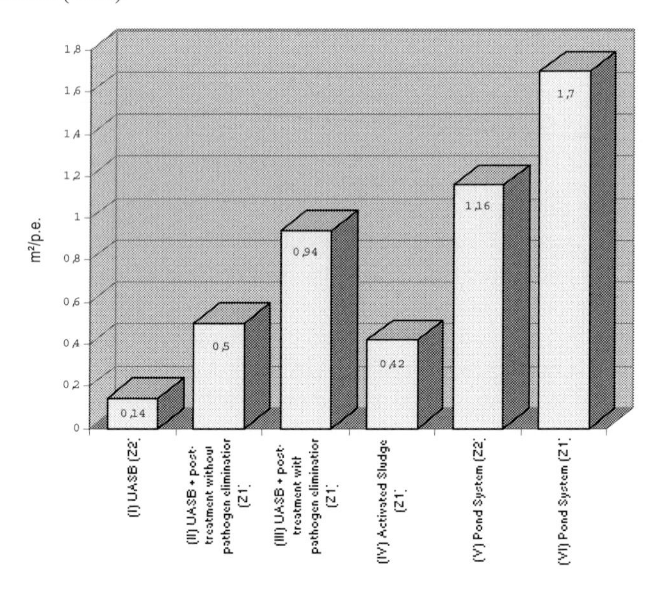

Figure 29.11. Land demand of different domestic wastewater treatment processes (in m^2/p.e.)

Figure 29.12. Decentralised 'UNITRAR' UASB reactor for municipal wastewater post-treatment in two lagoons (in series), followed by three parallel ponds, Lima, Peru. Top of picture: part of first lagoon on the left, three fishponds and second lagoon behind. (Design and construction: Universidad Nacional de Ingeniería, Lima; Monitoring by TBW within GTZ/TBW sectoral project 'Promotion of anaerobic technology').

The majority of municipal wastewater treatment plants in developing countries (see Figures 29.12 and 29.13) are built using public finance in the context of governmental or municipal disposal duties. There may be money from municipal budgets, national or international financing institutions or from multilateral institutions. The level of knowledge within the specific financing institutions is very often the decisive factor for the choice of technology. In the past, invitations for tenders did not always sufficiently consider anaerobic processes as an alternative.

29.3.2 Solid waste

Investment costs of anaerobic solid waste treatment are no higher than those for composting plants if the same environmental and emission standards are applied. Operational costs depend mainly on the costs of energy in the short term. Anaerobic solid waste treatment is presently more expensive than landfills if emissions are not accounted for – it may however possibly be more cost-efficient compared to sanitary landfills and when medium- and long-term environmental costs are taken into consideration. In any case, anaerobic solid waste treatment plants are cheaper than incineration plants.

Figure 29.13. Anaerobic treatment plant for agricultural (piggery) wastewater, Chiang Dao, Thailand. (Design and construction: Thai-German Biogas Programme; Chiang Mai University – GTZ).

A technology transfer to developing countries of the first generation of anaerobic solid waste treatment plants that were developed in Europe cannot be financed from present disposal fees, compost and energy sales alone. A number of technological and organisational adaptation measures are still needed to secure viability in developing countries. Cost reductions can be expected to result from further combination with agricultural fermentation technology.

29.4 FAVOURABLE FRAME CONDITIONS

The effectiveness of anaerobic or combined anaerobic-aerobic processes for the treatment of wastewater and waste is particularly high under specific frame conditions. A clearly defined political will to solve wastewater and waste problems from an ecological and economic view of the entire system is the most important precondition for success. Effective legislation, appropriate financing possibilities and a free choice of processes, as well as technical knowledge and training, are all indispensable.

Anaerobic processes are seldom successfully implemented without a minimum of technological knowledge and a reinforcing social and technical environment. Presently, there is little exchange of information and transfer of knowledge, in particular from south to north and north to south, and many national and international institutions and support agencies seem to be insufficiently aware of and informed about the potential, implications and regulatory needs for an increased application of anaerobic technologies. Hence, the importance of exemplary plants in operation, active institutions and sufficiently trained personnel for design and operation are of utmost importance for a successful implementation strategy.

29.4.1 Legal aspects

Experience has shown that solutions for wastewater and waste-related questions with respect to health care and the protection of natural resources are only sought if the legal framework is appropriate.

In most countries, a detailed legal framework with well defined discharge and emission standards for environmental polluting compounds already exists. Often there is a conflict between local and national regulations. In such cases, in general, the national law will prevail. Major problems are, however, the low enforcement and control of regulations, often due to limited financial means. In some situations corruption may interfere with the intention of the law. Another obstacle for effective law enforcement is a lack of monitoring equipment and qualified personnel within controlling agencies. Enforcement of legislation works best for newly built industries since they frequently need to comply with several regulations prior to putting a new plant into operation.

Anaerobic processes are taken into consideration by system-oriented decision makers than by those oriented towards specific cases or specific technologies. Nevertheless, rules and regulations for, for example, separate collection of wastes, landfills (covering and lining), sewerage systems (separate or mixed) or energy (feeding to the grid) may be decisive factors for the choice of technology.

For speedy dissemination, anaerobic waste and wastewater treatment favours integrated national and regional waste and wastewater treatment, energy conservation and climate change minimisation strategies that include comprehensive valorisation of the different benefits generated. This requires appropriate tendering procedures and legislation allowing for competition between centralised or decentralised (community-on-site) treatment options, and which target for a maximum (re)utilisation of the (by)products and possible cost reductions.

29.4.2 Organisational aspects

Privatisation of operations or the willingness of the public and private sectors to co-operate in plant management often ease a sustainable operation of public treatment facilities. Capacities of local governments to manage waste and wastewater collection and disposal, appropriate tariff systems that cover all running costs (including capital costs) and the willingness of the population to pay for these services influence the sustainability and durability of plant operation.

Except for the start-up of the reactor, the operation of anaerobic systems requires less attention than aerobic systems. Nevertheless, for anaerobic as much as for aerobic plants, skilled operators, well trained in the principles of anaerobic wastewater treatment, are necessary to avoid problems with malfunctioning reactors.

Anaerobic solid waste treatment requires a reliable organisation of the waste collection, preferably incorporating separation of the organic waste. The present situation with respect to collection and handling of solid waste in many developing countries is still poor and mechanical-biological treatment with anaerobic digestion appears to be one of the upcoming solutions. However, without drastic changes in the organisation of waste collection and in the general attitude and involvement of the public, this will not be possible. The existing collection and recycling activities of the so-called 'informal' sector cannot substitute a comprehensive collection and transport system. In many countries however, separation of solid waste at source (even in the household) is planned to become obligatory.

29.4.3 Sociological aspects

Surveys carried out in 16 countries reveal that environmental awareness is increasing on a broad scale, although present public awareness of environmental problems still permits the uncontrolled disposal of solid and liquid wastes.

29.4.3.1 Environment and health

Wastewater

In addition to water quality requirements (quality of the discharged wastewater and of the receiving water body), effects of anaerobic technologies on climate and soil have to be considered.

One advantage of anaerobic technology is its potentially positive effect on the climate if the biogas produced is utilised or flared, if utilisation cannot be realised (emitting CO_2 in the combustion process). In addition to the avoidance of diffuse methane emissions from, for example, wastewater swamps, greenhouse gas emissions are also prevented by substituting and reducing fossil fuel consumption.

Table 29.3. Emission behaviour of UASB + facultative pond, pond system and activated sludge process for municipal wastewater treatment. Source: Final report of GTZ/TBW sectoral project 'Promotion of anaerobic technology'

	UASB + facultative pond			Pond system	Activated sludge	
	+ gas utilisation	+ gas flare	no gas utilisation		no sludge digestion	+ sludge digestion, gas utilisation
CO_2 emissions kg/p.e.•a	–3	+8	+61	+8	+27	+1

The deposition of polluting substances on and into the soil, for example, from 'wild' wastewater swamps, lakes and channels can be prevented or reduced. Nutrients can be recirculated and the soil structure can be built up with the application of digested sludge.

Health and hygiene effects of successful municipal wastewater treatment are obvious. The sewerage system itself already has the effect of keeping critical wastewater flows away from the population. Both anaerobic and aerobic treatment processes reduce the pathogen content in wastewater. However, one-stage anaerobic processes disinfect wastewater to a lesser degree than combined processes. One-stage anaerobic processes only reach about 90% removal for bacteria (faecal coliforms), and 99% for helminth eggs and enteric viruses (Alaerts *et al.* 1990). Depending on the objective, a thorough elimination of pathogens (such as viruses and intestinal bacteria) is a point of concern, especially if the treated wastewater and the sludge are utilised in agriculture. From a hygienic point of view, the best removal efficiencies can be achieved by a combination of anaerobic and aerobic treatment steps if sufficiently long retention times are applied. Plant layouts with anaerobic treatment and post-treatment ponds can achieve pathogen removal rates of up to 99.9%, which is still slightly lower than required by WHO standards.

Solid waste

Due to the large amount of surplus energy produced in the form of biogas during the anaerobic digestion of solid waste, considerably larger amounts of fossil fuels can be substituted and greenhouse gas emissions can be avoided. In addition, methane emissions from uncontrolled landfills are avoided and leakage of harmful substances from these sites into the soil, ground and surface waters are reduced.

The lower land demand allows the operation within densely populated urban areas and even mega-cities, reducing transportation needs, thus having an additional positive effect on energy and climate. They can also be the starting point for an intensification of urban agriculture and tree-planting programmes.

Health risks and emissions, especially arising from open dumping and simple incineration facilities, are diminished due to the closed construction of anaerobic waste treatment plants. An improvement of the hygienic working conditions for the 'informal sector', whose members are active in small-scale recycling in numerous developing countries, can be achieved if this sector is integrated in programmes and projects.

Table 29.4. Emission behaviour of different solid waste treatment processes (exemplary calculated values). Source: Final report of GTZ/TBW sectoral project 'Promotion of anaerobic technology'

	Anaerobic digestion and aerobic post-treatment	Composting	Sanitary landfill
CO_2 equivalent in t per t volatile solids (VS)	–0.4* –0.16**	+0.09	+1.4***

with heat utilisation; ** without heat utilisation; *** high value due to methane losses despite landfill degasification

29.5 ANAEROBIC DESAR-EXAMPLES

In the following section, some treatment concepts are presented as examples for the successful implementation of anaerobic processes that can be accounted for as DESAR concepts.

29.5.1 The model project 'Living and working'

The model project 'Living and working' (Freiburg, Germany) and its innovative sanitary concept is a prime example for an anaerobic DESAR concept on a comparatively high technological level that is integrated in a comprehensive urban project for sustainable living (Design and Implementation: TBW GmbH; Frankfurt; Research: Fraunhofer Institut für Systemtechnik und Innovationsforschung (Fh-ISI, Karlsruhe); Client: Association for the promotion of ecological construction (Ökobau e.V., Freiburg); Co-funding: German Federal Environmental Foundation (DBU, Osnabrück)).

Buildings are constructed according to the passive house standard (insulation, controlled aeration etc.). A plant for the utilisation of thermal solar energy is installed, producing 32% of the total heat demand.

The black water is removed with vacuum toilets (Figure 29.14), using only small amounts of water. The separate collection of black and grey water enables the anaerobic digestion of night soil, urine, black water and household biowaste, which is preconditioned by grinding, in a biogas plant. The anaerobic treatment step takes place in a totally mixed reactor, from where the digested substrate is conveyed to a post-digestion tank with integrated gas storage. The final product is a liquid fertiliser which is kept in a storage tank, thus guaranteeing a defined retention time with the purpose of ensuring compliance with hygienic standards. The fertiliser is then directly transferred to agricultural application.

The produced biogas is utilised for cooking purposes, mixing in the digester and electricity generation. The grey water is conditioned for reuse as non-potable water (as toilet flush water or for garden irrigation purposes) by treatment in a filter bed plant.

In order to ensure sustainable and constant operation, the plant was designed to have low maintenance needs. Inhabitants' participation is however considerably higher than for conventional municipal biological treatment. On this scale of decentralisation, additional installations for waste separation cannot be part of the plant, so that separation has to be realised in the connected households of the building on a much higher level than is necessary for municipal collection and treatment of biowaste. For the Freiburg project, population participation is expected to be very good, as the participating households are known to be very 'ecologically oriented'. Future experience will show the participatory behaviour and constraints that should be expected in different social environments.

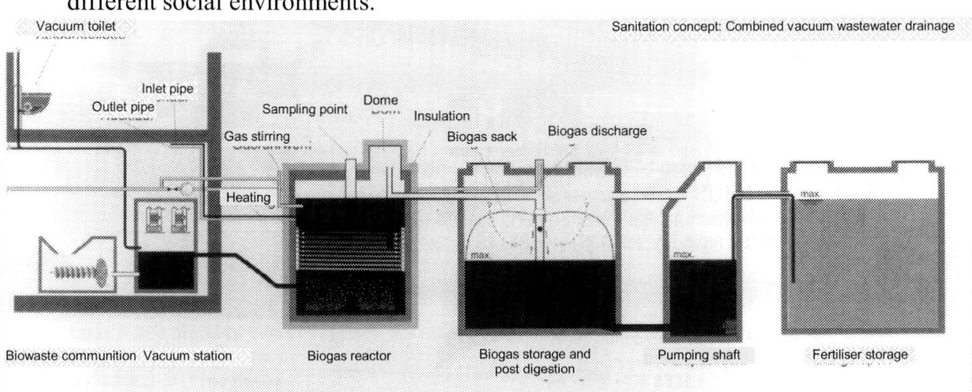

Figure 29.14. Sanitation concept using a combined vacuum drainage system

29.5.2 Biogas latrines

Biogas latrines are a very simple, but highly effective, treatment concept for toilet wastewater which has a great potential to significantly improve hygiene conditions, especially in rural areas. Due to their simplicity and because they can be implemented without being dependent on the existence of a sewerage system, they are generally suitable for integration into a DESAR concept.

The biogas latrine and the community biogas plant described in the following section are more advanced forms of the septic tank system, which consists of a closed tank where sedimentation takes place and settleable solids are retained. Retention time of the liquid is in the order of one day. Sludge is digested anaerobically in the septic tank, resulting in a reduced volume of sludge.

Septic tanks (Figure 29.15) are used for wastewater with a high percentage of settleable solids, typically for effluent from domestic sources (mostly black water, that is, toilet wastewater). Private households and enterprises in many communities, public buildings such as schools and hospitals currently use individual on-site and small-scale septic systems (up to about 50 households).

The biogas latrine (Figure 29.16) and the community biogas plants (Figure 29.17) are more appropriate and future-oriented since they collect the biogas, treat the effluent for longer and make use of the resulting water, nutrients and energy. Additional costs are socially easily covered by additional benefits to the household or community.

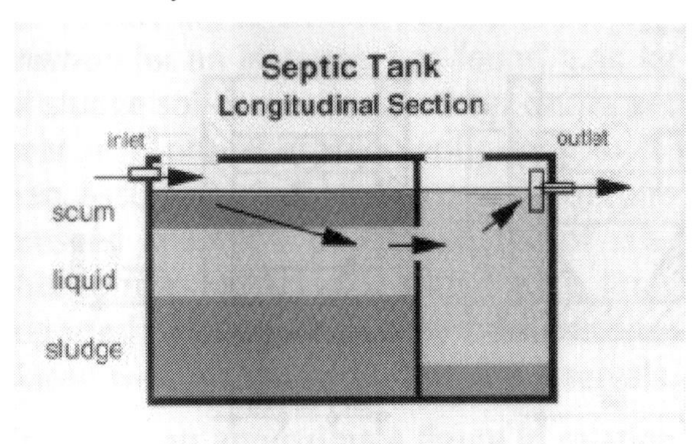

Figure 29.15: Schematic diagram of a septic tank.

In biogas latrines (Figure 29.16), toilet wastewater is collected in a biogas plant, where the organic substance is degraded by anaerobic bacteria. Household, agricultural and garden waste are usually added to increase gas yields to allow for household cooking with this renewable energy. Thus, the amount of fertiliser produced is also increased.

If livestock is available, then livestock buildings can additionally be connected directly to the plant; manure is sometimes collected from the fields and filled into the digester. If these plants are considered DESAR concepts, and built not only for fertiliser and energy reasons, then the cost-benefit relation improves considerably due to the saving on other solutions for the disposal of toilet effluent.

(a) (b)

Figure 29.16. (a) Biogas latrine under construction: dome biogas plant for domestic wastewater and organic kitchen and household waste (Ethiopia). (b) The same plant as in (a), shortly before completion.

29.5.3 Communal biogas plants

The same system as biogas latrines can be applied to institutional plants, where toilet water from flush toilets or from dry latrines enters a larger biogas plant. When building these communal biogas plants, a series of toilets is installed beyond the needs of the single institution alone. These can be used by other people as well and add to the amount of biomass available. Here, biomass from other sources such as kitchen, agricultural fields or industry can be utilised. Where livestock is available, manure is also added.

Constructing the necessary buffer tank (Figure 29.17) in the shape of a channel allows the effluent to be brought to the plantation, gardens or fields where it is needed, without having to organise any form of transportation. Pathogen removal here as well depends on the retention time, which again depends on the size of the plant and the dominant type of pathogens in the settlement. If application of the treated effluent is considered too risky – which applies in very few cases – percolation of the effluent into the underground can still be implemented.

Figure 29.17. Construction of a buffer tank in the shape of a channel, part of a 80 m^3 biogas plant for toilet wastewater from a school in Fitche, Ethiopia. The biogas produced is utilised in the canteen.

29.6 REFERENCES

Alaerts, G.J. *et al.* (1990) Feasibility of Anaerobic Sewage Treatment in Sanitation Strategies in Developing Countries. IHE-Report Series 20. International Institute for Hydraulic and Environmental Engineering (IHE), Delft, The Netherlands.

Baumann, W. and Karpe, H.J. (1980) Wastewater Treatment and Excreta Disposal in Developing Countries. GATE-Appropriate Technology Report. German Appropriate Technology Exchange (GATE), Eschborn, Germany

Euler, H., Müller, C. and Schroth, S. (1999) Anaerobe Vergärung im internationalen Vergleich. Contribution to 9th annual meeting of German Biogas Association, Weckelweiler, Germany.

GTZ/TBW Supraregional Sectoral Project (1998) Promotion of Anaerobic Technology for the Treatment of Municipal and Industrial Wastewater and Waste. Naturgerechte Technologien, Bau und Wirtschaftsberatung (TBW), Deutsche Gesellschaft für Technische Zusammenarbeit (GTZ), Eschborn, Germany.

Kellner, C., von Klopotek, F., Krieg, A. and Euler, H. (1997) Different systems and approaches to treat municipal solid waste – a state-of-the-art assessment.

Contribution to 'The Future of Biogas in Europe', Herning, Denmark, 7–10 September.

Model Project 'Living and working', Freiburg, Germany. Technology: and Fraunhofer Institut für Systemtechnik und Innovationsforschung (Fh-ISI, Karlsruhe); Client: Association for the promotion of ecological construction (Ökobau e.V., Freiburg); Promotion: German Federal Environmental Foundation (DBU, Osnabrück).

Naturgerechte Technologien, Bau und Wirtschaftsberatung (TBW) GmbH, Frankfurt (1996) Appropriate Technologies. Anaerobic technologies – a key technology for the future.

Naturgerechte Technologien, Bau und Wirtschaftsberatung (TBW) GmbH, Frankfurt (2000) Experiences with the-biocomp-process in the framework of the integrated municipal 'Rotenburg-Waste-Management Model', Germany. Contribution to International Symposium on Biogas Technology Development, Beijing, China, October 24–27.

Sasse. L. (1998) Decentralised Wastewater Treatment in Developing Countries. Bremen Overseas Research and Development Association (BORDA), Bremen, Germany.

Schulz, H. (1996) Biogas-Praxis. Grundlagen, Planung, Anlagenbau, Beispiele. Staufen, Germany. (In German.)

30

The micro and macro economic aspects of decentralized sewage treatment

M. von Hauff and P.N.L. Lens

30.1 INTRODUCTION

30.1.1 Economic activities and pollution

It has been estimated that it took the whole human history to grow to the 60 billion euros scale of the world economy in 1900 (EEA 1999). The world economy now grows by this amount every two years, and it was 39 trillion euros in 1998. It is the speed and scale of this economy that presents a threat to the integrity of the environment support system that underpins economic activity (Figure 30.1).

Edited by P. Lens, G. Zeeman and G. Lettinga. ISBN: 1 900222 47 7

Ecological services, unlike man-made services, are largely free. However, their value can depreciate, or may even disappear in case of overuse, for example, energy or materials (such as metals, minerals, forests). The exploitation of sources of energy and materials can still be controlled, by improvements in their utilization efficiency or by use of alternative products, such as plastics from biowaste. Other ecological services are much less easy to control, such as climate regulation and radiation protection from the ozone layer. It is not possible to replace the ozone layer or the climate regulatory systems with man-made capital, and their efficient functioning can fail once thresholds of 'loads' have been passed. These ecological services are not owned by anyone, nor do they have a price, which makes their preservation through market mechanisms impossible.

Industrialized nations are increasingly becoming aware of the fact that the continuing pollution of lakes and rivers leads to damage to a region or an economic area. Apart from the unpleasant smell and impact on the natural environment, pollution is increasingly perceived as unsightly. The quality of a location as a business site suffers if it is polluted. Companies avoid polluted locations and choose 'greener' ones.

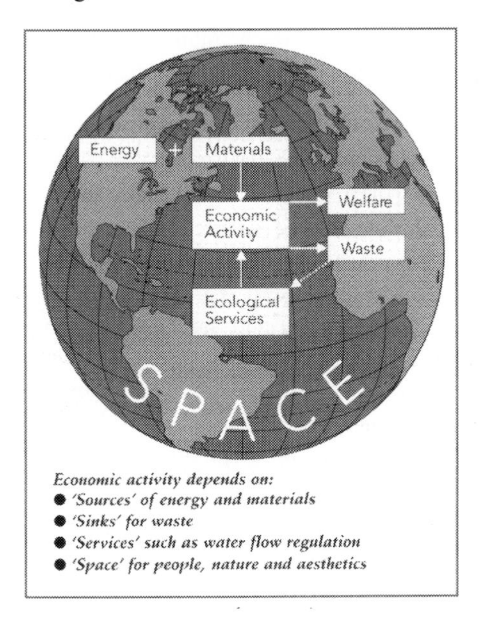

Figure 30.1. The four life support systems of the environment, essential to any economy (after EEA 1999).

30.1.2 The sanitation sector as an economic segment

In all economically advanced countries, the disposal and treatment of sewage is an important component of public infrastructure. Historically, the overriding reasons for investing in sanitation infrastructure were public health considerations. Man's direct contact with his faeces used to be a major cause of epidemics such as typhus, cholera and dysentery (Wilderer *et al.* 1998). For those affected, these diseases caused great suffering and misery (Appasamy and Lundqvist 1993). In material terms, this often resulted in serious difficulties for continued economic existence. For those affected, it means a loss of prosperity, which is reflected by a negative impact on a micro economic level. However, the social product, as an indicator of total economic activity (the macro economic level) does not statistically record this deplorable state of affairs.

The human and socio-economic costs of unmanaged and under-managed domestic waste are very high (Munasinghe 1992). In Peru, a recent cholera epidemic resulted in an estimated loss amounting to three times the expenditure on water and sanitation for the entire country over the preceding 10 years (Munasinghe 1992; Giles and Brown 1997). In India, the 1994 plague epidemic resulted in a loss of tourism revenue estimated at $US200 million. When considered from the point of view of the economics of public health, the lack of a centralized and/or decentralized sanitation infrastructure and inadequate disposal and treatment of sewage results in high health costs and has a negative impact on the productivity of the human capital. This in turn acts as an obstacle to the development of the economy as a whole and thus results in a loss of welfare (macro economic level).

To avoid these negative consequences, the disposal and treatment of sewage was declared a public duty. Apart from the disposal and treatment of sewage, the supply of water of sufficient quality is a complementary task (Serageldin 1995). In other words, water supply and sanitation cannot be considered in isolation. This contrasts with the current practice in many industrialized countries where drinking water is provided by private companies and not related to the public sewage treatment services. In many developing countries both the water supply and the sanitation infrastructure are still inadequate. Their installation represents a huge economic market.

30.2 COSTS OF PROVIDING ADEQUATE SANITATION

30.2.1 Wastewater

The initial capital costs of providing effective sanitation services can be high. The approximate cost of constructing sanitation systems ranges from US$75–

150 for a twin pit pour-flush latrine, to US$600–1200 for a conventional sewerage system (1990 prices) (Hardoy and Satterthwaite 1990). According to Grau (1994), countries with a per capita gross national product (GNP) of less than US$500 do not have the resources to construct treatment facilities and cannot maintain them (Niemczynowicz 1996). Additionally, the water resources consumed in some sanitation systems can be very high. In the developing world, flush toilets can consume 20–40% of the domestic water resources used in a sewered city (NRC 1981).

Most governments in developing countries are ready to admit that they lack the financial means to fill the need for water and sanitation. Moreover, historical, bilateral and multilateral funding accounts for less than 10% of total investment needs. Therefore, the need for private investment is imperative. In Latin America, the investment needs in water and sanitation have been estimated at about US$12 billion/year (Table 30.1). Similar estimates for Asia point to much greater needs, of about US$100 billion/year (Rivera 1996).

Table 30.1. Estimated annual investment needs in Latin America in US$ billion (1993 prices). After Rivera 1996

Country	Water supply	Sewerage and treatment	Rehabilitation	Total
Argentina	0.4	0.5	0.4	1.3
Brazil	1.2	2.6	0.9	4.7
Chile	0.1	0.2	–	0.3
Colombia	0.2	0.3	0.1	0.6
Mexico	0.5	1.2	0.3	2.0
Peru	0.3	0.2	0.2	0.7
Venezuela	0.2	0.2	0.1	0.5
Others	1.3	0.4	0.2	1.9
Total	4.2	5.6	2.2	12.0

Preventing pollution through engineered solutions can be inappropriate, as these often depend on high energy inputs, expert operator skills and continued maintenance expenditure (Edwards 1985; Boller 1997). The implementation of engineered solutions may also cause external and intangible ecological damage to adjacent ecosystems (Ahmad 1990). Any benefits that may result are often to the advantage of a local region, but often to the disadvantage of the larger society or environment (Yan and Ma 1991; Munasinghe 1992). The hygienic urban water supply and sewerage systems are now in question with regard to their environmental efficiency and sustainability and new alternatives must be found (Niemczynowicz 1993; Chapters 1 and 7 of this book).

30.2.2 Solid waste

The economic benefits of reusing human waste in agriculture can be realized at the farm level through supplementing the use of inorganic chemical fertilizers with reclaimed organic fertilizer derived from bio-waste (Furedy and Ghosh 1984). The benefits of reusing these organic wastes must also be measured against the cost of not doing so at both an economic and environmental level (Fahm 1980; Gardner 1998). However, the costs of implementing zero-discharge organic waste to agriculture recycling schemes may be inexpensive. Full-scale implementation of urban organic waste to agriculture systems could cost as little as US$5–6 million for a city of 1 million people (UN 1997).

30.3 ECONOMIC EVALUATION OF TECHNICAL MEASURES

30.3.1 Economic efficiency of technical measures

The economic efficiency of technical measures can be evaluated by both marginal cost analysis and water cost estimates.

30.3.1.1 Marginal cost analysis

Marginal cost analysis distinguishes the benefit-cost margin from additions to or changes in the system. The benefit-cost margin is the difference between the annual income (revenues from the water users) and the annual cost (operation and maintenance costs and the cost of buying potable water from a source). Operation and maintenance costs include personnel salaries, management expenses, operation fees (energy consumption and chemical costs) and maintenance costs (equipment repair and replacement).

Potable water and reclaimed water are sold to subscribers as a beneficial product. The investment will be paid back by the benefits in a certain number of years, i.e. the payback time C. When it is assumed that the sunk costs, i.e. the costs that occurred in the past, should not influence decisions on future actions, the payback time C can be calculated as (Asano 1998):

$$C = \text{Incremental capital investment/Incremental margin}$$

The incremental capital investment is the additional investment related to added equipment or systems. The incremental margin is the difference between the considered new scenario and the current situation.

30.3.1.2 Water cost estimates

The costs of a water system are generally paid by users in the form of charges. Therefore, it is important to consider the user's view of water costs. Lower costs will encourage users to accept the implementation of the system. On the other hand, the water system must be seen as a whole: some users may be affected by variations in water costs, even though they do not participate in the project directly. Thus, water costs help to estimate the economic feasibility of different scenarios.

Capital investment costs for water projects can be financed by the water supply association or by subsidies from, for example, ministries, water agencies or the county. Subsidies which do not need paying off can be deducted from the true capital cost. Loans need to be paid back by a certain date (typically 20 years) at an annual interest rate. Unlike the basis of marginal cost analysis, continuing debt payments for past investments are included in the water costs. For example, consumers will pay the remaining debt obligation for the potable distribution network, regardless of future actions.

Taking into account the different construction time of existing equipment and pipelines, different charge rates are developed to calculate the repayment of the remaining capital investment. For example, potential scenarios may have a 100% charge rate, while existing potable water supply systems might only have a 15% charge rate.

$$\text{Water cost} = \frac{(\sum_{j=1}^{m} \text{loans in capital investment } j \text{ x charge rate } j) \text{ x capital recovery factor} + \text{annual O \& M cost}}{\text{Annual water volume}}$$

$$\text{Capital recovery factor} = \frac{i(1+i)^n}{(1+i)^n - 1}$$

where i = interest rate (e.g. 8%), n = years of payment (e.g. 20 years), j = item of equipment or water project facility.

30.3.2 The effectiveness of financing instruments

The provision of sewage disposal systems requires investment, which can be allocated to the category of investment in environmental protection (Fehr 1992). From the point of view of the economy as a whole, investment in environmental protection makes sense if the benefit, measured in terms of the increase in social well-being and health, exceeds the cost of the investment in environmental protection. Positive effects on growth and employment are thus a positive by-product of, but not the rationale behind, investment in environmental protection.

Investment and consumption subsidies are used to help the poor to gain access to basic services. A closer review of the outcome of many subsidies in the sector reveals that the worthy objectives are seldom reached. Failure of subsidies to reach their intended objectives is in part due to a lack of transparency in their allocation. Subsidies are often indiscriminately assigned to support investment programmes, mostly capacity increases, which benefit more middle- and high-income families who already receive an acceptable service (Yepes 1996).

Cross-subsidies, prevalent in most developing countries, aim to foster consumption by the poor. Under this system, mostly domestic users pay tariffs below the average tariff while industrial and commercial users are charged at higher than average rates to subsidise the domestic users (Table 30.2). Cross-subsidies have adverse effects that are often not quantified or appreciated perhaps because regulators and utilities believe they are not substantial. Economic inefficiency occurs as some consumers receive water at a cost which is lower than the production value. Inefficiency also occurs as there are consumers who would pay for additional consumption at the real cost to provide it, but they are forced to reduce consumption or are priced out of the market.

Table 30.2. Selected cases of the magnitude of cross-subsidies (after Riviera 1996)

City/utility	Subsidised users (%)	Consumption subsidised (%)	Ratio highest/ lowest tariff	Subsidy transfer (% of total water billings)
Guayaquil/Ecuador	91	75	88	47
San Jose/Costa Rica	91	70	91	22
Chennai/India	92	72	16	53
Minsk/ Belarus	92	63	71	59

Cross-subsidies can also have an adverse impact on the financial position of the utility. At the low end of the tariff spectrum, tariffs often do not cover variable costs and therefore the utility will incur a net financial loss. Low tariffs may not even cover the cost of billing and collection, in which case the utility loses the interest to bill these consumers and the incentives to reduce water losses. At the high end of the tariff spectrum, subsiding consumers have the incentive to develop alternative sources of supply if the unit cost of these alternative supplies is lower than the tariff. When this situation occurs, these users disconnect from the public water system thus reducing the subsidising base which requires additional rates to be imposed on the remaining subsidising group. This sets a vicious circle in which the utility looses its most valuable consumers. Highly differentiated tariffs may also encourage corruption as users seek to be classified in a lower bracket.

Consumption subsidies often benefit upper-income domestic consumers substantially more than low-income ones as the former group consumes more water. In one Latin American city, for instance, upper-income consumers paying a volumetric – but still subsidised – 15 times higher rate than the poor income consumers, were receiving an average subsidy of US$380 per year while low-income ones receive only US$120 per year (Yepes 1992).

30.4 ECONOMIC ASSESSMENT OF ALTERNATIVE SEWAGE DISPOSAL SYSTEMS

30.4.1 Economic relevant framework conditions

The decision in favour of a particular sewage disposal system is influenced by a number of relevant framework conditions (von Hauff 1998). In industrialized and developing countries, for example, the framework conditions influencing the decision for centralized or decentralized sewage disposal systems are completely different. As well as comparing these two groups of countries, we also have to differentiate between the various regions of individual countries, depending on whether they are densely or less densely populated, for example. In other words, it is not possible to deduce any universal micro- or macro economic effects.

The difference between industrialized and developing countries is due to the following factors:

- Industrialized countries generally have an extensive public sewage disposal system, to which the majority of a country's regions and communities are connected. Authorities with qualified personnel are responsible for this system, and there is an efficient charging system. Given this situation, it can be asked under what conditions the establishment of a decentralized sewage disposal system in a community or region might make economic sense.
- Developing countries often have sewage disposal systems in cities only. However, these do not serve all parts of these cities, and are not designed to cover entire urban regions. Furthermore, in terms of both environment and health, the existing sewage disposal systems are often in an unsatisfactory state. One reason for this is that the authorities often do not have the necessary financial resources at their disposal, and lack the qualifications necessary to provide an

> ecologically acceptable sewage disposal system. This leads us to ask which sewage disposal system is economically and ecologically more efficient. This applies especially to rural regions where no sewage disposal system yet exists.

To a critical extent, demand for, and thus use of, the scarce resource water is determined by its price. There is without doubt broad consensus that, for poor people in developing countries in particular, sufficient supplies of existentially necessary water must be available at a price that is geared to the income of this population group. If this is not the case, then even decentralized sewage disposal systems will not be economically acceptable for this target group, in cities at least. In countries where the gap between rich and poor is wide, therefore, water charges have to be staggered.

Frequently, however, reality is a different matter. In a large number of cities 'dual systems' have developed, in which a proportion of the city's residents are served by subsidised town sewage collection and treatment facilities and water supply systems, whilst another section of the city is not incorporated into the town system and has instead developed a variety of on-site collective, or individual, treatment strategies through their own efforts (Johnstone 1997). As the guiding principle of sustainable development requires not only the twin pillars of ecology and economy, but also a more equitable society, this 'dual system' is not sustainable. One further problem in this context is that water supply is highly subsidised for demanders in many developing countries – a factor which has a negative effect on the water consumption if one considers how scarce this resource is.

It also has to be remembered that various sewage disposal systems generate varying levels of non-water related pollutant emissions. To set up and operate plants to treat water, for example, direct and indirect sources of energy are used. Their utilization conversion leads to the release of emissions and to the depletion of natural resources if non-renewable energy is used. They increase air pollution and thus result in social costs, as these negative external effects are not internalized. Where there is a rapid increase in air pollution, as can be observed in the conurbations of many developing countries, this aspect needs to be given special attention in the form of a comprehensive environmental impact assessment.

Moreover, the economic and ecological efficiency of different sewage disposal systems is not solely a matter of their technological sophistication. In this context, it is also important that the plants have been built properly and that their operation, including maintenance, has been taken care of. If breakdowns occur, skilled personnel must be available to remedy the faults at short notice. It is for this reason, for example, that technically second-best solutions (such as the use of appropriate technology in remote rural regions) may result in a greater degree of ecological and economic efficiency. They do not make high demands in terms of operation and maintenance: in other words, they usually work.

Finally, it has to be asked what incentives exist or have to be created in industrialized and developing countries to bring about demand for, and the construction of, decentralized sewage disposal systems, in rural areas in particular. It is strange that big markets exist for luxury products, such as cars and portable phones. Their negative side-effects, such as traffic jams and (lethal) accidents, are readily accepted by consumers. For some reason, no competitive market has developed for technologies to counteract the negative impact of our basic needs, that is, wastewater and solid waste production.

30.4.2 Micro economy of decentralized sanitation systems

From a micro economic point of view, there are suppliers and demanders of sewage disposal systems. Companies in the private sector will develop and produce more decentralized systems if they are legislatively and/or financially encouraged to do so by political actors or government policy. A good example are the costs associated with the current upgrade for nutrient removal of many centralized wastewater treatment plants, following the stricter effluents standards for nitrogen and phosphorous. Like the market for installations providing renewable energy, therefore, the market for decentralized systems is to a high degree determined politically. For their part, the demanders can be divided into:

- public-sector demanders such as local authorities, which have so far mainly been in the market for centralized sewage disposal systems, and
- private households, which are in the market for decentralized sewage disposal systems either as individual households or in conjunction with others (networks).

However, combinations of the above are also conceivable, with private households and a local authority co-operating. Operating models (built operate transfers, built operate own transfers) are also feasible, and are conceivable for decentralized solutions.

In all cases, these are investments for which a great deal of information is needed, and which thus give rise to information costs. Information must be gathered about the costs of different systems and their operating costs, about the level of technology that has been achieved and thus about ecological efficiency and, finally, about legal framework conditions such as licensing requirements. On the market for sewage disposal plants, however, demanders have to assume that the information available is incomplete. Such a situation means that decisions have to be made in a state of uncertainty, but that this uncertainty can be reduced by paying information costs.

30.5 THE FINANCIAL EVALUATION OF DECENTRALIZED TREATMENT

30.5.1 Decentralized sewage disposal in rural areas

This section will deal with the town of Geiselhöring, in Germany, where the construction of centralized plants for the disposal and treatment of sewage in rural area occasionally runs up against the limits of economic feasibility. This is especially true if, compared with more densely populated areas, the connection costs prove to be very high because of long connecting pipelines. This was the situation which prompted Marr and Steinle (1998) to prepare a pilot study for Geiselhöring, which looked at a potential alternative (micro-sewage treatment plants with biological post-treatment) to connection to a centralized sewage treatment plant.

Their most important findings were as follows:

- A comparison of the costs of the various alternatives shows that, in terms of investment costs, the decentralized solution using micro-sewage treatment is the less expensive variant. This is because savings are made in the area of the local sewerage system. Increasing competition on the market for micro-sewage treatment plants means that reductions in investment costs can be anticipated, thus making further cost savings possible.
- A comparison of total economic efficiency shows that, because of their relatively high operating costs and shorter depreciation periods, the annual costs of the micro-sewage treatment plants are as high as those of the centralized variants.
- When the decentralized solution is applied, citizens have to accept cuts in drainage standards.

It should also be mentioned that the local authority has no chance of forcing property owners to become connected to or to use the sewage treatment system. While voluntary solutions are conceivable, for example, as a result of negotiations between local authority and users, these will run into difficulties if users have to face additional costs and/or other charges. By law, the community is obliged to dispose of sewage effluent and excreta. In other words, the decision process for one of the alternatives is influenced by both economic and legal considerations. Moreover, when implementing decentralized alternatives, new issues and problems arise that are not found in centralized options. More specifically, the problem here is one of disposal, which has so far not been regulated by legislation.

30.5.2 Wastewater reclamation for irrigation

In Italy, the majority of water distributors are public authorities and water fees are usually not calculated on the basis of a metered consumption, but on the farm's irrigable surface (Nurizzo *et al.* 2000). Farmers in the large irrigation district in Northem Italy are charged with an irrigation fee of about 100 euros per ha per year when using spray irrigation, regardless the volumes of water effectively used. In case of corn, the cost of water is roughly 8% of the average corn price for farmers (Nurizzo *et al.* 2000). Using reclaimed water can considerably reduce the farmers' profit, as the amount of water will be metered and extra money has to be spent to ensure the quality of the water.

Nurizzo *et al.* (2000) present a cost estimate of different effluent polishing treatment steps for a plant with a flow of 50,000 m^3.d, representing a medium-sized Italian plant. The use of chlorine, UV radiation, peracetic acid addition and ozone as disinfectants prior to a nitrification-denitrification step was compared. Operation costs were calculated on the basis of the current energy prices (about 0.1 euro kWh^{-1}), chemicals, sludge disposal (about 50 euros t^{-1}) and labour. Maintenance costs have been evaluated as 0.5% of the building costs and 3% of the equipment costs on a yearly basis. Table 30.3 shows that the operation costs are strongly influenced by both the disinfection method and quality target. Reclaimed water distribution was not included in the calculation of this table.

Table 30.3. Increase in operation costs (amortization not included) of $FeCl_3$ contact filtration for different disinfection agents, expressed as a percentage of the standard nitro-denitro WWTP operation cost (Nurizzo *et al.* 2000)

Quality target	Contact filtration + NaClO	Contact filtration + UV	Contact filtration + PAA	Contact filtration + O_3
(*E. coli*/100 ml)	(%)	(%)	(%)	(%)
2.2	18.2	17.7	n.a.	60.6
23	16.0	15.4	64.4	41.6
100	15.3	14.1	42.2	33.1
1000	14.9	13.7	27.4	n.a.

n.a.: not applicable

30.5.3 Water reuse on Noirmoutier island (France)

Noirmoutier is an island located at the Atlantic coast and is a typical insular micro-environment. The main activities are agriculture, salt production, shellfish farms, fishing and tourism. The island's population increases from 9000 locals in the winter to 90,000–130,000 people in summer. Thus, its aquatic environment is very sensitive to water pollution, and the discharge of treated water into the sea is periodically forbidden in order to protect the aquatic industries and littoral areas. On the other hand, fresh water resources are virtually negligible on the island. Its main supply is potable water conveyed from a reservoir 70 km away and sold at a high price.

The four communities on the island have formed an association which is in charge of the water supply and wastewater collection, treatment, reuse and disposal. This association purchases the potable water and sells it to the consumers (Table 30.4). It also sells treated wastewater to farmers. Water irrigation has been implemented for many years in Noirmoutier. To date, reclaimed wastewater accounts for 80% of the agricultural irrigation in the north and 100% in the south. It prevents contamination of coastal areas by reducing waste disposal to the sea, while increasing available water resources.

Table 30.4. Average water prices in Noirmoutier, including the cost of subscription and meters (Xu *et al.* 2000)

Potable water (euro/m^3)				Reclaimed water (euro/m^3)
Purchased	Sold			Sold
	Domestic, hotel, ...	Landscape	Agriculture	Agricultural irrigation
0.60	4.57*	0.67	1.54	0.23–0.3

* = including the price for sewage treatment and disposal: 2.21 euros/m^3

Xu *et al.* (2000) applied an integrated technical-economic model to optimally manage water and wastewater treatment at the scale of a hydrologic unit. Their sub-model focuses on water engineering systems, including potable water production, distribution and consumption, and wastewater collection, treatment, disposal and reuse. Combined with an economic sub-model, water management schemes were assessed, taking into account economic and technical criteria and environmental impact. They calculated pay-back times and specific water costs for a number of scenarios that could improve the water utilization on the island, for example, variations in the amount of reclaimed irrigation water, agricultural versus landscape irrigation, inclusion of a desalination step for brackish water or seawater.

30.5.4 Decentralized solid waste treatment

Sonesson *et al.* (2000) examine the effects on the environment of different management systems for both solid waste and sewage. They used a system analysis to evaluate whether solid waste and blackwater treatment coupled to greywater treatment represented improvements for the environment and its sustainability. The system analysis was made using a substance-flow simulation model, OWARE, the organic waste research model. The system analysis was complemented with a life cycle assessment and an economic analysis.

For their case study and the assumptions made, anaerobic solid waste treatment resulted in the lowest environmental impact (Sonesson *et al.* 2000). Also, urine separation lowers the impact on the environment (less eutrophicating emissions of the treatment plant), although the amount of acidification increases after spreading. Table 30.5 gives the resource utilization of the different scenarios, indicating that composting shows the highest use of fossil fuels.

Table 30.5. Resource use for different solid waste processing schemes (Sonesson *et al.* 2000)

Resource	Incineration	Compost	Anaerobic digestion	Urine separation
Fossil fuel (TJ/year)	118	115	60	39
Woodchips*(TJ/year)	0	49	42	42
Total energy (TJ/year)	118	164	102	81
Phosphorus** (ton/year)	17	0	4	2

* used for supplementary heat production
** mineral fertilizer

The economic turnover for the different scenarios depends on several factors (Sonesson *et al.* 2000). In the study reported, waste transport is the largest cost of incineration and composting, while the treatment costs (investment into installation, more expensive waste transport) dominated anaerobic digestion costs. For both composting and anaerobic digestion, waste transport is more expensive as a separate collection system is needed. Transport and spreading of residuals were relatively minor costs in all three scenarios. When considering revenue, biogas as a vehicle fuel is the largest for anaerobic digestion, heat is the largest for incineration and organic fertilizer the largest for composting.

30.6 REFERENCES

Ahmad, A. (1990) Impact of human activities on marine environment and guidelines for its management: Environmentalist viewpoints. In *Recent Trends In Limnology* (eds V. Agrawal and P. Das), Delhi, pp. 49–60.

Appasamy, P. and Lundqvist, J. (1993) Water supply and waste disposal strategies for Madras. ***Ambio*** **22**(7), 442–448.

Asano, T. (1998) Wastewater reclamation and reuse. In *Water Quality Management Library* (eds W.W. Eckenfelder, J.F. Malina Jr. and J.W. Patterson), vol. 10, 260–261, Technomic, Lancaster, PA.

Boller, M. (1997) Small wastewater treatment plants – A challenge to wastewater engineers. *Wat. Sci. Tech.* **35**(6), 1–12.

Edwards, P. (1985) *Aquaculture: A Component of Low Cost Sanitation*. World Bank Technical Paper No. 36, Washington, DC.

EEA (1999) Societal developments and use of resources – Environment in EU at the turn of the century. Part II: Societal developments and use of resources, pp. 39–51.

Fahm, L. (1980) *The Waste of Nations: The Economic Utilisation of Human Waste in Agriculture*, Allanheld, Osmun & Co, Montclair, NJ.

Fehr, G. (1992) Entwicklung eines Bewertungsverfahrens zur Frage der zentralen oder dezentralen Abwasserreinigung im ländlichen Raum, Witten/Herdecke. (In German.)

Furedy, C. and Ghosh, D. (1984) Resource – conserving traditions and waste disposal: The garbage farms and sewage-fed fisheries of Calcutta. ***Conservation & Recycling,*** **7**(2–4), 159–165.

Gardner, G. (1998) *Recycling organic waste: From urban pollutant to farm resource.* Worldwatch Paper 135. State of the world 1998: A Worldwatch institute report on progress toward a sustainable society, W.W. Norton & Co., New York.

Giles, H. and Brown, B. (1997) 'And not a drop to drink'. Water and sanitation services in the developing world. *Geography* **82**(2), 97–109.

Grau, P. (1994) What Next? *Water Quality International* **4**, 29–32.

Hardoy, J. and Satterthwaite, D. (1990) Health and environment and urban poor. In *International Perspectives on Environment, Development and Health: Toward a Sustainable World* (eds G. Shahi, B. Levy, T. Kjellström and R. Lawrence), pp. 123–162, Springer-Verlag, New York.

von Hauff, M. (1998) Tendenzen und Perspektiven des Marktes für Umwelttechnik. In *Zukunftsmarkt Umwelttechnik?* (eds H.-D. Feser and M. von Hauff), Transfer Verlag, Regensburg.

Johnstone, N. (1997) Economic Inequality and the Urban Environment: The Case of Water and Sanitation. International Institute for Environment and Development, Discussion Paper 97-03, September.

Marr, G., and Steinle, E. (1998) Wirtschaftlichkeit und Machbarkeit dezentraler Abwasserentsorgung in ländlichen Ortsbereichen Beispiel Geiselhöring. In *Dezentrale Abwasserbehandlung für ländliche und urbane Gebiete* (eds P.A. Wilderer, E. Arnold, E. and D. Schreff), pp. 27–51, Munich.

Munasinghe, M. (1992) Water supply and environmental management. In *Studies in Water Policy and Management* (ed. C.W. Howe), pp. 163–195, Westview Press, San Francisco.

National Research Council (NRC) (1981) *Food, Fuel and Fertiliser from Organic Wastes*, National Academy Press, Washington, DC.

Niemczynowicz, J. (1993) New aspects of sewerage and water technology. *Ambio* **22**(7), 449–455.

Niemczynowicz, J. (1996) Megacities from a water perspective. *Water International* **21**(4), 198–205.

Nurizzo, C., Bonomo, L. and Malpei, F. (2000) Some economic considerations on wastewater reclamation for irrigation, with reference to the Italian situation. In Proceedings of the 3rd International Symposium on Wastewater Reclamation, Recycling and Reuse, 3–7 July, Paris, pp. 425–432.

Rivera, D. (1996) Private Sector Participation in the Water Supply tnd Wastewater Sector. Lessons From Six Developing Countries. World Bank, Washington, DC.

Serageldin, I. (1995) Water Supply, Sanitation, and Environmental Sustainability – The Financing Challenge, World Bank, Washington, DC.

Sonesson, U., Bjorklund, A., Carlsson, M. and Dalemo, M. (2000) Environmental and economic analysis of management systems for biodegradable waste. *Resour. Conserv. Recycl.* **28**, 29–53.

United Nations (UN) (1997) *Critical Trends: Global Change and Sustainable Development*, UN, New York.

Wilderer, P.A., Schreff, D. and Arnold, E. (1998) Dezentrale Abwasserentsorgung: Eine Herausforderung für die Zukunft. In *Dezentrale Abwasserbehandlung für ländliche und urbane Gebiete* (eds P.A. Wilderer, E. Arnold and D. Schreff), pp. 1–12, Munich.

Xu, P., Valette, F., Brissaud, F., Fazio, A. and Lazarova, V. (2000) Technical-economic modelling of integrated water management: wastewater reuse in a French island. In Proceedings of the 3rd International Symposium on Wastewater Reclamation, Recycling and Reuse, 3–7 July, Paris, pp. 417–424.

Yan, J. and Ma, S. (1991) The function of ecological engineering in environmental conservation with some case studies from China. In *Ecological Engineering for Wastewater Treatment* (eds C. Etnier and B. Guterstam), pp. 110–120, Bokskogen, Gothenburg, Sweden.

Yepes, G. (1992) Infrastructure maintenance of LAC. The cost of neglect and options for improvement. Water supply and sanitation sector. Vol. 3, Report 17, Regional Studies Program. LAC Technical Department, World Bank, Washington, DC.

Yepes, G. (1996) Do cross subsidies help the poor to benefit from water and water services? Lessons from a case study. Infrastructure note WS-18. TWU Department, World Bank.

Part VI

Architectural and urbanistic aspects of DESAR

31

Town planning aspects of the implementation of DESAR in new and existing townships

J. Kristinsson and A. Luising

31.1 INTRODUCTION

The road to permanent urban development appears to be hard to find. In modern town planning we have need of old, long-lost knowledge, of new inventions and of the introduction of light intelligent infrastructure. Only closed cycles for processes and use of material will, in the long run, make the urban environment permanent.

A complicating factor to the solution is the different levels of scale, but one has to think globally and act locally. Whilst a town is part of a total, worldwide

Edited by P. Lens, G. Lettinga and G. Zeeman. ISBN: 1 900222 47 7

process, it is becoming isolated from its former rural environment. Think globally, act locally (Duijvestein 1993).

31.1.1 DESAR on a world scale

When analysing and solving problems a scale level is usually indicated. This is the usual key to each approach. The durability of ecological systems requires insight into the relation between all scale levels (Figure 31.1). The scale level of 100,000 km is connected with the turbid atmosphere but also with cosmic radiation and the holes in the ozone layer at the magnetic poles. The extremes begin with dry cold and go via moderate temperatures and precipitation zones to tropical dry and wet areas. The scale of 10,000 km from the North or South Pole to the equator (that is, a quarter of the circumference of the earth) goes through completely different climatic zones. Each climate zone requires a different DESAR solution. Within this scale level we already find controversial differences in DESAR solutions in Europe. In some countries the subject is ignored because it is not socially acceptable and in other countries it is open to discussion in scientific circles. In the north and south DESAR is more open to discussion because there are no other solutions for waste disposal.

In the Netherlands, we live in a temperate climate zone with many opportunities for water closets, which the public considers the greatest invention since that of the wheel. The people live in a prosperous part of the world. In northern Europe and America, however, one kilo of water in the form of ice or snow is easier to come by than one litre of fresh water.

On the 1000 km scale of continents and large inland lakes the Baltic appears to be polluted to such an extent that it is an ecological disaster area without fish or birds. Also, in countries that border the Baltic, such as Finland, the glacier from the last ice age disappeared so recently that the average humus layer on the rocky ground is only 30 cm thick. When digging in the rocky soil to lay sewage pipes, running the risk that they will be quickly frozen over, dynamite – a lot of dynamite – is used, considering the highly-valued annual Nobel prizes. The further north one travels, the more attractive the self-sufficient house becomes. The Healthy House in Toronto in Canada, for example, has many installations in order to be self-sufficient. The cheapest sanitary systems are dry closets and aerobic septic tanks where possible.

At the 100km-scale level of countries and rivers, the reuse of water is a matter of course. Clean rainwater that falls on farming land between polluted rivers is drained into the North Sea in order to maintain the required water level for agricultural vehicles (Tjallingi 1996). The chance of floods caused by rivers bursting their banks is highly determined by rationally canalised waterways without any substantial water buffering. These river areas are becoming more and more important and therefore international consultation is necessary.

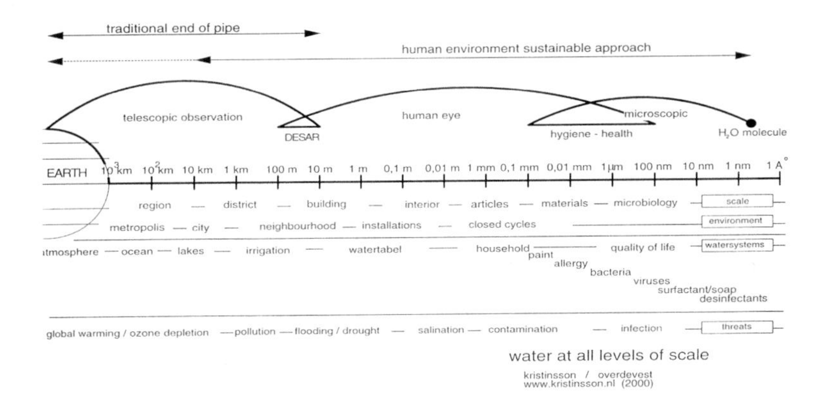

Figure 31.1. How to place DESAR on the logarithmic levels of scale.

At a scale level of 10 km the groundwater level is controlled and surface water is or is not drained off into the ground. This is also the scale level of sewage purification plants and canals. The closed cycle systems of DESAR send fertilisers and compost from the city to the rural outer areas where vegetables grow according to season (Mollison 1990). The lack of animal manure, trace elements and micronutrients leads in the long run to vegetable deficiency diseases such as potato sickness and an increasing susceptibility to pesticides.

The scale level of 1 km is the diameter of a small town in the Netherlands. All its inhabitants have a right to drinking water, even those living in a squat. There used to be a separate waterworks and power company and the local government took care for the urban infrastructure. Now there is a tendency to use a commercial company (the meter reader) to supply water and electricity and measure and collect payment for wastewater.

The contradiction is that a fixed annual amount of money is paid for domestic waste and for purification duties. It would be more sustainable if money was received for clean and sorted waste. The greater part of sorted waste can be brought back into the cyclical process as raw material. In my opinion it would be a good idea if the waste food collector of a decade ago came back to town with his potato peel basket.

At the 100 m scale level there is a residential street and a large building. The urban infrastructure is determined by the town council. To bring about optimal, that is, smaller, water concepts councillors need knowledge. At this scale level

B-quality water can be produced as rinse water and dishwater. To get rid of rainwater at ground level street drains and underground drainage is generally applied.

The 10 m scale level is that of a house. The self-sufficient house is on its way, but it is expensive and controversial. Leaving out the electricity/water/gas meter cupboard makes town planning more flexible because at present we still insert the meter cupboard within two metres of the front door. Power generation (12 volts) at the scale level of a house is expensive but feasible. This is the voltage of cars and pleasure boats. Water recycling has already been realised at a larger scale level. The sanitary system should be DESAR.

The 1 m scale level needs the type of robust DESAR we are waiting for. We are well aware that every continent has its own problems and its own solutions. What are they? Is the dry closet a vacuum closet, or is it an anaerobic or aerobic closet containing a septic tank, mineral wool purification, a helofyte filter or green algae as a purifying tropical pond? These are the questions that should be asked before implementing a simple durable DESAR system. There are many problems but the economy and politics are whimsical decision-makers.

The scale level of 0.1 metre, faeces, is a DESAR level which is not discussable. Therefore we are in the sad position of lacking the prior conditions to find architectural solutions for biological processes within closed cycles. We, as town planners, are waiting for a biologically durable solution. Gourmet experts tell us in great detail what to eat and drink and what not to – but there the story ends and what will happen four hours later is never mentioned.

The scale level of 0.01 metre and lower is entirely within the world of biologists, microbiologists and the sifters of health microbes.

31.1.2 The town within an ecological system

A growing number of people are gravely concerned about the decline of the environment. The majority of the world's inhabitants are preoccupied with their struggle for life. They don't possess the knowledge and do not have the same opportunity as those in the Western world to think about sustainability. If we want to express the ecological environment or the ecological burden in measurable units known to us, we do not have to look far to perceive that 'you cannot see the wood for the trees'. This imaginary wood, moreover, turns out not to be a forest but a jungle of different opinions and views, encompassing the whole world. The 'jungle' is due to lack of knowledge, lack of interest or because of economical and political reasons.

In order to view the environment as a whole, one has to be able to analyse the ecological system from a certain distance. The ecologist Tomásec (Tomásec 1979) divides the ecological system into three components (Figure 31.2): a technical component, containing everything that is man-made: buildings, roads, products; a biotic component, containing all living creatures, microorganisms, plants, animals; and an abiotic component, containing non-living elements: such as water/sea, ground, air, heat, light.

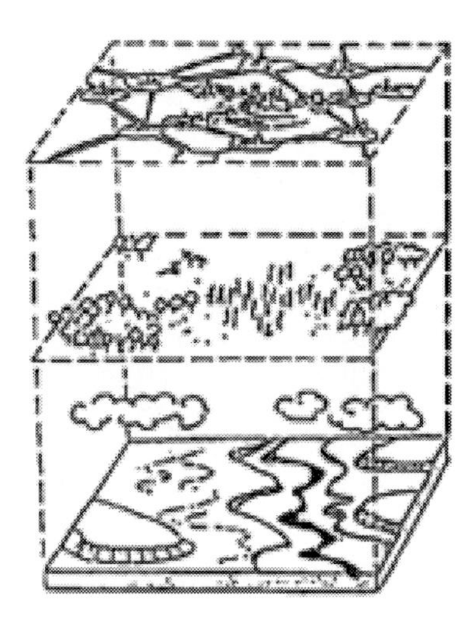

Figure 31.2. An ecosystem with three components, according to Tomásec.

The primary ecological problems we are facing today are exhaustion of resources, pollution of the earth and erosion caused by humanity. What is needed is an understanding of the various (material) flows and cycles.

Ecologist H.T. Odum shows the relationship of the town with its environment, loosely translated in Figure 31.3. This illustration shows clearly that unilateral influencing of the natural environment by the town/urban settlement will quickly lead to a wide-scale ecological problem.

Before starting to look for solutions using closed cycles, a technical engineer/architect can combine both illustrations. The image becomes clearer if we divide the a-biotic components of Tomasec into: (a) the a-biotic component, the earth and (b) the physical component, the infinite shell of the earth and the atmosphere of air, light and heat, liquid, sound, radiation, and so on.

In my analysis our ecological system consists of four components and appears, at first sight, more complicated (Figure 31.4(a)). Interactions between the components can be shown in flows of material, which clearly show the need of closed cycles (Figure 31.4(b)).

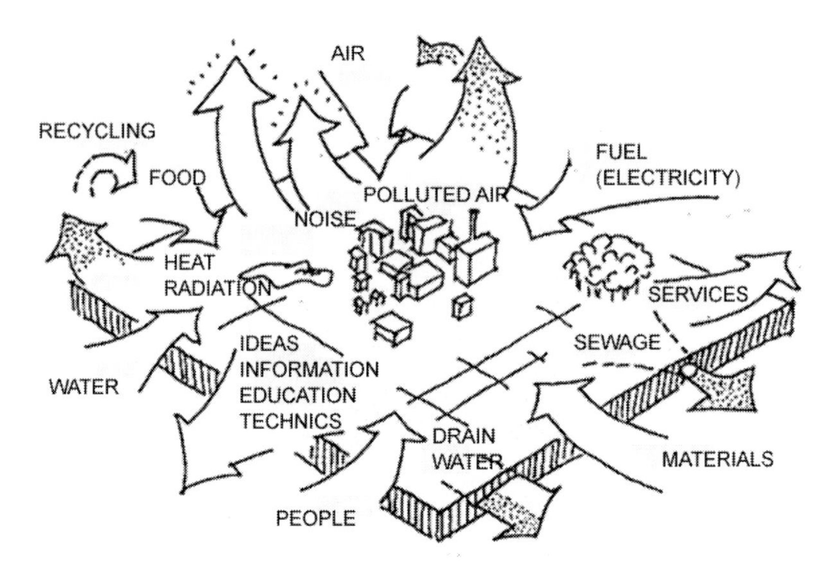

Figure 31.3. The town and its environment.

31.2 CLOSED CYCLES ARE SUSTAINABLE

The vulnerability of our ecological system becomes apparent, wherever exhaustion, pollution and erosion arise. In the long run only closed cycles are sustainable (Kristinsson 1997). The thickness of the atmosphere, in which we can breathe sufficient oxygen, is 7.5 km around the earth; this can be compared with a thin plastic foil around a soccer ball. As inhabitants of the earth we live in a vulnerable environment.

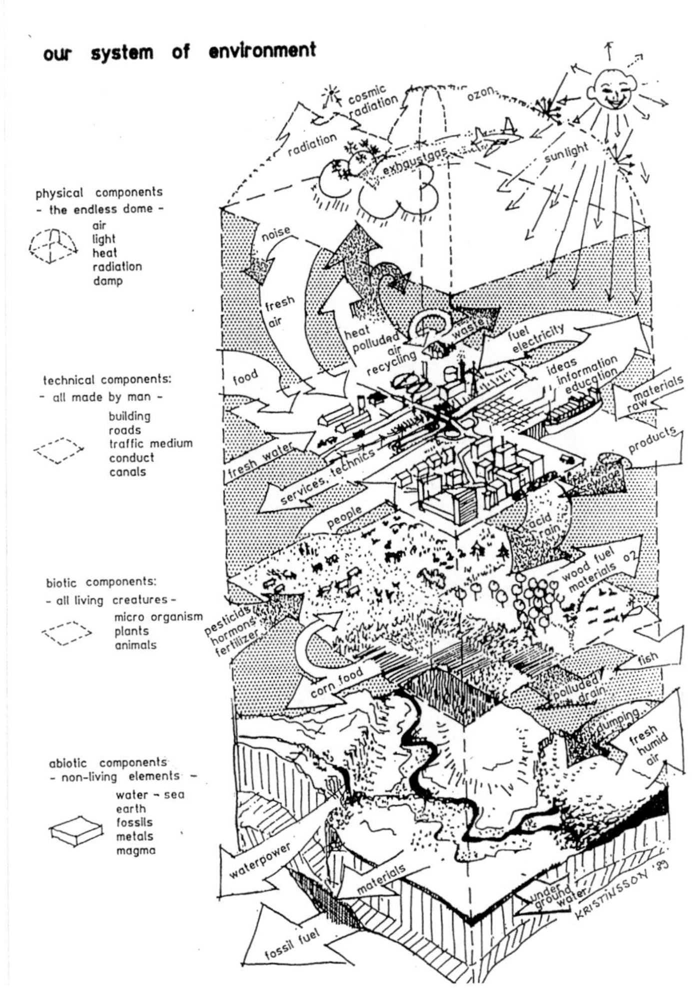

Figure 31.4(a). Our environment (Kristinsson 1994).

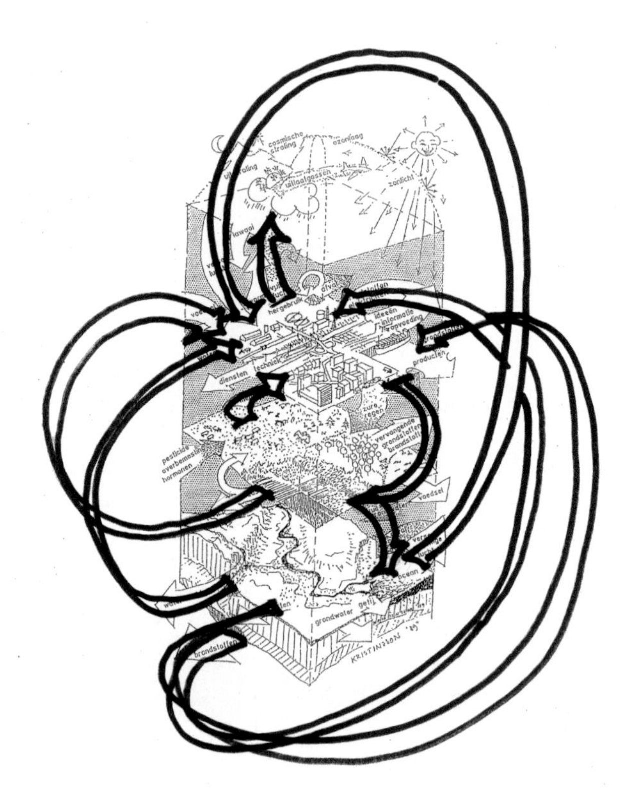

Figure 31.4(b). The flow of materials between components.

31.2.1 The water cycle on a large scale

In the Netherlands precipitation is 0.6–0.85 ml/year, almost entirely outside our influence. Within our influence, however, is evaporation, which we can affect by planting vegetation instead of, for instance, paving for road construction. One Dutch water board district hours (Jaarverslag Waterschap Salland) is proud of its ability to discharge temporary surplus rainwater from rivers into the North Sea within 36 hours. Flow of surface water is registered per district, but groundwater, once it enters an urban area, is difficult to record.

Finally we arrive at the loaded aspect of groundwater polluted by pesticides, hormones and topdressing in agricultural areas and on an urban level by rinse water as a carrier of drained-off faeces, waste matter from soap and kitchens and various chemical waste products. A very promising experiment is currently being carried out to compost vegetable, fruit and garden waste, and dry faeces, at very reasonable costs, at street level in a small residential area (Brabant 'RAZOB' waste treatment plant and De Twelf Ambachten, August 2000; see www waterbesparen.nl).

31.2.2 The water cycle on a small scale

Almost unnoticed we have arrived at the 'unmentionable' subject of the large-scale human refuse of DESAR, where most attention should be paid to small cycles.

When considering urban design, at the level of the residential area, it appears that opinions on surface water diverge widely. Half a century ago in town planning 5–6% of the surface was open water which fluctuated by 100–150 mm in heavy rain. The technological approach says that all surface water within the town should immediately be stored in a large-scale sewerage system and underground basins. In principle this is a mixed system of precipitation drainage (+ storm drains) and sewers in residential areas. If precipitation is extreme these outlets are opened into public, park-like open water, without making public the consequences for hygiene. Trying to increase the amount of surface water is not always successful. Economics can explain this. More money is invested in infrastructure, if there is a lot of it, and thus its maintenance increases. Building new large-scale installations are political priorities, unlike the search for optimal small-scale alternatives.

31.2.3 Sanitation solutions

What are the future sanitation solutions, and existing urban practice? The ditches along the Saramacca Street in Paramaribo on the Surinam River in Surinam have recently been filled up again because of the stench. Here, the tropical pond has not worked well enough as biological purification and the project has failed. One of the main differences between the present and the past is that in the past waste matter rotted away but that plastics and other packaging material used nowadays does not decay. This is an ecological change for the worse. The helofyte filters for sewage purification plants in hot countries may be a source of malaria mosquitoes and are therefore extremely questionable. We are currently facing a huge problem and the worst of it is that dirt, faeces, manure and waste matter are not discussable in our decent, respectable world.

Is it not possible for us, in a temperate climate zone, to dig a surrogate tropical pond that functions both in summer and in winter? During the dark winter months we lack sunlight and heat. Universities and research institutions are working on decentralised sanitation systems as a leap forward for urban infrastructures.

A different system would be that of a vacuum toilet with compact dung; a system whereby various standard bacilli can be rendered harmless by heating or ultraviolet (UV) light or by composting. In many Scandinavian summer cottages faeces is dried to dung by electricity, using a toilet made by Husquarna factories.

At our degree of latitude grey wastewater may be collected in helofyte filters within the residential area. Rockwool – a substrate from greenhouses – is used as a compact lightweight helofyten filter in other experimental small-scale sanitation projects. In the Netherlands, 15,000 tons of used rockwool/year are available for this purpose. Solutions will be different all over the world. Cold, rocky areas (such as Finland, Canada and Greenland) will need specially adapted DESAR. In 1999 a competition was held to design public housing and town planning in Greenland because, due to frost and adverse weather conditions, social housing is some 12 years behind schedule. Weather conditions have great influence on DESAR possibilities.

When taking a camping or sailing holiday we apparently have no problem with a different lifestyle using far less drinking water. We have to make do and therefore we become very inventive. Large cruise ships can in size be compared to a village. On board such a ship vacuum toilets using little water are generally used. When in harbour a collecting tank is used, which is emptied at sea.

31.2.4 Integrating the water and food cycles

The most fascinating DESAR is that used in a Russian spacecraft where the astronauts stay in space for a year. By using green algae the astronauts transfer their own faeces into protein as food. To my mind, green algae as food for breeding fish is a reasonable starting-point in reuse. As technicians, we may think: why can't we apply the high-tech systems of the space age in Earth's built-up areas in order to arrive at closed waste cycles? The question arises, why does the processing of faeces by means of green algae not occur in homes? If we cannot make food directly from faeces, the next best solution is composting dung as fertiliser to use in agriculture.

31.3 PUBLIC ACCEPTANCE OF DESAR

Public acceptance has very much to do with sanitation. What can we accomplish with the present robust technique within a Dutch urban area? What are our main choices in our 'water-closet world' that is not an experiment? By means of a small, narrow-gauged sewerage system, we have a socially acceptable solution using minimal 20-litres of B-quality flushing water per person per day to transport black waste (human faeces) to sewage treatment plants.

Cities in China have very decentralised sanitation. The old towns have no sewers and no flushing toilets. Every morning faeces is collected in the streets and for centuries it has been used as fertiliser for food processing. The smell is indescribable as the beautifully painted toiletpots are emptied. That must be a well-paid job!

Big cities such as Athens also have problems with their sewer systems. It is forbidden in Greece to put used toilet paper into the toilet because the small dimensions of the sewage pipes do not allow this to flush away. Every Greek toilet has a basket next to the toilet as for putting used toilet paper in.

For other reasons the Finns use small showers instead of toilet paper. Most Asians use water to clean their bottoms. It is a shock for them to find out that rich Anglo-Americans use the 'dirty' method of toilet paper.

31.4 IMPLEMENTING DESAR IN TOWNSHIPS

31.4.1 Closed water circuits within a residential suburb

Figure 31.5 shows a possible DESAR concept for a residential suburb. The black wastewater with all its nutrients should be processed in a small, warm algae tank as a small compact tropical pond lighted by direct sunlight. The light feeding device should be fresnel lenses to produce some 3000 lux (1 m/m^2) in the algae tank of 1–2m^3 per family. The green algae proteins should be given or sold to fish-breeders. A dry toilet or an aerobic septic tank for composting purposes is an alternative to this system.

An important aspect to the general public in changing sanitation systems is that living conditions should change as little as possible. Traditional flushing toilets use purified water from the algae tank. A final benefit is regaining sulphate from the algae tank, a substance that occurs rarely.

Grey wastewater is treated in wetlands. This is the greatest problem of this system in the long run because one has to burn or cut the reeds and rushes every year without being able to reuse them, thus wasting them.

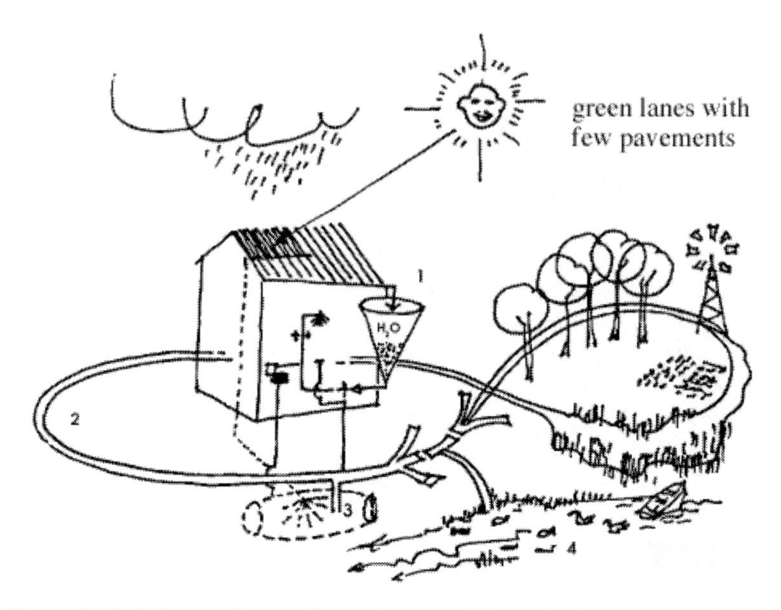

Figure 31.5. Scheme of mass fluxes in a possible sanitation concept. 1. Rainwater–sandfilter: B-quality consumption water. 2. Grey: wastewater, kitchen/washing: slow water circulation within the living area, helofyte filter, purification by reeds and rushes. 3. Black: wastewater and toilet faeces: processing proteins in green algae tanks lightened by glass fibre. 4. Open water as overflow to river or lake.

31.4.2 An ecological water philosophy for a new housing estate

When considering the new residential area of Kernhem Ede, near Wageningen in the Netherlands, it appears that various water concepts are readily available – with or without a financial or ecological basis. For various political reasons the concept with the largest infrastructure was chosen.

This section will explain the architectural concepts required to make satisfactory living quarters in this location. The authorities requested ideas for both public and private housing. First of all we designed a neighbourhood for new house-owners. Housing for elderly people as based on location, being only a short walk from amenities such as shops.

Surface water was an important architectural element in the allotments and thus was an extra feature we added to our plan. The groundwater level was suddenly drastically lowered by the reclamation of the Zuider Zee – now known as Northeast Polder – before the Second World War. It is the rule rather than the exception in the Netherlands that the groundwater level was about 10 m below the soil surface. In Kernhem attempts have been made, in spite of the fact that it is a built-up area, to raise the groundwater level by blocking the surface waterflow. This is illustrated in Figure 31.7 by the gardens which are laid out on various levels, and in the street where the water is running to a lower area.

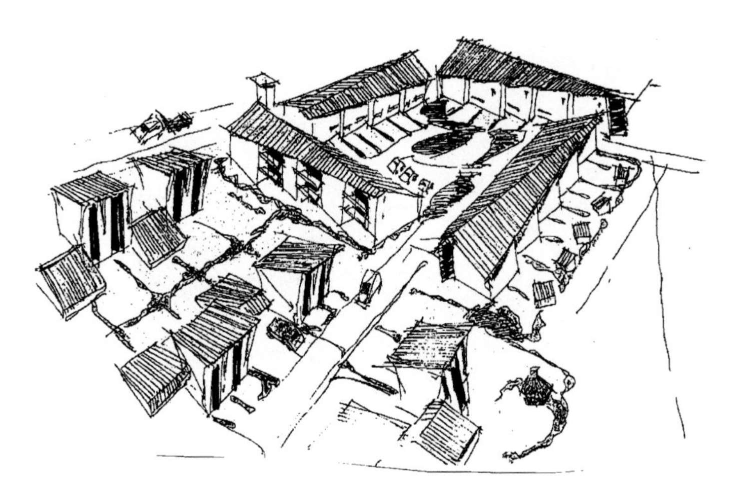

Figure 31.6. A senior citizen complex can also be attractive for young people.

In the Netherlands groundwater is often used as drinking water. Despite the fact that groundwater is getting scarcer, it is also used for flushing and washing purposes. Besides the evaporation that occurs because of this, groundwater is polluted by pesticides and over-manuring from infiltrating surface water. Thus a centuries-old drinking water supply has become tainted. A Dutch person uses 143 litres of drinking water per day on average, mainly for washing and flushing. The groundwater level is influenced negatively by this (Figure 31.8). Efforts are being made to find better water concepts with which people can retain their current standard of living (Figure 31.9).

There is an urgent need for a drastic reduction in the unnecessary use of drinking water. An important way of meeting this goal is to distinguish between drinkable and non-drinkable water. As one of the first residential quarters in the Netherlands, Kernhem is based upon an ecological water philosophy. In Kernhem two water qualities were proposed. This concept does not only concern house building; even the plan for the surrounding residential area is based upon a sustainable water system. In Kernhem, roads are placed beyond the side partition and rainwater slowly finds its way into the soil; and thus to the groundwater level. This measure will be followed elsewhere. For a well-functioning water system, so-called grease catching pits are necessary near car parks.

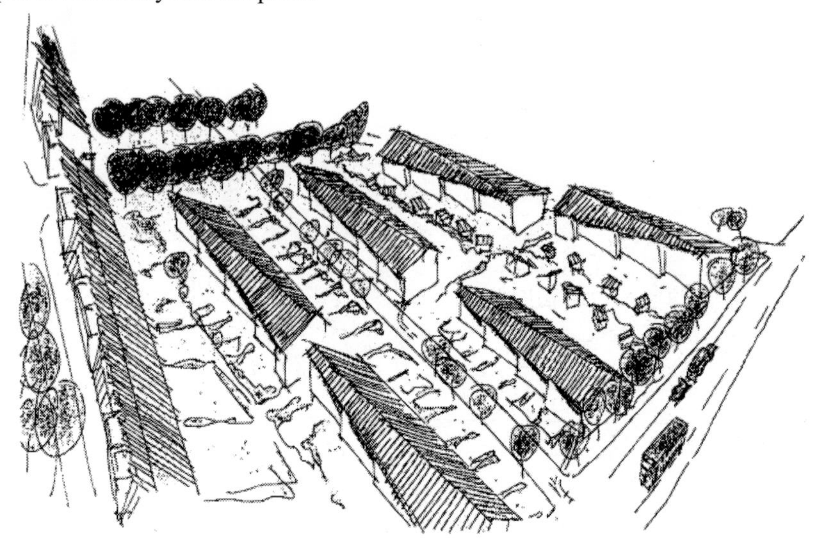

Figure 31.7. Raising the groundwater level by blocking the surface water flow.

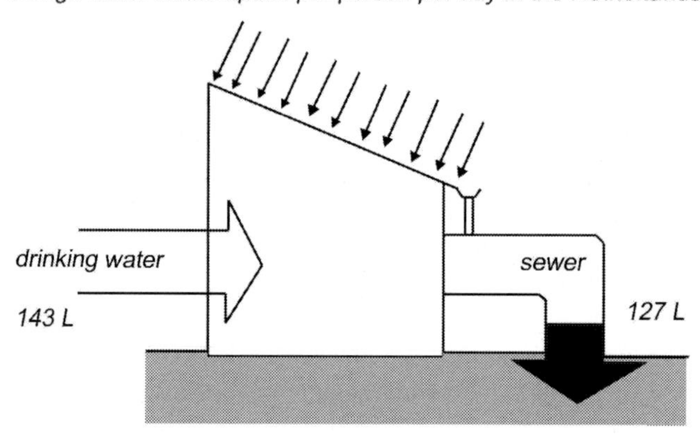

Figure 31.8. A traditional water concept.

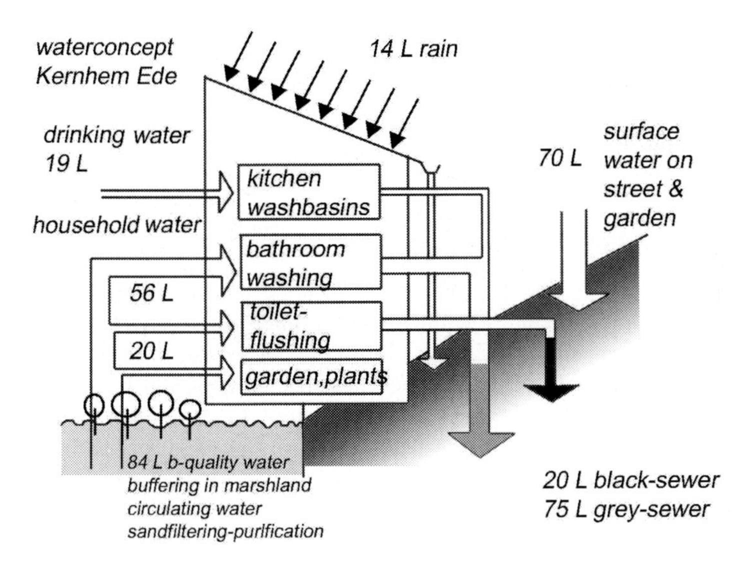

Figure 31.9. A water concept in Kernhem, Ede, the Netherlands.

In urban design the type of dwelling built depends on the natural landscape. Houses situated at the lowest part of the location will be built on poles. In this area we also find dike dwellings (split-level dwellings built along a dike). Further on, there are dwellings without crawl spaces or drainage, and dwellings higher up are supplied with dry crawl spaces, which may be used as storage space for rainwater bags.

Perhaps a new application of the rudiment could be to place bags in it for warm wastewater, to be used for heating the house. This water does not only concern drinking water but also, for instance, the cooling water of an electricity plant which is used for house heating in the change of energy to exergy. This is a completely different cycle of water use that fits into this water philosophy.

31.4.2.1 Water cycles in the estate

In the houses there will be a clear difference between drinking water and household water. Drinking water will be connected to the kitchen taps and the cold water taps. Washing and flushing will be done with household water. Thermic mixing taps will not be needed, because drinking water for drinking, brushing teeth and cleaning vegetables is cold water. This B-water is rainwater from roofs, streets, gardens and other cleaned surface water, reused from within the new housing estate.

In this project no attention is paid to small-scale sanitary systems in order not to mix the two aspects of water philosophy – reducing levels of drinking water is paramount. Attention will be paid, however, to the influence of dog and cat faeces in the residential area on household water quality. The reason for this is that the food chain of these animals very much resembles the human food chain.

31.4.2.2Adaptation to the surrounding

The design of the surrounding space in the new housing estate of Kernhem is based on the natural situation. The fluctuating groundwater level can be deduced from the tree species and other plants. In the higher part of the area (14–15 m above sea level) one finds dehydrated beeches. The groundwater level fluctuates between 2–4 m below the surface. This is the natural domain of pin and summer oak. In the middle part of the estate (11–12 m above sea level) with a groundwater level of 0.5–1 m below surface oak, there will be beech and elm.

To the west, at the foot of a hill and bordering the marshland during the wet spring months, the groundwater level is equal to the surface. The surface level is 10–11 m above sea level. This is appropriate for willows, alders and poplars. With natural afforestation the Lane of Kernhem will enjoy a rich diversity of tree species.

In essence three building types are perferable architecturally:

- for the central area, a dwelling type in east-west blocks with gardens at the front and back
- south-facing houses on a road parallel to the local road (the N224)
- `a new dwelling type, the 'noise wall-dwelling', between the main road (the A30) and the marshland which is evolving at the lowest western edge of Kernhem; these houses should keep annoying traffic noise out of the residential area.

31.4.3 Water purification systems

In view of household water purification in the marshland we suggested a plantation of reed and rush, near the banks and the pole dwellings. With a small watermill, circulation can be promoted. Experiences in the residential quarter of Morrapark in Drachten, the Netherlands (de Jong *et al.* 1994), built in 1992, prove that a very good water quality is achievable. There rainwater from the streets and roofs is caught and slowly pumped around. For the purifying capacity it is important that the water ponds are deep enough. With such a system, rainwater infiltration through a sand layer can probably be omitted and the household water system would be considerably simplified.

Realising this would have considerably shortened our search for durable water-collecting cycles. Social acceptance for this type of system appears to be different.

31.4.4 Designing a self-sufficient home integrated design 'peep-show of closed cycles'

In 1996 I tried to design a north/south cross-section of a self-sufficient home (Figure 31.10). This is self-sufficiency on a very small scale. DESAR is only a part of this private home concept. The advantage of a house without a meter cupboard is evident, although the issue is the integral design.

The concept is divided into four clear functional sections:

water	energy
transport	waste and food

In Figure 31.10, rainwater is preserved at gutter-height and in rainwater bags in the crawl space. Hollow, concrete foundation piles, filled with water, provide heat exchange with the soil. Black wastewater is drained into a tropical indoor algae pond in the floor of the veranda. Grey waste is drained into wetlands in the garden. The advanced energy supply has proved to function well in test projects.

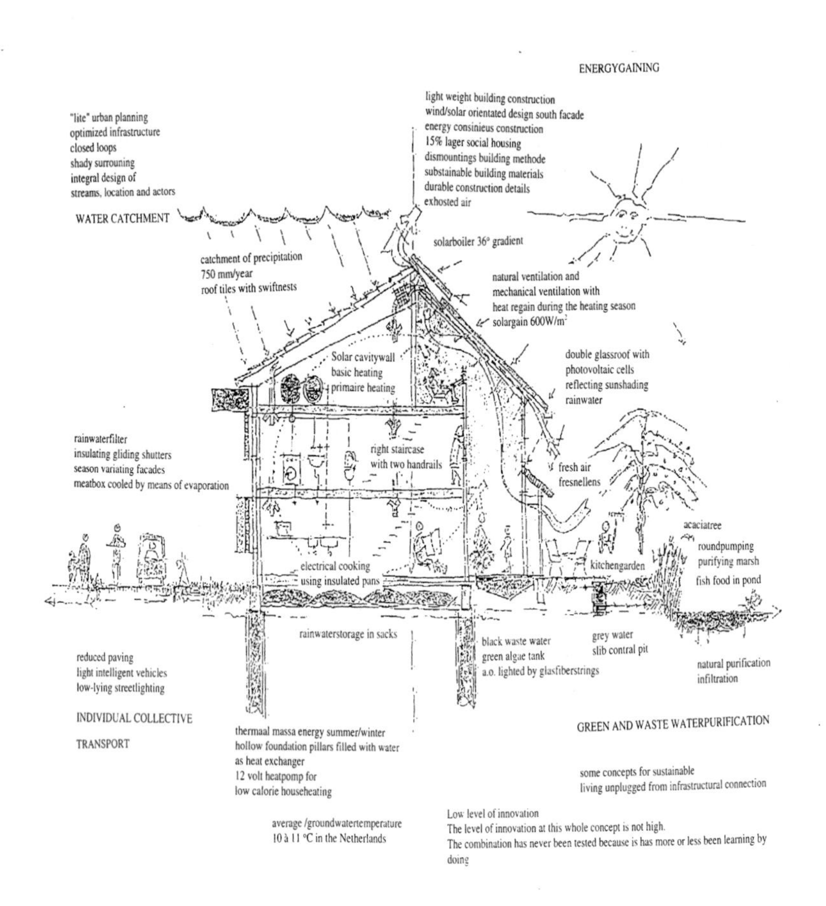

Figure 31.10. The start of a self-sufficient home.

31.5 INTEGRATION OF DECENTRALISED SANITATION SYSTEMS INTO A BUILT ENVIRONMENT

Today, wastewater flows to a central location to be purified (van der Graaf 1995). However, there are a few disadvantages to this form of sanitation. First, the wastewater has to be transported across long distances, thus large amounts of water and huge sewage systems are needed. Because of these long distances a large part of wastewater consists of water which is needed only to transport the waste to avoid clogging. Because of mixed sewage systems, flows with different degrees of pollution end up together in the same pipe. Finally, there is a loss of nutrients in this system. They are not reused, thus causing an impoverishment of the soil. In this case there is a one-way traffic where there should be a cycle. Figure 31.11, derived from Otterpohl *et al.* (1998), shows a wastewater cycle in which there is no loss of nutrients.

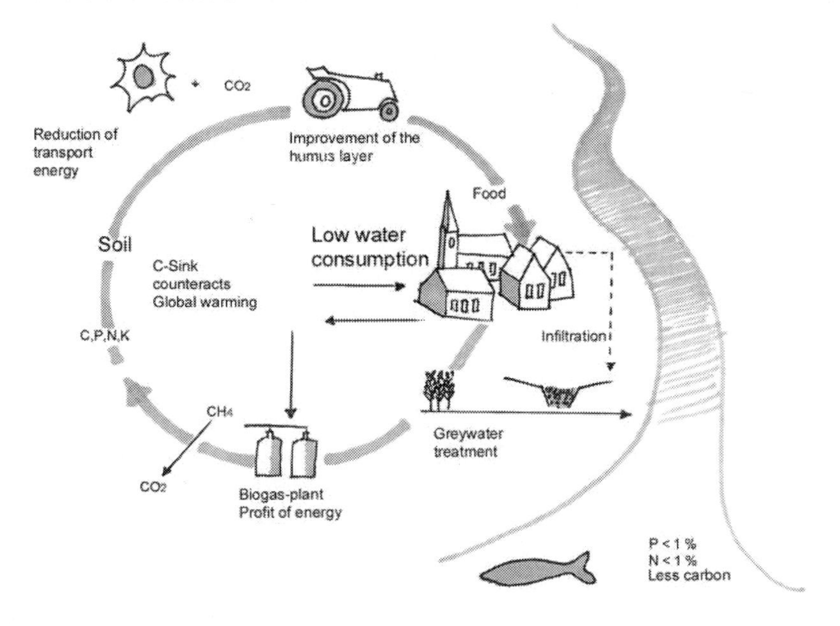

Figure 31.11. The closed wastewater cycle, derived from Otterpohl *et al.* (1998).

All the above-mentioned disadvantages of centralised sanitation systems point to one thing: a need for a new sanitation method. A system has to be developed in which the water cycle is respected. Several such systems are currently being developed and some have now been put into practice. In Lübeck, in Germany, a decentralised system has been applied to the Flintenbreite housing estate which accommodates about 300 people. The concept consists of a biogas installation which is connected to the houses' vacuum toilet systems. In this, installation biogas is produced and the slurry is reused as soil fertiliser. The area is not connected to sewage works. Grey water is led to wetlands. Another such system is in a housing complex in Freiburg, Germany. In this project another biogas installation was applied but this time in one building, in which

40 people live. There is a separate grey water cycle which leads to a sand filter. The water is directly available for reuse. Figure 31.12 shows closed wastewater cycles in a building.

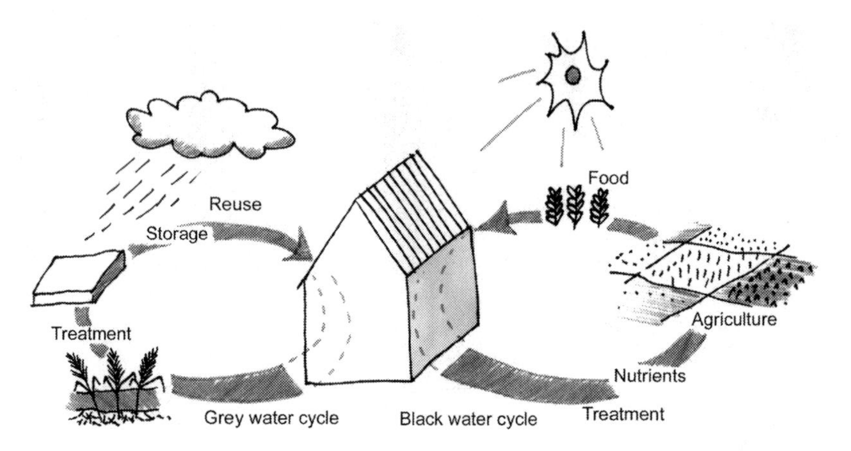

Figure 31.12. Closed wastewater cycles in buildings.

The change from centralised to decentralised sanitation systems does not only have consequences for the purification process. It means that more sanitation systems will be required which are closer to or even in a built-up area. Systems which respect the closed cycles will not only purify in another way but there has also to be a different design. They will be located in built-up areas and therefore changes may have to be made. The acceptance of a new form of sanitation close to built-up areas means the need for stringent rules concerning design and implementation. Besides sanitation, smell and visual hindrance must also be taken into account. It is likely that not only sanitation systems will change, but also buildings and areas which deal with centralised systems may also change. Now, no new buildings have separate rooms for treatment systems and buildings are designed in accordance with current sewage infrastructure.

Decentralised sanitation systems can be applied to wastewater flows of different sizes. The integration and application of these systems affect different abstraction levels in built-up areas. There are two levels, one of which is divided into two sub-levels:

- district level
- building level (building-part level and building-element level

The district level offers the largest possible application for decentralised systems. The application of a certain system influences the various abstraction levels. Dependent on the scale level of the application the implementation of the system will have more or less influence on the different levels of abstraction. For example, the integration of a large-scale wastewater system into a district influences factors such as district planning.

A decentralised system which treats the wastewater of a complete district will have more influence on the design and planning of a district than on design at building and lower levels. Vice versa, a system treating only a few people, that is, those located inside one building, has relatively little influence on design decisions at a district level.

Applications of different-sized systems all have identical effects on their surroundings and buildings. The next section offers a short overview of the characteristics of a decentralised system on different scale levels.

31.5.1 District level

By connecting more users to one system, a higher efficiency can be reached at lower costs. Because the systems grow larger with a growing number of users there will be a certain use of space in a district for a system. This must be kept in mind while planning designs. The possibility exists of combining systems with other district functions such as a community centre. Because wastewater is transported to a central location a sewage infrastructure is still needed although it will be less complicated. It is easier to construct because the volume of the streams is much smaller. The sewing pipes can have smaller dimensions and do not have to be located as deeply in the soil.

31.5.2 Building level

When a sanitation system is applied inside a building there are two different ways of doing this. The first is by reserving space in a room to house the system. The choice of location depends on the accessibility, supply and removal, safety-instructions and available space. This room can be combined with other installations in the building such as central heating.

Table 31.1. Advantages and disadvantages of the application of sanitation systems on building level

Disadvantages	Advantages
Occupation of space	No need for sewers
Design constrictions	Use of inner climate
Removal of fertiliser etc.	User has control and responsibility

The second option for integration is the installation of a system in a part of a building that already has another function; for example, in the foundations, roof construction, walls etc. This method of integration has consequences for the design of the element. In the first place it seems easier to place a system separate from its surroundings. It is interesting to combine functions when the system can use the properties of existing function. Aspects that can play a role in function are light, temperature, orientation and so on. The physical building properties of the element can also be used, such as insulation values, transparency, orientation and robustness. Beside these physical aspects, the building's user patterns can also be looked at. At certain times during the day, light and warmth are produced in a building. This energy may be used by the system. The life expectancy of the element and the system must be approximately the same to make the combination meaningful; this will also make the design technically more complex.

Table 31.2. Advantages and disadvantages of integration in elements

Disadvantage	Advantages
More complex design	Saving of space when the design is integrated No need of pipes

31.5.3 Types of decentralised sanitation systems

Apart from the application at different scale levels, a difference can be made between the ways in which systems function. Many sanitation principles can be put into practice to purify wastewater. To apply them in buildings a distinction can be made in another way, which relates to the integration of the system into the building. There are two systems:

- autonomous systems
- integrated systems

Autonomous systems do not use the properties of the surroundings in which they are placed. The systems come ready for use from the factory, are situated in an arbitrary place and put into use (Luising 1999). Integrated systems are the opposite, being completely integrated into their surroundings. To function they use the properties of their surroundings, such as temperature, light, and so on. When applied in a building, a physical climate is obtained in which transparency and insulation values can play a part. Important aspects of integrated systems are shown in Figure 31.13.

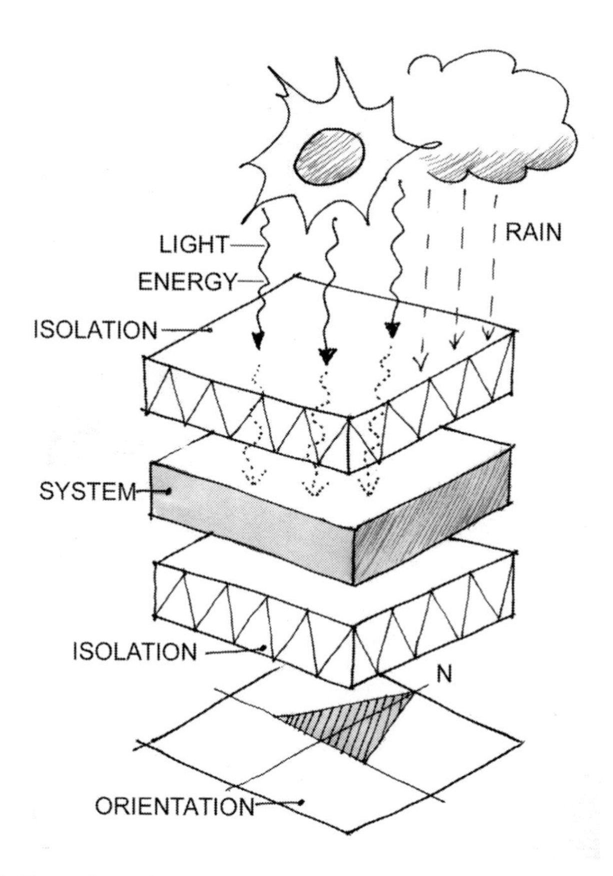

Figure 31.13. A schematic representation of a wastewater treatment.

To make further statements on all the aspects which can be discussed during the application of a decentralised sanitation system, research is being carried out into related aspects of integration into buildings (Luising 2000). Two systems have been designed that have to be integrated into a built-up area. To make a complete picture, two systems are chosen which differ from each other as much as possible. The first system, which is an autonomous large-scale application, is a DESAR system in which domestic wastewater streams are treated anaerobically and grey water is treated separately (DESAH 1999). For the second, an integrated small-scale system, an algae sanitation system has been chosen. Although algae have not yet been used on a small scale for domestic wastewater, there are some design aspects which can give more insight into integration aspects of decentralised sanitation systems (NOVEM 1993). The growth of algae depends on temperature and light intensity (Becker 1994). When this kind of system is integrated into its surroundings, it will have much influence on the design of these surroundings, for example, the orientation of windows in buildings, indoor humidity and on a larger scale the planning of build-up areas. The information obtained from this comparison will, in future, I hope lead to a better integration of human living with decentralised installations that respect natural resources.

31.6 REFERENCES

Becker, E.W. (1994) *Microalgae, Biotechnology and Microbiology,* Cambridge University Press, Cambridge.

DESAH (1999) Decentrale Sanitatie en Hergebruik op Gebouwniveau, Rapport KIEM EET project 98115, University of Wageningen, the Netherlands. (In Dutch.)

Duijvestein, K (1993) Ecologisch Bouwen. SOM, Dept of Architecture, University of Delft, the Netherlands. (In Dutch.) .

Van der Graaf, J.H.J.M. (1995) Behandeling van afvalwater, Dept. of Civil Engineering, University of Delft, the Netherlands. (In Dutch.) .

De Jong, T., Kristinsson, J. and Tjallingi, S.P. (1994) Aanzet Stadsrandvisie Drachten. KC-326.112.000, Rotterdam, the Netherlands. (In Dutch.) .

Kristinsson, J. (1994) Lecture, De Nieuwe Noodzakelijkheid. Dept of Architecture, University of Delft, the Netherlands. (In Dutch.)

Kristinsson, Architect and Engineering Office (1981, 1986, 1991, 1996) 3th, 4th, 5th and 6th editions, Deventer, the Netherlands.

Kristinsson, J. (1997) Inleiding integraal ontwerpen. Dept of Architecture, University of Delft, the Netherlands. (In Dutch.)

Luising, A.A.E. (1999) Studies on decentralised sanitation systems, Report, Department of Architecture, Delft University of Technology, the Netherlands .

Luising, A.A.E. (2000) Integratie van decentrale sanitatiesystemen in gebouwen. Rapport final thesis, Dept. of Architecture, University of Delft, the Netherlands. (In Dutch.)

Mollison, B. (1990) *Permaculture,* Island Press, Washington DC.

NOVEM (1993) Algen in de Nederlandse. Energiehuishouding. Novemrapport ROPD9283/d2. (In Dutch.)

Otterpohl, R. , Albold, A, Oldenburg, M. (1998) Internet Conference on Integrated Biosystems. http://www.ias.unu.edu/proceedings/icibs/oldenburg/index.htm

Tjallingi, S.P. (1996) De strategie van de twee netwerken, Den Haag, the Netherlands. (In Dutch.)

Tomásec, W. (1979) Die Stadt als Ökosystem: Überlegungen zum Vorentwürf Landschaftsplan Köln. *Landschaft & Stadt* **11**(2).

32

Architectural and urban aspects of the development and implementation of DESAR concepts

Jacob Schiere and Jørgen Løgstrup

32.1 INTRODUCTION

The purpose of this chapter is to move away from the often technical focus of DESAR to concentrate on the design, implementation and use of DESAR technology in architecture and urban town planning, and to discover how its implementation works in real life by looking at case studies.

Many of the readers of this book are experts in sanitation practice and policy. It is important that these readers relate to DESAR not only as a problem belonging to often contaminated and remote neighbourhoods (in analogy to Verhelst's statement (1986): 'The people are not only problems which need to

Edited by P. Lens, G. Zeeman and G. Lettinga. ISBN: 1 900222 47 7

be solved, but also mysteries which need to be explored, not only emptiness which need to be filled, but also fullness which can be discovered'), but also as something relevant to the home and life of the experts themselves. I stress here that the personal attachment of the experts' own daily life to DESAR challenges and solutions may be one basic condition for lowering the obstacles between policy-makers and policy-objects. A direct consequence of such an attitude is that talking and writing about DESAR should be more in terms of 'I' and 'we' and less in terms of 'him/her' and 'them'.

I refer here to a dialogue between Socrates and Polos, in which Socrates asks his opponent Polos to identify clearly his own personal position; 'whether he speaks the voice of the masses, or instead he dares to define who he is and how he thinks himself' (van Gelder 1960). These approaches should not contradict but should complement each other (Figure 32.1). Too easily we discard certain individual practices because they have no relevance to the bigger picture. Many seemingly opposing positions only contradict each other in certain limited contexts; they should instead be taken into free Socratian space in order to see what they have to offer each other. (One of Socrates' challenges was not to find out which of two opponents was right or wrong, but instead how they could come closer to true understanding of each other.)

Sanitation matters in today's world are too urgent to allow for one-sided approaches. One basic theme of this chapter is thus to validate the personal and specific approach as well as the general approach to sanitation.

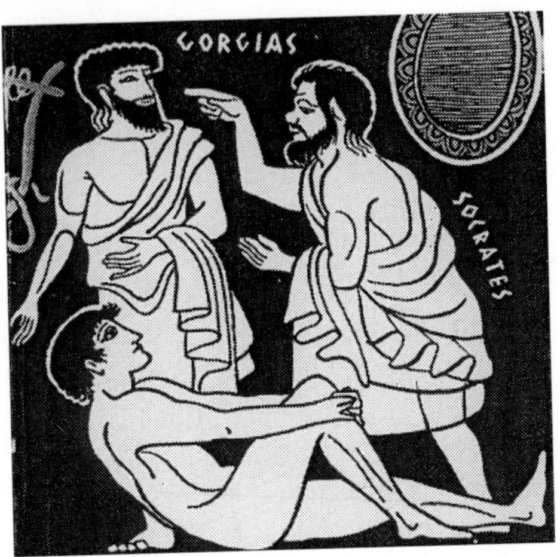

Figure 32.1. Socratian space. Source: van Gelder 1960.

It is thus relevant to clarify one of my personal biases. I believe that interaction between humankind and nature is not necessarily destructive, but instead can enhance and strengthen each other. Traditional Amazonian rainforest is partially so rich because of the careful and respectful presence and action of many thousand years of human presence (from extended and ongoing personal dialogue with Latin American ecologists and philosophers (including Ricardo Jordan and Claudio Tigre)). My conflicting personal experience between extremes of human creation and destruction has only reinforced this curiosity and belief in potential. Such curiosity is part of my drive into DESAR research.

A second basic theme is the challenge that DESAR and not DESAR can and should complement each other, depending on the context of the task to be completed. For that I also call on Socratian tradition 'not to ask who is right' but rather to figure out the truth together. It may require an architect's hand to undertake such a broad approach, because an architect blends all technical options and common sense into an image in which all of humanity can dwell in dignity (van Kranendonk 1980).

First, attention will be paid to authors' lives and practice (especially in the developing world). It may be of benefit to many experts to expose themselves to challenges and options from outwith their own familiar social and physical experience. Next, some basic conceptual perspectives, which have guided the author into his further understanding of DESAR, will be further discussed. Finally, attention will be paid to some authors and practice today, at the meeting of the developed and developing worlds. Some promising new developments and challenges will be presented.

- Author and reader are both object and subject in their professional practice of DESAR.
- DESAR can advance better as its practitioners develop a distinction between the individual and the general.
- DESAR can more easily be understood if its practitioners look outside their own schemes of thought and patterns of life.
- DESAR is to be set exactly in the meeting between changing perception on needs and options in the field of water and wastewater; hence DESAR goes beyond traditional feuds between conservative approaches (centralised and end-of pipe) and progressive sanitation practice (more decentralised and oriented to resource recovery).
- DESAR should be defined as a way to reduce friction in interactions between human action and the environment
- Architects (and urbanists in a broader sense) can help to set the context and challenges for DESAR.

Scheme 32.1. A summary of this chapter's introductory comments.

32.2 DESAR AND QUALITY OF LIFE

This section will elaborate on some of the basics of authors' lives and practice (especially in the developing world) through which we have entered the world of DESAR.

- DESAR practitioners should interact with the way poorer people live in their context of daily limitations and traditional sanitation practice. Such interaction is a reciprocal process, in which there is no place for imposition and in which openness and high expectations are imperative.
- DESAR should not only be promoted as a new practice; within the context of its new understanding it should also interact with and reinforce popular practice. Focus should be on transition and ongoing change.
- Legal and administrative procedures sometimes impose more limitations than protections on citizens. It is therefore imperative that DESAR distinguishes between the general and the individual; indeed, general development almost always feeds off individual understanding and courage.

Scheme 32.2. A summary of statements related to DESAR and quality of life.

Many 'sanitation experts' live middle class lives. More than the average person, we are confronted with technological advances and social backdrop in a setting of ecological change. As members of our individual families we live mainly in areas with sufficient potable water and centralised sewers. Many of us do not even know where and how our potable water is produced, where our wastewater flows to or how it is been treated.

As technicians we are close to advanced laboratories and computerised databases. As consultants we are called to look into sometimes quite different situations. We see people struggle for their daily quota of water and their human right to a clean environment. Such consultancies can bring us from Arctic snow to tropical deserts and jungles. We may have to walk far into peaceful valleys where pristine springs trickle from beautiful hillsides or we may have to search for water in destroyed and deserted areas, in vain searching for remaining water sources. We may even have to walk through minefields where war victims store their scarce water in second-hand pesticide barrels. Increasingly we are also becoming involved in areas struck (or vulnerable to being struck) by natural disaster. After our working day has ended, we may have to stay in expensive hotels where water abounds, or in remote cabins where we have to decide whether to wash our hands the same night or, because of the scarcity of water, wait and wash them tomorrow.

At least to the field expert, it is obvious that the issue of DESAR cannot be dealt with without asking questions about quality of life and differentiating the usual 'standards for quality'; hence it may prove difficult to discuss the issues of decentralised sanitation and reuse without asking ourselves urgent questions on the topic of sustainable development. I will use the term 'sustainable' here as a quest for balance between the three main components which are the social, material and natural well-being of humankind and its environment. Sustainability may be less of a goal and more of a path. The exploration of that path may oppose the ruling Western paradigm of consumerism and thus complicate the job of the DESAR expert.

32.2.1 A personal note on need and potential

Socrates asks his opponent Polos to make his personal position clear. Merleau-Ponti struggles with the same theme as he discovers that he is not an outsider to the social system, which he studies and redesigns (Simon 2000). So do I, as author of this chapter, feel the need to explain some of my personal background, even if only to make up for the lack of scientific proof of what I want to state.

After working in a commercial Dutch architects' practice I was challenged to help shape housing and people's environments, in a non-profit organisation, often under adverse conditions of poverty, ecological decay, natural disaster or even warfare. We tend to call such situations 'underdeveloped'. I came to call these situations 'overly complex'. Because water and hygiene represents such a basic need I was soon (even in my role as an architect/outsider) sucked into the issue of what in this book is called DESAR. I encountered challenges but also options to an extent and intensity I had not experienced before (Figure 32.2(a)–(c). Under such extraordinary circumstances, I came to see human and ecological misery fuse with the creative (born from necessity) conscience of housekeepers, technicians and scientists, occasionally and often unexpectedly producing realistic approaches for redefining need, means and answers (in that order!). In the most unexpected places I saw evidence of creativity and the positive energy of the common people. I was encouraged by the potential that existed and discouraged by the randomness of such efforts. My experience shows demand and potential, however, too many efforts are simply intuitive and out-of-context testimonies of rural times passed. However, in the stress and enormity of the task I was helping to carry out, I found that Western technology and economical context was often irrelevant, so that I could only help to strengthen local practices and, in a spirit of mutual learning, help in finding ways of translating such experiences into modern urban environments.

Figure 32.2(a) and (b). Examples of urban landscaping on difficult urban slopes (Villa Nueva Tegucigalpa, Honduras). Source: author.

Figure 32.2(c). Poor sanitation in Pantanal Managua, Nicaragua. Source: author.

I have already said that direct social context for every sanitation officer is his/her family, who provide a logic that is alien and often complementary to his/her office and laboratory experience. Out of necessity, my wife developed for our family 'the bucket-shower' (Figure 32.3), which involved cascading domestic water and reusing wastewater. A bath can be taken from a bucket-shower, standing in a tub. After that the runoff can be used to do the laundry, then to mop the floor and finally to irrigate your plants. It was fun for the children, thus adding the relevance of social components to technological innovation.

Figure 32.3. A bucket-shower as an example of need-driven practice of cascading domestic water use (Santa Maria Cauqué, Guatemala). Source: author.

In our water-scarce context the option of a dry composting toilet was soon considered. By building one, and by working out various improvements to it, I discovered that the original Guatemalan model (as copied from the poorly documented Vietnamese 'double vault' (Chongrak 1981)) was only promoted for hygienic purpose, to kill *Ascharis* and other disease vectors. However, it was our family's first priority to develop a practical waterless toilet for family use even if this was only making the best of poor options. The decision was actually more complex. The absolute scarcity of water but also the rental situation of our house, the subsoil in which we'd have to dig a pit latrine and our personal and technical curiosity were the main motivating factors behind our decision to use a composting toilet. The issue here is the complexity of our context vs. the apparent simple technological choice. The resulting and surprising simplicity and physical attraction were enough reason for neighbours to copy the idea (Figure 32.4). Although we had not intended to, we became heavily involved in this sanitary success story, even before we had decided on the second priority, 'What should we do with the compost and urine produced?' and later, 'What do we think of the resulting hygiene and sanitation?' (Cacerez 1987; Schiere 1989). (Respectful reference should be made here to CEMAT (Dr Armando Cacerez and Lic. Anamaria Xet) as well as to the Swiss EAWAG (Dr. Martin Strauss) who together made this an interesting experience.

Figure 32.4(a) and (b). An attractive and functional composting toilet in a middle class house (Cuernavaca, Mexico). Source: Esrey *et al.* 1998.

In conclusion I suggest that no matter how useful technological development may be, it should be complemented by other values, such as beauty and social cohesion.

32.3 A NON-WESTERN APPROACH TO AESTHETICS

This section will elaborate on some of the basic conceptual perspectives which have guided the author into his further understanding of DESAR.

- The traditional slogan 'Beauty, health and sustenance' brings to DESAR a more holistic and less restrictive focus.
- 'Balance' between risk and advantage provides for more realistic terms and expectation.
- DESAR requires a critical attitude towards ruling Western technological and consumerism, hence it should include in its objectives a constructive approach of sustainability and thus social and ecological cohesion.
- DESAR, in its interaction with today's sanitation challenges and options, complements traditional technological definition and pursuit. Therefore DESAR may prove a more difficult task than we as engineers are prepared to face.

Scheme 32.3. Summary statements for a non-Western approach.

As an architect I must speak on the issue of aesthetics. For the purpose of this DESAR statement I define aesthetics as 'creative friction'. Friction may be explained as 'making somebody think and/or respond'. As we respond to friction we basically have the choice of giving up, pushing even harder or reconsidering our position. Giving up may not be the most creative approach. Pushing harder may easily lead to the destruction of the matter at stake. Reconsidering may lead to growth or innovation. At best, development should be towards access. I talk about 'balance', but as we may see at the end of this section, the Western paradigm of development is not about balance. Aesthetics is not decoration but about discovering inherent beauty. As an engineer I would rather talk about logic than inherent beauty. The reader is now invited to make three steps into the principles of Western thought. DESAR is one of the expressions of that thought, but it carries with it encounters with other realities such as minefields and military camps, urban misery and the sometimes Stone Age practices of supposedly 'wise' mankind.

Today there is fast-growing acceptance of what is known as dry sanitation, or dry composting toilets. After several attempts (from Victorian London to revolutionary Vietnam) the advent of dry sanitation happened in the midst of civil war in the Mayan highlands of Guatemala and then became used in Mexico and beyond. Mayan culture is very much alive and has even worked its way into Western thinking (Figures 32.5 and 32.6). Strong development took place in the context of civil war within Mayan cultural understanding and may thus illustrate the relevance of non-Western components of understanding DESAR.

Figure 32.5. Traditional non-Western values and insights still inspire the Maya (Patzun, Guatemala). Source: author.

Let us now attempt to explain the concepts of aesthetics and balance from the Mayan point of view.

- Aesthetics: Mayan people thought that humankind was (at the third attempt) created out of corn (*ixim*). This story is found in the Popol Vuh, one of the ancient and authentic books of the Maya people and their cosmovision (Reinoso 1956). Somewhere in the endless and marvellous cycle of life, through decomposition, man returned to earth (*uleu*) and, out of *uleu*, *ixim* is recomposed. The *ixim* is then again sustenance to the body (mind and soul). I stress here that this perception is complementary and somewhat in conflict with the slogan of the DESAR summer course, which is 'economy, sociology and hygiene'. I call this extra factor 'aesthetics'; others call it 'wholeness'; Mayans call it *utzil* which means 'more than good'. Life and its components and processes such as DESAR, must look attractive through the architect's eye. At best, architects speak for the creativity that moves people. The concept of beauty depends on the onlooker. No expert is able to impose *utzil* alone; it can only be realised through interaction of the different actors and their environment. In Mayan thought there is inherent logic and beauty in seeing wastewater as a resource, simply one of the stages in the circle of life.
- Balance: the Mayan way of life seems to strive more for balance between good and bad than for the total eradication of bad. Hence one may not aim for the complete elimination of health risks but rather should aim to minimise and manage such risks (perhaps even turning them to your advantage). Such a search for the perfect solution may even be counterproductive. (The philosopher Ivan Ilich presented several critical studies of technological development and myth (Ilich 1978). A recent quote from a Dutch radio news broadcast stated that: 'For reasons of safety [traffic and so on] many parents take their children to school. There is evidence, however, that all this movement makes the school hour the most dangerous for children.) Let only one example challenge the mindset of the perfect; …the same unhygienic human 'waste' does by the wisdom of cyclical nature (we call that today 'urban agriculture') indeed provide for fertilisation of healthy growth of e.g. papaya and garlic, which have both proven to be excellent medicines against certain intestinal worms and parasites.

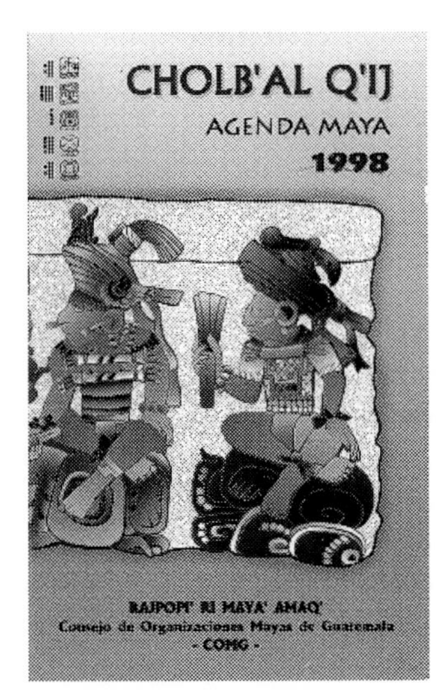

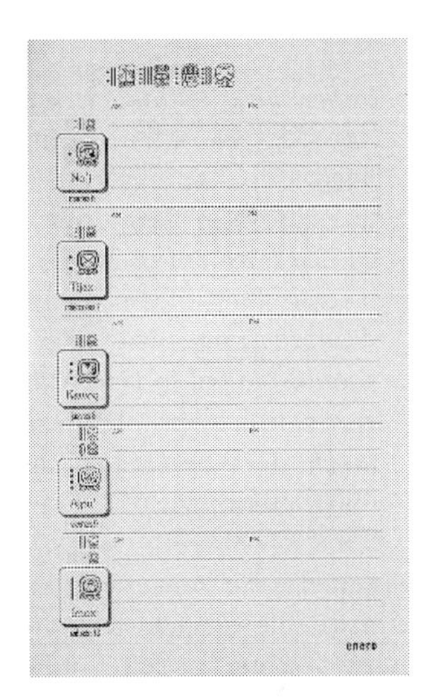

Figure 32.6(a) and (b). Ancient logic and insights are still shared today (with weeks of 13 and 20 days). Source: Maya agenda, COMG, Guatemala.

One complementary comment about utzil and balance is contained in Ellul's short statement on the misbalance of today's Western perception and action (Ellul 1990). Far from direct Mayan influence, he vocalises a criticism of observers who live and work within the Western development paradigm.

- *Concentrated advantage and diffuse disadvantage:* Ellul concludes that the Western emphasis on material goods tends to concentrate advantages and to diffuse disadvantages; in other words he argues that conventional and technological solutions to isolated problems tend to cause secondary and more diffuse problems somewhere else. I propose therefore to interpret 'sustainability' as the strive to close the gap of Ellul's contradiction, taking into account issues of socio-emotional and ecological cohesion.

All this leads us to one conclusive statement on DESAR. If DESAR pretends to respond also to the needs of those who have not been served by traditional sewer approaches, then practitioners and theoreticians should be aware that their task has just begun and that a strange new task lies ahead. Here I will not elaborate on the complex factors involved if we consider socio-economic realities and the structure of today's technological forces and powers (Schiere 1991). The following is a quality statement on the issue of approach:

- *Stimulating handicap:* implicitly related to previous points of my introduction is the potential of creative development out of real need. In line with Ellul's inherent misbalance of western technological development. I do emphasise the potential of what can be born out of need and what can be killed out of wealth. Isn't it true that much of DESAR was developed outside regular western experience under often un-western conditions? Respectful should we try to find proper balance with that rather non-western world full of problems and potentials. Sure they can teach us as much as that we can help them. They cannot be reduced to obedient objects of technical sanitation schemes.

I challenge here the suggestion that apparently romantic approaches may only be valid for the old and happy days dating to before urban misery and refugee camps. To me, at least, all this has become a practical inspiration while meeting adversity in life and work.

In conclusion to this philosophical section, I suggest that the Western socio-economic paradigm alone is too narrow to use as a base for DESAR practice.

32.4 ARCHITECTS' AND URBANISTS' INTERACTION WITH DESAR

Examples of images will be presented in this section. At first they may distract from other real, urgent and large-scale DESAR themes. It will be shown that on different levels major technical breakthroughs have been achieved. Once more these little examples may seem irrelevant to DESAR; however, it should not be forgotten that some fortunate and privileged people might through these examples become familiar with modern and hitherto unthought-of concepts of sanitation. This consciousness, in my belief, is one of the conditions which may contribute to the broader political and public acceptation of DESAR. Let me once more call upon Plato and Socrates. The small and individual perspective is as important as the general context. Their ancient struggle with political debate is also topical and relevant today. Progress is not to be expected from the political and technical debate alone; it needs also to be lived out among those who are the decision makers.

32.4.1 The interdisciplinary approach

DESAR can support and rebuild a sense of a 'healthy' and cohesive community in line with our Mayan concept of *utzil* and our understanding of Ellul's proposal for sustainability. The following examples may provide cross-cultural evidence that implementing DESAR has a potential beyond treating wastewater, involving all kind of experts, and of course local populations themselves. This listing may, by its context, stretch the understanding of the basic elements that make up DESAR; from the incredible beauty of dewdrops, spider webs, mosses and colourful fungi, passing by murky moments (such as black wastewater) all the way to the construction of hurricane-safe river banks.

32.4.2 Micro-ecology (Bad Kreuznach and Amersfoort)

Near Bad Kreuznach in Germany we find interesting interaction between minerals, water and air. Healthy mineral water from sources deep within the earth's crust trickle over large Gabion-like rock constructions looking like a huge open-air trickle-filter. (Gabion is basically a wall of rocks held together with wire mesh). Most fascinating aspects of the natural environment develop where water and air meet (Figure 32.7).

Near Amersfoort, in the Netherlands, physical barriers have been constructed to keep traffic noise out of the housing area (Figure 32.8(a) and (b)). These acoustic barriers have been finished with small Gabion (as in Kreuznach). It would be interesting to pump grey water from adjacent houses over and through these 'trickle filters' in order to provide irrigation to the inherently dry dike-tops.

Figure 32.7. The interaction of mineral water, rocks and air (near Bad Kreuznach, Germany). Source: city PR office.

Figure 32.8(a) and (b). Gabion-like constructions used with acoustic barriers could serve as trickle-filters for domestic grey water (Amersfoort, the Netherlands). Source: Sietzema.

32.4.3 Attitude, health and perception (Bad Kreuznach)

People from several social strata, each with respiratory problems, sit and stroll near the beautiful health walls as they relax in easy chairs (Figure 32.9) They recover by breathing the healthy water and air enriched with health-giving aerosols. Here, health and safety experts have investigated the risks and potentials of aerosol contamination and decontamination (see also the Amersfoort example). Recuperation also results from mental recovery, not simply as a result of inhaling the aerosols but due to relaxation and probably the presence of caring staff. Indeed, diseases of all kind are more prevalent in an unhealthy social environment, thus proving that DESAR is more than microbiology and oxygen demand.

32.4.4 Water, air and soil (outer space and Copenhagen)

Not only do astronauts travelling into outer space face the problem of human waste and wastewater, but they also live in an artificial and electrostatically contaminated environment, in a sense just like many modern office workers. NASA is thus interested in improving (using minimal means) the microclimate inside space shuttles in outer space. One of the authors of this chapter has developed, from the NASA experience, the concept of plants and their soil substrates filtering the 'waste' air just as they filter (or condition) wastewater (Figures 32.10 and 32.11).

Figure 32.9.The health wall trickle-filter, an environment of care and relaxation (Bad Kreuznach, Germany). Source: city PR office.

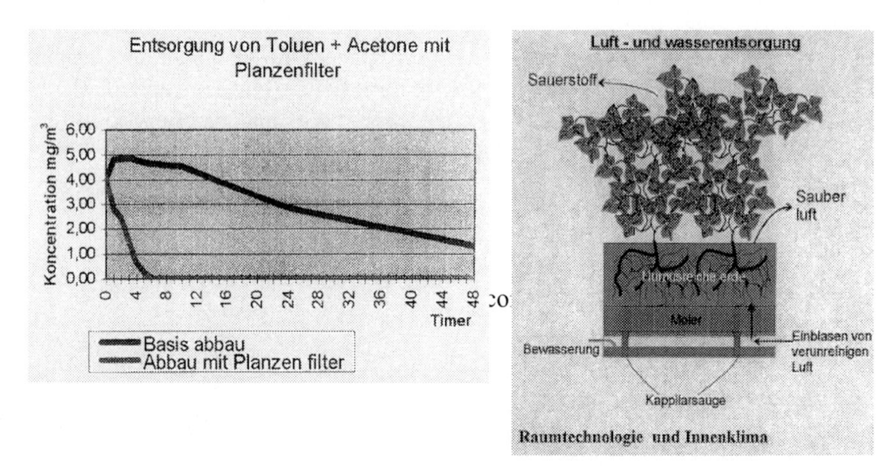

Figure 32.10(a) Soil substrates can bind and reduce toxic components from the air. (b). A combination of plants and soil can be designed into an effective air-conditioner. Source: DRT-TransForm files.

32.4.5 Domestic micro-application of integrated DESAR concepts

In experimental projects on renovating social housing projects in Kalmer (Sweden) and Alborg (Denmark)a similar art of combining quality treatment of water, air, noise, space and even storage of food has been practised. It is part of an architect's job to aim at the aesthetic integration of all aspects of our senses. In Alborg, experiments were first carried out with the combined treatment and recycling of indoor air and domestic grey water (Figure 32.11(a), (b)). The purpose of the project was to upgrade social housing and apartment buildings, in order to blend the old and the new. The Alborg experience should not be considered as the ideal of sustainable and cheap solutions, but rather as a first effort to blend new and unfamiliar concepts into new developments in indoor climate. The point here is the aesthetic aspect which even mini-DESAR can offer to middle class citisens.

In Kalmer (Sweden) the combination of trickling indoor grey water and kitchen air treatment makes an agreeable and calming background sound, thus setting a mood just as background music would (Figure 32.12).

Figures 32.11(a) and (b). Old apartment buildings rehabilitated into modern middle class apartments incorporating interesting features of urban ecology (Alborg, Denmark). Source: DRT-TransForm files.

Figure 32.12(a) and (b). Left: A mini-greenhouse sits in the window as an air- and domestic grey water conditioner, at the same time creating an visually pleasing environment (Alborg, Denmark). Source: DRT-Transform files. Right: Combined substrate and plant treatment improve office environments and their acoustic quality (Copenhagen, Denmark). Source: DRT-TransForm files.

Returning to the Alborg experience; the trickling fraction of the grey water-treatment for the community laundry (traditionally in the basement of apartment buildings) is designed to be functional in cooling down intake-air for big cooling cabinets, thus providing cool storage space for fresh vegetables and drinks (Figure 32.13). The point here is that this technology tends, through cellar-like storage, to enhance the community development aspect of sustainability because it enables the consumer to shop at local markets for fresh and local produce instead of having to depend on supermarket food stored in a small refrigerator.

The dreamers among DESAR practitioners can easily imagine expanding indoor water treatments, even with aquariums containing living fish, ready for an inviting meal for unexpected guests (Figure 32.14).

Figure 32.13(a) and (b).Left: A trickling filter enhancing a cooling system for large indoor cooling cabinets. Right: An indoor cooling cabinet fed by trickle-filter cooling energy (Copenhagen, Denmark). Source: DRT-TransForm files.

Figure 32.14. An indoor aquarium as a component of decentralised domestic grey water treatment. Source: DRT-TransForm files.

32.4.6 DESAR on-the-spot treatment and reuse (Lima, San Marcos, Hichtum and Nieuwersluis)

In Lima, in Peru, a local wastewater engineer Ing. Alejandro Vinces recently brought together a well-to-do neighbourhood and convinced the people not to discharge their wastewater into the central sewer pipe, but to use it (after basic treatment) for irrigation of the beach area. In fact, Lima is built in the middle of a dry desert. A huge living and green curtain, full of flowers at certain times of the year, now softens the otherwise arid microclimate and improves the visual experience of the residential area. The area is coincidentally called 'Mira Flores' (see the flowers). We can well believe that the lush, green appearance of the area means that its inhabitants stay at home more often instead of travelling in search of rare green areas. Moreover, because of their active participation in the process, neighbours have become acquainted with each other. In fact, this wastewater engineer has only improved upon local farmers' practice which is to cut into urban sewer pipes and irrigate their fields using urban effluent.

Around the Salvadoran refugee camps in arid southern Honduras in the early 1980s, working for United Nations High Commission on Refugees (UNHCR), I participated in similar experiences. Wastewater provided food, work and socio-economic cohesion and satisfaction even for those desperate refugees in that war situation (Figure 32.15). Attempts were made to partly use wastewater to irrigate the slopes. A peculiar irony in this case was the difficulties experienced in obtaining medium- or long-term funds for investment. First, it is widely thought that refugees will not stay in one area for a long time and they are thought to be unwilling to build communities within their camp setting. The Salvadoran war was about issues of social perception and justice, which may have been one of the major reasons that social reality in the camps allowed for social–ecological perceptions and subsequent actions.

The second problem was the official political statement that the war would soon be over. As a matter of fact, the type of integrated development which was practised in the camps triggered similar schemes in neighbouring native communities around the camps.

In part of the rural Dutch province of Fryslân a village community has recently refused to have a centralised sewer system. Their resistance is challenging because it exposes the limits of urban sanitation options in the context of a rural environment. The canals and wetlands around the village provide the inhabitants with more natural wetland treatment capacity than is needed for regular wastewater runoff. The hope is that a low density natural treatment system can be developed together with the production of appropriate biomass such as willow trees or Robinia. These trees can then be harvested yearly by the community for use as firewood, fencing etc.

Contrary to the rural reality in Fryslân there may also be relevance to DESAR in the much more urban environment of the western part of the Netherlands. In a rare rural spot of Utrecht, enclosed by a highway, a high-speed railway, an international waterway and several essential polder dikes a certain Dutch entrepreneur wants to develop his industrial area into a less polluting eco-park (Figure 32.16). Surrounding physical barriers cut this area off from access to a centralised sewer. The plan is to develop an option in which all wastewater could be transformed into protective, decorative and attractive bird-hosting biomass on the land. However, up to now there has been no provision for legal experiments within the Dutch administrative framework.

Figure 32.15. Arid slopes turned into fertile gardens (Mesa Grande refugee camp, San Marcos, Honduras). Source: author.

Figure 32.16. Even in the Netherlands some areas may be cut off from access to conventional centralised sewer lines (Nieuwersluis, the Netherlands). Source: Mr Peter van Bolhuis, Pandeon Productions, Westervoort.

32.4.7 Storm-water and erosion defence (Managua and Choluteca)

I present only two more challenges. They deserve to be taken seriously by forums as DESAR. These two examples reflect a sense of urgency and potential beyond what we usually work with in the Dutch lowlands or in Danish apartment houses. They also defy accepted philosophy on health risks. I detected them during the past two years while I was involved in urban renovation and disaster management in Central America. Both cases are impossible without community participation or community awareness. They offer the potential for strengthening and healing communities by giving them purpose and a common interest.

Managua makes an excellent case for sustainable and DESAR water management simply because of its confinement to a relatively small watershed of less than 20 km across. The soil is very sandy and only some areas are 'protected' aquifers used for drinking water. During the last decades (because of earthquakes and war) many people have established extreme shanty towns around old Managua, often where urban planning would not allow construction due to the area being ecologically and geologically unstable. Administrators tend to structure urban renewal efforts according to the logic of highways and traffic lights. As an architect I work more with the logic of spaces and visual highlights. As a member of an IDB consultancy team I proposed, beyond visual identification, the consideration of a third major criterion, namely the logic of watersheds and sewers. Although I oppose urban design as a function of the briefest possible sewer connection from house to street, I try to take maximum advantage of natural drop by cascading and oxygenating the water while it is running downhill.

At the same time I suggested making use of the sewer effluents for irrigation of urban green areas in order to stabilise critical slopes and to soften the urban climate. Remembering the Amersfoort gabions, I suggested using the 'creative friction' between eroding runoff streams over the steep slopes, with the aim of eventually integrating at least the grey water problem into reinforcing the same slopes (Figure 32.17). I call this urban terracing and it hints at the concept of urban agriculture. In my view, urban agriculture includes flowers, fruit and shade. Giving economical and aesthetic advantage, measures of physical protection may also provide essential elements to public acceptance.

Figure 32.17. Grey water could be used for the stabilisation of urban slopes and improving the microclimate by planting flowers and vegetables (Tegucigalpa, Honduras). Source: author.

In Managua, pit latrines were the main method of black water disposal. However (in order to save the pit latrines for black water only) grey water runs through the streets and playgrounds where children play (Figure 32.18(a) and (b)). Whether on slopes or flat areas, the grey water could be used much more creatively without much expense.

In 1998 Hurricane Mitch brought great devastation to Central America. The destruction of the environment was almost beyond imagination. Here I mention only the valley of Choluteca in Honduras. The course of the main river has changed dramatically, while making modern bridges useless (Figure 32.19). Over a great distance this river destroyed the scarce fertile land or left metres of sand and rock on top of fertile land. In Choluteca the resulting flood washed away the sewage treatment plant. It will take many years to build another one in its place.

As a DESAR practitioner, I question whether the conventional approach of sewer treatment makes sense (if used in this arid desert climate) with so much open space and unprotected river banks.

First, I suggest that wastewater, after minimal pre-treatment, should be used as fertigation (fertigation is a combination of fertilisation and irrigation) for rapid-growing (and regrowing) deep-rooting trees. The potential damage that a further storm could do will then be restrained by the 'living fence', and as an additional advantage firewood can be harvested and people will be able to enjoy a cooler and dust-free microclimate. An enriched natural habitat for birds and other animals will develop by itself, balancing threats (such as the common fear of snakes) with benefits (such as the harvesting of firewood, the sound of birdsong and the smell of blossom). Second, I suggest exploring the potential wealth of harvesting different qualities of water for different purposes from the destroyed riverbed. Huge reservoirs of rock and sand now provide an interesting storage of water and also show a creative friction within the logic of nature.

Figure 32.18(a) and (b). Grey-water being used around the house (Pantanal, Managua, Nicaragua). Source: author.

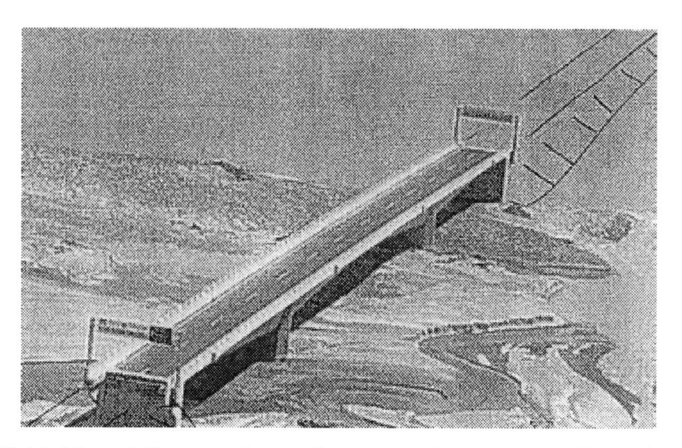

Figure 32.19. Natural disasters change the course of nature, sometimes also rendering obsolete human engineering such as this bridge (Choluteca, Honduras). Source: Prensa's memorial, San Pedro Sula, 1998.

To an architect's mind there is *utzil* to balance. Wastewater is an essential component in the play of beauty, health and sustenance.

32.5 REFERENCES

Cacerez, A. (1987) Primer Seminario –Taller Nacional sobre Letrinas Aboneras Secas Familiares, Guatemala. (In Spanish.)
Chongrak, P. (1981) *Human Faeces and their Utilization,* AIT, Bangkok.
Ellul, J. (1990) *The Technological Bluff*, Erdmann, Grand Rapids.
Esrey, S. *et al.* (1998) *Ecological Sanitation,* SIDA, Stockholm. (A general, very worthwhile resource on the subject.)
Ilich, I. (1978) *Towards a History of Needs*, Pantheon Books, New York
Reinoso, D. (1956) Popol Vuh, Tradiciones Antiguas del Quiché, Guatemala. (In Spanish.)
Ross, G.D. (1999) Cities feeding people. Report 27, Community-based technologies for domestic wastewater treatment and reuse for urban agriculture, IDRC, Ottawa. (A general, very worthwhile resource on the subject.)
Schiere, J.J. (1989) LASF, Una Letrina para la Familia,.MCC, Santa Maria Cauqué. (In Spanish.)
Schiere, J.J. (1991) *Beyond Technology*, MCC occasional paper, Akron.
Simon, C. (2000) Westerse Mantra's XVII, Maurice Merlau-Ponti, *Filosofie* magazine. (In Dutch.)
Van Gelder, J. (1960) *Plato: Gorgias en Socrates*, Bert Bakker, The Hague, the Netherlands. (In Dutch.)
Van Kranendonk, A. (1980) Het Analytische en Compositorische in de Architectuur, OUP, Delft. (In Dutch.)
Verhelst, T. (1986) Het recht anders te zijn, Unistad, Antwerpen. (In Dutch.)

Index

Decentralised Sanitation and Reuse

Integrated Environmental Technology Series

The *Integrated Environmental Technology Series* addresses key themes and issues in the field of environmental technology from a multidisciplinary and integrated perspective.

An integrated approach is potentially the most viable solution to the major pollution issues that face the globe in the 21st century.

World experts are brought together to contribute to each volume, presenting a comprehensive blend of fundamental principles and applied technologies for each topic. Current practices and the state-of-the-art are reviewed, new developments in analytics, science and biotechnology are presented and, crucially, the theme of each volume is presented in relation to adjacent scientific, social and economic fields to provide solutions from a truly integrated perspective.

The *Integrated Environmental Technology Series* will form an invaluable and definitive resource in this rapidly evolving discipline.

Series Editor

Dr. Ir. Piet Lens, Sub-department of Environmental Technology, The University of Wageningen, P.O. Box 8129, 6700 EV Wageningen, The Netherlands. (piet.lens@algemeen.mt.wag-ur.nl)

Forthcoming titles in the series include:

Closing Industrial Cycles: *Challenges for the integration of biotechnology*
Technologies to Treat Phosphorous Pollution: *Principles and engineering*
Anaerobic Environmental Biotechnology
Biofilms: *Analysis, prevention and utilisation*